# COMPARATIVE HIGH PRESSURE BIOLOGY

T0187946

# COMPARATIVE HIGH PRESSURE BIOLOGY

*Editor*
**PHILIPPE SÉBERT**

Orphy-EA 4324
Université de Brest-UEB
6, Avenue Le Gorgeu 29238 Brest Cedex 3
France

CRC Press
Taylor & Francis Group
Boca Raton London New York

CRC Press is an imprint of the
Taylor & Francis Group, an **informa** business

A SCIENCE PUBLISHERS BOOK

First published 2010 by Science Publishers Inc.

Published 2019 by CRC Press
Taylor & Francis Group
6000 Broken Sound Parkway NW, Suite 300
Boca Raton, FL 33487-2742

© 2010, Copyright Reserved
CRC Press is an imprint of Taylor & Francis Group, an Informa business

First issued in paperback 2019

No claim to original U.S. Government works

ISBN 13: 978-0-367-45240-7 (pbk)
ISBN 13: 978-1-57808-638-2 (hbk)

This book contains information obtained from authentic and highly regarded sources. Reasonable efforts have been made to publish reliable data and information, but the author and publisher cannot assume responsibility for the validity of all materials or the consequences of their use. The authors and publishers have attempted to trace the copyright holders of all material reproduced in this publication and apologize to copyright holders if permission to publish in this form has not been obtained. If any copyright material has not been acknowledged please write and let us know so we may rectify in any future reprint.

Except as permitted under U.S. Copyright Law, no part of this book may be reprinted, reproduced, transmitted, or utilized in any form by any electronic, mechanical, or other means, now known or hereafter invented, including photocopying, microfilming, and recording, or in any information storage or retrieval system, without written permission from the publishers.

For permission to photocopy or use material electronically from this work, please access www.copyright.com (http://www.copyright.com/) or contact the Copyright Clearance Center, Inc. (CCC), 222 Rosewood Drive, Danvers, MA 01923, 978-750-8400. CCC is a not-for-profit organization that provides licenses and registration for a variety of users. For organizations that have been granted a photocopy license by the CCC, a separate system of payment has been arranged.

**Trademark Notice:** Product or corporate names may be trademarks or registered trademarks, and are used only for identification and explanation without intent to infringe.

**Visit the Taylor & Francis Web site at**
**http://www.taylorandfrancis.com**

**and the CRC Press Web site at**
**http://www.crcpress.com**

Cover Illustration reproduced by courtesy of M. Castellini and Y. Grossman

Library of Congress Cataloging-in-Publication Data
Comparative high pressure biology / editor, Philippe
    Sébert.
       p. cm.
    Includes bibliographical references and index.
    ISBN 978-1-57808-638-2 (hardcover)
1. High pressure biochemistry. 2. High pressure biology.
    3. High pressure chemistry. I. Sébert, Philippe.
QP517.H53C66 2009
572'.43--dc22
                                              2009032722

# Foreword

It should be noted that hydrostatic pressure is an environmental factor as well as a thermodynamical factor. In thermodynamics pressure, as temperature does, can modify enzyme kinetics, membrane fluidity and more generally structure and/or function of macromolecules and their interactions. Clearly, high pressure is potentially able to modify all the biological process which, in return, makes pressure an efficient tool. In the same way, as an environmental factor, pressure naturally concerns most of the species living on the earth due to the fact that about 80% of the biomass lives at depth, mainly in the oceans which represent about 70% of the earth surface. To the species living under pressure, we must add the species (birds, mammals and thus Man) which episodically are submitted to relatively high pressures for feeding, reproduction, exploration, working activities and also leisure (for these species pressure gas effects must also be considered).

From the molecules to the overall organism, via the cells and unicellular species, invertebrates and vertebrates, ectotherms and endotherms, more than 30 specialists have participated in this book in order to offer a panorama of their topic as complete as possible. The objective of this book is to present the state of art in terms of comparative high pressure biology. Consequently, not only does this book provide the more recent results in each of its chapter but suggests new directions for research. I hope that the diversity in the approaches and the numerous animal species studied at different levels of organization make this 'pressure book' interesting for biologists, whatever their research topic, and attractive for adventurous young researchers looking out for a fascinating world.

Finally, I would like to thank all the contributors who offer through this book, many interesting results, ideas and views useful for any researcher or teacher.

**Philippe SÉBERT**

# Contents

## PRESSURE AND LIVING ORGANISMS

## PRESSURE AND MAN

# List of Contributors

**Abraini, J.H.**
Université de Caen UMR 6232, CI-NAPS, Centre CYCERON, BP 5229, Boulevard Becquerel, 14074 Caen cedex, France.
E-mail: abraini@cyceron.fr

**Amérand, A.**
ORPHY EA4324, Universite Europeenne de Bretagne, Universite de Brest, UFR Sciences et Techniques, France.
E-mail: aline.amerand@univ-brest.fr

**Aviner, B.**
Department of Physiology, Faculty of Health Sciences,Ben-Gurion University of the Negev, Beer-Sheva, 84105, Israel.
E-mail: benaviner@gmail.com

**Bartlett, D.**
8750 Biological Grade, 4405 Hubbs Hall, Marine Biology Research Division, Center for Marine Biotechnology and Biomedicine, Scripps Institution of Oceanography, University of California, San Diego 92093-0202, La Jolla California, USA.
E-mail: dbartlett@ucsd.edu

**Blumelhuber, G.**
Sonnenweg 16, 85084 Reichertshofen, Germany.
E-mail: info@immobrau.com

**Bode, C.**
Institute of Biophysics and Radiation Biology, Semmelweis University, H-1444 Budapest P.O.Box 263.
E-mail: Csaba.Bode@eok.sote.hu

**Castellini, M.**
School of Fisheries and Ocean Sciences, University of Alaska Fairbanks, Fairbanks, Alaska, USA 99775.
E-mail: mikec@ims.uaf.edu

**Chevalier-Lucia, D.**
Assistant Professor, Department of Agro-resources and Biological and Industrial Processes, Mixed Research Unit of Agro-polymer Engineering and Emerging Technologies, Université Montpellier 2, Sciences et Techniques, Place E.Bataillon, 34095 Montpellier, France.
E-mail: Dominique.Chevalier-Lucia@univ-montp2.fr

**Damasceno-Oliveira, A.**
CIMAR-LA/CIIMAR, Universidade do Porto, Rua dos Bragas, 289 4050-123 Porto, Portugal.
E-mail: aol@ciimar.up.pt

**Daniels, S.**
Wales Research and Diagnostic Positron Tomography Imaging Centre School of Medicine, Cardiff University, Cardiff CF14 4XN, UK.
E-mail: danielss@cf.ac.uk

**Dumay, E.**
Professor, Department of Agro-resources and Biological and Industrial Processes, Mixed Research Unit of Agro-polymer Engineering and Emerging Technologies, Université Montpellier 2, Sciences et Techniques, Place E.Bataillon, 34095 Montpellier, France.
E-mail: Eliane.Dumay@univ-montp2.fr

**Fraser, P.J.**
Institute of Biological and Environmental Sciences, Zoology Department, College of Life Sciences and Medicine, Aberdeen University, Tillydrone Avenue, Aberdeen, UK, AB24 2TZ.
E-mail: p.fraser@abdn.ac.uk

**Friedrich, O.**
School of Biomedical Sciences, University of Queensland, Skerman Bldg, QLD 4072, St Lucia, Brisbane, Australia.
E-mail: o.friedrich@uq.edu.au

**Grossman, Y.**
Feldman Chair for Neurophysiology, Department of Physiology, Faculty of Health Sciences, Zlotowki Center for Neuroscience, Ben-Gurion University of the Negev, Beer-Sheva, 84105, Israel.
E-mail: ramig@bgu.ac.il

**Jammes, Y.**
UMR MD2 P2COE, Faculté de Médecine, Université de la Méditerranée, France.
E-mail: jammes.y@jean-roche.univ-mrs.fr

**Jebbar, M.**
Laboratoire de Microbiologie des Environnements extremes, UMR 6197 (UBO, Ifremer, CNRS), Universite de Brest, France.
E-mail: mohamed.jebbar@univ-brest.fr

**Kato, C.**
Extremobiosphere Research Center, Japan Agency for Marine-Earth Science and Technology, 2-15 Natsushima-cho, Yokosuka 237-0061, Japan.
E-mail: kato_chi@jamstec.go.jp

**Lange, R.**
INSERM U710, Université Montpellier 2, Place Eugene Bataillon 34095 Montpellier cedex5, France.
E-mail: reinhard.lange@inserm.fr

**Lavoûte, C.**
Université dela Méditerranée et Institut de Médecine Navale di Service de Santé des Armées, UMR-MD2—Physiologie et physiopathologie en condition d'oxygenation extrême, Institut de Neurosciences Jean Roche, Faculté de Médecine Nord, Bd P. Dramard, 13015 Marseille France.
E-mail: cecile.lavoute@univmed.fr

**Le Péchon, J-C.**
Ingénieur Conseil, 94, rue de Buzenval, 75020 Paris, France.
E-mail: hyperbar@club-internet.fr

**López-Pedemonte, T.**
Assistant Professor, Departamento de Ciencia y Tecnologia de los Alimentos, Facultad de Quimica—Universidad de la República (UDELAR), Avenida General Flores 2124 – Montevideo, Uruguay.
E-mail: tlopez@fq.edu.uy

**Marchal, S.**
INSERM U710, Université Montpellier 2, Place Eugene Bataillon, 34095 Montpellier cedex5, France.
E-mail: stephane.marchal@inserm.fr

**Moisan, C.**
ORPHY EA4324, Université Européenne de Bretagne, Université de Brest, UFR Sciences et Techniques, France.
E-mail: christine.moisan@univ-brest.fr

**Mor, A.**
Department of Physiology, Faculty of Health Sciences, Ben-Gurion University of the Negev, Beer-Sheva, 84105, Israel.
E-mail: morami12@gmail.com

**Oger, P.**
Laboratoire de Science de la Terre, UMR 5570 CNRS-ENSL-UCBL 46, allée D'Italie, F-69364 Lyon, France.
E-mail: poger@ens-lyon.fr

**Péqueux, A.J.R.**
Ecophysiology Unit, University of Liège 22,Quai Van Beneden B-4020 Liège, Belgium.
E-mail: A.Pequeux@ulg.ac.be

**Prieur, D.**
Laboratorie de Microbiologie des Environnements extrémes, UMR 6197 (UBO, Ifremer, CNRS), Université de Brest, France.
E-mail: daniel.prieur@univ-brest.fr

**Risso, J.J.**
Université de la Méditerranee et Institut de Médecine Navale du Service de Santé des Armées, UMR-MD2—Physiologie et physiopathologie en condition d'oxgénation extrême IMNSSA, BP 84, 83800 Toulon Armées, France.
E-mail: j.jrisso@imnssa.net

**Rostain, J.C.**
Université de la Méditerranée et Institut de Médecine Navale du Service de Santé des Armees, UMR-MD2—Physiologie et physiopathologie en condition d'oxgénation extrême, Institut de Neurosciences Jean Roche, Faculté de médicine Nord, Bd P.Dramard 13015 Marseille, France.
E-mail: jean-claude.rostain@univmed.fr

**Sébert, P.**
ORPHY–EA4324, Université Européenne de Bretagne—UBO 6, Avenue Le Gorgeu, CS 93837, 29238 Brest Cedex3 France.
E-mail: Philippe.sebert@univ-brest.fr

**Siebenaller, J.F.**

Department of Biological Sciences, Louisiana State University, Baton Rouge, Louisiana 70803 USA.
E-mail: zojose@1su.edu

**Smeller, L.**

Semmelweis University, Dept. Biophysics and Radiation Biology, Tuzolto u. 37-47, Budapest IX, H-1444 Pf 263.
E-mail: laszlo.smeller@eok.sote.hu

**Thom, S.R.**

Professor of Emergency Medicine, Chief, Hyperbaric Medicine, University of Pennsylvania, 3620 Hamilton Walk, Philadelphia, PA 19104 USA.
E-mail: sthom@mail.med.upenn.edu

**Tolgyesi, F.G.**

Institute of Biophysics and Radiation Biology, Semmelweis University, H-1444 Budapest P.O.B. 263.
E-mail: Ferenc.Tolgyesi@eok.sote.hu

**Torrent, J.**

INSERM U710, Université Montpellier 2, Place Eugene Bataillon, 34095 Montpellier cedex5, France.
E-mail: joan.torrent@inserm.fr.

**Winter, R.**

TU Dortmund University, Physical Chemistry 1, Biophysical Chemistry, Otto-Hahn Str. 6, D-44227 Dortmund, Germany.
E-mail: roland.winter@tu-dortmund.de

# PRESSURE AND
# CELL COMPONENTS

# Protein Kinetics under High Pressure

*Stéphane Marchal, Joan Torrent* and *Reinhard Lange*

## INTRODUCTION

High pressure affects protein structures at the secondary, tertiary and quaternary level (Gross and Jaenicke, 1994). Depending on the protein and on incubation conditions, exposure to pressure may thus convert a protein from α-helical to β-sheet structure, induce protein unfolding, or dissociate or aggregate proteins (Smeller et al., 2006). Of course, these effects are very important for a variety of applications, such as maintaining or optimizing the structural integrity of enzymes used as bio-catalysts under extreme conditions of temperature and pressure, developing new and safer food products, or controlling folding/misfolding reactions of proteins associated to systemic or neurodegenerative diseases. (Lange, 2006).

In the last decades, considerable progress was made in developing analytical instruments for recording protein structural changes under high pressure. The interest of using pressure as structure perturbation parameter is i) it provides complementary thermodynamic information to the more conventional parameters, such as temperature or chemical agents; ii) pressure is a relatively mild perturbant: irreversible structural changes, such as those induced by heat, can generally be avoided; iii) its use does not lead to complications such as specific interactions of protein with chemical denaturants.

However, this progress was mainly restricted to the study of pressure dependent reactions under equilibrium conditions. Only in the last decade, the idea of studying protein reaction kinetics under pressure became popular.

INSERM U710, Université Montpellier 2, Place Eugène Bataillon 34095 Montpellier cedex5, France.
E-mail: stephane.marchal@inserm.fr, joan.torrent@inserm.fr, reinhard.lange@inserm.fr

The underlying theory is simple, though. The pressure dependence of the rate constant (k) of an elementary process yields the activation volume, $\Delta V^{\ddagger}$, which is equal to the difference in volumes of the kinetic transition state, $V^{\ddagger}$, and the ground state, $V_g$. This relation is described in equations 1 and 2:

$$(\partial \ln k / \partial p)_T = -(\partial \Delta G^{\ddagger} / \partial p)_T / RT = -\Delta V^{\ddagger} / RT \qquad (1)$$

$$\Delta V^{\ddagger} = V^{\ddagger} - V_g \qquad (2)$$

where p is the pressure, T the temperature, R, the gas constant and $\Delta G^{\ddagger}$ the activation free energy. Hence, any chemical reaction—and *a fortiori* one involving protein structural changes— is accelerated (for a negative value of $\Delta V^{\ddagger}$) or decelerated (for a positive value of $\Delta V^{\ddagger}$) by pressure depending on the relative volume of the kinetic transition state with respect to that of the ground state (Mozhaev et al., 1996). These differences in volume are due to transient changes of protein packing and hydration in the course of the reaction (Royer, 2005). As an example, a value of $\Delta V^{\ddagger}$ of 69 mL mol$^{-1}$, found for invertase (Morild, 1981), gives rise to a more than 100-fold acceleration for a pressure increase from 0.1 to 200 MPa.

In practice, protein kinetics under high pressure have been studied mainly by two methods, using high pressure stopped flow and pressure jump devices. Here we will illustrate the use of these experimental set-ups for studying the mechanism of enzyme activity, of ligand binding to hemoproteins, and of protein folding. As our team is specialized in these approaches, most of the examples will stem from our laboratory in Montpellier.

## EXPERIMENTAL

### High pressure stopped flow technique

The high pressure stopped-flow technique (Fig. 1)  was introduced by Heremans  and further developed by  Balny's group (Balny et al., 1984, 1987) and Merbach (Pappenberger et al., 2000). It allows mixing, within less then 10 ms, of two solutions under defined pressure (up to 200 MPa) and temperature (between −20 and +40°C). The reactant solutions (5 mL) are kept in glass syringes. The mixing bloc and observation cell (1 cm optical path) are made of Teflon. The apparatus is equipped with quartz windows for absorbance and fluorescence measurements and interfaced via optical fibers to a BioLogic light source and detection system.

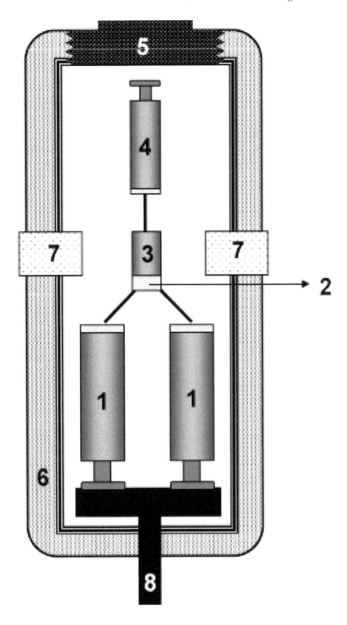

**Figure 1.** Schematic representation of a high pressure stopped-flow apparatus. The actual mixing device, consisting of mixing syringes (1), mixing chamber (2), observation cell (3), and waste syringe (4), is introduced via an opening screw (5) into a silicon oil containing cylindrical high pressure cell (6), equipped with quartz windows (7). Mixing of the reactants contained is induced by a pneumatically driven upward movement of a central piston (8), pushing upward the pistons of the mixing syringes.

## Pressure-jump device to induce protein structural relaxations

Figure 2 gives a schematic representation of a pressure-jump device connected to a fluorescence spectrometer, as it is used our laboratory. A limitation of this technique is the pressure-induced variation of temperature. During a pressure-jump, the temperature of the sample is heated (for upward jumps) or cooled (for downward jumps) by about 0.1°C for a pressure jump amplitude of 10 MPa. In order to minimize these temperature variations, experiments are therefore usually limited to pressure jump amplitudes of 50 MPa (Font et al., 2006b)

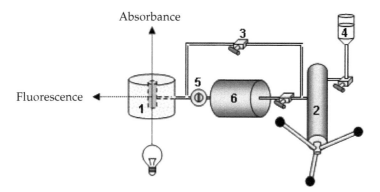

**Figure 2.** Schematic representation of the pressure jump device. The high pressure optical cell (1) allows the sample cell to be cooled or heated at a set temperature, and operates at a pressure of several hundred MPa (up to 700 MPa) without leakage, even at sub-zero temperatures. It holds a quartz sample cell, which is sealed with a polyethylene stretch, maintained by a rubber O-ring. This high pressure cell, equipped with sapphire windows, is easily interfaced to conventional spectrophotometers and fluorimeters. Pressure is generated by a manual pump (2) and controlled by HP valves (3). The pressure vector, water (4), is mediated through stainless steel capillaries. A home made pressure-jump device (Torrent et al., 2006a) can be connected to the high pressure cell. Pressure jumps are carried out by opening an electrically driven pneumatic valve (5) localized between the high pressure optical cell and the ballast tank (6). The pressure-jumps consist of sudden changes of pressure (up to 100 MPa) within a pressure range of 0.1–600 MPa.

## THE USE OF HIGH PRESSURE TO STUDY THE MECHANISM OF ENZYME REACTIONS

The main problem for studying enzyme reactions under high pressure is the generally high reactivity of enzymes; the conversion of a substrate into a product starts immediately after mixing the enzyme with its substrate. If this mixing is done at atmospheric pressure, the time lap that is necessary to raise the pressure of the mixture precludes observation of early reaction stages. Moreover, in the case of rapidly reacting enzymes, the reaction may be even finished before the end of the pressurization time.

A convenient means to overcome this difficulty is given by the stopped-flow technique under high pressure (see Fig. 1). This approach allows keeping enzyme and substrate separated during the pressurization time. They are mixed only once the final pressure has been reached. An alternative technical solution, optimized for steady state enzyme kinetics, has been developed by (Hoa et al., 1990). The use of high pressure stopped-flow devices for studying enzyme activity is illustrated by the following three examples.

Frequently, enzyme catalyzed chemical reactions are carried out by oligomeric proteins. However, one may ask, what is the role of the oligomeric states? And are the monomers equally active as dimmers or tetramers? These questions have been dealt with by Kornblatt and co-workers for the case of yeast enolase (Kornblatt et al., 1998). This dimeric enzyme catalyzes the interconversion of 2-phosphoglyceric acid and phosphoenolpyruvate. Raising pressure induces the dissociation of the protein into monomers. This information could be deduced from a fourth derivative UV absorbance spectroscopic investigation (Lange et al., 1996) under pressure. Furthermore, as shown by stopped-flow results, the activity of the enzyme decreased as a function of pressure. As shown in Fig. 3, these two processes occurred concomitantly, suggesting that monomers do not retain native-like structures.

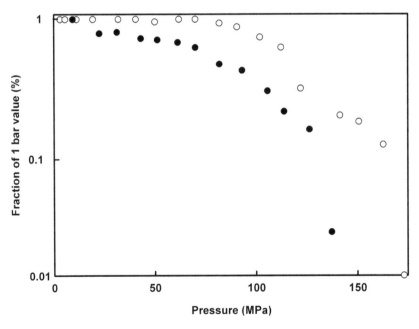

**Figure 3.** Effects of pressure on the quaternary structure and activity of yeast enolase. The quaternary structure (full dots) and the enzyme activity (open circles) are expressed as percentage of the respective values at atmospheric pressure. Adapted from (Kornblatt et al., 1998).

This study also showed that pressure-induced dissociation of yeast enolase in the presence of magnesium involves the loss of bound metal ion Mg(II). Interestingly, when magnesium was replaced by manganese, the application of pressure did not lead to a loss of the metal ion. Moreover, the pressure-induced monomers of Mn-bound enolase remained active and showed a native-like structure. Figure 4 resumes the effects of pressure on magnesium and manganese-bound yeast enolase (Kornblatt et al., 2004).

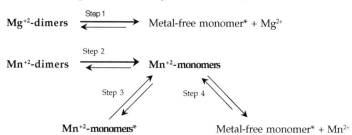

(Species in bold are enzymatically active; monomers and monomers* indicate different conformations of the monomer)

**Figure 4.** Effects of hydrostatic pressure on yeast enolase. Adapted from (Kornblatt et al., 2004).

Carboxypeptidase from the thermophilic archaebacterion *Sulfolobus solfataricus* is another example of successful application of high pressure to study enzyme activity. In the absence of glycerol and beta-mercaptoethanol, this otherwise heat-stable enzyme undergoes thermal inactivation at 50°C. However, this loss of activity can be inhibited when the enzyme is maintained at 200 MPa (Bec et al.,1996). As shown in Fig. 5, at higher temperatures, even higher pressures (up to 400 MPa) can be used to drastically reduce the thermal inactivation rate. This property of high pressure to stabilize proteins against thermal inactivation is interesting for the use of enzymes in biotechnological applications under high temperature conditions.

Using the same carboxypeptidase, pressure was used for investigating the role of individual non-covalent interactions in enzymatic reactions (Occhipinti et al., 2006). This carboxypeptidase turned out to be ideally suited for this investigation due to its very broad substrate specificity. This enzyme can remove any amino acid, with the exception of praline, from short peptides and N-blocked amino-acids, irrespective of the blocking group. This allowed a first comparative analysis of the effects temperature and pressure have on the interaction of an enzyme with a broad range of substrates. Furthermore, the pressure and temperature dependence of the enzyme kinetic parameters (turnover number and Michaelis constant) could be used to dissect the individual non-covalent interactions in substrate binding. Together with molecular docking experiments performed on a 3D

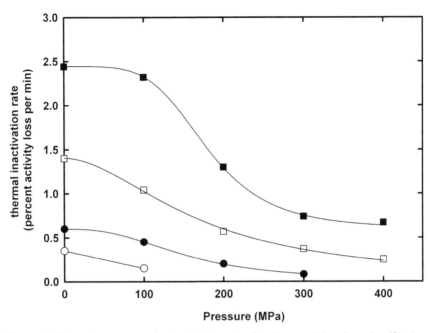

**Figure 5.** Effect of pressure on the inactivation rate of carboxypeptidase from *S. solfataricus* at different temperatures. Adapted from (Bec et al., 1996).

model of the enzyme, it turned out that both hydrophobic and energetic interactions (i.e., stacking and van der Waals) contribute most strongly to substrate binding.

As shown in Fig. 6, a strong correlation was observed between the volume changes associated with $K_m$ ($\Delta V_{Km}$) and $k_{cat}$ ($\Delta V^{\ddagger}_{kcat}$). Taking into account the kinetic reaction model of ordered Bi Bi system, describing $K_m$ and $k_{cat}$ in terms of individual rate constants, this correlation indicates that the pressure dependence of the kinetic substrate binding constant does not depend on the nature of the substrate. However, as shown in Fig. 7, the thermodynamic parameters describing the temperature dependence of the enzyme-substrate interaction and those describing its pressure dependence were not correlated. This suggests that—in contrast to the pressure dependence—the temperature dependence of the kinetic substrate binding constant depends on the nature of substrate.

## LIGAND BINDING TO HEMOPROTEINS

In the field of hemoproteins, one of the key features of the puzzling mechanism of heme-type monooxygenases has been to identify the structural features responsible for binding and subsequent activation of molecular

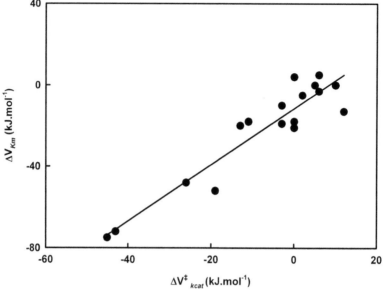

**Figure 6.** Distribution of the volume changes associated with the Michaelis constants ($\Delta V_{Km}$) versus the activation volumes ($\Delta V^{\ddagger}_{kcat}$) of the catalytic step for different N-blocked amino acids used as substrates. Reprinted with permission from (Occhipinti et al., 2006).

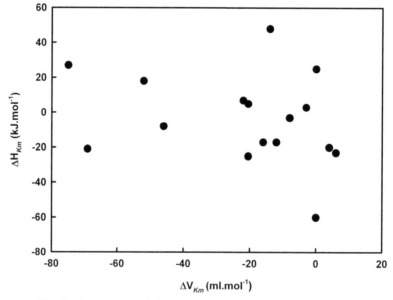

**Figure 7.** Distribution of the enthalpy changes associated with the Michaelis constants ($\Delta H_{Km}$) versus the volume changes associated with the Michaelis constants ($\Delta V_{Km}$) for different N-blocked amino acids used as substrates. Reprinted with permission from (Occhipinti et al., 2006).

oxygen. The best-studied monooxygenase is cytochrome P-450, the generic name for a multigene family of heme-thiolate proteins that catalyze the oxidation of a variety of exogenous, as well as endogenous, substrates (Makris et al., 2002; Makris et al., 2003; Sligar, 1999). Closely related to P450 is nitric oxide synthase (NOS), a two-domains protein that consists of a cytochrome P450 reductase (Bredt et al., 1991) domain and an oxygenase domain that resembles cytochrome P450 (Marletta, 1994; McMillan et al., 1992; Stuehr and Ikeda-Saito, 1992; White and Marletta, 1992). For both classes of proteins, the heme moiety in its active form contains a pentacoordinated high-spin iron ($Fe^{III}$) with an axial cysteine thiolate ligand. Oxygen can bind to the heme iron as a sixth ligand. This step is the essential prerequisite for its enzymatic function. Generally, observation of these binary complexes is very difficult because of their restricted stability in the timescale of measurement at ambient temperature (Gorren et al., 2000; Lipscomb et al., 1976; Peterson et al., 1972; Sligar et al., 1974). Besides oxygen, other small ligands can bind. For example, carbon monoxide (CO) induces a characteristic red shift of the Soret band to 450nm. Due to the chemical stability of the ferrous iron -CO complex, the latter is often used as a model for a better understanding of the interaction of cytochrome P-450 and NOS with oxygen. An example for a stopped-flow kinetics of CO-binding under high pressure to cytochrome P450 is shown in Fig. 8.

Several studies reported the importance of thiolate as a fifth proximal heme-iron ligand on $O_2$ activation for cytochrome P450. Using high pressure stopped-flow kinetic experiments, Lange et al., 1994 have determined the activation volume for CO binding in several isoforms of cytochrome P450 in the absence of a substrate and compared the results with those obtained with histidine ligated hemoproteins such as haemoglobin and myoglobin. They pointed out that the pressure effect depended on the nature of the proximal axial heme ligand. Markedly negative values of the transition state volume were determined for histidine ligand proteins indicative of protein conformational changes and/or solvation processes of the transition state. In contrast, small positive activation volume changes led the authors to the conclusion that CO-binding transition state was only slightly affected by pressure for thiolate ligated proteins (Table 1). A first explanation of the origin of this behaviour was to suggest the presence of the negatively charged sulphur from cysteine ligand that produces specific electronic properties.

Reconsidering this conclusion, Jung et al. (2002) showed that CO-binding in cytochrome P450cam complexed with substrate analogues of camphor carrying methyl groups was also characterized by negative activation volumes in contrast to the results observed for the protein in the presence or in the absence of natural substrates (Jung et al., 2002). From these observations, the authors suggested that the accessibility of the protein heme centre can be modulated by substrate binding. The positive sign of the

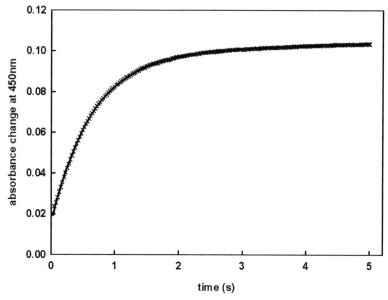

**Figure 8.** Time course of the CO binding to cytochrome P-450 BM3 after stopped flow mixing at 150 MPa. Kinetics were started by mixing equal volumes (60μl) of the reduced form of enzyme and CO solution in a thermostated high-pressure cell placed in Aminco DW2 spectrophotometer operating in dual wavelength mode. The experiments were set-up with one syringe filled with enzyme and the other one containing the CO-saturated solution (1mM). Enzyme and gas-saturated solutions were prepared as described by (Gorren et al., 2000; Lange et al., 2001). The monoexponential fit is shown by the solid line through data points.

**Table 1.** Activation volume changes of the CO binding to various hemoproteins.

| Enzyme | Ligand | $\Delta V^{\ddagger}$ (ml.mol$^{-1}$) | References |
|---|---|---|---|
| P-450scc | S⁻ (cys) | 2 ± 2 | (Lange et al., 1994) |
| P-450LM2 | S⁻ (cys) | 3 ± 2 | (Lange et al., 1994) |
| P-450LM3 | S⁻ (cys) | 6 ± 5 | (Lange et al., 1994) |
| P-450cam | S⁻ (cys) | 4.6 ± 2.0 (slow phase) | (Jung et al., 2002) |
|  |  | 10.2 ± 2.1 (fast phase) |  |
| P-450cam + 1R-camphor | S⁻ (cys) | −19.6 ± 0.9 (slow phase) | (Jung et al., 2002) |
|  |  | −13.2 ± 0.8 (slow phase) |  |
| P-450cam + norcamphor | S⁻ (cys) | 7.6 ± 2.0 (fast phase) | (Jung et al., 2002) |
| P-450cam + norbornane | S⁻ (cys) | 8.4 ± 0.8 (fast phase) | (Jung et al., 2002) |
| chloroperoxidase | S⁻ (cys) | 1 ± 2 | (Lange et al., 1994) |
| lactoperoxidase | N (his) | −10 ± 3 | (Lange et al., 1994) |
| Horseradish peroxidase | N (his) | −24 | (Balny and Travers, 1989) |
| myoglobin | N (his) | −9 | (Hasinoff, 1974) |
|  |  | −18.8 | (Adachi and Morishima, 1989) |
| hemoglobin | N (his) | −3 | (Hasinoff, 1974) |
|  |  | −36 | (Hasinoff, 1974) |

activation volume for CO binding is rather indicative for solvent accessibility and flexibility of the protein than for diffusion-controlled CO binding or for the specific electronic structure of the thiolate proximal group.

With regard to the CO binding properties of cytochrome P450, it was surprising that the closely related NOS did not show the same behaviour. Indeed, using stopped-flow kinetic approach, Lange et al. (2001) ( pointed out that the rate of CO binding was strongly decreased by the presence of substrate. However, application of high pressure induced a strong increase of the rate constant, up to a value similar to that in the absence of substrate. In contrast, for cytochrome P450 BM3, a protein structurally homologous to NOS, CO-binding to the heme iron was not affected by pressure, whether in the absence or in the presence of substrate. Altogether, these observations indicated that the substrate binding site in BM3 would be large and flexible while that of NOS would be more rigid and narrow. Thus, the effect of pressure could be attributed to an expulsion of substrate of the active centre of NOS.

After instrumental optimization of the high pressure stopped-flow apparatus and a very careful design of anaerobic experimental conditions, it became possible to record oxygen binding kinetics under high pressure.

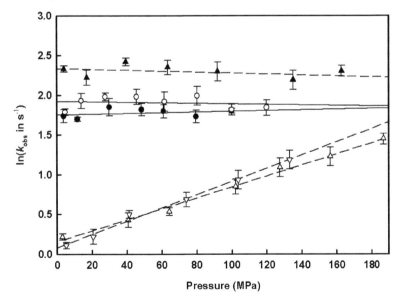

**Figure 9.** Effect of pressure on the rate of CO binding by BM3 (circles) and neuronal NOS (triangles). High pressure stopped-flow kinetics were carried out at 25°C in the absence (solid symbols) and in the presence of 100μM palmitate (open circles) with cytochrome P450 BM3 and with NOS, in the presence of 200μM substrate, L-arginine (open triangles up) or in the presence of both 200μM L-arginine and an excess of 25μM cofactor BH4 (open triangles down) for nNOS (Lange et al., 2001).

Accordingly, Marchal et al., 2003 successfully investigated the kinetics of CO and oxygen-binding to heme-thiolate proteins, the F393H mutant of cytochrome P450 BM3 and the endothelial NOS under pressure. These proteins were used as models because of their specific chemical properties, the formation of a stable ferrous-oxy complex for the former and the absence of a reductase domain for the latter, preventing the reduction of the oxycomplex by electron transfer. For BM3, rates and activation volumes of oxygen-binding matched the values measured for CO binding. In contrast, these properties differed for NOS, suggesting different binding mechanisms of CO and oxygen in NOS. Moreover, these results indicated that CO binding does not seem to be an appropriate probe for studying oxygen-binding.

The analysis of the pressure dependence of the oxygen binding rates of NOS revealed biphasic profiles. A closer analysis of these profiles, in the presence of substrates, L-arginine and hydroxyl-arginine, and the cofactor tetrahydropterin or its inactive analogs, led to interesting results which could be interpreted by the transient formation of a new form of ferrous-heme centre for NOS (see reviews (Gorren et al., 2005; Marchal et al., 2005)). These results were corroborated in more recent rapid-scanning stopped-flow experiments performed at low temperature (Marchal et al. 2004) which

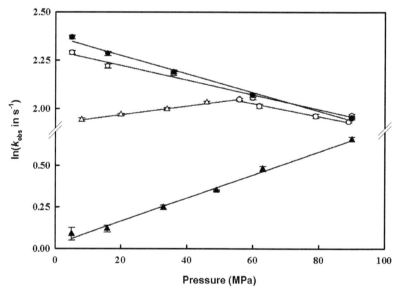

**Figure 10.** Pressure-dependence of ligand binding to the ferrous heme centre of P450 BM3 (circles) and eNOSoxy (triangles). CO binding rates (solid symbols) and $O_2$ binding rates (open symbols) have been measured by rapid mixing of reduced proteins in the presence of substrate (60μM arachidonate for BM3) and inactive cofactor (0.5mM L-arginine, 50μM amino-BH2 for eNOS) with 0.5mM CO and $O_2$, respectively, in appropriate buffer at pH 7.4 and at 4°C (Marchal et al., 2003).

showed that oxygen-binding to the reduced endothelial nitric oxide synthase (eNOS) resulted in two distinct species differing in their Soret and visible absorbance maxima, and in their capacity to exchange oxygen by CO.

Altogether, ligand-binding studies using high pressure which stopped flow technology appears to be informative for the characterization of elementary steps in the complex mechanism of several enzymatic systems.

## PROTEIN FOLDING AND UNFOLDING

Protein folding/unfolding is a complex reaction that can be described properly only in terms of multidimensional energy surfaces (Dill and Chan, 1997; Socci et al., 1998). Determining the path and rate of this reaction remains a major experimental challenge. Nonetheless, there have been significant advances in experimental techniques. Apart from computational approaches, the p-jump method appears as an interesting tool. Over the past 15 years, the pressure-jump induced relaxation kinetics has yielded many significant insights in this research field, especially towards the understanding of folding and unfolding mechanisms (Desai et al., 1999; Font et al., 2006b; Herberhold et al., 2003; Mohana-Borges et al., 1999; Panick et al., 1999; Panick and Winter, 2000; Pappenberger et al., 2000; Tan et al., 2005; Torrent et al., 2006b; Vidugiris et al., 1995; Woenckhaus et al., 2001), and the structure of folding transition states (Font et al., 2006a; Jacob et al., 1999; Mitra et al., 2007; Perl et al., 2001).

The p-jump approach can be used to study both the unfolding reaction (by upward p-jump—pressurization) and the folding reaction (by downward p-jump—depressurization). These studies monitor the structural changes through relaxation of the protein's spectroscopic properties. The data obtained, with a dead-time on the order of milliseconds, exhibit single- or multiple-exponential time-decay profiles, showing, respectively, that proteins can fold in a two-state process (between the native state and the unfolded state) or in an scenario involving intermediate states.

Generally, protein folding and unfolding reactions are triggered by rapid flow techniques, involving turbulent mixing with high concentrations of a chemical denaturant, and by laser-induced temperature-jump (T-jump). Although, the pressure-jump technique is used much less often, it offers several advantages: (1) pressure propagates rapidly so that sample inhomogeneity is a minor problem. (2) Pressure-jumps can be performed bi-directionally, i.e., by upward and downward p-jumps of magnitude between 10 and 100 MPa. (3) In the case of fully reversible structural changes of the sample, identical p-jumps can be repeated to allow for an averaging of the data over several jumps. (4) Pressure allows the activation volumes for folding and unfolding reactions to be determined, a parameter that informs about the hydration of the transition state.

Here, we discuss the application of this method to several model proteins: 33-kDa and 23-kDa proteins from photo-system II, a variant of the green fluorescent protein, and a fluorescent variant of ribonuclease A, and a set of seven hydrophobic variants of this protein. As illustrated by the examples shown thereafter, the pressure-jump relaxation method permits, on one hand, to reveal path dependent protein folding/unfolding reactions, which are hidden in equilibrium unfolding studies and barely accessible by other techniques; and on the other hand, to explore the folding transition state ensemble of a protein, giving direct information about its state of hydration.

## Path dependent protein folding/unfolding reactions

Depending on the model protein and the experimental conditions, significant discrepancies were observed when comparing upward and downward p-jump induced kinetics. In fact, the recent landscape theories of protein folding suggest that individual proteins within a large ensemble may follow different routes in conformation space from the unfolded state toward the native state, and vice versa. The results are in line with a more complex free energy landscape involving multiple pathways.

### *The 33-kDa and 23-kDa proteins from photo-system II*

The thermodynamically predicted equivalency of upward and downward pressure-jump induced relaxation kinetics for a typical two-state protein was observed for the 33-kDa model. The situation becomes more complex with the 23-kDa protein. At 45°C, the relaxation profiles are fast, occurring on the timescale of seconds. Hence, the apparent rates for the folding and unfolding reaction are comparable. However, upon decreasing the temperature to 20°C, and further on to 10°C, increasing differences in the apparent reaction rates are revealed. At 10°C, refolding is about 20 times slower than unfolding.

### *A GFP variant*

Intrigued by the precedent results, we extended our study to other proteins. The GFP variant used appeared to be a very suitable model. At pressures below the overall breakdown of the secondary structure of the native protein, partial protein unfolding results in a quenching of the fluorescence of its co-factor. The associated unfolding kinetics occurs on the millisecond to second time-scale, and is accelerated at high pressure, indicating that the activation volume is negative. In Fig. 11, the observed relaxation rates, $k_{obs}$, are compared for upward and downward pressure jumps within a broad pressure range, from 150 to 500 MPa. We note that the observed rate constants $k_{obs}$ are not the

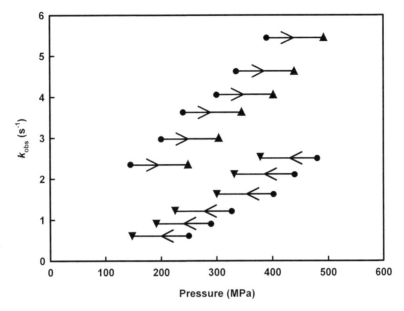

**Figure 11.** Pressure-induced shift of observed rate constants (inverse of relaxation times) for rsGFP after p-jumps of 100 MPa amplitude at pH 5.5 and 5°C. Triangle symbols (▲) and (▼) are the $k_{obs}$ values determined in the pressurizing (right-handed arrow) and depressurizing (left-handed arrow) direction, respectively. The length of the arrows refers to the p-jump amplitude. Figure adapted from (Herberhold et al., 2003).

same for forward and backward pressure-jumps, i.e., the process under consideration is path-dependent, indicating that p-jumps in opposite directions appear to induce kinetics involving different transition states.

## A Fluorescent variant of ribonuclease A

Again, significant deviations from the expected symmetrical protein relaxation kinetics were observed. Whereas downward p-jumps always resulted in single exponential kinetics, the kinetics induced by upward p-jumps was biphasic in the low pressure range and monophasic at higher pressures (Fig. 12). The relative amplitude of the slow phase decreased as a function of both pressure and temperature. At higher temperatures (50°C), only the fast phase remained. These results were interpreted within the framework of a two dimensional energy surface containing a pressure- and temperature-dependent barrier between two unfolded states differing in the isomeric state of the Asn-113-Pro-114 bond. Analysis of the activation volume of the fast kinetic phase revealed a temperature-dependent shift of the unfolding transition state to a larger volume. The observed compensation of this effect by glycerol offered an explanation for its protein stabilizing effect.

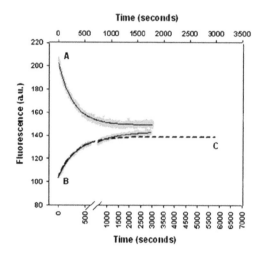

**Figure 12.** p-Jump induced Y115W RNase A folding and unfolding relaxation kinetics at 30°C. Comparison of upward and downward p-jumps led to identical final pressures. The kinetics after a downward p-jump (upper x-axis) is fitted by a single exponential (A), that after an upward p-jump (lower x-axis) by a double (B) and a single exponential (C, dashed line). Figure adapted from (Font et al., 2006).

## The folding transition state ensemble of a protein. The case of hydrophobic variants of ribonuclease A

Investigation of the transition state ensemble is experimentally extremely challenging, since—due to its transient nature—it can not be directly observed. However, some characteristics of this state can be deduced from kinetic experiments, especially by applying the Φ-analysis (Fersht, 2004; Raleigh and Plaxco, 2005), which involves measuring the folding kinetics and equilibrium thermodynamics of variant proteins containing amino acid substitutions. Since experiments under pressure allow in determining reaction and activation volume changes (and hence the change in hydration of the protein upon unfolding), using the βp-value analysis, the relative hydration of the TSE with respect to the folded and unfolded states can be evaluated. This is illustrated by the following example.

The role of hydrophobic interactions in the pressure-folding transition state of ribonuclease A was investigated using the Φ-value and βp-value methods. To this aim, the ratio between the folding activation volume and the reaction volume (βp-value), which is an index of the compactness or degree of hydration of the transition state, was calculated. The results obtained indicate a pressure-folding transition state for the main hydrophobic core of RNase A that is approximately halfway between the native and unfolded states. The results showed that the TSE was a relatively uniformly expanded form of the folded structure presenting a weakened hydrophobic core. This strongly suggests a pressure-folding pathway following a nucleation-condensation mechanism. In addition, a direct comparison of the nature of the transition state inferred from pressure-induced folding studies and the results of the protein engineering method was presented. As shown in Fig. 13, a good correlation was found between the $\Phi$-values and the $\Delta\beta p$-values.

The stage is now well set for more detailed and further work addressing these questions, but also for looking at more complex questions, such as the study of the folding reaction of oligomeric proteins as well as for studies of protein aggregation and the fibrillation phenomena. Of particular interest may be the application of this method to proteins which naturally fold into different thermodynamically stable conformations, such as prion proteins. In fact, for such proteins, identification of distinct path-dependent folding/unfolding reactions might help defining strategies aimed at selecting or modulating particular folding routes.

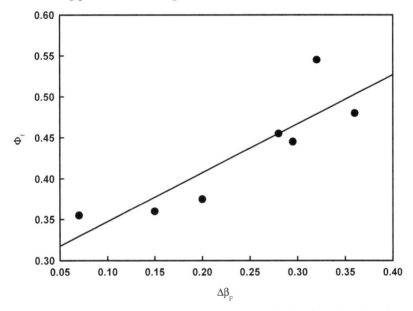

**Figure 13.** Correlation between the fractional $\Phi$-values and the $\Delta\beta p$-values of each variant. The solid line shows the best fit to a linear equation (r = 0.85). Reprinted with permission from Font et al., 2006a.

## CONCLUSION

In this chapter we have analyzed the effects of pressure on several types of protein reactions. The necessity to incorporate pressure as a variable in the thermodynamic analysis of protein kinetics and for getting a closer insight into the structural features of their reaction mechanism have been shown.

Without changing pressure, it seems indeed very difficult to characterize transient interactions of proteins with their hydration shell.

Sometimes people who are less acquainted with pressure, object that pressure variations of some hundred MPa are not in the 'physiological range'—meaning that using such approaches, albeit intellectually interesting, might be void of practical utility. They are wrong for two reasons. First, there are many organisms that do indeed live under high pressure conditions (cf. Chapter "Pressure and Living Organisms" in this book). Second, pressure must be considered as a perturbation parameter—just as is temperature, pH, chemical composition of the solvent, etc. The only difference to pressure is that parameters like temperature or pH have been used for a long time as protein perturbation tools. Though, noone is questioning the use of temperature variations for determining energetic parameters. However, using pressure as a variable is still sometimes considered as something 'exotic'. But isn't real scientific progress possible only by taking 'exotic' paths, thus permitting to overcome traps and barriers of much-tramped paths?

## REFERENCES

Adachi, S. and I. Morishima. 1989. The effects of pressure on oxygen and carbon monoxide binding kinetics for myoglobin. A high pressure laser flash photolysis study. J. Biol. Chem. 264: 18896–18901.

Balny, C. and F. Travers. 1989. Activation thermodynamics of the binding of carbon monoxide to horseradish peroxidase. Role of pressure, temperature and solvent. Biophys Chem. 33: 237–244

Balny, C. and J.L. Saldana and N. Dahan. 1984. High-pressure stopped-flow spectrometry at low temperatures. Anal Biochem. 139: 178–189.

Balny, C. and J.L. Saldana and N. Dahan. 1987. High-pressure stopped-flow fluorometry at subzero temperatures: application to kinetics of the binding of NADH to liver alcohol dehydrogenase. Anal Biochem. 163: 309–315.

Bec, N. and A. Villa, P. Tortora, V.V. Mozhaev, C. Balny, R. Lange, E.V. Kudryashova and C. Balny. 1996. Enhanced stability of carboxypeptidase from *Sulfolobus solfataricus* at high pressure. Biotechnol. Letts. 18: 483–488.

Bredt, D.S. and P.M. Hwang, C.E. Glatt, C. Lowenstein, R.R. Reed and S.H. Snyder. 1991. Cloned and expressed nitric oxide synthase structurally resembles cytochrome P-450 reductase. Nature 351: 714–718.

Desai, G. and G. Panick, M. Zein, R. Winter and C.A. Royer. 1999. Pressure-jump studies of the folding/unfolding of trp repressor. J. Mol Biol. 288: 461–475.

Dill, K.A. and H.S. Chan. 1997. From Levinthal to pathways to funnels. Nat Struct Biol. 4: 10–19.

Fersht, A.R. 2004. Relationship of Leffler (Bronsted) alpha values and protein folding Phi values to position of transition-state structures on reaction coordinates. Proc Natl Acad Sci. USA 101: 14338–14342.

Font, J. and A. Benito, R. Lange, M. Ribo and M. Vilanova. 2006a. The contribution of the residues from the main hydrophobic core of ribonuclease A to its pressure-folding transition state. Protein Sci. 15: 1000–1009.

Font, J. and J. Torrent, M. Ribo, D.V. Laurents, C. Balny, M. Vilanova and R. Lange. 2006b. Pressure-jump-induced kinetics reveals a hydration dependent folding/unfolding mechanism of ribonuclease A. Biophys J. 91: 2264–2274.

Gorren, A.C. and N. Bec, A. Schrammel, E.R. Werner, R. Lange and B. Mayer. 2000. Low-temperature optical absorption spectra suggest a redox role for tetrahydrobiopterin in both steps of nitric oxide synthase catalysis. Biochemistry 39: 11763–11770.

Gorren, A.C. and M. Sorlie, K.K. Andersson, S. Marchal, R. Lange and B. Mayer. 2005. Tetrahydrobiopterin as combined electron/proton donor in nitric oxide biosynthesis: cryogenic UV-Vis and EPR detection of reaction intermediates. Methods Enzymol. 396: 456–466.

Gross, M. and R. Jaenicke. 1994. Proteins under pressure. The influence of high hydrostatic pressure on structure, function and assembly of proteins and protein complexes. Eur J. Biochem. 221: 617–630.

Hasinoff, B.B. 1974. Kinetic activation volumes of the binding of oxygen and carbon monoxide to hemoglobin and myoglobin studied on a high-pressure laser flash photolysis apparatus. Biochemistry 13: 3111–3117.

Herberhold, H. and S. Marchal, R. Lange, C.H. Scheyhing, R.F. Vogel and R. Winter. 2003. Characterization of the pressure-induced intermediate and unfolded state of red-shifted green fluorescent protein—a static and kinetic FTIR, UV/VIS and fluorescence spectroscopy study. J. Mol Biol. 330: 1153–1164.

Heremans, K. and J. Snauwaert and J. Rijkenberg. 1980. Activation and reaction volumes for redox reactions of horse-heart cytochrome c with inorganic reagents. Rev. Sci. Instrum. 51: 806–808.

Hoa, G.H. and G. Hamel, A. Else, G. Weill and G. Herve. 1990. A reactor permitting injection and sampling for steady state studies of enzymatic reactions at high pressure: tests with aspartate transcarbamylase. Anal Biochem. 187: 258–261.

Jacob, M. and G. Holtermann, D. Perl, J. Reinstein, T. Schindler, M.A. Geeves and F.X. Schmid. 1999. Microsecond folding of the cold shock protein measured by a pressure-jump technique. Biochemistry 38: 2882–2891.

Jung, C. and N. Bec and R. Lange. 2002. Substrates modulate the rate-determining step for CO binding in cytochrome P450cam (CYP101). A high-pressure stopped-flow study. Eur J. Biochem. 269: 2989–2996.

Kornblatt, M.J. and R. Lange and C. Balny. 1998. Can monomers of yeast enolase have enzymatic activity? Eur J. Biochem. 251: 775–780.

Kornblatt, M.J. and R. Lange and C. Balny. 2004. Use of hydrostatic pressure to produce 'native' monomers of yeast enolase. Eur. J. Biochem. 271: 3897–3904.

Lange, R. (ed.). 2006. <u>Proteins under High Pressure</u>. Bioch. Biophys. Acta—Proteins and Proteomics Elsevier. Amsterdam.

Lange, R. and I. Heiber-Langer, C. Bonfils, I. Fabre, M. Negishi and C. Balny. 1994. Activation volume and energetic properties of the binding of CO to hemoproteins. Biophys J. 66: 89–98.

Lange, R. and N. Bec, V.V. Mozhaev and J. Frank. 1996. Fourth derivative UV-spectroscopy of proteins under high pressure. II. Application to reversible structural changes. Eur Biophys J. 24: 284–292.

Lange, R. and N. Bec, P. Anzenbacher, A.W. Munro, A.C. Gorren and B. Mayer. 2001. Use of high pressure to study elementary steps in P450 and nitric oxide synthase. J. Inorg Biochem. 87: 191–195.

Lipscomb, J.D. and S.G. Sligar, M.J. Namtvedt and I.C. Gunsalus. 1976. Autooxidation and hydroxylation reactions of oxygenated cytochrome P-450cam. J. Biol. Chem. 251: 1116–1124.

Makris, T.M. and R. Davydov, I.G. Denisov, B.M. Hoffman and S.G. Sligar. 2002. Mechanistic enzymology of oxygen activation by the cytochromes P450. Drug Metab Rev. 34: 691–708.

Makris, T.M. and I.G. Denisov and S.G. Sligar. 2003. Haem-oxygen reactive intermediates: catalysis by the two-step. Biochem. Soc Trans. 31: 516–519.

Marchal, S. and H.M. Girvan, A.C. Gorren, B. Mayer, A.W. Munro, C. Balny and R. Lange. 2003. Formation of transient oxygen complexes of cytochrome p450 BM3 and nitric oxide synthase under high pressure. Biophys. J. 85: 3303–3309.

Marchal, S. and A.C. Gorren, M. Sorlie, K.K. Andersson, B. Mayer and R. Lange. 2004. Evidence of two distinct oxygen complexes of reduced endothelial nitric oxide synthase. J. Biol. Chem. 279: 19824–19831.

Marchal, S. and A.C. Gorren, K.K. Andersson and R. Lange. 2005. Hunting oxygen complexes of nitric oxide synthase at low temperature and high pressure. Biochem. Biophys. Res. Commun. 338: 529–535.

Marletta, M.A. 1994. Approaches toward selective inhibition of nitric oxide synthase. J Med Chem. 37: 1899–1907.

McMillan, K. and D.S. Bredt, D.J. Hirsch, S.H. Snyder, J.E. Clark and B.S. Masters. 1992. Cloned, expressed rat cerebellar nitric oxide synthase contains stoichiometric amounts of heme, which binds carbon monoxide. Proc. Natl Acad. Sci. USA 89: 11141–11145.

Mitra, L. and K. Hata, R. Kono, A. Maeno, D. Isom, J.B. Rouget, R. Winter, K. Akasaka, B. Garcia-Moreno and C.A. Royer. 2007. V(i)-value analysis: a pressure-based method for mapping the folding transition state ensemble of proteins. J. Am Chem. Soc. 129: 14108–4109.

Mohana-Borges, R. and J.L. Silva, J. Ruiz-Sanz and G. de Prat-Gay. 1999. Folding of a pressure-denatured model protein. Proc. Natl Acad. Sci. USA 96: 7888–7893.

Morild, E. 1981. The theory of pressure effects on enzymes. Adv. Protein Chem. 34: 93–166.

Mozhaev, V.V. and R. Lange, E.V. Kudryashova and C. Balny. 1996. Application of high hydrostatic pressure for increasing activity and stability of enzymes. Biotechnol Bioeng. 52: 320–331.

Occhipinti, E. and N. Bec, B. Gambirasio, G. Baietta, P.L. Martelli, R. Casadio, C. Balny, R. Lange and P. Tortora. 2006. Pressure and temperature as tools for investigating the role of individual non-covalent interactions in enzymatic reactions *Sulfolobus solfataricus* carboxypeptidase as a model enzyme. Biochim. Biophys. Acta 1764: 563–572.

Panick, G. and R. Winter. 2000. Pressure-induced unfolding/refolding of ribonuclease A: static and kinetic Fourier transform infrared spectroscopy study. Biochemistry 39: 1862–1869.

Panick, G. and G.J. Vidugiris, R. Malessa, G. Rapp, R. Winter and C.A. Royer. 1999. Exploring the temperature-pressure phase diagram of staphylococcal nuclease. Biochemistry 38: 4157–4164.

Pappenberger, G. and C. Saudan, M. Becker, A.E. Merbach and T. Kiefhaber. 2000. Denaturant-induced movement of the transition state of protein folding revealed by high-pressure stopped-flow measurements. Proc. Natl Acad. Sci. USA 97: 17–22.

Perl, D. and G. Holtermann and F.X. Schmid. 2001. Role of the chain termini for the folding transition state of the cold shock protein. Biochemistry 40: 15501–15511.

Peterson, J.A. and Y. Ishimura and B.W. Griffin. 1972. Pseudomonas putida cytochrome P-450: characterization of an oxygenated form of the hemoprotein. Arch Biochem. Biophys. 149: 197–208.

Raleigh, D.P. and K.W. Plaxco. 2005. The protein folding transition state: what are Phi-values really telling us? Protein Pept Lett. 12: 117–122.

Royer, C.A. 2005. Insights into the role of hydration in protein structure and stability obtained through hydrostatic pressure studies. Braz J. Med. Biol. Res. 38: 1167–1173.

Sligar, S.G. 1999. Nature's universal oxygenases: the cytochromes P450. Essays Biochem. 34: 71–83.

Sligar, S.G. and J.D. Lipscomb, P.G. Debrunner and I.C. Gunsalus. 1974. Superoxide anion production by the autoxidation of cytochrome P450cam. Biochem. Biophys. Res. Commun. 61: 290–296.

Smeller, L. and  H. Roemich and R. Lange. 2006. Proteins under high pressure. Biochim. Biophys. Acta 1764: 329–330.

Socci, N.D. and  J.N. Onuchic and P.G. Wolynes. 1998. Protein folding mechanisms and the multidimensional folding funnel. Proteins 32: 136–158.

Stuehr, D.J. and M. Ikeda-Saito. 1992. Spectral characterization of brain and macrophage nitric oxide synthases. Cytochrome P-450-like hemeproteins that contain a flavin semiquinone radical. J. Biol. Chem. 267: 20547–20550.

Tan, C.Y. and C.H. Xu, J. Wong, J.R. Shen, S. Sakuma, Y. Yamamoto, R. Lange, C. Balny and K.C. Ruan. 2005. Pressure equilibrium and jump study on unfolding of 23-kDa protein from spinach photosystem II. Biophys. J. 88: 1264–1275.

Torrent, J. and C. Balny and R. Lange. 2006a. High pressure modulates amyloid formation. Protein Pept. Lett. 13: 271–277.

Torrent, J. and J. Font, H. Herberhold, S. Marchal, M. Ribo, K. Ruan, R. Winter, M. Vilanova and R. Lange. 2006b. The use of pressure-jump relaxation kinetics to study protein folding landscapes. Biochim. Biophys Acta. 1764: 489–496.

Vidugiris, G.J. and J.L. Markley and C.A. Royer. 1995. Evidence for a molten globule-like transition state in protein folding from determination of activation volumes. Biochemistry 34: 4909–4912.

White, K.A. and M.A. Marletta. 1992. Nitric oxide synthase is a cytochrome P-450 type hemoprotein. Biochemistry 31: 6627–6631.

Woenckhaus, J. and R. Kohling, P. Thiyagarajan, K.C. Littrell, S. Seifert, C.A. Royer and R. Winter. 2001. Pressure-jump small-angle x-ray scattering detected kinetics of staphylococcal nuclease folding. Biophys. J. 80: 1518–1523.

# 2

# Protein Folding and Aggregation under Pressure

*László Smeller*

## INTRODUCTION

### The protein folding problem

The folding problem, namely how a freshly synthesized polypeptide chain reaches its secondary, tertiary and sometimes quaternary structure belonging to the native state, has been studied widely in the last decades (Pain, 2000; Dobson, 2003; Gruebele, 2002, 2005; Royer, 2008). The field was initiated by Anfinsen who unfolded the bovine pancreatic ribonuclease and found that after dialyzing the denaturant out, the protein folded back and showed enzymatic activity. He concluded that the primary structure, which remains intact in the denatured state, unequivocally determines the native conformation of the protein (Anfinsen, 1973). This principle is known today as the Anfinsen dogma.

It is clear now that Anfinsen was fortunate with this experiment because his protein did not misfold, which is common nature for proteins, but his conclusion gave rise to a new area of research.

A theoretical approach by Levinthal led to a paradox. If one estimates the number of conformations, which are available for an average sized protein, using a simple approximation of 2 rotamers (rotational states) pro residue, the total number of possible conformations is an astronomic number ($\sim 10^{30}$ for a protein of 100 residues). Assuming that the protein needs only 10 ps to try one conformation, the time to sample the whole conformational space is much longer than the age of the universe ($4 \; 10^{11}$ years, while the

Semmelweis University, Dept. Biophysics and Radiation Biology, Tûzoltó u. 37–47, Budapest IX. H-1444 Pf 263.
E-mail: laszlo.smeller@eok.sote.hu

universe is c.a. $10^{10}$ years old). It is clear that proteins do not fold by a random search in the conformational space. The above calculation, ( known today as Levinthal's paradox) led to the idea of directed folding. Levinthal himself used such a calculation to support his idea of a directed folding, rather than believing in the random walk-type folding. According to this idea the possible choices in the folding path are restricted by the previous folding steps. This is visualized by the so called funnel model. (Fig. 1.) The funnel model was invented in order to visualize the nonrandom nature of the process (Bryngelson et al., 1995). The conformation of the protein backbone can be determined by two angles called dihedral angles ($\phi$, $\psi$). This spans a 200 dimensional conformational space for a protein of 100 residues. For characterization of the complete structure one needs further coordinates, giving a conformational space with several hundred or thousand dimensions. Each point in this space represents a unique conformation with certain energy. Plotting the energy versus the coordinates of the conformational space one gets the so called energy landscape. In order to visualize the energy landscape one reduces on the plot the number of the conformational coordinates to two, resulting in a three-dimensional plot seen in Fig. 1. According to the funnel hypothesis this plot has one global minimum. The side of the funnel is a rugged surface with several local minima, representing intermediate states with different lifetime.

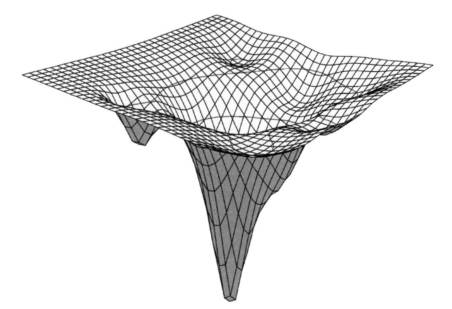

**Figure 1.** The folding funnel.

It has to be mentioned that the folding process *in vivo* often starts co-translationally. This means that fractions at the N-terminus of the protein start to fold before the completion of the synthesis of the C-terminus. Beside this, there are a number of helper proteins, called chaperones which facilitate the folding process. They also prevent aggregation of the freshly synthesized polypeptide in the crowded environment of the cell. A separate chapter of this book is dedicated to the effect of pressure on chaperones (Tölgyesi and Böde, 2009).

## Misfolding and aggregation. Conformational diseases

Folding of the polypeptide chain into the native conformation is not always successful. There are intermediate conformations on the folding pathway where intermolecular interactions can become favorable, which leads to aggregation and misfolding of the protein. These aggregates can be deposited either in the cell or in the extracellular space. Such insoluble tissue deposits of water soluble proteins can be observed in several serious diseases (like Alzheimer, diabetes II, Creuzfeld-Jacob). In these diseases partially folded or misfolded proteins form amyloid fibers, which can lead to tissue death (Booth et al., 1997). The mechanism of the formation of the fibers is not yet fully understood. In mature fibers proteins are arranged in antiparallel β-sheet structure in such a way, that the polypeptide chains are perpendicular to the direction of the fiber (Fink, 1998; Dobson, 2003). The proteins attached to each other with hydrogen bonds typical in antiparallel β-sheets.

The 'conformational diseases' described above usually appear when the protein, which is aggregating, is present in a form with a reduced stability. The most probable reason for this is that the formation of the amyloid fibers is preceded by a partial unfolding of the protein. The destabilization of the native structure can be a result of a mutation (like in the case of lysozyme, two mutants of which known to cause familial amyloidosis), fragmentation or can be a consequence of interactions with other proteins (like in case of Creuzfeld-Jacob syndrome). Kendrick et al. (1998) showed that the aggregation of human interferon is preceded by the transient expansion of the molecule, which suggests that proteins should be partially unfolded or destabilized in order to be able to aggregate.

More than a dozen proteins and protein families are known today which cause conformational diseases due to aggregation of their metastable conformations (Carrell and Gooptu, 1998; Meersman and Dobson, 2006). These proteins already have (intramolecular) β-strands, which were thought of as precursors of the intermolecular structuring, but according to the current view all proteins can form fibers under appropriate conditions (Fändrich et al., 2001).

Formation of fibrous aggregates is however the last step of a complex pathway. Aggregation prone protein chains usually form amorphous aggregates first, followed by profilaments and fibers. The formation of fibrous aggregates is usually very slow, it takes several days *in vitro*, but *in vivo* it can even take years. The stability of the aggregates increases and the well packed fibers are then the most resistant ones both against chemical and physical effects.

## PRESSURE INDUCED UNFOLDING

Pressure is an intensive thermodynamic parameter, equally important to temperature. It is not surprising, that pressure has pronounced effect on the folded–unfolded equilibrium of proteins (Heremans and Smeller, 1998). The first observation about the unfolding effect of pressure was published by Bridgman (1914). He pressurized the white of the egg and noticed that after the pressure cycle it had "an appearance much like that of a hard boiled egg". Of course this time neither the structure nor the folding process of the protein was known, the observation could not be interpreted in the way we can see the case now. The next milestone was done by Brandts et al. ( 1970) and Hawley (1971), who performed a systematic investigation of the unfolding of ribonuclease and chimotrypsinogen as function of pressure and temperature. Hawley also presented a short thermodynamic theory describing his observations. This is known today as the elliptic phase diagram of the proteins. The theory predicts an elliptic region on the pressure-temperature plane, inside which the native state is more stable than the unfolded, while outside of the ellipse the unfolded state is more stable (Fig. 2). The theory was based on the equilibrium thermodynamic description of a system containing only two states, namely the folded and unfolded states.

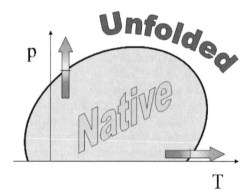

**Figure 2.** The elliptic phase diagram of the proteins (for further explanations, see text).

The Gibbs free energy difference

$$\Delta G = G_{\text{unfolded}} - G_{\text{folded}}$$

between the unfolded and folded state was written using the first and second order terms of $p$ and $T$. The $\Delta G = 0$ transition curve is mathematically a second order curve, because terms up to the second derivatives of $\Delta G$ versus $T$ and $p$ were taken into account. This curve is an ellipse, in a certain range of the thermodynamic parameters, like the change in heat capacity ($\Delta C_p$), compressibility ($\Delta\beta$) and thermal expansion ($\Delta\alpha$) during the unfolding. However later it was pointed out that the Hawley theory has several serious limitations: 1) it assumes that $\Delta C_p$, $\Delta\beta$, $\Delta\alpha$ are independent of the temperature and pressure, 2) the two-state system approach is too simple, there are intermediate and aggregated states, which have also to be taken into account. While the appearance of the third order terms due to the temperature and pressure dependence of $\Delta C_p$, $\Delta\beta$, $\Delta\alpha$ does not change the elliptic shape drastically (Smeller and Heremans, 1997), the new states will change the phase diagram completely. The extension of the Hawley-diagram will be discussed later in this chapter.

There were several experimental investigations performed on a number of proteins to determine the actual shape of the elliptic curve (Zipp and Kauzmann, 1973; Heinisch et al., 1995; Zhang et al., 1995; Takeda et al., 1995; Panick et al., 1999; Torrent et al., 2001; Winter, 2002). The typical pressure for the protein unfolding is in the range of 5–10 kbar (0.5–1 GPa) at room temperature (Smeller, 2002). Some pressure sensitive proteins can denature at lower pressure values, even at 1–2 kbar (Ruan et al, 2003). The pressure studies of aqueous protein solutions are limited by the freezing of water, which is slightly above 1GPa at room temperature. There are also extremely stable proteins, such as the heat shock protein MjHSP16.5 which was isolated from a thermophile (Tölgyesi et al., 2004).

The Hawley theory is a completely universal, phenomenological theory, without using any specific feature of proteins. It is not surprising therefore, that one can find similar elliptic phase diagrams for other systems, like liquid crystals (Cladis, 1988), polymers (Kunugi et al., 1997; Kato, 2005), and starch (Rubens and Heremans, 2000; Baks et al., 2008).

## FOLDING EXPERIMENTS

The folding experiments all start with a sample, containing the aqueous solution of the folded protein. That is why the experiment must start with an unfolding step. The following (re)folding step is the actual subject of the investigation. Consequently there are two important technical parameters, the method of the unfolding (the nature of the unfolding 'agent' ) and the spectroscopic technique used in the observation of the folding process.

The most widely used technique is the so called stopped flow technique (Table 1), where the protein is denatured by a chemical agent, usually urea or guanidine hydrochloride (GuHCl), and the solution is diluted by fast mixing of buffer and the denatured protein containing solution (Volk, 2001). If the concentration of the denaturant decreases below a critical limit, folding starts and can be observed by any appropriate spectroscopic methods. It hast to be mentioned, that stopped flow experiments are possible under pressure, (Heremans et al., 1980) but these equipments are used in study of enzyme kinetics rather than protein folding under pressure (Masson and Balny, 2005).

Alternatively one can unfold the protein by high pressure and the folding can be observed after a fast decrease of the pressure. This is known as the pressure-jump technique, which will be discussed later in detail.

The fastest method to initiate protein folding or unfolding is the temperature jump (Table 1). In this case a fast heating of the sample is induced by a Nd-YAG laser pulse, which is Raman shifted in a hydrogen gas cell, in order to obtain a pulse with the wavelength of 1.9 μm, which is absorbed by the water (falls in the range of the water vibration overtone). This way the sample is heated in its whole volume, avoiding delay and inhomogeneity coming from slow heat conduction. The typical temperature jump is 15°C which can be reached in 10 ns, resulting in a heating rate over $10^9$ K/s. Since this jump can only be performed in the direction of the heating, the subject of the experiments is limited to those proteins that can be cold-unfolded above the freezing point of the solvent (Ballew et al., 1996a and b; Osvath et al., 2003).

Because of its high sensitivity fluorescence spectroscopy is used as a detection method in most cases. Detection of the fluorescence of intrinsic chromophores, mostly that of tryptophan residues is widely used (Liu and Gruebele, 2007). Circular dicroism or infrared spectroscopy provide an alternative to fluorescence, but their time resolution is considerably lower. In Fourier transform infrared spectroscopy (FTIR) there is a special technique called 'step-scan', with a resolution in the range of ns, (Meilleur et al., 2004) but it can only be used with samples that can be unfolded-refolded in a reproducible way several thousand times, which is not typical for proteins.

The folding time of the various secondary and tertiary structure elements is significantly different (Bieri and Kiefhaber, 1999). It is not surprising that the characteristic folding time relates strongly to the size of the motif. The fastest folding times are achieved by small loops. Alpha helices formed by a nucleation process, require at least one turn formed at the beginning. Building up a beta sheet needs an order of magnitude longer time. The characteristic folding times of different polypeptide structures are presented in the Fig. 3.

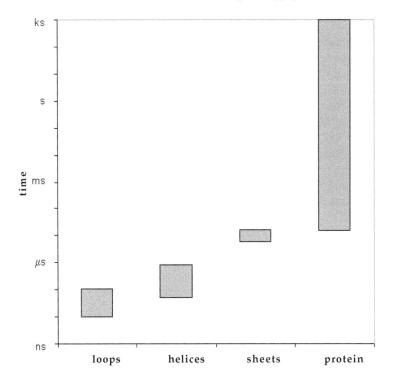

**Figure 3.** Characteristic folding times of different motifs.

## PRESSURE JUMP TECHNIQUE

Pressure jump experiments have several advantages over the classical stopped flow technique. First of all, the pressure can be released completely, while the concentration of the chemical denaturant can never be diluted to zero. Unlike the temperature jump, pressure jump can be performed in both directions; one can observe both the unfolding and refolding kinetics (Ruan et al., 2003).

For applying pressure one can use two basically different methods. In the first approach one uses a strong pressure container, known as a thick wall cylinder, equipped with special windows usually made of sapphire for spectroscopic studies (Sherman and Stadmuller, 1987). The second approach is called diamond anvil cell (DAC), which compresses a very small sample between two diamond anvils (Jayaraman, 1986). There are several variants of these two main types with various types of pumps and windows for the cylindrical cell, or driving systems for the DAC.

The speed of the pressure jump is limited by the technique by which the pressure is created (Table 1). For pressure systems with the thick wall cylinder type pressure cell a pneumatic driven valve is used to change the pressure. In this way both positive and negative pressure-jumps can be achieved with the amplitude of 100 MPa in the pressure range from 0.1 to 600 MPa. The dead-time of the setup is about 5–10 ms (Woenckhaus, 2000; Winter, 2002; Ruan et al., 2003; Tan et al., 2006). Another group uses piezoelectric actuator for fast pressure jump (Jacob et al., 1999; Pearson et al., 2002). The cell used in this study is very simple; it contains a sapphire cylinder, with elastic sealing at the ends. The two ends of the cylinder are directly attached to a piezoelectric actuator and a pressure transducer, respectively. The first version of that cell could perform a pressure jump of a few bars (Clegg and Maxfield, 1976), but the design has been improved, the current limit is about hundred bars (few 10 MPa). This pressure step is performed within 50–100 µs (Jacob et al., 1999).

**Table 1.** Time ranges of the different techniques used in the folding studies.

| Method | Time scale (sec) |
| --- | --- |
| Stopped flow | $10^{-3}$–$10^2$ |
| Pressure jump | $5\ 10^{-5}$–$10^3$ |
| Laser T-jump | $10^{-8}$ –$10^{-3}$ |

A simple pressure jump system for FTIR spectroscopy has been recently constructed for pressures up to 10 MPa using the pumps of a high pressure liquid chromatograph (Schiewek et al., 2007) which resulted in a few ms decay time of the pressure.

For the study of the slow process of protein folding, or for the study of the aggregation, one can simply release the pressure by opening a valve by hand, or even turning the screw pump in a backward direction.

To my knowledge there are no fast pressure jump experiments published for the DAC technique. Several studies have been performed in DAC, however, on slow aggregation processes that followed the pressure release.

One needs to mention the adiabatic heating-cooling effect (depending on the direction of the jump) that accompanies pressure jumps. This has to be taken into account when the pressure jump is fast. The adiabatic heating-cooling causes a temperature change of less than 2°C/kbar (Winter, 2002). If the studied process is fast, one needs fast p-jump, and there is no time for temperature relaxation. The temperature, where the folding process occurs is different from the one before the p-jump, as a consequence of the adiabatic heating-cooling. Studying slow processes, it is possible to decrease the pressure slow enough to avoid significant temperature change due to the adiabatic effect. The first few seconds are in this case unuseable even if one uses fast p-jump, because the temperature is not constant in this range.

## PRESURE-JUMP PROTEIN FOLDING EXPERIMENTS

A piezo-activated submicrosecond folding device was used by Jacob et al. (1999) to study the folding of the cold shock protein of *Bacillus subtilis*. This is a small (66 residue long) beta barrel protein, with no helices, showing a pure two-state folding behavior, which made it ideal for these studies. The limited amplitude of the pressure jump (up: $3 \rightarrow 160$ bar or down: $160 \rightarrow 3$ bar) made it necessary to use additional chemical denaturant. The pressure jump caused 6% change in the fluorescence signal if the jump was applied at the middle point of the unfolding curve, i.e. at 1.5 M GuHCl. Comparing this change with the 50% fluorescence decrease upon complete unfolding of the protein achieved by 3 M GuHCl, the pressure induced equilibrium change was estimated as 78% – >54%. These values resulted in a rough estimate $\Delta V = 60 \pm 20 \text{cm}^3/\text{mol}$ of the unfolding volume change. In a systematic study of the wild type and three mutant proteins they measured the folding and unfolding rate constants, repeating the pressure jump experiment at several temperatures. These experiments allowed the determination of the thermodynamic parameters of the unfolding ($\Delta G$, $\Delta H$, $\Delta S$, $\Delta C_p$) and the corresponding activation parameters ($\Delta G^{\#}$, $\Delta H^{\#}$, $\Delta S^{\#}$, $\Delta C_p^{\#}$). In the three mutants one of the three phenylalanine residues of the central beta-sheet region was substituted by alanine. These substitutions did not change the overall properties of the activated state relative to the unfolded state. The activated state was found to be a native- type one.

Woenckhaus et al. (2000) investigated the folding and unfolding kinetics of a small single-domain protein called staphylococcal nuclease (SNase) using small-angle X-ray scattering. This protein was proven to fold reversibly, and served as a model protein in several previous stopped flow studies. In contrast to denaturant and pH-jump studies, pressure-jump folding was found to have a single exponential behavior (Vidugiris et al., 1995, Panick et al., 1998, 1999; Woenckhaus et al., 2001). There was a big difference between the folding and unfolding rates of SNase. While the folding after the jump from 400 MPa to 80 MPa proceeded with a short (4,5 s time constant, the unfolding at 300MPa had an almost 200 times longer time constant (14 min). The similarity of the folding activation volume (89 cm³/mol) and the equilibrium folding volume change (81 cm³/mol) led the authors to the conclusion that the transition state is highly native-like.

Pressure jump experiments on the 33 kDa protein of spinach photosystem (Ruan et al., 2001) showed folding and unfolding relaxation times in the same order of magnitude (100 and 125 s respectively). In a separate study the same group investigated the effect of sucrose and glycerol on the stability and kinetics of this protein (Ruan et al., 2003). All the rate constants decreased, but in both cases the unfolding rate was influenced more than the refolding one. It was proven earlier, that glycerol can lower the unfolding volume change (Oliviera et al., 1994) leading to the stabilization of the folded state against pressure perturbation.

## MISFOLDING, AGGREGATION AND PRESSURE

### Pressure sensitivity of aggregates and oligomers

Protein aggregates have a small surface compared to the same number of separate monomers, which means that the hydration shell is also smaller. This is the reason why these aggregates cannot form under pressure and tend to dissociate under pressure (Silva and Weber, 1988; Cheung et al., 1998; Gorovits and Horowitz, 1998; Smeller et al., 1999; Meersman et al., 2005).

In their seminal paper Silva and Weber (1988) reported the dissociation of the brome mosaic virus into subunits. If the pressure during the treatment reached 140 MPa the monomer units formed amorphous aggregates instead of re-associating into virus particles. This behavior was explained by the conformational drift theory, i.e. that the dissociated monomer units slightly changed their conformation, which prevented from forming a virus particle again. Gaspar et al. (2001) compared the stability of enveloped and nucleocapsid particles of the alpha virus Mayaro using hydrostatic pressure. Pressure dissociated the capsid particles but could not dissociate the ones with an envelope.

Pressure-dissociation of several olygomer proteins were observed, among them hemoglobin (Schay et al., 2006) and the tetramer form of transthyretin (Ferraro -Gonzales, 2000). In case of transthyretin, the pressure induced a conformation, in which the tryptophan residue become exposed to the solvent. Although only a small fraction of monomers was observed, the tetramer which appeared after the pressure cycle had different conformation and increased aggregation capability (Ferraro-Gonzales, 2000). These aggregated conformations were characterized by Cordeiro et al. (2006) using FTIR spectroscopy.

An interesting application of the pressure sensitivity of aggregates is the pressure assisted protein recovery (refolding) from inclusion bodies. (St John et al., 1999; Lee et al., 2006). Inclusion bodies are protein aggregates formed in the bacteria during biotechnological expression of hardly soluble proteins. Pressure can dissociate these aggregates, and facilitate the correct refolding of the protein under appropriate conditions (Randolph et al., 2002).

Infrared spectroscopy is a suitable tool for the simultaneous study of the conformation of the protein and its aggregation during and after pressure treatment. Myoglobin has been shown to have an increased aggregation tendency after a pressure unfolding-refolding cycle (Smeller et al., 1999). This was explained by a highly populated intermediate state, which was folded or aggregated competitively. Refolding of horseradish peroxidase, in contrary, did not populate aggregation prone intermediate states, unless, it was modified, in a way that decreased its stability (Smeller et al., 2003). Substitution of the heme group by a fluorescent porphyrine, or binding of a substrate did not lead to aggregation. Stronger destabilization, however, like the removal of the heme, cleavage of SS-bridges or $Ca^{2+}$-removal, led to a metastable intermediate, and finally to aggregated states.

Similarly to myoglobin and destabilized horseradish peroxidases, lysozyme was also found to form amorphous aggregates after an unfolding pressure cycle. Subdenaturing pressure treatment did not induce aggregation, which means that the intermediate state forms only after the unfolding of the protein. This conformation is metastable under pressures lower than the denaturation limit (i.e. inside the elliptic shape on Fig. 2).

These experiments can be summarized by an energy landscape depicted in the Fig. 4.

At the end of this section it should be noted, that the term 'pressure induced aggregation' can cause confusion. It could give the impression that pressure induces aggregation, as it is inducing unfolding, i.e. the proteins aggregate *under* pressure. This is however not the case, because if one reads these publications carefully, the aggregation is detected after

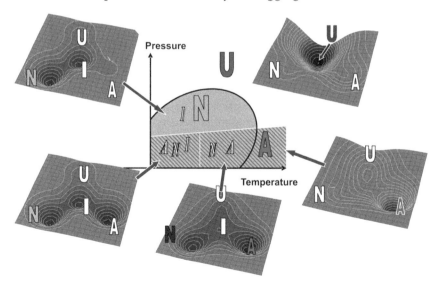

**Figure 4.** Pressure–temperature phase diagram of lysozyme and the corresponding Gibbs-free energy landscapes. Open letters show metastable states, corresponding to local minima on the landscape, while closed letters indicate stable thermodynamic states, which belong to global minima.
A: Aggregated; I: Intermediate; N: Native; U: unfolded.
(Reprinted from Smeller et al. (2006) with permission from Elsevier)

pressure release. The pressure treatment converts the protein into an aggregation prone state, which is aggregating only *after* the pressure treatment. The name 'pressure treatment induced aggregation' would be a more precise name for this phenomenon.

## Pressure sensitivity of fibrous protein aggregates

Contrary to amorphous aggregates and olygomers, mature fibers are quite pressure resistant. Dirix et al. (2005) studied the 105–115 fragment of transthyretin using FTIR spectroscopy and AFM (atomic force microscopy). While the early (amorphous) aggregates could be dissociated by pressures as high as 220 MPa, the mature fibrils were resistant to 1.3 GPa. This supports the idea that fiber formation needs more flexible structure, but when the fibers are formed, their close packed nature makes them resistive against most physical and chemical intervention (Meersman and Dobson, 2006).

The high pressure tolerance of the matured fibrils observed in conformational diseases, limits the possible use of pressure as preventive treatment for foodstuff against the proliferation of the conformational diseases (Dubois et al., 1999; Heindl et al., 2006). Inactivation of the pressure resistant subpopulation of highly infectious prion subpopulations needed three cycles of pressurization at 800 MPa and at 60 C, (Heindl et al., 2008) conditions which would denature almost all proteins.

Radovan et al. (2008) studied a 37-amino acid peptide hormone, called islet amyloid polypeptide (IAPP), which is the principal component of the islet amyloid in type II diabetes mellitus. The deposition of these amyloids in the extracellular matrix of beta-cells, leads to islet cell death. The fragments 1–19 and 1–29 were resistant to pressure treatment, which was explained by a more densely packed aggregate structure with less void volume and strong cooperative hydrogen bonding. They also observed that the 30–37 amino acids region of the peptide is less densely packed, leading to pressure sensitivity of the full-length IAPP.

Sawaya et al. (2007) observed a 'steric zipper' like structure formed by segments of various amiloid proteins. This very efficiently packed, low volume structure can also be a key to the understanding of high pressure resistivity.

Protofibrils of intrinsically denatured disulfide-deficient variant of hen egg white lysozyme have been found to have increased volume and compressibility (Akasaka et al., 2007). The volume increase during protofilament formation was found to be 570 ml/mol, which is a few times larger than the typical volume change during pressure unfolding. This can explain the increased pressure sensitivity of the nonmature aggregates in accordance with the FTIR studies on TTR peptide (Dirix et al., 2005).

## CONCLUSION

High pressure can unfold proteins which consequently can be used for (re)folding studies. During the refolding, metastable states can be populated, which have increased aggregation propensity. The aggregation of proteins has a multistep nature, ending in very stable fibrous aggregates, which play a role in many serious conformational diseases, like BSE, CJD, Alzheimer's and Parkinson diseases. Mild pressure can reverse the amorphous aggregation, but the mature fibers are even more stable against pressure than the native folded structure. The following scheme summarizes the possible conformations of the proteins including folding / unfolding and aggregation pathways and the pressure effect on the conformational changes.

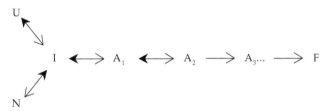

U: unfolded, I: intermediate, $A_1 A_2$, $A_3$: aggregated to different extent, F> fibrous aggregate, N: Native pressure promotes the transitions in the direction of the full arrow.

As it can be seen from the scheme, pressure can be used to guide the protein on the folding-unfolding-misfolding pathway. Also some thermodynamic parameters, that can only be determined from pressure studies also have relevance in the chemical-biochemical processes occurring at ambient pressure.

## ACKNOWLEDGEMENT

The author thanks Sz. Osvath for useful discussions and for carefully reading the manuscript. Financial support of OTKA 49213 grant is also acknowledged.

## REFERENCES

Akasaka, K. and A.R.A. Latif, A. Nakamura, K. Matsuo, H. Tachibana and K. Gekko. 2007. Amyloid protofibril is highly voluminous and compressible. Biochemistry 46: 10444–10450.

Anfinsen, C.B. 1973. Principles that govern the folding of protein chains. Science 181: 223–230.

Baks T. and M.E. Bruins, A.E.M. Janssen and R.M. Boom. 2008. Effect of pressure and temperature on the gelatinization of starch at various starch concentrations. Macromolec. 9: 296–304.

Ballew, R.M. and J. Sabelko and M. Gruebele. 1996a Direct observation of fast protein folding: The initial collapse of apomyoglobin Proc. Natl. Acad. Sci. USA 93: 5759–5764.

Ballew, R.M. and J. Sabelko, C. Reiner and M. Gruebele. 1996b. A single-sweep, nanosecond time resolution laser temperature-jump apparatus. Rev. Sci. Inst. 67: 3694–3699.

Bieri, O. and T. Kiefhaber. 1999. Elementary steps in protein folding. Biol. Chem. 380: 923–929.

Booth, D.R. and M. Sunde, V. Bellotti, C.V. Robinson, W.L. Hutchinson, P.E. Fraser, F.N. Hawkins, S.M. Dobson, S.E. Radford, D.D.F. Blake and M.B. Pepys. 1997. Instability, unfolding and aggregation of human lysozyme variants underlying amyloid fibrillogenesis. Nature 385: 787–793.

Bridgman, P.W. 1914. The coagulation of albumen by pressure J. Biol. Chem. 19: 511–512.

Brandts J.F. and R.J.Oliveira and C. Westort. 1970. Thermodynamics of protein denaturation. Effect of pressure on the denaturation of ribonuclease A, Biochemsistry 9: 1038–1047.

Bryngelson, J.D. and J.N. Onuchic, N.D. Socci and P.G. Wolynes. 1995. Funnels, Pathways and energy landscape of protein folding—A syntesis. Proteins- Struct Funct. Gen. 21: 167–195.

Carrell, R.W. and B. Gooptu. 1998. Conformational changes and disease—serpins, prions and Alzheimer's. Curr. Opin. Struct. Biol. 8: 799–809.

Cheung C.Y. and D.J. Green, G.J. Litt and J.A. Laugharn, Jr. 1998. High pressure mediated dissociation of immune complexes demostrated in model systems Clin. Chem 44: 299–303.

Cladis, P.E. 1988. A 100 year perspective of the reentrant nematic phase. Mol. Cryst. Liq. Cryst. 165: 85–121

Clegg, R.M. and B.W. Maxfield. 1976. Chemiscal kinetic studies by a new small pressure perturbation method. Rew. Sci Instrum. 47: 1383–1393.

Cordeiro, Y. and J. Kraineva, M.C. Suarez, A.G. Tempesta, J.W. Kelly, J.L. Silva, R. Winter and D. Foguel. 2006. Fourier Transform Infrared Spectroscopy Provides a Fingerprint for the Tetramer and for the Aggregates of Transthyretin. Biophys. J. 91: 957–967.

Dirix, C. and F. Meersman, C.E. MacPhee, C.M. Dobson and K. Heremans. 2005. High Hydrostatic Pressure Dissociates Early Aggregates of TTR105–115, but not the Mature Amyloid Fibrils. J. Mol. Biol. 347: 903–909.

Dobson, C.M. 2003. Protein folding and misfolding. Nature: 426, 884–890.

Dubois, J. and A.A. Ismail, S.L. Chan and Z. Ali-Khan. 1999. Fourier transform infrared spectroscopic investigation of temperature- and pressure-induced disaggregation of amyloid A. Scand. J. Immunol. 49: 376–380.

Fändrich, M. and M.A. Fletcher and C.M. Dobson. 2001. Amyloid fibrils from muscle myoglobin. Nature 410: 165–166.

Ferraro-Gonzales, A.D. and S.O. Souto, J.L. Silva and D. Foguel. 2000. The preaggregated state of an azloigogenic protein: Hydrostati pressure converts native transthyretin into amzloidogenic state. Proc. Natl. Acad. Sci. USA 97: 6445–6450.

Fink, A.L. 1998. Protein aggregation: folding aggregates, inclusion bodies and amyloid Folding and Design 3: R9–R23.

Gaspar, L.P. and A.F. Terezan, A.S. Pinheiro, D. Foguel, M.A. Rebello and J.L. Silva. 2001. The metastable state of nucleocapsids of enveloped viruses as probed by high hydrostatic pressure. J. Biol. Chem. 276: 7415–7421.

Gorovits, B.M. and P.M. Horowitz. 1998. High hydrostatic pressure can reverse aggregation of protein folding intermedieates and facilitate acquisition of native structure. Biochemistry 37:6132–6135.

Gruebele, M. 2002. Protein folding: the free energy surface. Curr Opin. Struct. Biol. 12:161–168.

Gruebele, M. 2005. Downhill protein folding: evolution meets physics. C.R. Biologies 328: 701–712.

Hawley, S.A. 1971. Reversible pressure-temperature denaturation of chymotrypsinogen. Biochemistry 10: 2436–2442.

Heindl, P. and A.F. Garcia, P. Butz, E. Pfaff and B. Tauscher. 2006. Protein conformation determines the sensibility to high pressure treatment of infectious scrapie prions. BBA-Proteins Proteomix 1764: 552–557.

Heindl, P. and A. Fernandez-Garcia, P. Butz, B. Trierweiler, H. Voigt, E. Pfaff and B. Tauscher. 2008. High pressure/temperature treatments to inactivate highly infectious prion subpopulations. Innov. Food Sci. Emerging Techn. 9: 290–297.

Heinisch, O. and E. Kowalski, K. Goossens, J. Frank, K. Heremans, H. Ludwig and B. Tauscher. 1995. Pressure effects on the stability of lipoxygenase: Fourier transform infrared spectroscopy and enzyme activity studies. Z. Lebensm. Unters. Forsch. 201: 562–565.

Heremans, K. and L. Smeller. 1998. Protein Structure and Dynamics at High Pressure. Biochim. Biophys. Acta 1386: 353–370

Heremans, K. and J. Snauwaert and J. Rijkenberg. 1980. Stopped-flow apparatus of fast reactions in solution under high pressusre. Rev. Sci. Insturm. 51: 806–808.

Jacob, M. and G. Holtermann, D. Perl, J. Reinstein, T. Schindler, M.A. Geeves and F.X. Schmid. 1999. Microsecond folding of the cold shock protein measured by a pressure-jump technique. Biochemistry 38: 2882–2891.

Jayaraman, A. 1986. Ultrahigh pressures. Rev. Sci. Instrum 57: 1013–1031.

Kato, E. 2005. Thermodynamic study of a pressure-temperature phase diagram for poly (N-isopropylacrylamide) gels. J. Appl. Polymer Sci. 97: 405–412.

Kendrick, B.S. and J.F Carpenter, J.L. Cleland and T.W. Randolph. 1998. A transient expansion of the native state precedes aggregation of recombinant human interferon. Proc. Natl Acad. Sci. USA 95: 14142–14146.

Kunugi, S. and K. Takano, N. Tanaka, K. Suwa and M. Akashi. 1997. Effects of pressure on the behavior of the thermoresponsive polymer poly(N-vinylisobutyramide) (PNVIBA). Macromol. 30: 4499–4501.

Lee, S.H. and J.F. Carpenter, B.S. Chang, T.W. Randolph and Y.S. Kim. 2006. Effects of solutes on solubilization and refolding of proteins from inclusion bodies with high hydrostatic pressure. Protein Sci. 15: 304–313.

Liu, F. and M. Gruebele. 2007. Tuning lambda(6–85) towards downhill folding at its melting temperature. J. Mol. Biol. 370: 574–584.

Masson, P. and C. Balny. 2005. Linear and non-linear pressure dependence of enzyme catalytic parameters BBA 1724: 440–450.

Meersman, F. and C.M. Dobson. 2006. Probing the pressure-temperature stability of amyloid fibrils provides new insights into their molecular properties. Biochim. Biophys. Acta 1764: 452–460.

Meersman, F. and L. Smeller and K. Heremans. 2005. Extending the pressure-temperature state diagram of myoglobin. Helv Chim. Acta 88:546–556.

Meilleur, F. and J. Contzen, D.A.A. Myles and C. Jung. 2004. Structural stability and dynamics of hydrogenated and perdeuterated cytochrome P450cam (CYP101). Biochemistry 43: 8744–8753.

Oliveira, A.C. and L.P. Gaspar, A.T. Dapoian and J.L. Silva. 1994. ARC repressor will not denature under pressure in the absence of water. J. Mol. Biol. 240: 184–187

Osvath, Sz. and J.J. Sabelko and M. Gruebele. 2003.Tuning the Heterogeneous Early Folding Dynamics of Phosphoglycerate Kinase. J. Mol. Biol. 333: 187–199.

Pain, R.H. 2000. Mechanism of Protein Folding Oxford University Press, Oxford.

Panick, G. and R. Malessa, R. Winter, G. Rapp, K.J. Frye and C.A. Royer. 1998. Structural characterization of the pressure-denatured state and unfolding/refolding kinetics of staphylococcal nuclease by synchrotron small-angle x-ray scattering and Fourier-transform infrared spectroscopy. J. Mol. Biol. 275: 389–402.

Panick, G. and J.A. Vidugiris, R. Malessa, G. Rapp, R. Winter and C. Royer. 1999. Exploring the Temperature-Pressure Phase diagram of Staphylochoccal Nuclease. Biochemistry 38: 4157–4164.

Pearson, D.S. and G. Holtermann, P. Ellison, C. Cremo and M.A. Geeves. 2002. A novel pressure-jump apparatus for the microvolume analysis of protein-ligand and protein–protein interactions: its applications to nucleotide binding to skeletal-muscle and smooth-muscle myosin subfragment-1. Biochem J. 366: 643–651.

Radovan, D. and V. Smirnovas and R. Winter. 2008. Effect of pressure on islet amyloid polypeptide aggregation: Revealing the polymorphic nature of the fibrillation process. Biochemsitry 47: 6352–6360.

Randolph, T.W. and M. Seefeldt and J.F. Carpenter. 2002. High hydrostatic pressure as a tool to study protein aggregation and amyloidosis. BBA Proteins Struct. Mol. Enzimol. 1595: 224–234.

Royer, C.A. 2008. The nature of the transition state ensemble and the mechanisms of protein folding. Arch. Biochem. Biophys. 469: 34–45.

Ruan, K.C. and C.H. Xu, T.T. Li, R. Lange and C. Balny. 2003. The thermodynamic analysis of protein stabilization by sucrose and glycerol against pressure-induced unfolding—The typical example of the 33-kDa protein from spinach photosystem II. Eur J. Biochem. 270: 1654–1661.

Ruan, K.C. and C.H. Xu, Y. Yu, J. Li, R. Lange, N. Bec and C. Balny. 2001. Pressure-exploration of the 33-kDa protein from the spinach photosystem II particle Eur. J. Biochem. 268: 2742–2750.

Rubens, P. and K. Heremans. 2000. Pressure-temperature gelatinization phase diagram of starch: An in situ fourier transform infrared study. Biopolymers 54: 524–529.

Sawaya, M.R. and S. Sambashivan, R. Nelson, M.I. Ivanova, S.A. Sievers, M.I. Apostol, M.J. Thompson, M. Balbirnie, J.J.W. Wiltzius, H.T. McFarlane, A.Ø. Madsen, C. Riekel and D. Eisenberg. 2007. Atomic structures of amyloid cross-beta spines reveal varied steric zippers. Nature 447: 453–457.

Schiewek, M. and M. Krumova, G. Hempel and A. Blume. 2007 Pressure jump relaxation setup with IR detection and millisecond time resolution Rew. Sci. Instrum. 78: 045101.

Sherman, W.F. and A.A. Stadmuller. 1987. Experimental Techniques in High Pressure Research, John Wiley & Sons Ltd. Chichester.

Silva, J.L. and G. Weber. 1988. Pressure induced dissociation of brome mosaic virus. J. Mol. Biol. 1999: 149–159.

Schay, G. and L. Smeller, A. Tsuneshige, T. Yonetani and J. Fidy. 2006 Allosteric effectors influence the tetramer stability of both R- and T-states of hemoglobin A.J. Biol. Chem. 281: 25972–25983.

Smeller, L. 2002. Pressure-temperature phase diagrams of biomolecules. Biochim. Biophys. Acta 1595: 11–29.

Smeller, L. and K. Heremans. Some thermodynamic and kinetic consequences of the phase diagram of protein denaturation, pp. 55–58 In: K. Heremans (ed.). 1997. High Pressure Research in Bioscience and Biotechnology, Leuven University Press, Leuven.

Smeller, L. and P. Rubens and K. Heremans. 1999. Pressure effect on the temperature-induced unfolding and tendency to aggregate of myoglobin. Biochemistry 38: 3816–3820.

Smeller, L. and F. Meersman, J. Fidy and K. Heremans. 2003. High pressure FTIR study of the stability of horseradish peroxidase. Effect of heme substitution, ligand binding, $Ca^{++}$ removal and reduction of the disulfide bonds. Biochemistry 370: 859–866.

Smeller, L. and F. Meersman and K. Heremans. 2006. Refolding studies using pressure: The folding landscape of lysozyme in the pressure–temperature plane Biochim. Biophys. Acta 1764: 497–505.

St. John, R.J. and J.F. Carpenter and T.W. Randolph. 1999. High pressure fosters protein refolding from aggregates at high concentrations.Proc. Natl. Acad. Sci. USA 96: 13029–13033.

Takeda, N. and M. Kato and Y. Taniguchi. 1995 Pressure and thermally-induced reversible changes in the secondary structure of ribonuclease A studied by FT-IR spectroscopy. Biochemistry 34: 5980–5987.

Tan, C.Y. and C.H. Xu and K.C. Ruan. 2006. Folding studies of two hydrostatic pressure sensitive proteins. BBA-Proteins Proteomics 1764: 481–488.

Torrent, J. and P. Rubens, M. Ribo, K. Heremans and M. Vilanova. 2001. Pressure versus temperature unfolding of ribonuclease A: An FTIR spectroscopic characterisation of 10 variants at the carboxy-terminal site. Protein Sci. 10: 725–734

Tölgyesi, F. and Cs. Böde, L. Smeller, K.K. Kim, K. Heremans and J. Fidy. 2004. Pressure activation of the chaperone function of small heat-shock proteins. Cellular and Molecular Biology 50: 361–369.

Tölgyesi, F.G. and Cs. Böde. 2009. Pressure and heat shock proteins. *Chapter in this book.*

Vidugiris, G.J.A. and J.L. Markley and C.A. Royer. 1995. Evidence for a molten globule-like transition state in protein folding from determination of activation volumes. Biochemistry 34: 4909–4912.

Volk, M. 2001. Fast initiation of peptide and protein folding processes. Eur. J. Org. Chem. 14: 2605–2621.

Winter, R. 2002. Synchrotron X-ray and neutron small-angle scattering of lyotropic lipid mesophases, model biomembranes and proteins in solution at high pressure. Biochim. Biophys. Acta—Prot. Struct. Mol. Enzymol. 1595: 160–184.

Woenckhaus, J. and R. Kohling, R. Winter, P. Thiyagarajan and S. Finet. 2000 High pressure-jump apparatus for kinetic studies of protein folding reactions using the small-angle synchrotron x-ray scattering technique. Rew. Sci. Instrum. 71: 3895–3899.

Woenckhaus, J. and R. Kohling, P. Thiyagarajan, K.C. Littrell, S. Seifert, C.A. Royer, and R. Winter. 2001. Pressure-Jump Small-Angle X-Ray Scattering Detected Kinetics of Staphylococcal Nuclease Folding. Biophys. J. 80: 1518–1523.

Zhang, J. and X. Peng, A. Jonas and J. Jonas. 1995. NMR study of the cold, heat, and pressure unfolding of ribonuclease A. Biochemistry 34:8361–8641.

Zipp, A. and W. Kauzmann. 1973. Pressure Denaturation of Metmyoglobin, Biochemistry 12: 4217–4228.

# 3

# Pressure and Heat Shock Proteins

*Ferenc G. Tölgyesi[1] and Csaba Böde[2]*

## INTRODUCTION

At physiological conditions the three-dimensional structure of the proteins are uniquely encoded in the amino acid sequence (Anfinsen, 1973; Watters and Baker, 2004). However, most of the proteins are not able to find this structure without help of other proteins. Heat Shock Proteins (HSPs) or chaperones are the members of a protein family which act in different stages of protein folding, they protect damaged proteins and prevent them from aggregation and they assist at refolding. Chaperones thus have an important role in the stress response of the cell. The members of the HSP family are usually categorized by similar functions and structural motifs (Table 1), the family is then divided into HSP100, HSP 90, HSP70, HSP60, HSP33 groups and small Heat Shock Proteins (sHSP).

**Table 1.** Heat shock proteins and their supposed role.

| Family | Example | Supposed role | Reference |
|---|---|---|---|
| HSP100 | ClpB | dissociation of protein aggregates | (Goloubinoff et al., 1999) |
| | ClpA | protease complex | (Ishikawa et al., 2001) |
| HSP90 | HtpG | protection against aggregation | (Buchner, 1999) |
| HSP70 | DnaK | protein folding | (Bukau and Horwich, 1998) |
| HSP60 | GroEL | protein folding | (Grallert and Buchner, 2001) |
| HSP33 | HSP33 | oxidative stress | (Jakob et al., 1999) |
| sHSP | α-crystallin | protection against aggregation | (Haslbeck, 2002) |
| | MjHSP16.5 | | |

[1] Institute of Biophysics and Radiation Biology, Semmelweis University, H-1444 Budapest P.O.B. 263.
E-mail: Ferenc.Tolgyesi@eok.sote.hu
[2] Institute of Biophysics and Radiation Biology, Semmelweis University, H-1444 Budapest P.O.B. 263.
E-mail: Csaba.Bode@eok.sote.hu

## EFFECT OF PRESSURE ON THE HEAT SHOCK PROTEIN PRODUCTION OF CELLS

As HSPs generally play a key role in the stress response of cells, it might be expected that HSPs are also activated at stress conditions imposed by elevated pressure. Whereas higher order structure of proteins is affected only in the pressure region above 100 MPa (Gross and Jaenicke, 1994; Smeller, 2002) cell life is already perturbed in the lower pressure region (Lammi et al., 2004). It is known that high continuous hydrostatic pressure (below 100 MPa) inhibits protein synthesis in general (Hildebrand and Pollard, 1972; Schwarz and Landau, 1972). On the contrary, there are several observations that cells express higher amounts of HSPs at elevated pressures. It has been demonstrated, e.g. that accumulation of heat shock protein 70 occurs under high pressure (30 MPa) independent of the general inhibition of protein synthesis (Kaarniranta et al., 1998).

Moderate elevated pressures occur in living organism as well, e.g. in the eye. A recent work of Salvador-Silva studied the effect of pressure on small heat-shock protein expression in human optic nerve head astrocytes (Salvador-Silva et al., 2001) , thus modelling the stress induced by elevated intraocular pressure. Their experiments detected increased synthesis of Hsp27 at ~0.01 MPa (60 mmHg), but interestingly no change in alpha B-crystallin synthesis. In the pressurized cells, Hsp27 was condensed in large granules around the nucleus. As it was presumed earlier (Quinlan, 2002) that these results suggest that sHSPs may have an important role in stabilizing the cytoskeletal network.

Complexity of pressure effects on cell life is demonstrated by a recent work that investigated the role of HSP genes in the growth and survival of Saccharomyces cerevisiae under high hydrostatic pressure at 25 and 125 MPa (Miura et al., 2006). Interesting and somewhat surprising results about the roles of HSPs in cell survival were obtained. It was shown that cells of strain BY4742 were capable of growth at moderate pressure of 25 MPa and the expression of six HSPs (HSP78, HSP104, HSP10, HSP32, HSP42, and HSP82) was upregulated between 2–4-fold. However, the loss of 1 of the 6 genes did not markedly affect growth at 25 MPa, while on the contrary the loss of HSP31 impaired high-pressure growth. Extremely high pressure of 125 MPa decreased the viability of the wild-type cells to 1% of the control level. Notably, the loss of HSP genes other than HSP31 enhanced the survival rate by about 5 fold at 125 MPa, suggesting that the cellular defensive system against high pressure could be strengthened upon the loss of the HSP genes.

## EFFECT OF PRESSURE ON THE STRUCTURE AND FUNCTION OF HEAT SHOCK PROTEINS

The application of hydrostatic pressure in general shifts equilibria toward the more compact state; it accelerates reactions with a negative activation volume and slows down reactions with positive activation volume.

The structure of proteins are determined by covalent, ionic, hydrophobic and hydrogen bonding. The energy input in a usual compression experiment is too small to influence covalent bonding. Ion pairs are destabilized by hydrostatic pressure, since separated charges arrange water molecules in their vicinity more densely than bulk water. Thus the overall volume change favours the dissociation of ionic interactions under pressure. Similarly, the exposure of hydrophobic groups to water leads to a hydrophobic solvation layer which is assumed to be more densely packed. Formation of hydrogen bonds in biomacromolecules is connected to a negligibly small reaction volume. As a result hydrostatic pressure affects mainly the tertiary and quaternary structure of proteins: low pressures cause elastic response of the protein, moderate pressures will promote the dissociation of oligomeric proteins and higher pressures will unfold proteins (Gross and Jaenicke, 1994; Meersman et al., 2006; Smeller et al., 2002).

The HSP family is very diverse, it contains proteins with various structural motifs and function. Most members of the HSP family have a complex structure, the functional protein molecules have a oligomeric structure. This oligomeric structure may play an important role in the protein function (Haslbeck, 2002). Thus it can be expected that their structure and function will be influenced by pressure considerably. It also means that high pressure methods might be a unique tool to understand the structure-function relationships of oligomeric HSPs.

To our knowledge only some small heat-shock proteins, i.e. $\alpha$-crystallin, HSP16.5 and HSP26, and GroEL, a member of the HSP60 family had been studied so far under high pressure.

### Effect of Pressure on Small Heat Shock Proteins

#### *About Small Heat Shock Proteins*

Small heat-shock proteins (sHSPs) constitute the most variable family of molecular chaperones (Haslbeck, 2002), they can be found in all kingdoms of life (Haslbeck et al., 2005). The possible main role of the sHSPs is to prevent the aggregation of unfolded substrate proteins, and maintain them in a folding-competent state (Kim et al., 1998b). They do not promote the refolding of their substrates, their role is the prevention of the protein aggregation, but they can cooperate with other members of the HSP family

(Mogk et al., 2003). Although the members of the sHSP family are diverse, they all have a similar structural motif, the 'α-crystallin domain', named after the mammalian eye lens protein α-crystallin. The presence of this 80-amino acid length structural motif defines the whole sHSP family (de Jong et al., 1998), It has been shown in several *in vivo* and *in vitro* experiments, that the small heat-shock proteins play an important role in stress response and protect several cellular functions, for instance stabilizing the cytoskeletal network(Carra et al., 2005; Kappe et al., 2001).

Most sHSPs are active in an oligomeric form. The monomer (subunit) size of sHSPs varies between 12 and 42 kDa, these subunits usually form huge oligomers. The oligomerization is shown to be a prerequisite of proper chaperone function (Giese and Vierling, 2002; Kokke et al., 1998; Lentze et al., 2003). The oligomerizational states of sHSPs are different: the individual members of the family exhibit different stress responses and they have different functional quaternary structure (Haley et al., 2000). With high pressure technique three sHSPs have been studied so far: eye lens α-crystallin (Bode et al., 2003; Radlick and Koretz, 1992; Skouri-Panet et al., 2006), yeast HSP26 (Skouri-Panet et al., 2006) and HSP 16.5 from the extremophile Methanococcus Jannaschii (Tolgyesi et al., 2004).

*Alpha-Crystallin.* The oligomeric protein α-crystallin is a main vertebrate eye lens structure protein which is responsible for the transparency of the eye lens (Tardieu, 1998). Beside its structural role in the eye, a decade ago it was proved to be a small heat-shock protein (Horwitz, 1992). As a small-heat shock protein its possible role is to prevent the aggregation of unfolding substrate proteins and maintaining them in a folding competent state. The chaperone-like activity of α-crystallin has been demonstrated with numerous systems among others citrate synthase and alpha-glucosidase (Jakob et al., 1993), which suggests the importance of α-crystallin in the whole human body, and indicates that α-crystallin and sHSPs protect a wide range of cellular functions (Basha et al., 2004).

α-Crystallin exists in the eye lens as a heterooligomeric complex built up by two homologous proteins: αA and αB crystallin. The size of the oligomers is roughly 800 kDa (Horwitz, 2003), but the oligomer population is rather heterogeneous (Raman and Rao, 1997). Due to the lack of its crystal structure, the oligomer structure is not known. Only a cryo-EM image has been published so far, which reveals a spherical structure with an inner hole (Horwitz, 2003), similar to other sHSPs with known crystal structure (Kim et al., 1998a; Stamler et al., 2005; van Montfort et al., 2001). While αA-crystallin can be found only in the eye lens, αB-crystallin is expressed in several tissues and its role for the whole body proved to be essential (Horwitz, 2003).

*Methanococcus Jannaschii HSP16.5 (MjHSP16.5).* MjHSP16.5, a hyperthermophile small heat-shock protein from archebacteria

*Methanococcus Jannaschii* confers protection against thermally induced stress when overexpressed in *Esherichia coli* cells (Kim et al., 2003b), and it prevents the thermally or chemically induced aggregation of certain substrate proteins *in vitro* (Bova et al., 2002; Kim et al., 2003a). MjHSP16.5 is rather ineffective chaperone at room temperature (Bova et al., 2002), but its potency sharply increases above 60–65°C and it can function as an effective chaperone with selected substrates at temperatures of 70°C and greater (Kim et al., 1998c).

The structure of MjHSP 16.5 has been determined by X-ray crystallography (Kim et al., 1998a), it is a homomeric complex of 24 subunits arranged in octahedral symmetry. The overall shape is a hollow sphere with an external diameter of around 12 nm and an internal diameter of around 6.5 nm with 14 small 'windows' on the surface. Electron microscopy revealed a similar structure as in the crystalline state (Kim et al., 1998c). The monomer contains 10 β-strands, one of which contacts a neighbouring subunit. In the case of this sHSP, the dimer is the fundamental building block of the oligomeric assembly.

*Yeast HSP26.* In S. cerevisiae, HSP26 is the best-characterized member of the sHSP family (Haslbeck et al., 1999). At physiological conditions, HSP26 forms a large monodisperse oligomer of 24 subunits of 23.7 kDa (Bentley et al., 1992). After heat shock it dissociates into dimers (Haslbeck et al., 1999; Stromer et al., 2004) In vitro chaperone assays performed at different temperatures showed that the dissociation of the HSP26 complex at heat shock temperatures is a prerequisite for efficient chaperone activity. Binding of non-native proteins to dissociated HSP26 produced large globular assemblies with a structure that appeared to be completely reorganized relative to the original HSP26 oligomers (Haslbeck et al., 1999). In this complex one monomer of substrate is bound per HSP26 dimer. As yet it remains unclear whether the dimeric or the oligomeric form has the chaperone activity in vivo (Franzmann et al., 2005; Stromer et al., 2004).

### Effect of High Pressure on the Function and Structure of Small Heat Shock Proteins

It is widely accepted that the quaternary structure of the sHSPs, the oligomeric organization, plays an important role in the chaperone function. However, the nature of this role is not clear yet (Franzmann et al., 2005; Haslbeck et al., 1999; Stromer et al., 2004). While some sHSPs turn into smaller oligomers on increasing temperature and it is stated that a certain degree of dissociation is a prerequisite of their chaperone activity (Gu et al., 2002; Haslbeck et al., 1999), the oligomeric size of α-crystallin and MjHSP16.5 increases at higher temperatures, and it was hypothesized that their chaperone activity correlates with a higher degree of association (Burgio et al., 2000; Kim et al., 2003a).

We earlier reported  the enhancement of chaperone potency of α-crystallin by high pressure induced dissociation (Bode et al., 2003), and similar results were obtained for MjHSP16.5 (Tolgyesi et al., 2004). Here we note that in our experiments TRIS and BES buffers were used, which are known to be only marginally pressure dependent (Neuman et al., 1973).

In both cases the chaperone activity of the pressure-treated proteins was compared with the control. Pressure treatment means that the samples were exposed to high pressure (several 100 MPa for 15–20 min), and the chaperone-activity was measured with an aggregation assay immediately after returning to ambient pressure. In the case of the α-crystallin to separate the effect of pressurization from the temperature effects, a non-thermal aggregation phenomenon was used, i.e. the aggregation of insulin B chains after reduction of insulin by DTE. The experiments showed that pressurized α-crystallin prevented aggregation of insulin B chains to a greater extent, than the control. The induced activity depends on the applied pressure. Figure 1a shows that higher pressures, up to 340 MPa, induce higher activity. Chaperone activity was defined as described  by (Bode et al., 2003) It has also been shown that the enhanced protective ability is  lost slowly with time. In the case of MjHSP 16.5 a rhodanese aggregation assay was used, and the pressure dependence and time-decay of the chaperone-activity enhacement had not been studied. Pressure treatment at 300 MPa also enhanced the chaperone activity of MjHSP16.5 significantly (Fig. 1a).

Enhancement of chaperone activity was found to correlate well with the dissociation of the oligomers under pressure treatment (Fig. 1b). Dissociation was evidenced by decreasing light scattering on increasing pressure (shown in Fig. 1b), and by decreasing the 1688 $cm^{-1}$ band area in FTIR spectra. (This band is assigned to the hydrogen bonds in intermolecular β contacts.) Furthermore, dissociation is expected to open new hydrophobic surfaces which were hidden in subunit contacts before for the fluorescent probe ANS (8-anilinonaphthalene-1-sulphonic acid). It was  found that for both sHSPs the fluorescent intensity of ANS increased significantly under pressure indicating higher availability of hydrophobic patches.  It was suggested by us that this feature can have a role in the enhanced chaperone activity. Tryptophan fluorescence experiments also demonstrated the loosening in the molecular structure around tryptophan residues, i.e. a red shift in the tryptophan spectrum, which might accompany the dissociation of the oligomers. In the case of MjHSP16.5, dissociation was followed in addition by native gel electrophoresis under high pressure. A homogeneous oligomer population was detected on the electroforegrams at ambient pressure, corresponding to previous observations which suggest the MjHSP16.5 exists in a 24-mer form. At high pressures above 50 MPa inhomogeneity could be observed, and a Fergusson analysis showed a dissociation of the 24mers probably to 12mers at 100 MPa (unpublished results).

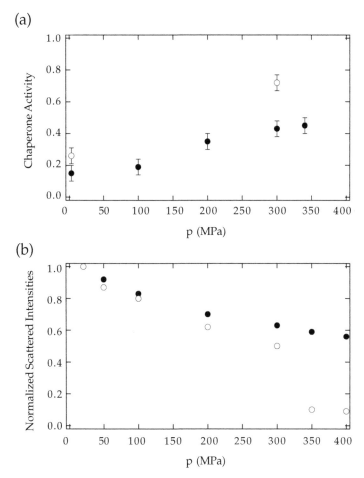

Figure 1. (a) Chaperone activity of α-crystallin (filled circles) and of MjHSP16.5 (open circles) after pressure treatment (15–20 min). The chaperone activity was measured by insulin aggregation assay for α-crystallin, and by rhodanese aggregation assay for MjHSP16.5, and activity values were calculated as described by Bode et al. (2003). (b) Dissociation of α-crystallin oligomers (filled circles) and of MjHSP16.5 oligomers (open circles) under pressure treatment. Dissociation was evidenced by decreasing light scattering on increasing pressure.

Similar to chaperone activity enhancement, structural changes were also found to be reversible. The structure relaxation after pressurization showed two phases. In the first phase, dissociated monomers reassociate quickly into smaller oligomers as seen from light scattering, FTIR and Trp-ANS energy transfer data. The time constant of this process is in the range of minutes. In the second slower phase oligomers grow further until equilibrium by subunit exchange. The time constant of this phase was 33 h, roughly the same as the characteristic time of subunit exchange. The relaxation time of the functional relaxation (2 h) was between that of the fast and slow structural rearrangements.

Two more papers studied α-crystallin under high pressure. J. Koretz and her co-workers studied the effect of pressure on the α-crystallin oligomers up to 80 MPa with turbidity measurements and electronmicroscopy technique. Their observations revealed a slight incerease in the oligomer size in this pressure range (Radlick and Koretz 1992). Similar observations were published by Skouri-Planet from small-angle X-ray scattering experiments (Skouri-Panet et al., 2006). Skouri-Panet unfortunately carried out high pressure experiments for α-crystallin in phosphate buffer. The $pK_a$ values of this buffer is known to depend not negligibly on hydrostatic pressure, its pH shifts roughly 0.5 pH unit at 100 MPa to the acidic direction. As a consequence the pH value of their sample was shifted e. g. at 300 MPa from the original value 6.8 to 5.5, which is near to the isoelectric point of aA sbunits (5.3) and it is beyond the isoelectric point of αB subunits (6.8). We also observed the size increase and the aggregation of α-crystallin at around this pH value in our separate experiments when α-crystallin was subjected to pH shock (unpublished results). Radlick and Koretz 1992 used imidazole buffer which is appropriate for high pressure experiments. We have no explanation for the discrepancy between their and our observations.

None of the two papers (Radlick and Koretz, 1992; Skouri-Panet et al., 2006) studied the effect of pressure on the chaperone activity of α-crystallin.

It should be mentioned that we studied the effect of high hydrostatic pressure on the secondary structure of α-crystallin and that of MjHSP16.5 in a wide pressure range up to 1600 MPa. Both tryptophan flourescence and FTIR spectroscopy demonstrated that the secondary structure of α-crystallin is not affected until 400 MPa. Higher pressures, however cause a secondary structural disorder at around 800 MPa. MjHSP 16.5 is more stable than α-crystallin, its secondary stucture damaged only at pressures above 1600 MPa.

According to the general expectation, dissociation was also observed for yeast HSP26 under high pressure by Skouri-Panet et al., 2006 . In these experiments a pressure-insensitive buffer, i.e. Tris buffer was used in contrast to their α-crystallin experiments. Small-angle X-ray scattering data suggested a fully reversible pressure dissociation of HSP26 oligomers to dimers between 100 and 175 MPa without the detectable presence of any intermediate oligomerization state. In this pressure range no subunit denaturation could be observed. No functional studies were carried out for pressure perturbed HSP26 in this work.

## Effect of Pressure on the Bacterial Chaperonin GroEL

Horowitz and his coworkers studied the effect of pressure on the structure and function of the chaperonin GroEL in detail, and they performed some investigations for its cochaperonin GroES and for the single ring mutant of GroEL, the so called SR1 too (Gorovits and Horowitz, 1995; Panda and Horowitz, 2002; Panda and Horowitz, 2004; Panda et al., 2001; Panda et al., 2002). Here, we present GroEL, GroES and SR1 briefly and then summarize the main findings of Professor Horowitz and his coworkers at first in a short subsection and then in more detail.

### *About GroEL*

The chaperonins are a class of molecular chaperones consisting of double-ring assemblies that carry out an ATP-dependent protein folding to the native state. They are present in all three kingdoms of life: GroEL resides in the bacterial cytoplasm. GroEL, together with some other chaperonins is a thermally inducible heat shock protein.

GroEL is a ≈ 800 kDa cylindrical homo-oligomer formed of 14 subunits arranged in two heptamer rings. Each ring forms a large central cavity for the binding of unfolded or newly synthesized proteins. It appears that GroEL has a wide range of substrates. After substrate binding this central cavity is closed by a 'cap' , i.e. the co-chaperonin, GroES that binds in an ATP-dependent fashion. Folding proceeds in this cage, subsequently GroES dissociates and the native-like protein is released.

GroEL subunits (547 amino acids, 57 kDa) are divided into three domains. The apical domain faces the solvent, forms the opening to the central channel, and contains the GroES and peptide binding site. The highly helical equatorial domain contains the ATP binding site. The hingelike intermediate domain links the apical and equatorial domains and transfers the ATP-induced conformational changes between the equatorial and the apical domain. ATP plays an important role both as an energy source and as an allosteric effector.

The co-chaperonin, GroES is a single heptameric ring of ≈10 kDa subunits, with a dome-shaped architecture. GroES cycles on and off GroEL in a manner allosterically regulated by the GroEL ATPase activity. Nonnative protein, exposing hydrophobic amino acid residues, binds with the highest affinity to the nucleotide-free state of GroEL. Binding of ATP and of GroES then induces a structural conversion of the inner GroEL surface from hydrophobic to hydrophilic and generates an enclosed chamber with approximate dimensions of 8 nm in diameter and 8.5 nm in height. As a result, bound protein is transiently displaced into this cage and allowed to fold. The enclosure time of ≈10 s reflects the time required for the hydrolysis of the 7 ATP molecules in the GroES bound ring (the cis-ring) of GroEL. Following hydrolysis, GroES is triggered to dissociate by ATP binding to the trans GroEL ring. At this point, folded protein leaves GroEL, whereas incompletely folded states are rapidly recaptured for another folding attempt.

The presence of $Mg^{2+}$ and $K^+$ is also necessary for the GroEL-assisted folding cycle.

There are still questions concerning the precise role of the GroEL cavity in assisting folding: e. g. the role played by the individual heptameric rings of the double-ring GroEL. The mutation of four residues that prevent the major contacts between the two rings of GroEL produces the single-ring mutant termed SR1. It has been shown that it also supports the folding of rhodanese in the presence of GroES, ATP or ADP, and KCl (Ellis, 2005; Horwich et al., 2007; Krishna et al., 2007; Radford, 2006; Tang et al., 2006).

## *Effect of High Pressure on the Function and Structure of GroEL in Summary*

GroEL and SR1 were purified from *Escherichia coli* cells. 50 mM Tris-HCl buffer (pH 7.8) was used throughout in the experiments. Tris buffer is appropriate for high pressure experiments because of its small $pK_a$ dependence upon hydrostatic pressure (Neuman et al., 1973).

As one can expect from the general effect of pressure, all of the above presented oligomers, GroEL, GroES and SR1 dissociate into monomers in the high pressure range of 50–250 MPa. It is interesting, however, that only monomers and oligomers were produced at any given pressure, no intermediate was observed. There are differences between the double-ring GroEL and its single-ring mutant SR1: SR1 is less stable and it dissociates at lower pressures than GroEL. The extent of dissociation depends on the applied pressure in the case of the GroEL tetramer. In contrast, SR1 undergoes complete dissociation into monomers at all pressures in the range of 75–300 MPa. The dissociation kinetics is pressure dependent with a biphasic character for both proteins, nevertheless SR1 dissociates faster than GroEL. Parallel to dissociation the chaperone function of GroEL is lost.

GroEL and SR1 are subject to conformational drift during pressurization. As a result, the monomers reassociate back to the oligomer only very slowly ($t_{1/2}$ of 150 h at 25 °C for GroEL), or not at all (SR1) after depressurization. The function of GroEL is also regained slowly after depressurization, namely before the monomers fully reassemble. Unlike GroEL and SR1, the conformation of GroES monomers seems not to alter during pressurization since they do reassociate readily on depressurization.

Salts ($MgCl_2$ and KCl) stabilize GroEL and decrease its dissociation rates up to a certain concentration. Unlike GroEL, dissociation of SR1 and that of GroES are independent of salt concentrations.

On the contrary, ATP or ADP destabilize both GroEL and the GroEL-GroES complex, but hardly SR1. Dissociation rates of GroEL are increased and the reassociation rates following depressurization also increase, but only if the ligands are already present during pressurization.

## More Details of the Effect of Pressure

The dissociation of the oligomeric proteins during pressurization was followed in the studies of Horowitz and his group mainly by static light scattering. The intensity of scattered light is proportional to the product of the molecular weight and concentration of a given type protein. The larger the particles (the oligomers), the stronger the scattering, in spite of a decreasing number concentration of the scattering particles. For example, the scattering by a solution of fully dimerised protein is twice as strong than that of a solution of monomers of a twice higher number concentration. In static light scattering experiments the effect of pressure on the optical properties (e.g. the refractive index) of the sample is generally disregarded as it was the case in the studies reviewed here. For investigation of the kinetics of the dissociation process it is thus sufficient to observe directly the time changes in the scattering light intensity (Banachowicz, 2006).

In some experiments for GroEL bisANS was used to probe the accessibility of hydrophobic surfaces. BisANS fluorescence intensity is known to increase significantly on binding to hydrophobic regions of proteins. The increase of ANS fluorescence during pressurization signalled increased exposure of hydrophobic surfaces due to dissociation of the oligomers and conformational alterations of the subunits.

Neither of these two methods can, however, give exact information on the state of dissociation. For this purpose native gel electrophoresis and ultracentrifugation were used, but only after depressurization and only for GroEL. The sedimentation coefficient of <5 S measured for GroEL after a pressure treatment of 200 MPa (for 30 min) and subsequent relaxation at atmospheric pressure (for 2 h at 5°C) in comparison to the 20 S value for untreated GroEL showed that the sample after pressure treatment contained no tetradecamers only monomers. Native gel electrophoresis results were in accordance to sedimentation analysis results.

It is interesting, that no intermediate species were found in the pressure treated samples, only full oligomers and monomers. It could mean that there is no hierarchy in the organization of the oligomers, the subunits play equivalent roles. This is, however, difficult to accept in case of the double-ring GroEL where two levels of organization seem to exist. The possible explanation is, that the two rings are bound together more strongly than the subunits inside the rings. Thus, each ring dissociates into monomers immediately when the double-ring falls apart.

For the sake of comparison of the dissociation of GroEL with that of SR1, the apparent $p_{1/2}$ values (the pressure at which the dissociation process shows one-half of the total intensity change in the scattered light) can be compared. They are 70 MPa and 145 MPa for SR1 and GroEL, respectively. It seems that interring bonds give higher stability to GroEL. These $p_{1/2}$ values support the previous hypothesis, i. e. the two rings are bound together more strongly than the subunits inside the rings. GroES was investigated only at 250 MPa, at which full dissociation can be suspected from light scattering measurements.

Gorovits and Horowitz (1995) calculated the standard volume change on association from the dissociation curves of GroEL. They obtained the value of 99 ml/mol of monomer which can be compared with the values in the range of 70–125 ml/mol for several oligomeric or aggregated proteins (Gross and Jaenicke, 1994; Randolph et al., 2002).

The kinetics of dissociation at fixed pressures were predominantly biexponential both for GroEL and for SR1. In general SR1 dissociation was faster. The fast and the slow rate constants of the biexponential processes were in the order of magnitude of $10^{-3}$ 1/s and $10^{-4}$ 1/s, respectively. The faster rate increased with pressure both for GroEL and for SR1 showing a negative activation volume of dissociation process. The activation volumes at zero pressure ($\Delta V_0^{\neq}$) were calculated for SR1, the values are between –65 and –95 ml/mol at different temperatures. At 20°C $\Delta V_0^{\neq}$ = –95 ml/mol for SR1, and –85 ml/mol for GroEL (calculated for this chapter from Fig. 5 of (Panda et al., 2001)). The nonavailability of $\Delta V_0^{\neq}$ data for the dissociation of oligomeric proteins limits the discussion of the magnitude of this parameter. Royer (2002) has reviewed the volume changes upon unfolding of a number of proteins and reported that values were in the range of –51 to –114 ml/mol.

The slow phase of the dissociation process was attributed to pressure independent spontaneous dissociation on the basis that they observed dissociation process also at atmospheric condition when SR1 samples were diluted and this spontaneous dissociation had a similar rate constant to that of the slow rate constant of pressure dissociation (Panda and Horowitz, 2004).

The chaperone function of GroEL was tested by rhodanese folding intermediates. Untreated GroEL is capable of capturing rhodanese folding intermediate on dilution from concentrated urea. Pressure-pretreated GroEL was not able to capture rhodanese at all immediately after depressurization, but it slowly regained this ability due to the reassociation of the subunits (see later).

After depressurization the monomers of GroEL or SR1 were not ready to reassociate in contrast to monomers obtained by urea dissociation. For SR1, Panda et al. (2002) were unable to detect any reassociated heptamer even after storing the dissociated monomers for 1 month for any of the several reaction conditions. Gorovits et al., 1995 had shown that the reassociation of monomers from GroEL tetramer dissociation was extremely slow with a $t_{1/2}$ of 150 h at 25°C. This reveals that the dissociated monomers undergo conformational drift, making them unsuitable for reassociation, and this also happened when the oligomer was a single-ring mutant instead of a double ring. The authors could not find a technique such as gel filtration or circular dichroism to distinguish between the conformations of the monomers generated by chaotropes and high pressure. However it is sure, that thiol oxidation as the cause of irreversibility can be ruled out. Namely one might suspect that the irreversibility of the dissociation process could be attributed to disulphide formation since there are three cysteines in each monomer of the GroEL. But GroEL monomers showed in no way higher propensity for reassociation in the presence of DTT than without DTT.

Unlike GroEL and SR1, GroES monomers readily associated on depressurization.

As it was already mentioned, GroEL regained its ability to capture rhodanese folding intermediates after depressurization before the complete reassociation of the subunits. This observation demonstrates clearly that the tetradecameric structure is not an absolute prerequisite for this part of the chaperone function of GroEL.

At the end the authors investigated the effect of ions and that of nucleotides on the pressure dissociation. The presence of $MgCl_2$ and KCl stabilizes GroEL up to 300 MPa. At 250 MPa, e. g. both, the fast and the slow, dissociation rates decreased exponentially with the increase in KCl concentration and reached a minimum at ≈75 mM. A plausible interpretation of this observation is the formation of salt bridges that are broken in a kinetically controlled process leading to slower observed dissociation rates. Similarly, the rates decreased with the increase in $MgCl_2$ concentration and reached a minimum at ≈3 mM $MgCl_2$. Unlike the GroEL tetramer, the observed rates for SR1 dissociation are independent of the concentrations of $MgCl_2$ and KCl in the studied range. It appears that the salt bridges mentioned above are important only for the double ring, they stabilize only the inter-ring interactions. $MgCl_2$ has no effect on either the dissociation of GroES heptamer or the reassociation of the monomers.

The functionally related ligands, ATP or ADP destabilize the quaternary structure of GroEL. Their presence during pressurization shifts the dissociation transition of GroEL to lower pressures and increases the rates of both dissociation and reassociation following depressurization. The addition of ATP after depressurization has little effect on the reassociation of monomers, indicating that the sites of ATP binding are either not available or are not competent. In contrast, the dissociation process of the SR1 heptamer was not dramatically different in the presence of these ligands from that in the buffer alone.

The GroEL-GroES complex is very stable in the range of 100–250 MPa. However, the addition of ADP destabilizes the complex, which dissociates completely at 150 MPa.

## CONCLUSION

High pressure studies have physiological relevance up to about 100 MPa. Application of even higher pressures can be considered as an additional tool to explore protein structure and function. The studies reviewed in this chapter demonstrate clearly that high pressure technique can really provide new insights into structure and chaperone function of heat-shock proteins. These can be summarized as follows.

The oligomeric HSPs (three small heat-shock proteins, i. e. α-crystallin, HSP16.5 and HSP26, and three members of the HSP60 family, i. e. GroEL, SR1 and GroES) studied so far under high pressure showed dissociation to various extents in the pressure range up to 400 MPa. Upon depressurization they are more or less ready to reassociate with the exception of SR1, the single ring mutant of GroEL. Some conformational changes of the subunits namely may already appear in this pressure range, the extent and type of which may be different for the individual HSPs. Conformational drift under high pressure influences ability for reassociation upon depressurization, making reassociation process slow or impossible at all. The complete loss of secondary structure, however, occurs only at higher pressures.

Dissociation influences the chaperone function of HSPs under study. The chaperone function of small heat-shock proteins is simply to bind damaged proteins. Partial dissociation of sHSP oligomers seems to enhance this passive function. GroEL on the contrary loses its function after complete dissociation, but a part of its function, i. e. the passive binding of folding intermediates is regained upon partial reassociation.

As for the physiologically relevant relatively narrow pressure range, the structural changes outlined above begin already in this pressure range, even if they are at the limit of observability. Partial dissociation, penetration of some water molecules into the interior of protein molecules results in stronger dynamics and in a moderate loosening of the structure. These changes may be beneficial for some protein functions such as the chaperone activity of small heat-shock proteins. The answer of living organisms to elevated pressures is of course more complex and as yet little is known about it. As the usual answer to a stress condition, the production of heat-shock proteins seems to increase under high pressure. Whether it really helps the survival of cells or not, remains to be seen .

## ACKNOWLEDGEMENTS

The authors would like to thank Dr. László Smeller for a critical reading of the chapter.

## REFERENCES

Anfinsen, C.B. 1973. Principles that govern the folding of protein chains. Science 181: 223–230.

Banachowicz, E. 2006. Light scattering studies of proteins under compression. Biochim. Biophys. Acta 1764: 405–413.

Basha, E. and G.J. Lee, L.A. Breci, A.C. Hausrath, N.R. Buan, K.C. Giese and E. Vierling. 2004. The identity of proteins associated with a small heat shock protein during heat stress in vivo indicates that these chaperones protect a wide range of cellular functions. J. Biol. Chem. 279: 7566–7575.

Bentley, N.J. and I.T. Fitch and M.F. Tuite. 1992. The small heat-shock protein Hsp26 of Saccharomyces cerevisiae assembles into a high molecular weight aggregate. Yeast 8: 95–106.

Bode, C. and F.G. Tolgyesi, L. Smeller, K. Heremans, S.V. Avilov and J. Fidy. 2003. Chaperone-like activity of alpha-crystallin is enhanced by high-pressure treatment. Biochem. J. 370: 859–866.

Bova, M.P. and Q. Huang, L. Ding and J. Horwitz. 2002. Subunit exchange, conformational stability, and chaperone-like activity function of the small heat-shock protein 16.5 from Methanococcus Jannaschii. J. Biol. Chem. 277: 38468–38475.

Buchner, J. 1999. Hsp90 & Co.—a holding for folding. Trends Biochem. Sci. 24: 136–141.

Bukau, B. and A.L. Horwich. 1998. The Hsp70 and Hsp60 chaperone machines. Cell 92: 351–366.

Burgio, M.R. and C.J. Kim, C.C. Dow and J.F. Koretz. 2000. Correlation between the chaperone-like activity and aggregate size of alpha-crystallin with increasing temperature. Biochem. Biophys. Res. Commun. 268: 426–432.

Carra, S. and M. Sivilotti, A.T. Chavez Zobel, H. Lambert and J. Landry. 2005. HspB8, a small heat shock protein mutated in human neuromuscular disorders, has in vivo chaperone activity in cultured cells. Hum. Mol. Genet. 14: 1659–1669.

de Jong, W.W. and G.J. Caspers and J.A. Leunissen. 1998. Genealogy of the alpha-crystallin—small heat-shock protein superfamily. Int. J. Biol. Macromol. 22: 151–162.

Ellis, R.J. 2005. Chaperomics: in vivo GroEL function defined. Curr. Biol. 15: R661–3.

Franzmann, T.M. and M. Wuhr, K. Richter, S. Walter and J. Buchner. 2005. The activation mechanism of Hsp26 does not require dissociation of the oligomer. J. Mol. Biol. 350: 1083–1093.

Giese, K.C. and E. Vierling. 2002. Changes in oligomerization are essential for the chaperone activity of a small heat shock protein in vivo and in vitro. J. Biol. Chem. 277: 46310–46318.

Goloubinoff, P. and A. Mogk, A.P. Zvi, T. Tomoyasu and B. Bukau. 1999. Sequential mechanism of solubilization and refolding of stable protein aggregates by a bichaperone network. Proc. Natl. Acad. Sci. USA 96: 13732–13737.

Gorovits, B.M. and P.M. Horowitz. 1995. The chaperonin GroEL is destabilized by binding of ADP. J. Biol. Chem. 270: 28551–26556.

Grallert, H. and J. Buchner. 2001. Review: a structural view of the GroEL chaperone cycle. J. Struct. Biol. 135: 95–103.

Gross, M. and R. Jaenicke. 1994. Proteins under pressure. The influence of high hydrostatic pressure on structure, function and assembly of proteins and protein complexes. Eur. J. Biochem. 221: 617–630.

Gu, L. and A. Abulimiti, W. Li and Z. Chang. 2002. Monodisperse Hsp16.3 nonamer exhibits dynamic dissociation and reassociation, with the nonamer dissociation prerequisite for chaperone-like activity. J. Mol. Biol. 319: 517–526.

Haley, D.A. and M.P. Bova, Q.L. Huang, H.S. McHaourab and P.L. Stewart. 2000. Small heat-shock protein structures reveal a continuum from symmetric to variable assemblies. J. Mol. Biol. 298: 261–272.

Haslbeck, M. 2002. sHsps and their role in the chaperone network. Cell Mol Life Sci. 59: 1649–1657.

Haslbeck, M. and S. Walke, T. Stromer, M. Ehrnsperger, H.E. White, S. Chen, H.R. Saibil and J. Buchner. 1999. Hsp26: a temperature-regulated chaperone. Embo. J. 18: 6744–6751.

Haslbeck, M. and T. Franzmann, D. Weinfurtner and J. Buchner. 2005. Some like it hot: the structure and function of small heat-shock proteins. Nat. Struct. Mol. Biol. 12: 842–846.

Hildebrand, C.E. and E.C. Pollard. 1972. Hydrostatic pressure effects on protein synthesis. Biophys. J. 12: 1235–1250.

Horwich, A.L. and W.A. Fenton, E. Chapman and G.W. Farr. 2007. Two families of chaperonin: physiology and mechanism. Annu. Rev. Cell Dev. Biol. 23: 115–145.

Horwitz, J. 1992. Alpha-crystallin can function as a molecular chaperone. Proc. Natl. Acad. Sci. USA 89: 10449–10453.

Horwitz, J. 2003. Alpha-crystallin. Exp. Eye. Res. 76: 145–153.

Ishikawa, T. and F. Beuron, M. Kessel, S. Wickner, M.R. Maurizi and A.C. Steven. 2001. Translocation pathway of protein substrates in ClpAP protease. Proc. Natl Acad. Sci. USA 98: 4328–4333.

Jakob, U. and M. Gaestel, K. Engel and J. Buchner 1993. Small heat shock proteins are molecular chaperones. J. Biol. Chem. 268: 1517–1520.

Jakob, U. and W. Muse, M. Eser and J.C. Bardwell. 1999. Chaperone activity with a redox switch. Cell 96: 341–352.

Kaarniranta, K. and M. Elo, R. Sironen, M.J. Lammi, M.B. Goldring, J.E. Eriksson, L. Sistonen and H.J. Helminen. 1998. Hsp70 accumulation in chondrocytic cells exposed to high continuous hydrostatic pressure coincides with mRNA stabilization rather than transcriptional activation. Proc. Natl. Acad. Sci. USA 95: 2319–2324.

Kappe, G. and P. Verschuure, R.L. Philipsen, A.A. Staalduinen, P. Van de Boogaart, W.C. Boelens and W.W. De Jong. 2001. Characterization of two novel human small heat shock proteins: protein kinase-related HspB8 and testis-specific HspB9. Biochim. Biophys. Acta 1520: 1–6.

Kim, D.R. and I. Lee, S.C. Ha and K.K. Kim. 2003a. Activation mechanism of HSP16.5 from Methanococcus jannaschii. Biochem. Biophys Res. Commun 307: 991–998.

Kim, K.K. and R. Kim and S.H. Kim. 1998a. Crystal structure of a small heat-shock protein. Nature 394: 595–599.

Kim, K.K. and H. Yokota, S. Santoso, D. Lerner, R. Kim and S.H. Kim. 1998b. Purification, crystallization, and preliminary X-ray crystallographic data analysis of small heat shock protein homolog from Methanococcus jannaschii, a hyperthermophile. J. Struct. Biol. 121: 76–80.

Kim, R. and K.K. Kim, H. Yokota and S.H. Kim. 1998c. Small heat shock protein of Methanococcus jannaschii, a hyperthermophile. Proc. Natl. Acad. Sci. USA 95: 9129–9133.

Kim, R. and L. Lai, H.H. Lee, G.W. Cheong, K.K. Kim, Z. Wu, H. Yokota, S. Marqusee and S.H. Kim. 2003b. On the mechanism of chaperone activity of the small heat-shock protein of Methanococcus jannaschii. Proc. Natl. Acad. Sci. USA 100: 8151–8155.

Kokke, B.P. and M.R. Leroux, E.P. Candido, W.C. Boelens and W.W. de Jong. 1998. Caenorhabditis elegans small heat-shock proteins Hsp12.2 and Hsp12.3 form tetramers and have no chaperone-like activity. FEBS Lett. 433: 228–232.

Krishna, K.A. and G.V. Rao and K.R. Rao. 2007. Chaperonin GroEL: structure and reaction cycle. Curr. Protein Pept. Sci. 8: 418–425.

Lammi, M.J. and M.A. Elo, R.K. Sironen, H.M. Karjalainen, K. Kaarniranta and H.J. Helminen. 2004. Hydrostatic pressure-induced changes in cellular protein synthesis. Biorheology 41: 309–313.

Lentze, N. and S. Studer and F. Narberhaus. 2003. Structural and functional defects caused by point mutations in the alpha-crystallin domain of a bacterial alpha-heat shock protein. J. Mol. Biol. 328: 927–937.

Meersman, F. and C.M. Dobson and K. Heremans. 2006. Protein unfolding, amyloid fibril formation and configurational energy landscapes under high pressure conditions. Chem. Soc. Rev. 35: 908–917.

Miura, T. and H. Minegishi, R. Usami and F. Abe. 2006. Systematic analysis of HSP gene expression and effects on cell growth and survival at high hydrostatic pressure in Saccharomyces cerevisiae. Extremophiles 10: 279–284.

Mogk, A. and C. Schlieker, K.L. Friedrich, H.J. Schonfeld, E. Vierling and B. Bukau. 2003. Refolding of substrates bound to small Hsps relies on a disaggregation reaction mediated most efficiently by ClpB/DnaK. J. Biol. Chem. 278: 31033–31042.

Neuman Jr., R. C. and W. Kauzmann and A. Zipp. 1973. Pressure dependence of weak acid ionization in aqueous buffers. J. Phys. Chem. 77: 2687–2691.

Panda, M. and P.M. Horowitz. 2002. Conformational heterogeneity is revealed in the dissociation of the oligomeric chaperonin GroEL by high hydrostatic pressure. Biochemistry 41: 1869–1876.

Panda, M. and P.M. Horowitz 2004. Activation parameters for the spontaneous and pressure-induced phases of the dissociation of single-ring GroEL (SR1) chaperonin. Protein J. 23: 85–94.

Panda, M. and J. Ybarra and P.M. Horowitz. 2001. High hydrostatic pressure can probe the effects of functionally related ligands on the quaternary structures of the chaperonins GroEL and GroES. J. Biol. Chem. 276: 6253–6259.

Panda, M. and J. Ybarra and P.M. Horowitz. 2002. Dissociation of the single-ring chaperonin GroEL by high hydrostatic pressure. Biochemistry 41: 12843–12849.

Quinlan, R. 2002. Cytoskeletal competence requires protein chaperones. Prog. Mol. Subcell Biol. 28: 219–233.

Radford, S.E. 2006. GroEL: More than Just a folding cage. Cell 125: 831–833.

Radlick, L.W. and J.F. Koretz. 1992. Biophysical characterization of alpha-crystallin aggregates: validation of the micelle hypothesis. Biochim. Biophys. Acta 1120: 193–200.

Raman, B. and C.M. Rao. 1997. Chaperone-like activity and temperature-induced structural changes of alpha-crystallin. J. Biol. Chem. 272: 23559–23564.

Randolph, T.W. and M. Seefeldt and J.F. Carpenter. 2002. High hydrostatic pressure as a tool to study protein aggregation and amyloidosis. Biochim. Biophys Acta 1595: 224–234.

Royer, C.A. 2002. Revisiting volume changes in pressure-induced protein unfolding. Biochim. Biophys Acta 1595: 201–209.

Salvador-Silva, M. and C.S. Ricard, O.A. Agapova, P. Yang and M.R. Hernandez. 2001. Expression of small heat shock proteins and intermediate filaments in the human optic nerve head astrocytes exposed to elevated hydrostatic pressure in vitro. J. Neurosci. Res. 66: 59–73.

Schwarz, J.R. and J.V. Landau. 1972. Hydrostatic pressure effects on Escherichia coli: site of inhibition of protein synthesis. J. Bacteriol. 109: 945–948.

Skouri-Panet, F. and S. Quevillon-Cheruel, M. Michiel, A. Tardieu and S. Finet. 2006. sHSPs under temperature and pressure: the opposite behaviour of lens alpha-crystallins and yeast HSP26. Biochim. Biophys. Acta 1764: 372–383.

Smeller, L. 2002. Pressure-temperature phase diagrams of biomolecules. Biochimica Et Biophysica Acta—Protein Structure and Molecular Enzymology 1595: 11–29.

Smeller, L. and F. Meersman, J. Fidy and K. Heremans. 2002. Intermolecular interactions of proteins under pressure. Aggregation, dissociation, chaperoning. High Pressure Effects in Chemistry, Biology and Materials Science 208–2: 19–24.

Stamler, R. and G. Kappe, W. Boelens and C. Slingsby. 2005. Wrapping the alpha-crystallin domain fold in a chaperone assembly. J. Mol Biol. 353: 68–79.

Stromer, T. and E. Fischer, K. Richter, M. Haslbeck and J. Buchner. 2004. Analysis of the regulation of the molecular chaperone Hsp26 by temperature-induced dissociation: the N-terminal domain is important for oligomer assembly and the binding of unfolding proteins. J. Biol. Chem. 279: 11222–11228.

Tang, Y.C. and H.C. Chang, A. Roeben, D. Wischnewski, N. Wischnewski, M.J. Kerner, F.U. Hartl and M. Hayer-Hartl. 2006. Structural features of the GroEL-GroES nano-cage required for rapid folding of encapsulated protein. Cell 125: 903–914.

Tardieu, A. 1998. alpha-Crystallin quaternary structure and interactive properties control eye lens transparency. Int. J. Biol. Macromol. 22: 211–217.

Tolgyesi, F.G. and C. Bode, L. Smeller, D.R. Kim, K.K. Kim, K. Heremans and J. Fidy. 2004. Pressure activation of the chaperone function of small heat shock proteins. Cell Mol Biol. (Noisy-le-grand) 50: 361–369.

van Montfort, R.L. and E. Basha, K.L. Friedrich, C. Slingsby and E. Vierling. 2001. Crystal structure and assembly of a eukaryotic small heat shock protein. Nat. Struct. Biol. 8: 1025–1030.

Watters, A.L. and D. Baker. 2004. Searching for folded proteins in vitro and in silico. Eur. J. Biochem. 271: 1615–1622.

# 4

# Pressure Perturbation of Artificial and Natural Membranes

*Roland Winter*

## INTRODUCTION

The interest in also using pressure as a thermodynamic and kinetic variable, along with temperature and chemical potential (or activity) has been largely increasing in physico-chemical studies of biological systems in recent years (Mishra and Winter, 2008; Winter and Jonas, 1999; Silva et al., 2001; Gross and Jaenicke, 1994; Heremans and Smeller, 1998; Balny et al., 1992; Winter, 2003a; Daniel et al., 2006). To probe the concept of any energetic description and the resultant set of parameters necessary to provide a general explanation of the phase behavior of biomolecular systems, one needs to scan the appropriate parameter space experimentally. To this end, pressure dependent studies have also proven to be a very valuable tool. High hydrostatic pressure (HHP) acts predominantly on the spatial (secondary, tertiary, quaternary and supramolecular) structures of biomacromolecules. Besides the general physico-chemical interest in using high pressure as a tool for understanding the structure, phase behavior and energetics of biomolecules, HHP is also of biotechnological (e.g., for high pressure food processing) and physiological (e.g., for understanding the physiology of deep-sea organisms living in cold and high-pressure habitats, which exist at pressures up to about 1 kbar (0.1 MPa = 1 bar, 1 GPa = 10 kbar)) interest (Balny et al., 1992; Winter, 2003a; Daniel et al., 2006; Hausmann and Kremer, 1994; Wharton, 2002; Bartlett, 2002). Pressure stress affects all levels of cellular physiology including metabolism, membrane physiology, transport, transcription and translation. Interestingly, the biological membrane seems

TU Dortmund University, Physical Chemistry I, Biophysical Chemistry, Otto-Hahn Str. 6, D-44227 Dortmund, Germany.
E-mail: roland.winter@tu-dortmund.de

to be one of the most pressure sensitive biological systems. In this chapter, we discuss results of studies of the effects of pressure on the structure and phase behavior of lyotropic lipid mesophases, natural and model biomembrane systems as well as pressure effects on the interaction of peptides and drugs with membranes. In fact, at a more empirical level, there exists a quasi-pharmacologial aspect of pressure in which it is used to perturb membrane-drug interactions. High pressure biophysical studies generally call for unique methods, which have largely been developed in recent years. The principle designs can be found elsewhere (Heremans and Smeller, 1998; Balny et al., 1992; Winter, 2003a; Winter, 2002; Winter, 2003b; Woenckhaus et al., 2000; Arnold et al., 2003; Jonas, 1991; Akasaka, 2006). In this chapter, we focus only on basic concepts and results.

## THEORETICAL ASPECTS OF LIPID SELF-ASSEMBLY AND PHASE TRANSITIONS

The amphiphilic properties of lipid molecules lead to self-aggregation in water solution, so that their hydrocarbon chains are pointing away from contact with water, while the hydrophilic headgroups are hydrated. The main driving force behind this self-assembly is the hydrophobic effect, which on its own would lead to a macroscopic phase separation (one polar and one non-polar phase), but which is prevented owing to the requirement that the polar headgroups should be in contact with water. Instead, packing restrictions of the lipid molecules in the aggregate structures formed have to be fulfilled (Cevc, 1993; Lipowski and Sackmann, 1995; Seddon, 1990). A useful concept for a qualitative understanding of the mesophase behavior of amphiphilic systems is based on a consideration of the shape of the lipid molecules (Fig. 1a). The self-assembly of lipid molecules can be rationalized by a dimensionless packing parameter, $P$, defined by the ratio $v(al)^{-1}$, where $v$ is the molecular volume, $l$ is the molecular length, and $a$ is the molecular area at the hydrocarbon-water interface. When the packing parameter $P \approx 1$ (cylindrical-like molecules), there are optimal conditions for the formation of a bilayer structure. For $P > 1$, the molecules are wedgeshaped and the lipid monolayer prefers to curve towards the water region, and, for example, an inverse hexagonal ($H_{II}$) phase may form (e.g., for phosphatidylamines at high temperatures) (Figs. 1, 2) (Cevc, 1993; Lipowski and Sackmann, 1995; Seddon, 1990).

From a more general point of view, the molecular organization within a lipid aggregate can be understood in terms of a balance of attractive and repulsive forces acting at the level of the lipid polar headgroups and non-polar acyl chains (Fig. 1b). Within the headgroup (lipid/water) region there is an effectively attractive force $F_{lipid/water}$, which arises from the unfavorable contact of the hydrocarbon chains with water (the hydrophobic effect).

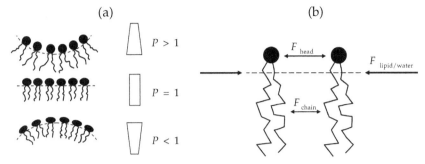

**Figure 1. (a)** Possible sponateous curvatures of a lipid monolayer arising from the distribution of lateral forces within the headgroup and acyl chain regions, including the corresponding packing parameter $P$ of the molecular building blocks: negative ($P > 1$), zero ($P = 1$), and positive ($P < 1$). **(b)** Illustration of the balance of lateral forces across a lipid monolayer.

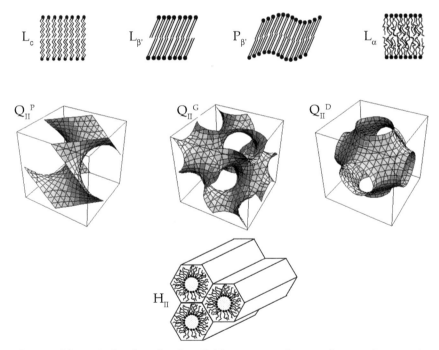

**Figure 2.** Schematic drawing of various lipid-water mesophases and states of aggregation adopted by membrane lipids ($L_c$, lamellar crystalline; $L_{\beta'}$, $P_{\beta'}$, lamellar gel; $L_\alpha$, lamellar liquid-crystalline (fluid-like); $Q_{II}{}^G$ (space group Ia3d, number 230), $Q_{II}{}^P$ (space group Im3m, number 229), $Q_{II}{}^D$ (space group Pn3m, number 224), inverse bicontinuous cubics of different space group—the cubic phases are represented by the G, D, and P minimal surfaces, which locate the midplanes of fluid lipid bilayers; $H_{II}$, inverse hexagonal).

Attractive contributions from hydrogen bonding, as in the case of phosphatidylethanolamines, may also be present. The repulsive headgroup pressure $F_{head}$ is the result of hydrational, steric, and - for charged headgroups - electrostatic contributions. For the acyl chain region, the attractive van der Waals interactions among the $CH_2$-groups are compensated by the repulsive lateral pressure, $F_{chain}$, owing to entropy-driven thermally activated dihedral angle isomerizations. In a membrane bilayer at equilibrium, the various lateral pressures are balanced. An imbalance of attractive and repulsive forces at the level of the headgroup and acyl chains within a given lipid monolayer yields a spontaneous curvature, $H_0$. Clearly, the two monolayers cannot be simultaneously at a free energy minimum with regards to their intrinsic or spontaneous curvature. It follows that a symmetrical bilayer composed of nonlamellar forming lipids is under a condition of curvature elastic stress. If sufficiently large, this curvature free energy associated with the membrane lipid-water interface then leads to the formation of nonlamellar (cubic or $H_{II}$) phases with more favorable curvature energy (Fig. 2, (Cevc, 1993; Lipowski and Sackmann, 1995; Seddon, 1990; Hyde et al., 1997; Luzzati, 1997)). The spontaneous curvature increases, for instance, if the lateral chain pressure increases due to an increase of trans/gauche isomerisations of the acyl chains at high temperatures, or if the level of headgroup hydration decreases (e.g., due to $Ca^{2+}$ adsorption at the polar/apolar interface).

It is generally assumed that the nonlamellar lipid structures, such as the inverted hexagonal ($H_{II}$) and bicontinuous cubic ($Q_{II}$) lipid phases, are also of signifiant biological relevance. Fundamental cell processes, such as endo- and exocytosis, membrane recycling, protein trafficking, fat digestion, membrane budding, fusion and enveloped virus infection, involve a rearrangement of biological membranes where nonlamellar lipid structures are probably involved. Furthermore, the cubic phases can be used as controlled-release drug carriers and crystallization media for proteins (Hyde et al., 1997).

Until now, no full theoretical description of the lyotropic lipid phase behavior exists, though some progress has been made in recent years. Hence, a phenomenological concept is often used that utilizes a small set of parameters, neglecting the precise chemical nature of the lipid molecules. Following Helfrich (1978), the surface curvature energy contribution associated with amphiphile films is described in terms of curvature elastic parameters: the spontaneous mean curvature $H_0$, the mean curvature modulus $k_m$, and $k_G$, the Gaussian curvature modulus. Applying differential geometry, the surface energy per unit area for small curvatures is given by (Cevc, 1993; Lipowski and Sackmann, 1995; Seddon, 1990; Hyde et al., 1997).

$$g_{curv} = 2\kappa_m \left\langle (H - H_0)^2 \right\rangle + \kappa_G \left\langle K \right\rangle \tag{1}$$

In differential geometry, two principal curvatures at a given point of a surface measure, how the surface bends by different amounts in different directions at that point. If we cross a given surface at a point under consideration with planes that are perpendicular to the surface and are in the principal directions, then the principle curvatures are the curvatures of the two lines of intercepts between the planes and the surface. The radii of these two circular fragments, $R_1$ and $R_2$, are called the principle radii of curvature, and their inverse values, $C_1 = 1/R_1$ and $C_2 = 1/R_2$, are referred to as the two principle curvatures. $H = (C_1+C_2)/2$ is the mean interfacial curvature, which is equal to half the sum of the principal curvatures $C_1$ and $C_2$. $K = C_1C_2$ is the Gaussian curvature at the interface, given by the product of the principal curvatures $C_1$ and $C_2$ of the interface (e.g., $C_1 = C_2 = 0$ for a planar bilayer and $C_1 < 0$, $C_2 = -C_1 > 0$ (i.e., $H = 0$, $K < 0$) for a hyperbolic saddle surface, as observed in bicontinuous cubic phases (Fig. 2). The spontaneous mean curvature $H_0$ is the mean curvature the lipid aggregate would wish to adopt in the absence of any constraints, and $\kappa_m$ tells us what energetic cost there would be for deviations away from this (typically 10–20 $k_BT$ for phospholipids). Thus, to minimize the free energy of curvature, it is advantageous to have $H$ close to $H_0$.

Besides the curvature energetic contribution, there will be other energetic contributions. Due to the desire to fill all the hydrophobic volume by the amphiphile chains (due to the hydrophobic effect), there will be a contribution quantifying an eventual packing frustration. For example, for a phospholipid monolayer to form a cylinder of radius $R_0 = 1/H_0$, there will also be a non-zero packing energy (Seddon, 1990; Hyde et al., 1997). In particular, the acyl chains of the lipid molecules must stretch to fill the interstitial regions between the cylindrical aggregates building up the $H_{II}$ phase (see Fig. 2). Thus, the forces of curvature and packing may oppose each other, a situation which is referred to as frustration. It can, therefore, be concluded that the smaller the radius of the water cylinder, the smaller are the hydrophobic interstices and the easier it is for the lipid acyl chains to elongate to fill the volume. This explains why the $H_{II}$ phase tends to form at low water contents. Similarly, it can be understood why increasing the acyl chain length also favors the $H_{II}$ phase formation. The curvature elastic energy is believed to be the crucial term governing the stability of nonlamellar phases and the ability of lipid membranes to bend, in particular at high levels of hydration.

## LIPID MESOPHASES AND MODEL BIOMEMBRANE SYSTEMS

### Lamellar lipid bilayer phases

Lyotropic lipid mesophases are organized systems formed by amphiphilic molecules, mostly phospholipids, in the presence of water. They exhibit a rich structural polymorphism, depending on their molecular structure, water content, pH, ionic strength, temperature and pressure (Cevc, 1993; Lipowski and Sackmann, 1995; Seddon, 1990; Hyde et al., 1997; Luzzati, 1997; Winter and Czeslik, 2000; Winter and Köhling, 2004; Czeslik et al., 1998; Winter et al., 2007; Hammouda and Worcester, 1997; Bernsdorff et al., 1996; Bernsdorff et al., 1997; Seemann and Winter, 2003; Reis et al., 1996). The basic structural element of biological membranes consists of a lamellar phospholipid bilayer matrix (Fig. 2). Not only is the entire cell membrane very complex, containing a large variety of different lipid molecules and a large body of proteins performing versatile biochemical functions, but also the simplest lipid bilayer consisting of only one or two kinds of lipid molecules already exhibits a very complex phase behavior as discussed below. In excess water, saturated phospholipids often exhibit two thermotropic lamellar phase transitions, a gel-to-gel ($L_{\beta'}$-$P_{\beta'}$) pretransition and a gel-to-liquid-crystalline ($P_{\beta'}$-$L_{\alpha}$) main transition at a higher temperature, $T_m$ (Fig. 2). In the fluid-like $L_{\alpha}$ phase, the acyl chains of the lipid bilayers are conformationally disordered, whereas in the gel phases, the chains are more extended and ordered. In addition to the thermotropic phase transitions, a variety of pressure-induced phase transformations have been observed (Winter and Czeslik, 200; Winter and Köhling, 2004; Czeslik et al., 1998; Winter et al., 2007; Hammouda and Worcester, 1997).

The main effect of pressure is to diminish the void volume between the lipid's acyl chains by an intermolecular rather than an intramolecular compression. Hence, the compression of the bilayer as a whole is anisotropic, lateral shrinking being accompanied by a decrease in thickness due to a straightening of the acyl chains. Because the average end-to-end distance of disordered hydrocarbon chains in the $L_{\alpha}$-phase is smaller than that of ordered (all-trans) chains, the bilayer becomes thinner during melting at the $P_{\beta'}/L_{\alpha}$-transition, even though the lipid volume increases. This is demonstrated in Fig. 3a which shows the temperature dependence of the specific lipid volume $V_L$ of DMPC (a $C_{14}$ double-chain phospholipid, see list of abbreviations) in water. The change of $V_L$ near 14°C corresponds to a small volume change in course of the $L_{\beta'}$-$P_{\beta'}$ transition. The main transition at $T_m$ = 23.9°C for this phospholipid is accompanied by a well pronounced 3% change in volume, which is mainly due to changes of the chain cross-sectional area, because the chain disorder increases drastically at the transition. Figure 3b exhibits the pressure dependence of $V_L$ at a temperature above $T_m$, e.g. 30°C. Increasing pressure triggers the phase transformation from the $L_{\alpha}$ to the gel phase, as can be seen from the rather abrupt decrease of the lipid volume at 270 bar. The volume change $\Delta V_m$ at the main transition decreases slightly with increasing temperature and pressure along the main transition line. Other thermodynamic parameters have been determined as well. It appears that the coefficient of isothermal compressibility of the $P_{\beta'}$ gel phase is substantially lower than

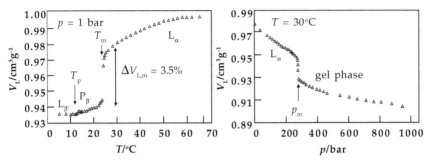

**Figure 3.** Effect of (a) temperature and (b) pressure (at $T = 30°C$) on the partial lipid volume $V_L$ of DMPC bilayers. The gel-to-gel ($L_{β'}$ to $P_{β'}$) and gel-to-fluid ($L_α$) lamellar phase transitions at temperatures $T_p$ and $T_m$, respectively, are accompanied by a small and larger volume expansion, respectively.

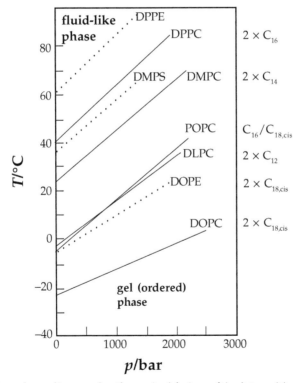

**Figure 4.** $T$, $p$-phase diagram for the main (chain-melting) transition of different phospholipid bilayer systems. The fluid (liquid-crystalline) $L_α$-phase is observed in the low-pressure, high-temperature region of the phase diagram; the ordered gel phase regions appear at low temperatures and high pressures, respectively. The Clapeyron slopes of the gel-to-fluid transition lines of the phosphatidylcholines are drawn as full lines, those of the phospholipids with different headgroups as dashed lines. The acyl chains of the various phospholipids are denoted on the right-hand side of the figure.

that of the liquid-crystalline phase (typically, $\kappa_T(P_{\beta'}) \approx 5 \cdot 10^{-5}$ bar$^{-1}$ and $\kappa_T(L_\alpha)$ $\approx 13 \cdot 10^{-5}$ bar$^{-1}$) (Seemann and Winter, 2003). Whereas the lateral compressibility of the lipid chains is rather high, a slight lateral compression is observed for the polar headgroups, only. Hence, this will have little effect on the nature of the electrical double layer in the interfacial region, but pressure would generally be expected to favor ionization of polar groups and electrostriction of the surrounding solvent water.

Upon compression, the lipids adapt to volume restriction by changing their conformation and packing. A common slope of ~22°C/kbar has been observed for the gel-fluid phase boundary of this and other saturated phosphatidylcholines as shown in Fig. 4. Using the Clapeyron relation describing first-order phase transitions, $dT_m/dp = T_m \Delta V_m / \Delta H_m$, the positive slope can be explained by an endothermic enthalpy change, $\Delta H_m$, and a volume increase, $\Delta V_m$, for the gel-to-fluid transition, which have indeed been found in direct measurements of these thermodynamic properties (Landwehr and Winter, 1994). Similar transition slopes have been determined for the mono-*cis*-unsaturated lipid POPC, the phosphatidylserine DMPS, and the phosphatidylethanolamine DPPE. Only the slopes of the di-*cis*-unsaturated lipids DOPC and DOPE have been found to be markedly smaller. The two *cis*-double bonds of DOPC and DOPE lead to very low transition temperatures and slopes, as they impose kinks in the linear conformations of the lipid acyl chains, thus creating significant free volume in the bilayer so that the ordering effect of high pressure is reduced. Hence, in order to remain in a physiological relevant, fluid-like state at high pressures, more of such *cis*-unsaturated lipids are incorporated into cellular membranes of deep sea organisms, a phenomenon called homeoviscous adaption (Cossins and Macdonald, 1986).

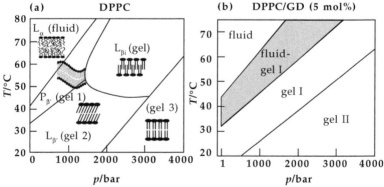

**Figure 5. (a)** *T, p*-phase diagram of DPPC bilayers in excess water (besides the Gel 1 (P$_{\beta'}$), Gel 2 (L$_{\beta'}$) and Gel 3 phase, an additional crystalline gel phase (L$_c$) can be induced in the low-temperature regime after prolonged cooling which is not shown here). **(b)** Phase diagram of DPPC-gramicidin D (GD) (5 mol%) in excess water as obtained from SAXS and FT-IR spectroscopy data.

As seen in Fig. 4, pressure generally increases the order of membranes, thus mimicking the effect of cooling. But we note that applying high pressure can lead to the formation of additional ordered phases, which are not observed under ambient pressure conditions, such as a partially interdigitated high pressure gel phase, $L_{\beta i}$, found for phospholipid bilayers with acyl chain lengths $\geq C_{16}$ (Winter and Czeslik, 2000; Hammouda and Worcester, 1997). To illustrate this phase variety, the results of a detailed SAXS/SANS and FT-IR spectroscopy study of the *p,T*-phase diagram of DPPC in excess water are shown in Fig. 5a. At much higher pressures as shown here (data not shown), further ordered gel phases appear, differing in the tilt angle of the acyl chains and the level of hydration in the headgroup area (Winter and Czeslik, 2000; Czeslik et al., 1998).

## Cholesterol effect and heterogeneous lipid bilayers

Cholesterol is an important sterol, which constitutes up to 50% of animal cell membranes. Due to both its amphiphilic character and size (intermediate between that of the shortest and longest acyl chains, $C_{14}$ and $C_{24}$), it is inserted in phospholipid membranes and is able to regulate the effects of external stress on the physical state of membranes. Cholesterol thickens a liquid-crystalline bilayer and increases the packing density of the lipid acyl-chains in a way that has been referred to as a 'condensing effect'. Increasing cholesterol concentration leads to a drastic reduction of the main transition enthalpy, $\Delta H_m$, until at cholesterol contents higher than 30 mol% the main transition vanishes. For phospholipid-cholesterol lipid mixtures, a rather complex phase behavior is found for lower sterol concentrations (Cevc, 1993; Lipowski and Sackmann, 1995; Krivanek et al., 2008). Measurements of the acyl-chain orientational order of the lipid bilayer system by measuring the $^2$H-NMR spectra or the steady-state fluorescence anisotropy $r_{ss}$ of the fluorophore TMA-DPH clearly demonstrate the ability of cholesterol and other plant or bacterial sterols to efficiently regulate the structure, motional freedom and hydrophobicity of biomembranes (Winter and Jonas, 1999; Winter, 2003b; Bernsdorff et al., 1996; Bernsdorff et al., 1997). The pressure dependencies of $r_{ss}$ of TMA-DPH labeled DPPC and DPPC/cholesterol are shown in Fig. 6. $r_{ss}$ of pure DPPC at $T = 50°C$ increases slightly up to about 400 bar, where the pressure-induced liquid-crystalline to gel phase transition starts to take place. Since $r_{ss}$ essentially reflects the mean order parameter of the lipid acyl-chains, these results indicate that increased pressures cause the chain region to be ordered in a manner similar to that which occurs upon decreasing the temperature. Addition of increasing amounts of cholesterol leads to a drastic increase of $r_{ss}$ values in the lower pressure region, whereas the corresponding data at higher pressures in the gel-like state of DPPC are slightly reduced. For concentrations above ~30 mol%

cholesterol, the main phase transition can hardly be detected any more. At a concentration of 50 mol% cholesterol, $r_{ss}$ values are found to be almost independent of pressure. Incorporation of the sterol also significantly changes the expansion coefficient and isothermal compressibility of the membrane (Seemann and Winter, 2003), and it significantly increases the hydrophobicity and hence decreases the water permeability of the bilayer (Bernsdorff et al., 1996; Bernsdorff et al., 1997). An increase in pressure up to the 1 kbar range is much less effective in suppressing water permeability than cholesterol embedded in fluid DPPC bilayers at high concentration levels. These data and further FT-IR spectroscopic pressure studies (Reis et al., 1996) clearly demonstrate the ability of sterols to efficiently regulate the structure, motional freedom and hydrophobicity of membranes, so that they can withstand even drastic changes in environmental conditions, such as in temperature or external pressure.

$T, p$-phase diagrams of binary mixtures of saturated phospholipids have been determined as well. They are typically characterized by lamellar gel phases at low temperatures, a lamellar fluid phase at high temperatures, and an intermediate fluid-gel coexistence region. The narrow fluid-gel

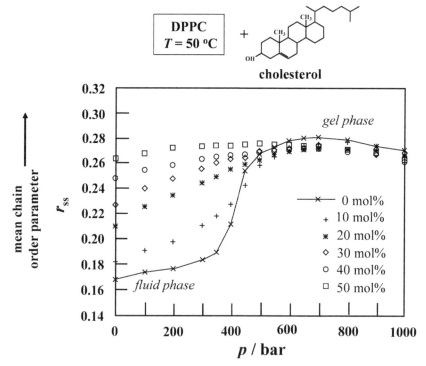

**Figure 6.** Pressure dependence of the steady-state fluorescence anisotropy $r_{ss}$ of TMA-DPH in DPPC unilamellar vesicles at different cholesterol concentrations (at $T = 50°C$).

coexistence region in the DMPC(di-C$_{14}$)–DPPC(di-C$_{16}$) system indicates a nearly ideal mixing behavior of the two components (isomorphous system). In comparison, the coexistence region in the DMPC(di-C$_{14}$)–DSPC(di-C$_{18}$) system is broader and reveals pronounced deviations from ideality. With increasing pressure, the gel-fluid coexistence region of the binary lipid systems is shifted toward higher temperatures. A shift of about 22°C/kbar is observed, similar to the slope of the gel-fluid transition line of the pure lipid components. The *p,T*-phase diagrams of a series of other mono- and two-component lipid mixtures  exhibiting also nonlamellar phases  have been investigated in recent years as well ((Winter, 2002; Winter and Czeslik, 2000) and refs. therein).

Studies were also carried out on the phase behavior of cholesterol-containing ternary lipid mixtures, generally containing an unsaturated lipid like DOPC and a saturated lipid like sphingomyelin (SM) or DPPC (Munro, 2003). Such lipid systems are supposed to mimic distinct liquid-ordered lipid regions, called 'rafts', which seem to be also present in cell membranes. They are thought to be important for cellular functions such as signal transduction and the sorting and transport of lipids and proteins (Munro, 2003; Janosch et al., 2004; Nicolini et al., 2006b). Recently, we determined the liquid-disordered/liquid-ordered phase coexistence region of POPC/SM/Chol (1:1:1) model raft mixtures, which extends over a rather wide temperature range, and an overall fluid phase is reached at rather high temperatures (above ~50°C), only (Nicolini et al., 2006c). Upon pressurization at ambient temperatures (20–40°C), an overall (liquid- and solid-) ordered membrane state with high ordering of their acyl-chains is reached at pressures of about 1–2 kbar. A similar behavior has been observed for the model raft mixture DOPC/DPPC/Chol (1:2:1) (Jeworrek et al., 2008) (Fig. 7). Interestingly, in this pressure range of ~2 kbar, ceasing of membrane protein function in natural membrane environments has been observed for a variety of systems (Winter et al., 2007; Chong et al., 1985; Kato et al., 2002; Powalska et al., 2007; de Smedt et al., 1979), which might be correlated with the membrane matrix reaching a physiologically unacceptable overall ordered state at these pressures. Moreover, many bacterial organisms have been shown to completely lose activity at these pressures.

Recently, Nicolini et al. (2006a)  used a technique that allows us to visualize morphological changes of the membrane of giant more-component unilamellar vesicles (GUVs) upon pressure perturbation. Under these conditions, the bending rigidity and the line tension, which control the domain structure of heterogeneous vesicles, and the differential area of the opposing monolayers in the lipid vesicle are influenced by pressure-induced changes in lipid molecular packing and shape. It has been shown by two-photon excitation fluorescence microscopy on these multidomain GUVs that budding and fission of daughter vesicles may occur, and this at

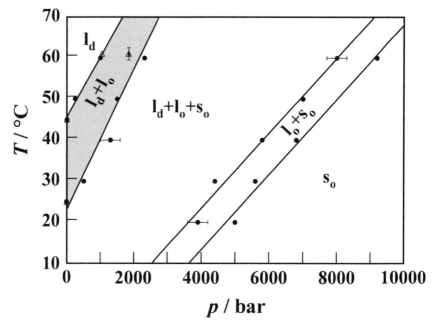

**Figure 7.** *p,T*-phase diagram of the ternary lipid mixture DOPC/DPPC/Chol (1:2:1) in excess water as obtained from the FT-IR spectroscopy (•) and SAXS (Δ) data. Selected error bars are shown which indicate the uncertainties in the pressure and temperature determination and phase assignments. The $l_d+l_o$ two-phase coexistence region is marked in gray.

surprisingly low pressures of 200–300 bar, already. Theses budding processes are not directly related to phase transitions to an overall ordered conformational state of the lipid membrane, which occur at much higher pressures (see above). The topological changes of the lipid vesicles are irreversible and exhibit a different behavior depending if the pressure is increased or decreased. Hence, this scenario provides a further mechanism that may be responsible for disruption of natural membranes upon compression.

The results discussed in this chapter demonstrate that biological organisms are able to modulate the physical state of their membranes in response to changes in the external environment by regulating the fractions of the various lipids in a cell membrane differing in chain length, chain unsaturation or headgroup structure ( 'homeoviscous adaption' ). Moreover, nature has further means to regulate membrane fluidity, such as by changing the membrane concentration of its sterols, by a lateral redistribution of their various lipid components and ordered/fluid domains or by the various proteins embedded in the membrane. In fact, many studies have demonstrated that membranes are significantly more fluid in barophilic and/or psychrophilic species. This is principally a consequence of an increase in the unsaturated to saturated lipid ratio. Interestingly, in both atmospheric and high-pressure adapted species, the plasma membrane appears to play a key role in the defense against pressure shocks. Notably, transitions between different local membrane structures, induced by physical changes in the cell environment (such as by compression), are also thought to act as signals to trigger particular signal transduction cascades.

**Ions and amphiphilic drugs.** The lipid conformation, membrane structure and phase behavior is also influenced by the ionic strength of the surrounding solvent (in particular for lipids with charged headgroups) and incorporation of amphiphilic molecules, such as anesthetics and drugs (Winter and Jonas, 1999; Winter, 2003a; Winter and Czeslik, 2000). For example, small amphiphilic molecules like anesthetics interact with membranes and decrease the gel-to-liquid-crystalline transition temperature of lipid bilayers. They induce a volume expansion which has the opposite effect of HHP and so they antagonize the effect of HHP on membrane's fluidity and volume, making membranes more fluid and expanded. The application of HHP to membrane-anesthetic systems reverses the effect and may even result in the expulsion of the drug to the aqueous environment (Winter and Jonas, 1999; Winter, 2003a; Winter et al., 1991; Böttner and Winter, 1993; Eisenblätter et al., 2000).

### Effects of peptide incorporation

Certainly, the incorporation of peptides and proteins leads to a drastic change of the structure and phase behavior of the membrane system. As an example, we describe the effect of the incorporation of the model channel peptide gramicidin D (GD) on the structure and phase behavior of phospholipid bilayers of different chain-lengths (Zein and Winter, 2000; Eisenblätter and Winter, 2005; Eisenblätter and Winter, 2006). Gramicidin is highly polymorphic, being able to adopt a wide range of structures with different topologies. Common forms are the dimeric single-stranded right-handed $\beta^{6.3}$-helix with a length of 24 Å, and the antiparallel double-stranded $\beta^{5.6}$-helix, being approximately 32 Å long. For comparison, the hydrophobic fluid bilayer thickness is about 30 Å for DPPC bilayers, and the hydrophobic thickness of the gel phases is larger by 4–5 Å. Depending on the GD concentration, significant changes of the lipid bilayer structure and phase behavior were found. These include disappearance of certain gel phases formed by the pure lipid bilayer systems, and the formation of broad two-phase coexistence regions at higher GD concentrations. And *vice versa,* the peptide conformation is also influenced by the lipid environment. Depending on the phase state and lipid acyl chain length, GD adopts at least two different types of quaternary structures in the bilayer environment, a double helical 'pore' and a helical dimer 'channel' . When the bilayer thickness changes at the gel/fluid main phase transition of DPPC, the conformational equilibrium of the peptide also changes. In gel-like DPPC bilayers, the equilibrium of the GD species in the lipid bilayer is shifted in favor of the longer double helical configuration. Hence, not only the lipid bilayer structure and *T*, *p*-dependent phase behavior drastically depends on the polypeptide concentration, but the peptide conformation (and hence function) can be also significantly influenced by the lipid environment. No pressure-induced unfolding of the polypeptide is observed up to 10 kbar. For large integral and peripheral proteins, however, pressure-induced changes in the physical state of the membrane may lead to a weakening of protein-lipid interactions as well as protein dissociation (see below).

### Dynamical properties

Little is known about pressure effects on the dynamical properties of lipid bilayers at elevated pressures (Winter, 2003b; Jonas, 1991). Of particular interest is the effect of pressure on lateral diffusion, which is related to biological functions such as electron transport and membrane-associated signalling processes. Pressure effects on the lateral self diffusion coefficient *D* of DPPC and POPC vesicles have been studied by Jonas et al., 1991. The lateral diffusion coefficient of DPPC in the liquid-crystalline phase decreases by about 30% from 1 to 300 bar at 50°C. A further 70% decrease in the *D*-value occurs at the pressure-induced $L_\alpha$ to gel phase transition. Compared to the lateral diffusion, the rotational dynamics of lipids and small amphiphilic molecules in membranes seems to be less influenced by high pressure (Bernsdorff et al., 1996; Bernsdorff et al., 1997). Phospholipid flip-flop and lipid transfer between membranes is also markedly slowed down by high pressure (Macdonald, 1992; Mantulin et al., 1984).

### Nonlamellar lipid phases

For a series of lipid molecules, also nonlamellar lyotropic phases are observed as thermodynamically stable phases or as long-lived metastable phases after special sample treatment (Seddon, 1990; Hyde et al., 1997; Luzzati, 1997) (Fig. 2). Lipids, which can adopt a hexagonal phase, are present at substantial levels in biological membranes, usually with at least 30 mol% of the total lipids. Some lipid extracts, such as those from archaebacteria (*S. solfataricus*), even exhibit cubic liquid-crystalline phases (Hyde et al., 1997; Luzzati, 1997). Fundamental cell processes, such as endo- and exocytosis, fat digestion, membrane budding and fusion, involve a rearrangement of biological membranes where such nonlamellar lipid structures are probably involved, but also static cubic structures (cubic membranes) might occur in biological cells (Hyde et al., 1997).

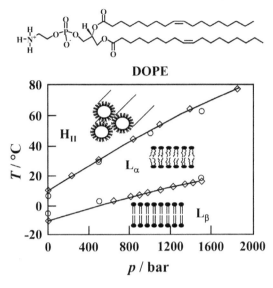

**DOPE**

**Figure 8.** $T$, $p$-phase diagram of DOPE in excess water.

For example, contrary to DOPC which shows a lamellar $L_\beta$-$L_\alpha$ transition only (Fig. 4), the corresponding lipid DOPE with ethanolamine as headgroup exhibits an additional phase transition from the lamellar $L_\alpha$ to the nonlamellar, inverse hexagonal $H_{II}$ phase at high temperatures (Fig. 8). As pressure forces a closer packing of the lipid chains, which results in a decreased number of *gauche* bonds and kinks in the chains, both transition temperatures, of the $L_\beta$-$L_\alpha$ and the $L_\alpha$-$H_{II}$ transition, increase with increasing pressure. The $L_\alpha$-$H_{II}$ transition observed in DOPE/water and also in egg-PE/water (egg-PE is a natural mixture of different phosphatidylethanolamines) is the most pressure-sensitive lyotropic lipid phase transition found to date. The reason why this transition has such a strong pressure dependence is due to the strong pressure dependence of the chain length and volume of its *cis*-unsaturated chains. Generally, at sufficiently high pressures, hexagonal and cubic lipid mesophases give way to lamellar structures as they exhibit smaller partial lipid volumes.

## Archaebacterial membranes

In order to survive under extreme conditions, the membranes and the proteins of the extremophilies have to be rather different from those of eukaryotes and most prokaryotes. Nature has evolved lipids of a particular molecular structure in order to provide the membranes of the extremophiles with proper physical conditions to support their biological activity. A most fascinating way of developing lipid-based strategies to survive under

extreme conditions is found in the kingdom of the archaebacteria that live
in deep-ocean hot volcanic vents, in very salty water, in the acid guts of
animals, and under harsh chemical conditions with high levels of, e.g.,
methane or sulfur. The chemistry of their fatty-acid chains is based on poly-
isoprene, which forms so-called phytanyl chains. These chains are saturated
and have methyl groups sticking out from the chain at every fourth carbon
atom. The phytanyl chains are connected to the glycerol backbone by ether
bonds rather than ester bonds. Often, the isoprene units are even cyclized to
form rings of five carbon atoms on the chain. The ends of the two phytanyl
chains of a di-ether lipid of this type can even be chemically linked to the
corresponding ends of another lipid of the same type, forming so-called
tetraether lipids (bolalipids). As yet little is known about the physical
properties of membranes made of these lipids. However, it can be surmised
that the isoprenic character of the hydrocarbon chains, in particular,  the
case of cyclization along the chain, provide for sufficient chain disorder
and fluidity. The bolalipids are expected to provide additional mechanical
stability under harsh chemical conditions. Recently, the conformation and
phase behavior of bipolar tetraether liposomes composed of the polar lipid
fraction E (PLFE) isolated from the thermoacidophilic archaeon *Sulfolobus
acidocaldarius* have been studied (Chong et al., 2003; Chong et al., 2005).
Interestingly, these lipids also exhibit some polymorphism. Apparently, the
presence of branched methyl groups, cyclopentane rings, ether linkages,
and membrane spanning structures is compatible with the formation of a

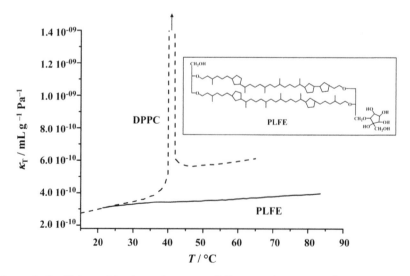

**Figure 9.** Coefficient of isothermal compressibility, $\kappa_T$, of PLFE membranes (grown at
$T = 65°C$) compared to that of DPPC bilayers as a function of temperature. PLFE contains
two types of dibiphytanyldiglycerol tetraethers; one type is depicted in the inlet of the figure.

variety of lamellar phases with rather stiff chains and high packing density, and the PLFE vesicles have unusually high thermal, chemical, and mechanical stability. The relatively small volume and enthalpy changes involved in the phase transitions help to explain why PLFE liposomes are remarkably thermally stable and are rather rigid and tightly packed. As an example, Fig. 9 shows the coefficient of isothermal compressibility of PLFE membranes as a function of temperature in comparison to that of an orthodox phospholipid bilayer undergoing the pre- and main transition. The data clearly show those the PLFE membranes have low compressibilities comparable to that of ordered gel phases of phospholipid bilayers over the whole temperature range covered (Sulc et al., 2008).

## BIOLOGICAL AND RECONSTITUTED MEMBRANES

The cytoplasmic membrane is a complex heterogeneous aggregate structure that can be disturbed by rather low pressures in its structure and function. The integrity and functionality of natural membranes are vital for the cell, e.g., for energy generation, transport, signalling processes, and maintenance of osmotic pressure and intracellular pH. Although some ion transporters are unaffected or even activated upon mild compression, certain other channels and pumps are inactivated at moderate pressures. However it has generally been observed, that at sufficiently high pressures of several kbar, membrane protein function ceases, and integral and peripheral proteins may even become detached from the membrane when its bilayer is sufficiently ordered by pressure, and depolymerization of cytoskeletal proteins may be involved as well (Mcdonald, 1992).

In a more detailed study, the influence of hydrostatic pressure on the activity of $Na^+$, $K^+$-ATPase enriched in the plasma membrane from rabbit kidney outer medulla was studied using a kinetic assay that couples ATP hydrolysis to NADH oxidation ((Powalska et al., 2007) and refs. therein). $Na^+$, $K^+$-ATPase is a heterodimer consisting of three subunits: an $\alpha$-subunit, with a molecular mass of about 110 kDa, a highly glycosylated $\beta$-subunit, with a molecular mass of about 50 kDa, and a $\gamma$-subunit with a molecular mass of about 12 kDa. The $\alpha$-subunit contains the catalytic center required for ATP hydrolysis. It contains the ATP binding and phosphorylation domains, as well as essential amino acids for the binding of sodium, potassium, and some inhibitors. The $\beta$-subunit has a structural function and is important for the membrane insertion of the enzyme. A small hydrophobic protein, the $\gamma$-subunit, has been shown to associate with the $\alpha$-subunit of $Na^+$, $K^+$-ATPase in a tissue-specific manner, although its function is still unclear. The data shown in Fig. 10 reveal that the activity $k$ of $Na^+$, $K^+$-ATPase is inhibited by pressures below 2 kbar. The plot of $\ln k$ *vs.* $p$ revealed an apparent activation volume of the pressure-induced inhibition reaction which amounts to $\Delta V^{\neq} = 47$ mL mol⁻¹. At higher pressures, exceeding 2 kbar, the enzyme is inactivated irreversibly, in agreement with observations from Chong et al. (1985) and de Smedt et al. (1979) who concluded that the decrease in the fluidity of the membrane caused by increased pressure might hinder the conformational changes that accompany the reaction steps, and thus decrease the rate of the overall reaction. On the other hand, it was suggested by Kato et al. (2002) that high pressures up to 2 kbar might lead to a dissociation of the subunits of the Na⁺, K⁺-ATPase. In this study it was suggested that the activity of the enzyme shows at last three step changes induced by pressure: At pressures below and around 1 kbar, a decrease in the fluidity of the lipid bilayer and a reversible conformational change in the transmembrane protein is induced, leading to functional disorder of the membrane associated ATPase activity. Pressures of 1–2 kbar cause a reversible phase transition and the dissociation or conformational changes in the protein subunits, and pressures higher that 2200 bar irreversibly destroy the membrane structure due to protein unfolding and interface separation. Pressure dissociation of water-soluble oligomeric proteins in this pressure range is well documented (Mishra and Winter, 2008; Winter, 2003a). To be able to explore the effect of the lipid matrix on the enzyme activity, the Na⁺, K⁺-ATPase was also reconstituted into various lipid bilayer systems of different chain lengths, conformation, phase state and heterogeneity including model raft mixtures. Interestingly, in the low-pressure region, around 100 bar, a significant increase of the activity was observed for the enzyme reconstituted into DMPC and DOPC bilayers. We found that the enzyme activity decreases upon further compression, reaching zero activity around 2

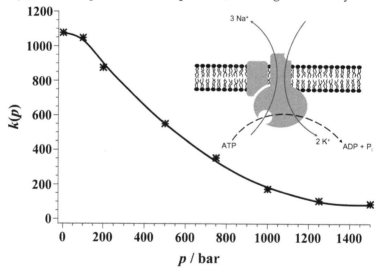

**Figure 10.** Activity $k$ (in arbitrary units) of Na⁺, K⁺-ATPase as measured using an enzymatic assay at selected pressures and $T = 37°C$.

kbar for all reconstituted systems measured. Pressures above about 2.2 kbar, irreversibly change the protein conformation, probably due to dissociation and partial unfolding of the subunits (Powalska et al., 2007).

A similar behavior has been found for the chloroplast ATP-synthase. This enzyme is a $H^+$/ATP-driven rotary motor in which a hydrophobic multi-subunit assemblage rotates within a hydrophilic stator, and subunit interactions dictate alternate-site catalysis. To explore the relevance of these interactions for catalysis Souza et al. (2004) used hydrostatic pressure to induce conformational changes and/or subunit dissociation, and the resulting changes in the ATPase activity and oligomer structure were evaluated. Under moderate hydrostatic pressures (up to 800 bar), the ATPase activity increased by 1.5-fold, which did not seem to be related to an increase in the affinity for ATP, but rather seemed to correlate with an enhanced turnover induced by pressure. The activation volume determined for the ATPase reaction was found to be $-23.7$ mL $mol^{-1}$. Higher pressures of up to 2 kbar lead to dissociation of the enzyme. At these high pressures, dissociation seemed to impair the contacts needed for rotational catalysis.

The effect of pressure was also measured on the HorA activity of *Lactobacillus plantarum* (Ulmer et al., 2002). HorA is an ATP-dependent multi-drug-resistance transporter of the ABC family conferring hop resistance to beer spoilage bacteria. Changes were determined in the membrane composition of *L. plantarum* induced by different growth temperatures and their effect on the pressure inactivation, and a temperature-pressure phase diagram was constructed for the *L. plantarum* membranes that could be correlated with the respective kinetics of high pressure inactivation. Upon pressure-induced transitions to rigid (e.g., gel-like) membrane structures, i.e., at pressure around 500–1500 bar for temperatures between 20 and 37°C, fast inactivation of HorA was observed. Evidence was provided that the composition and the phase behavior of cytoplasmic membranes before and during high pressure treatment, respectively, affect vital membrane bound transporters.

The effect of pressure and influence of the lipid matrix on lipid-protein interactions was also studied for the multidrug resistance protein LmrA, which was expressed in *Lactococcus lactis* and functionally reconstituted in different model membrane systems (Periasamy et al., 2008; Teichert, 2008). These systems were composed of DMPC, DOPC, DMPC+10 mol% cholesterol (Chol), and the model raft mixture DOPC:DPPC:Chol (1:2:1). Teichert showed that a sharp pressure-induced fluid-to-gel phase transition without the possibility for lipid sorting, such as for DMPC bilayers, has a drastic inhibitory effect on the LmrA activity (Teichert, 2008). Otherwise, an overall fluid-like membrane phase over the whole pressure-range covered, with suitable hydrophobic matching, such as for DOPC, prevents the membrane protein from total high pressure inactivation even up to 2 kbar. Also the systems exhibiting thicker membranes with higher lipid order parameters, such as DMPC/10 mol% cholesterol and the model raft mixture, exhibit significant pressure stabilities at high pressure. What seems to be clear is that an efficient packing with optimal lipid adjustment to prevent (also pressure-induced) hydrophobic mismatch might be a particular prerequisite for the homodimer formation and hence function of LmrA.

Recently, pressure has also been found to be of interest to study perturbations of signalling events, such as ToxR/ToxS interaction and phospholipase activation by G-proteins (Linke et al., 2008; Scarlata, 2005). Scarlata revealed perturbations of phospholipase PLβ-Gβγ association caused by the Gα(GDP) subunit in the membranous context that where not observable by atmospheric association measurements (Scarlata, 2005). PLCβ membrane binding was stable throughout the 1–2000 bar range, Gβγ only at high concentrations, whereas Gα(GDP) dissociated from membranes above 1000 bar.

Pressure experiments on ion channels at high pressure are still scarce as well. They revealed that high pressures below 1000 bar affect the kinetics of gating (generally, a retardation is observed due to a positive activation volume, possibly owing to a conformational change following a change in dipole moment) but not the conductance of the channel (Macdonald, 2002). More recent data will be discussed in a separate chapter of this book.

## CONCLUSION

The study of high-pressure effects on membrane proteins and in particular on ion channels and signalling processes is still in its infancy, but there are sufficient published experiments now to encourage further work in this growing area. What is already clear is that the membrane physical-chemical effects markedly influence the lipid-protein interaction, activity and pressure stability of the membrane protein. It also seems to be clear that the specific nature of the membrane protein (e.g., oligomeric assembly, a required dimerization reaction, etc.) plays a significant role in its pressure-dependent activity and stability as well.

## ACKNOWLEDGEMENTS

Financial support from the Deutsche Forschungsgemeinschaft (DFG) is gratefully acknowledged.

## ABBREVIATIONS

GD gramicidin D, HHP high hydrostatic pressure, DMPC 1,2-dimyristoyl-*sn*-glycero-3-phosphatidylcholine (di-$C_{14:0}$), DMPS 1,2-dimyristoyl-*sn*-

glycero-3-phosphatidylserin (di-C$_{14:0}$), DPPC 1,2-dipalmitoyl-*sn*-glycero-3-phosphatidylcholine (di-C$_{16:0}$), DPPE 1,2-dipalmitoyl-*sn*-glycero-3-phosphatidylethanolamine (di-C$_{16:0}$); DSPC 1,2-distearoyl-*sn*-glycero-3-phosphatidylcholine (di-C$_{18:0}$), DOPC 1,2-dioleoyl-*sn*-glycero-3-phosphatidylcholine (di-C$_{18:1,cis}$), DOPE 1,2-dioleoyl-*sn*-glycero-3-phosphatidylethanolamine (di-C$_{18:1,cis}$), DEPC 1,2-dielaidoyl-*sn*-glycero-3-phosphatidylcholine (di-C$_{18:1,trans}$), POPC 1-palmitoyl-2-oleoyl-*sn*-glycero-3-phosphatidylcholine (C$_{16:0}$,C$_{18:1,cis}$), egg-PE egg-yolk phosphatidylethanolamine, Chol cholesterol, SM sphingomyelin, DSC differential scanning calorimetry, SAXS (WAXS) small(wide)-angle X-ray scattering, FT-IR Fourier-transform infrared, NMR nuclear magnetic resonance, AFM atomic force microscopy.

## REFERENCES

Akasaka, K. 2006. Probing conformational fluctuation of proteins by pressure perturbation. Chem. Rev. 106: 1814–1835.

Arnold, M.R. and R.H. Kalbitzer and W. Kremer. 2003. High-sensitivity sapphire cells for high pressure NMR spectroscopy on proteins. J. Magn. Reson. 161: 127–131.

Balny, C. and R. Hayashi, K. Heremans and P. Masson (eds.). 1992. High Pressure and Biotechnology, Colloque Inserm, Vol. 224. John Libbey Eurotext Ltd, Montrouge.

Bartlett, D.H. 2002. Pressure effects on in vivo microbial processes. Biochim. Biophys. Acta 1595: 367–381.

Bernsdorff, C. and A. Wolf and R. Winter. 1996. The effect of temperature and pressure on structural and dynamic properties of phospholipid/sterol mixtures-a steady-state and time-resolved fluorescence anisotropy study. Z. Phys. Chem. 193: 151–173.

Bernsdorff, C. and A. Wolf, R. Winter and E. Gratton. 1997. Effect of hydrostatic pressure on water penetration and rotational dynamics in phospholipid-cholesterol bilayers. Biophys. J. 72: 1264–1277.

Böttner, M. and R. Winter. 1993. The influence of the local anesthetic tetracaine on the phase behaviour and the thermodynamic properties of phospholipid bilayers. Biophys. J. 65: 2041–2046.

Cevc, G. (ed.). 1993. Phospholipids Handbook. Marcel Dekker, New York.

Chong, P.L.-G. and P.A.G. Fortes and D.A. Jameson. 1985. Mechanisms of inhibition of (Na,K)-ATPase by hydrostatic pressure studied with fluorescent probes. J. Biol. Chem. 260: 14484–14490.

Chong, P.L.-G. and M. Zein, T.K. Khan and R. Winter. 2003. Structure and conformation of bipolar tetraether lipid membranes derived from thermoacidophilic archaeon *sulfolobus acidocaldarius* as revealed by small-angle x-ray scattering and high-pressure FT-IR spectroscopy. J. Phys. Chem. B107: 8694–8700.

Chong, P.L.-G. and R. Ravindra, M. Khurana, V. English and R. Winter. 2005. Pressure perturbation and differential scanning calorimetric studies of bipolar tetraether liposomes derived from the thermoacidophilic archaeon sulfolobus acidocaldarius. Biophys. J. 89: 1841–1849.

Cossins, A.R. and A.G. Macdonald. 1986. Homeoviscous adaptation under pressure. III. The fatty acid composition of liver mitochondrial phospholipids of deep-sea fish. Biochim. Biophys. Acta 860: 325–335.

Czeslik, C. and O. Reis, R. Winter and G. Rapp. 1998. Effect of high pressure on the structure of dipalmitoylphosphatidylcholine bilayer membranes: a synchrotron-x-ray diffraction and FT-IR spectroscopy study using the diamond anvil technique. Chem. Phys. Lipids 91: 135–144.

Daniel, I. and P. Oger and R. Winter. 2006. Origins of life and biochemistry under high pressure conditions. Chem. Soc. Rev. 35: 858–875.

de Smedt, H. and R. Borghgraef, F. Ceuterick and K. Heremans. 1979. Pressure effects on lipid-protein interactions in (Na⁺ K⁺)-ATPase. Biochim. Biophys. Acta 556: 479–489.

Eisenblätter, J. and R. Winter. 2005. Pressure effects on the structure and phase behavior of phospholipid-polypeptide bilayers—a synchrotron small-angle x-ray scattering and $^2$H-NMR spectroscopy study on DPPC-gramicidin lipid bilayers. Z. Phys. Chem. 219: 1321–1345.

Eisenblätter, J. and R. Winter. 2006. Pressure effects on the structure and phase behavior of DMPC-gramicidin lipid bilayers—a synchrotron SAXS and $^2$H-NMR spectroscopy study. Biophys. J. 90: 956–966.

Eisenblätter, J. and A. Zenerino and R. Winter. 2000. High pressure $^1$H-NMR on model biomembranes: a study of the local anaesthetic tetracaine incorporated into POPC lipid bilayers. Magn. Reson. Chem. 38: 662–667.

Gross, M. and R. Jaenicke. 1994. Proteins under pressure: The influence of high hydrostatic pressure on structure, function and assembly of proteins and protein complexes. Eur. J. Biochem. 221: 617–630.

Hammouda, B. and D. Worcester. 1997. Interdigitated hydrocarbon chains in C20 and C22 phosphatidylcholines induced by hydrostatic pressure. Physica B 241–243: 1175–1177.

Hausmann, K.and B.P. Kremer (eds.). 1994. Extremophile. Mikroorganismen in ausgefallenen. Lebensräumen, VCH, Weinheim.

Helfrich, W. 1978. Z. Naturforsch. 33: 305–315.

Heremans, K. and L. Smeller. 1998. Protein structure and dynamics at high pressure. Biochim. Biophys. Acta 1386: 353–370.

Hyde, S. and S. Andersson, K. Larsson, Z. Blum, T. Landh, S. Lidin and B.W. Ninham. 1997. The language of shape. The role of curvature in condensed matter: physics, chemistry and biology. Elsevier, Amsterdam.

Janosch, S. and C. Nicolini, B. Ludolph, C. Peters, M. Völkert, T.L. Hazlet, E. Gratton, H. Waldmann and R. Winter. 2004. Partitioning of dual-lipidated peptides into membrane microdomains—lipid sorting vs. peptide aggregation. J. Am. Chem. Soc. 126: 7496–7503.

Jeworrek, C. and M. Pühse and R. Winter. 2008. X-ray kinematography of phase transformations of three-component lipid mixtures: a time-resolved synchrotron x-ray scattering study using the pressure-jump relaxation technique. Langmuir 24: 11851–11859.

Jonas, J. (ed.). 1991. High Pressure NMR. Springer-Verlag, Berlin.

Kato, M. and R. Hayashi, T. Tsuda and K. Taniguchi. 2002. High pressure-induced changes of biological membrane. Study on the membrane-bound Na⁺/K⁺-ATPase as a model system. Eur. J. Biochem. 269: 110–118.

Krivanek, R. and L. Okoro and R. Winter. 2008. Effect of cholesterol and ergosterol on the compressibility and volume fluctuations of phospholipid-sterol bilayers in the critical point region: a molecular acoustic and calorimetric study. Biophys. J. 94: 3538–3548.

Landwehr, A. and R. Winter. 1994. High-pressure differential thermal analysis of lamellar to lamellar and lamellar to non-lamellar lipid phase transitions. Ber. Bunsenges. Phys. Chem. 98: 214–218.

Linke, K. and N. Periasamy, M. Ehrmann, R. Winter and R.F. Vogel. 2008. Influence of high pressure on the dimerization of ToxR, a protein involved in bacterial signal transduction. Appl. Environ. Microbiol. 74: 7821–7823.

Lipowski, R. and E. Sackmann (eds.). 1995. Structure and dynamics of membranes. Volumes 1A and 1B, Elsevier, Amsterdam.

Luzzati, V. 1997. Biological significance of lipid polymorphism: the cubic phases. Curr. Opin. Struct. Biol. 7: 66–668.

Macdonald, A.G. 1992. Effects of high hydrostatic pressure on natural and artificial membranes. Pp. 67–75. *In*: C. Balny, R. Hayashi, K. Heremans and P. Mason (eds.) High Pressure and Biotechnology Colloque INSERM, John Libbey Eurotext Ltd..

Macdonald, A.G. 2002. Experiments on ion channels at high pressure. Biochim. Biophys. Acta 1595: 387–389.

Mantulin, W.W. and A.M. Gotto and H.J. Pownall. 1984. Effect of hydrostatic pressure on the transfer of a fluorescent phosphatidylcholine between apolipoprotein-phospholipid recombinants. J. Am. Chem. Soc. 106: 3317–3319.

Mishra, R. and R. Winter. 2008. Cold- and Pressure-Induced Dissociation of Protein Aggregates and Amyloid Fibrils. Angew. Chem. Int. Ed. 47: 6518–6521.

Munro, S. 2003. Lipid rafts: elusive or illusive? Cell 115: 377–388.

Nicolini, C. and A. Celli, E. Gratton and R. Winter. 2006a. Pressure tuning of the morphology of heterogeneous lipid vesicles: a two photon-excitation fluorescence microscopy study. Biophys. J. 91: 2936–2942.

Nicolini, C. and J. Baranski, S. Schlummer, J. Palomo, M.L. Burgues, M. Kahms, J. Kuhlmann, S. Sanchez, E. Gratton, H. Waldmann and R. Winter. 2006b. Visualizing association of N-Ras in lipid microdomains: influence of domain structure and interfacial adsorption. J. Am. Chem. Soc. 128: 192–201.

Nicolini, C. and J. Kraineva, M. Khurana, N. Periasamy, S. Funari and R. Winter. 2006c. Temperature and pressure effects on structural and conformational properties of POPC/SM/cholesterol model raft mixtures—a FT-IR, SAXS, DSC, PPC and laurdan fluorescence spectroscopy study. Biochim. Biophys. Acta 1758: 248-258.

Periasamy, N. and H. Teichert, K. Weise, R.F. Vogel and R. Winter. 2008. Effects of temperature and pressure on the lateral organization of model membranes with functionally reconstituted multidrug transporter LmrA. Biochim. Biophys. Acta (in press).

Powalska, E. and S. Janosch, E. Kinne-Saffran, R.K.H. Kinne, C.F.L. Fontes, J.A. Mignaco and R. Winter. 2007. Fluorescence spectroscopic studies of pressure effects on $Na^+$, $K^+$-ATPase reconstituted into phospholipid bilayers and model raft mixtures. Biochemistry 46: 1672–1683.

Reis, O. and R. Winter and T.W. Zerda. 1996. The effect of high external pressure on DPPC-cholesterol multilamellar vesicles—a pressure-tuning Fourier-transform infrared spectroscopy study. Biochim. Biophys. Acta 1279: 5–16.

Scarlata, S. 2005. Determination of the activation volume of $PLC\beta$ by $G\beta\gamma$-subunits through the use of high hydrostatic pressure. Biophys. J. 88:2867–2874.

Seddon, J.M. 1990. Structure of the inverted hexagonal ($H_{II}$) phase, and non-lamellar phase transitions of lipids. Biochim. Biophys. Acta 1031: 1–69.

Seemann, H. and R. Winter. 2003. Volumetric properties, compressibilities and volume fluctuations in phospholipid-cholesterol bilayers. Z. Phys. Chem. 217: 831–846.

Silva, J.L. and D. Foguel and C.A. Royer. 2001. Pressure provides new insights into protein folding, dynamics and structure. Trends Biochem. Sci. 26: 612–618.

Souza, M.O. and T.B. Creczynski-Pasa, H.M. Scofano, P. Gräber and J.A. Mignaco. 2004. High hydrostatic pressure perturbs the interactions between $CF_0F_1$ subunits and induces a dual effect on activity. Int. J. Biochem. Cell Biol. 36: 920–930.

Sulc, M. and R. Winter and P.L.-G. Chong. 2008 (in preparation).

Teichert, H. 2008. Behaviour of membrane transport proteins under high hydrostatic pressure, M.S. Thesis, Technische Universität München, München, Germany.

Ulmer, H.M. and H. Herberhold, S. Fahsel, M.G. Gänzle, R. Winter and R.F. Vogel. 2002. Effects of pressure-induced membrane phase transitions on inactivation of HorA, an ATP—dependent multidrug resistance transporter, in lactobacillus plantarum. Appl. Environ. Microbiol. 68: 1088–1095.

Wharton, D.A. 2002. Life at the Limits. Organisms in Extreme Environments. Cambridge University Press, Cambridge.

Winter, R. 2002. Synchrotron x-ray and neutron small-angle scattering of lyotropic lipid mesophases, model biomembranes and proteins in solution at high pressure. Biochim. Biophys. Acta 1595: 160–184.

Winter, R. (ed.). 2003a. High Pressure Bioscience and Biotechnology II. Springer-Verlag, Heidelberg.

Winter, R. 2003b. High pressure NMR studies on lyotropic lipid mesophases and model biomembranes. Ann. Rep. NMR Spectr. 50: 163–200.

Winter, R. and J. Jonas (eds.). 1999. High Pressure Molecular Science. Kluwer Academic Publishers, NATO ASI E 358, Dordrecht.

Winter, R. and C. Czeslik. 2000. Pressure effects on the structure of lyotropic lipid mesophases and model biomembrane systems. Z. Kristallogr. 215: 454–474.

Winter, R. and R. Köhling. 2004. Static and time-resolved synchrotron small-angle x-ray scattering studies of lyotropic lipid mesophases, model biomembranes and proteins in solution. J. Phys.: Condens. Matt. 16: S327–S352.

Winter, R. and M.-H. Christmann, M. Böttner, P. Thiyagarajan and R. Heenan. 1991. The influence of the local anaesthetic tetracaine on the temperature and pressure dependent phase behaviour of model biomembranes. Ber. Bunsenges. Phys. Chem. 95: 811–820.

Winter, R. and D. Lopes, S. Grudzielanek and K. Vogtt. 2007. Towards an understanding of the temperature/pressure configurational and free-energy landscape of biomolecules. J. Non-Equilib. Thermodyn. 32: 41–97.

Woenckhaus, J. and R. Köhling, R. Winter, P. Thiyagarajan and S. Finet. 2000. High pressure-jump apparatus for kinetic studies of protein folding reactions using the small-angle synchrotron x-ray scattering technique. Rev. Sci. Instrum. 71: 3895–3899.

Zein, M. and R. Winter. 2000. Effect of temperature, pressure and lipid acyl-chain length on the structure and phase behaviour of phospholipid-gramicidin bilayers. Phys. Chem. Chem. Phys. 2: 4545–4551.

# 5

# High Pressure and Food Conservation

*Eliane Dumay,[1] Dominique Chevalier-Lucia[2] and Tomás López-Pedemonte[3]*

## INTRODUCTION

Nowadays consumer demand for food products do not only refer to the microbiological and toxicological safety, but also require the guarantee for other key characteristics such as preserved nutritional and sensorial quality, extended shelf-life and the convenience of use. In addition, food manufacturers have to integrate actual considerations linked to the environmental requirements, reduction of energy consumption and regulation evolution. Besides, as a consequence of market globalization, the food industry and especially small and medium-sized enterprises (SME's) have to demonstrate innovative dynamics in launching products with high added-value and developing technological 'niches' as new market opportunities. Thermal processing which has a long tradition in food preservation is unfortunately often accompanied by degradation of some food sensorial and nutritional properties. Even if heat treatments can be optimized, such as short-time-high-temperature treatments able to inactivate pathogens and spores while limiting vitamin loss, some deteriorative side-

[1] Professor, Department of Agro-resources & Biological and Industrial Processes, Mixed Research Unit of Agro-polymer Engineering and Emerging Technologies, Université Montpellier 2, Sciences et Techniques, Place E. Bataillon, 34095 Montpellier, France. E-mail: Eliane.Dumay@univ-montp2.fr

[2] Assistant Professor, Department of Agro-resources & Biological and Industrial Processes Mixed Research Unit of Agro-polymer Engineering and Emerging Technologies, Université Montpellier 2, Sciences et Techniques, Place E. Bataillon, 34095 Montpellier, France. E-mail: Dominique.Chevalier-Lucia@univ-montp2.fr

[3] Assistant Professor, Departamento de Ciencia y Tecnología de los Alimentos, Facultad de Química—Universidad de la República (UDELAR), Avenida General Flores 2124– Montevideo, Uruguay. E-mail: tlopez@fq.edu.uy

effects cannot be completely avoided with regard to fresh-like color and flavor, and product functionality. The development of novel technologies including high hydrostatic pressure (HHP) for food processing and preservation complies with these requirements. Minimal processing is designed to limit processing impacts on nutritional and sensory quality while preserving food products and limiting the use of 'chemical' additives. HHP is a physical non-thermal process able to reduce/inactivate spoilage and pathogenic microorganisms at cold or mild temperatures while preserving food quality attributes and saving energy with a low impact on the environment. HP treatments at 4–20°C and 400–600 MPa are nowadays implemented at an industrial scale (Tonello 2008). Such treatments lead to food product pasteurization. More recently, combined high pressure-high temperature treatments (or pressure-assisted sterilization) that take advantage of adiabatic heating during pressure build up to achieve 105–120°C for few seconds or minutes (starting from product temperatures of 60–90°C) are investigated at the laboratory scale to assess spores inactivation (Meyer et al., 2000; Matser et al., 2004; Ahn, 2007).

The first studies dealing with pressure-induced effects on foods date back to the end of the XIX[e] century and the early of the XX[e]. These pioneering and promising studies reported the effects of high pressure on microorganisms in milk and vegetables (Hite, 1899; Hite et al., 1914), and also on egg albumen coagulation (Bridgman, 1914). Bridgman (1912) reported 5 crystalline forms of ice (type I to IV, and VI) while actually more than 10 ice polymorphs are known. The phase diagram of water as a function of pressure and temperature has retained attention this last decade for potential food applications. On the basis of results obtained from grape and apple juices by Hite et al. (1914), Cruess (1924) suggested the preservation of fruit juices using HHP whose application is nowadays implemented at an industrial scale. Basset and Macheboeuf (1932, 1933) and Basset et al. (1933) studied pressure-induced inactivation of microorganisms, spores and enzymes, and assessed the immunological properties of toxins after HP treatment, all those topics that actually remain of great interest. They also observed the pressure-induced denaturation/gelation of horse serum proteins (Basset et al., 1933) and also of gelatin (Talwar et al., 1952, 1953). From the 1960's numerous studies started on pressure-induced unfolding and denaturation of protein such as ovalbumin (Suzuki, 1960), β-casein (Payens and Heremans, 1969), metmyoglobin (Zipp and Kauzmann, 1973). Pressure has been used as a tool by biochemists and biophysicists to follow (most of time using spectroscopy under pressure) protein structural changes and mechanisms of enzyme activity, to understand organism adaptation to extreme conditions of pressure and temperature using protein mutants, to study mechanisms of protein unfolding/refolding or misfolding, catching intermediate states (Smeller, 2002; Winter, 2002) which is not possible to observe using heating and more recently, to follow protein aggregation pathways (e.g., aggregation of prion protein) (Randolph et al., 2002; El Moustaine et al., 2008). All these studies constitute a useful scientific background for the food technologist in order to assess/predict effects of HP processing on food enzymes and proteins.

Eighteen years ago HP technology was transferred from material science to food technology with the first machines built at the pilot scale (1–2 liter capacity) for food science laboratories, then at an industrial scale. From the 1990's, HP food research activity has been widely supported in EU through national and European projects, also in Japan and much less in the USA. In a decade, various kinds of food and biological materials have been processed by HHP. The first pressurized foods appeared in the market in Japan in 1990: it was jam (fruit-based product with pH < 4.5) launched in the market by the Meidi-Ya food factory, then followed by fruits jellies, fruit juices, rice cakes, sauces, dairy preparations and fish products (Cheftel, 1995). After a little while, HP-processed jams and juices were also introduced in Europe, avocado paste and shell oysters in USA, sliced cooked ham in Spain. Actually, most of the commercially available products processed by HHP are launched in the American, Canadian and Australian markets, much less in the European market, probably due to the difficulties that the Novel Food Regulation opposed to innovation in Europe. Effectively, the Regulation 258/97/EC for novel foods and food ingredients that was introduced in 1997 by the European Commission defines novel foods as foods and food ingredients that have not been used for human consumption to a significant degree within the Community before 15 May 1997. Foods commercialized in at least one Member State before the entry in force of the Regulation on Novel Foods on 15 may 1997, are on the EU market under the principle of 'mutual recognition'. In order to ensure the highest level of protection of human health, novel foods must undergo a safety assessment before being placed on the EU market. Only those products considered to be safe for human consumption are authorized for marketing. Companies that want to place a novel food on the EU marked need to submit their application in accordance with Commission Recommendations 97/618/EC that concerns the scientific information and the safety assessment report required. Foods or food ingredients to which a production process not currently used (such as HHP) has been applied may follow a simplified procedure, only requiring notifications from the Company, when they are considered by a national food assessment body as 'substantially equivalent' to existing foods or food ingredients, as regards to their composition, nutritional value, metabolism, intended use and the level of undesirable substances contained therein (e.g., toxic or allergenic compounds). The European Commission has recently adopted (January 2008) a proposal to revise the Novel Foods Regulation with a view to improve the access of new and innovative foods to the EU market, while maintaining a high level of consumer protection and ensuring food safety.

Among the novel (or emerging) technologies offering potential industrial developments for food products, HP processing is probably one of the most advanced with machines operating at an industrial scale (capacity up to 400 L, 600 MPa). Large scale equipments have been developed by firms such as ACB, ABB, Flow International Corporation, National Forge, Avure Technologies, Hyperbaric, etc. At least 124 industrial HP machines are operating in the world including 71 equipments in the USA, 26 in Europe, 22 in Asia, 5 in Oceania (Tonello, 2008). HP processing is a semi-continuous process which necessitates HP vessels of high capacity, pumps, pressure intensifiers, HP valves, HP tubing and seals, lubricants, specific stainless-steels able to resist pressure as well as low (or high) temperatures (for combined pressure-temperature treatments). Pressure is generated by hydro-pneumatic or hydro-electric pumps which need energy for the pressure build up, but the pressure maintained does not require much additional energy. For industrial developments, horizontal machines with rapid closing/opening systems, one way loading/ unloading, and equipped with multiple pumps and intensifiers are preferred to facilitate access and maintenance, and allow sufficient production capacities. The processing cost ranges between 0.080 and 0.2 € /kg of product depending on the applied HHP processing and the equipment capacity. The investment cost varies between 500–2000 k € depending of the equipment capacity (Tonello, 2008). HHP machines are designed to treat foods preferably in their packaging. But few data are actually available on the possible interactions that HHP could induce between food and packaging.

## EFFECTS OF HIGH PRESSURE PROCESSING ON FOOD MATRICES AND CONSTITUENTS

Two general scientific principles are of direct interest for food applications: (1) the principle of Le Chatelier, which indicates that any phenomenon (phase transition of water or lipids, chemical reaction, change in macromolecule configuration) accompanied by a decrease in volume is enhanced by an increase in pressure (and *vice versa*), (2) the isostatic rule, according to which pressure is transmitted in a uniform and quasi-instantaneous manner throughout the whole biological sample (Cheftel, 1992, 1995). According to the first principle and depending on the value of the reaction volume ($\Delta V^0$) and the activation volume ($\Delta V^{\#}$), some (bio-)chemical reactions will be favored by pressure, others will never happen under pressure. In the range of pressure levels involved to treat foods or biological materials (well below 1 GPa) and at low or moderate temperatures (< 40°C), covalent bonds are not affected (Mozhaev et al., 1996;

Winter, 2002) and small molecules such as vitamins, aromas, pigments are particularly not or slightly affected (Oey et al., 2008a, b) in contrast to heating. Some enzymatic reactions are promoted, others are not. As a consequence of the isostatic rule, the pressure experienced by the food is independent of its volume, shape or physical state. The pressurization process time is therefore independent of the sample volume, in contrast to thermal processing (Cheftel, 1992). In the food area, isostatic high pressure treatment refers to a hydrostatic process that exposes to pressure the food product hermetically sealed in a flexible packaging (i.e., indirect compression). The pressure transmitting medium (PTM) in contact with the packaged food products is usually water or alcool-water mixtures for processing at negative temperatures. Isostatic high pressure differs from dynamic high pressure involved in high-pressure homogenization processing. In this latter process, a fluid under pressure is forced through a small orifice of some micrometer width, the HP valve gap. The resulting strong pressure gradient between the inlet and outlet of the HP valve generates intense shear forces and laminar extensional stress before and in the valve gap, simultaneously to cavitation and impact at the valve gap outlet, plus short-life heating phenomena (Grácia-Juliá et al., 2008). Such shearing effects are not induced by isostatic high pressure, since pressure is uniformly and instantly applied throughout the sample. Foods containing high proportion of water, little gas and little fat display the same compressibility as that of water. The volume reduction of water varies from 4% at 100 MPa to 15% at 600 MPa and 22°C. This volume reduction corresponds to an increase in the water density (Cheftel, 1992; Farkas and Hoover, 2000). The quasi-adiabatic compression of water (or aqueous solutions) causes a temperature increase of 2–3°C per 100 MPa (at ambient temperature) that must be taken into account for process designing. Such increase in temperature depends on the rate of compression, the respective initial food sample and PTM temperatures, the respective composition of PTM and food product (Kolakowsky et al., 2001). Expansion during the pressure release is accompanied by a decrease in PTM and sample temperatures of the same order of magnitude than the temperature increase during compression. Heat of compression has to be considered when using prototypes or industrial equipments for which the rate of compression is rapid (~300 MPa/min or more). Since heat exchanges exist between the food samples and the PTM, and between the PTM and the metal walls of the HP vessel, the heat of compression can be minimized or in the opposite used to reach a desired quite constant temperature during the pressure holding (if the vessel walls have been previously equilibrated at the expected temperature). Temperature is an important parameter to control during HP processing in the case of large scale equipments (i.e., temperature of metal walls, PTM and product; temperature distribution in the HP vessel; ratio of PTM to food product in the HP vessel). Pressure gauges and thermocouples are used to follow the pressure and temperature history of the processed products and predict/ guarantee the processing impact on microbial inactivation and possible (bio)chemical changes. Good temperature homogeneity in the HP vessel during HP processing will effectively guarantee the expected microbial inactivation (Hartmann et al., 2003). Phase transition phenomena of mainly water or lipids induced by pressure are accompanied by sample temperatures changes (temperature jumps in the case of crystal formation or temperature drops for crystal melting) that must be considered in food process designing.

The interest in the properties of water at high pressure and low temperature increased in these last 15 years (Makita, 1992; Kalichevsky et al., 1995; Knorr et al., 1998; Cheftel et al., 2000, 2002; Le Bail et al., 2002; Urrutia Benet et al., 2004; Dumay et al., 2006; Urrutia et al., 2007). The phase transitions of water and water characteristics (latent heat, density, viscosity) are influenced by pressure (Schlüter et al., 1998). The pressure-temperature phase diagram of water indicates that water remains in a liquid state down to – 22°C at 207 MPa, probably because pressure opposes the volume increase accompanying the formation of ice (type I). At higher pressure levels other types of ice are formed (III, V, VI) that have been also explored in the food area (see next section).

Phase transitions phenomena may also affect food lipids. Melting temperatures of triglycerides increases by 10–20°C per 100 MPa and pressure enhances the formation of the denser and most stable crystals (Cheftel, 1992). Nevertheless few data are available for food lipids. The melting point of triglycerides is increased by about 14°C per 100 MPa in cacao butter (Yasuda and Mochizuki, 1992). An increase in the fat crystalline state of HP-treated dairy cream (firstly heated at 60°C to liquefy fat globules, then pressurized at 450 MPa and 25°C for 30 min) was observed after pressure release as compared to the untreated sample (Dumay et al., 1996). This result agreed with the previous observations of Buccheim and Abou El Nour (1992) for milk fat. More studies are needed to assess the crystalline state of lipids depending on the size of emulsion droplets, and also to assess the further changes in crystal forms during storage after HP processing of emulsions or bulk fat. However, processing at 450 MPa and 10 or 25°C of dairy creams (without additives) or oil-in-water (O/W) emulsions prepared with water, peanut oil or triolein and dairy proteins did not induce changes in the fat globules or oil droplet sizes (Dumay et al., 1996). Interestingly, processing at 450 MPa and 25 or 40°C of O/W emulsions stabilized by whey proteins (mainly β-lactoglobulin) induced an increase in emulsions viscosity (25°C) or emulsion gelation (40°C). Hence such HP processing did not break down or coalesce oil/fat particles in any model emulsions because of the isostatic principle, but did change the rheological behavior of emulsions stabilized by pressure-sensitive proteins. It may thus be possible to induce desirable texture in food or cosmetic emulsions, simultaneously to microbial inactivation. Appropriate packaging needs to be considered. The pressure-induced inactivation of microorganisms is partly explained by the physical state of lipids in bio-membranes. The temperature- and pressure-dependent structure and phase behavior of model membrane systems such as dipalmitoyl-phosphatidyl-choline bilayer membranes have been investigated, showing various domains from the liquid-crystalline to gel phases (Luzzati, 1997; Czeslik et al., 1998; Winter, 2002).

Numerous data have been gathered concerning the effects of high pressure on food macromolecules, both proteins and polysaccharides: milk proteins, animal and fish muscle proteins, starch and pectins, and mixture of proteins and polysaccharides. Enzymatic kinetics in buffers, vegetable extracts or milk have been extensively studied and modelled with a view to assess the product stability during HP processing and further storage (Hendrickz et al., 1998; Ludikhuyze et al., 2001). Such enzymatic models may also serve as pressure-temperature-time indicators for HP processing of foods (Claeys et al., 2003). Effects of high pressure processing on milk and dairy products have been extensively studied to assess pressure-induced structural changes in proteins, milk enzyme inactivation and microbial load reduction, changes in milk coagulation properties (acid or rennet gel formation) and cheese production (Trujillo et al., 2002; López-Pedemonte 2007a, b). Pressure-induced unfolding and aggregation of milk globular proteins dispersed in water, buffers or milk have been studied to compare with heating (Funtenberger et al., 1997; Cheftel and Dumay, 1998; Tewari 1999; Huppertz et al., 2006; Considine et al., 2007). The pressure sensitivity of globular proteins in bovine milk has been stated as follows: β-lactoglobulin > α-lactalbumin > bovine serum albumin > immunoglobulins, so that it could be envisaged to extend the shelf-life of colostrum rich in immunoglobulins by HP processing. HP-induced inactivation of endogenous milk enzymes such as alkaline phosphatase (ALP), γ-glutamyl transferase, lactoperoxydase, plasmin have been studied to be used as internal process markers for pathogenic strains (Rademacher and Hinrinchs, 2002). Pressurization of β-lactoglobulin in the presence of food anionic polysaccharides such as sodium alginate, carraghenate, high methoxy or low methoxy pectins induced phase separation between protein and polysaccharide, and the formation of original gelled structures and textures: depending on the polysaccharide/protein ratio, mixed gels with honeycomb structures or protein microparticulation could be obtained (Dumay et al.,1999). HP treatment of milk induces casein micelle dissociation/re-association phenomena as determined by photon correlation spectroscopy and microscopy (Needs et al., 2000; Huppertz et al., 2004, Regnault et al., 2004). Original milk micelles were dissociated into neo-micelles of half their initial sizes by processing up to 300 MPa at 9 or 20°C for 15 min (Regnault et al., 2004). This reduction in micelle size explains the pressure-induced decrease in milk turbidity already noticed by Hite (1899) long ago. These dissociation/re-association phenomena could be envisaged as a tool to encapsulate bio-molecules of interest. Hereto, HP processing may be used for food protein engineering. Temperature associated to HHP is an important parameter to consider when processing foods, even at cold or mild temperatures since both pressure and temperature influence the protein native state through their effects on hydrophobic and hydrogen bonds, and electrostatic interactions. For example, pressure-induced denaturation of β-lactoglobulin as evaluated after pressure release decreases after pressurization at low or negative temperature (Kolakowski et al., 2001). On the contrary, milk micelle dissociation is enhanced by pressurization at low temperatures (Regnault et al., 2004). Pressure-induced changes in protein immune-reactivity and allergenicity of known allergens have to be investigated to compare with heating.

Effects of high pressure on muscle proteins, meat and fish products have been extensively investigated in view of meat/fish muscle preservation and refrigerated storage, microbial inactivation and meat tenderization. Post-mortem tenderization is caused by enzymatic degradation of proteins in myofibrils, extracellular matrix and proteins involved in linkages between myofibrils or with sarcolemma (Cheftel and Culioli, 1997; Delbarre-Ladrat, 2006). Enzymatic systems such as cathepsins and calpains have been especially studied, as well as phosphorylases and $Ca^{++}$-dependent ATPase involved in the regulation of glycogen breakdown. The modifications of muscle ultrastructure are strongly dependent on the time post-mortem (pre- or post-rigor) when pressure is applied. They are more extensive when pressure is applied in combination with heat. Actin and myosin are pressure sensitive from relatively low pressure levels (150–200 MPa). Pressure induces actomyosin depolymerization then protein aggregation and gelation at higher pressure levels. Pressure-induced changes in meat color and myoglobin have also been investigated under vacuum, in the presence of air or $O_2$ (Carlez et al., 1995). Pressurized meat becomes pink at 200–350 MPa (10°C for 10 min), then gray-brown at higher pressure levels (400–500 MPa). Meat discoloration results from whitening effects in the range 200–350 MPa due to globin denaturation, then oxidation of ferrous to ferric metmyoglobin at or above 400 MPa. It seems that pressurization at low temperature and below 200 MPa may protect meat color. These color changes are not observed after pressurization of raw or cooked ham ( which is commercialized).

Effects of high pressure processing on vegetable and fruit based products have been recently reviewed by Oey et al. (2008a, b). On the basis of the bibliographic data  available mainly for water soluble vitamins (ascorbic acid, folate, B vitamins), it can be stated that high pressure processing at ambient/moderate temperatures maintains the vitamin content of fruit and vegetable based products. Vitamin stability during processing and further storage depends on the initial oxygen content in the product, possible existence of anti- or pro-oxidants in the food matrix and storage temperature (Rovere et al., 1996; Oey et al., 2008a). By contrast, degradation of water soluble vitamin in vegetable or meat based products is increased by combination of extreme pressure and temperature since effects of pressure and temperature on food constituents are governed by activation volume (pressure) and activation energy (temperature) (Van den Broeck et al., 1998; Touakis, 1998; Butz et al., 2004, 2007). Carotenoid stability in juice and puree during HP processing at moderate temperature and further storage at 4°C is not or slightly affected (Fernandez Garcia et al., 2001; Qui et al., 2006). Pressure influences the extraction yields of carotenoids and flavones and consequently anti-oxidant capacity of fruit or vegetable processed juices and purees (Sanchez-Moreno et al., 2004a, b). Human studies have shown that HP-processed orange juice and vegetable soup still possess their vitamin C bioavaibility and their potentiality to decrease human inflammatory biomarkers (Sanchez-Moreno et al., 2004a; Oey et al., 2008b). HP processing at low or moderate temperature has a limited effect on pigments responsible for the color of fruits and vegetables (Oey et al., 2008a). In many cases the green color becomes even more intense (Krebers et al., 2002). However some color changes may happen during further storage due to incomplete enzymatic inactivation or oxidation phenomena (Oey et al., 2008a). Combining high pressure and high temperature above 50°C affects chlorophyll stability (Butz et al., 2002) but in that case no further changes were noticed under storage (Krebers et al., 2002). Storage at 4°C and the presence of ascorbic acid in the food matrix limit color changes during storage. Effects of pressure on fruit and vegetable texture vary depending on the pressure level, pressurization time, the processing temperature and the type of product (Ludikhuyze and Hendrickx, 2001). Changes in plant cell walls and therefore in plant-based food texture during ripening and food processing are mainly related to pectin structure and composition. Pressure and temperature may affect pectin degrading enzymes such as pectinmethylesterase (PME), polygalacturonase (PG), pectate lyases (PL) and other unknown enzymes (Rastogi et al., 2008; Sila et al., 2008). PME can be regarded as heat-sensitive and pressure-resistant enzyme. PG is easily inactivated by pressure at moderate temperature. Increase in PME activity has been reported after pressurization at 300–500 MPa and 50–60°C (Sila et al., 2008). HP processing in combination with mild temperature (60°C) has been proposed as pre-treatment to reduce further cooking softening for calcium soaked carrots via the formation of cross-linked pectin by divalent ions after pectin demethylation (Sila et al., 2007). Effect of pressure on cell walls has been explored by microscopy examination, and the redistribution of water at the sub-cellular level by RMN. Higher pressures than 100 MPa induced membranes rupture and water redistribution in parenchyma tissues of delicate fruits such as strawberries (Marigheto et al., 2004). Other vegetables rich in fiber such as green beans resist pressure effects better.

## HIGH PRESSURE-LOW TEMPERATURE PROCESSING OF FOOD SAMPLES

As shown on the phase diagram of water (Fig. 1a–f), pressure markedly influences ice-water transitions, and high pressure technology offers distinct potentialities for preserving food materials at low temperature. Conventional freezing at 0.1 MPa (atmospheric pressure) is admitted as an efficient preservation method for foods, minimizing microbial or enzymatic activities during frozen storage due to water immobilization at the solid state, but inducing no marked inactivation of most detrimental microorganisms (Dykes and Moorhead, 2001; Novak and Juneja, 2003). But the formation of ice I crystals during freezing at atmospheric pressure is responsible for most of the damages that biomaterials and food structure may suffer. The larger specific volume of ice I crystals compared with liquid water is responsible for cell breakage and tissue disruption. During freezing at atmospheric pressure a freezing front forms and proceeds from the surface to the interior of the sample. Extracellular water freezes more readily than intracellular water because of its lower ionic and solute concentration. The kinetic of freezing (or thawing, the reverse processing) influences the quality of food products. It is well known that slow freezing rates induce the formation of large extracellular crystals (ice I) causing tissue mechanical damage and drip loss during thawing. Besides, water crystallization associated to an increase in extracellular concentration of solutes favors cell dehydration and death through osmotic plasmolysis and membrane damage. Conversely, rapid freezing enhances nucleation and the formation of both smaller extra- and intracellular crystals, little translocation of water and reduced drip loss during thawing. But intracellular crystals promote cell death through disruption of the cytoplasm gel and organelles (Devine, 1996; Reid, 1998).

According to the phase diagram of water, different pathways (Fig. 1a–f) can be followed at high pressure and low temperature with or without phase transitions (liquid/solid or solid/solid) and subsequent changes in the physical state of HP-processed biological samples. Several reviews (Cheftel et al., 2000, 2002, 2003; Le Bail et al., 2002; Urrutia Benet et al., 2004; Dumay et al., 2006) have detailed these possibilities of processing food samples in the high pressure-low temperature (HP-LT) domain. They include water phase transitions that may affect the viability of undesirable microorganisms present in the processed foods and also food structures.

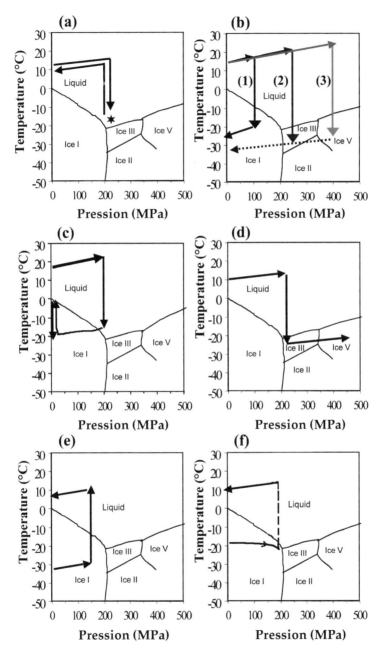

**Figure 1.** Schematic pathways for high pressure-low temperature processes. (a) Sub-zero cooling without ice crystal formation. (b) Pressure-assisted freezing to ice I (1), ice III (2) or ice V (3). (c) Pressure-shift freezing. (d) Pressure-induced freezing. (e) Pressure-assisted thawing of ice I. (f) Pressure-induced thawing. For further explanations, see text.

**Process without phase transition: subzero cooling without ice crystal formation (Fig. 1a)**

According to the phase diagram, water remains in the liquid state up to 207 MPa and down to –22°C. A food sample may be pressurized, e.g., to 112, 200 or 300 MPa then cooled down under pressure to about –10, –21°C or –18°C, respectively, without ice crystal formation. Cooling without freezing could be used for example to extend the shelf-life of high value biological samples at pressure below 200 MPa to limit denaturing effects of pressure on protein materials. Cooling without freezing may also be used to study the cold-induced changes of biomaterials at sub-zero temperature without addition of co-solvent or cryo-protectant.

**Freezing at a constant pressure or pressure-assisted freezing (Fig. 1b)**

An unfrozen food sample may be pressurized then cooled under a constant pressure down to a temperature below that of the corresponding freezing/melting point, leading to the formation of ice I, or even ice III, V or VI, depending on the pressure-temperature conditions. Since ice III (or V or VI) is denser than ice I, damage to sample microstructure may be minimized. The driving force of the process is the temperature gradient between the cooling medium and the sample. Depending on the overall processing goal, the process can be ended by releasing the pressure down to 0.1 MPa, or followed by pressure-assisted thawing under the same pressure then final pressure release at a temperature well above the freezing/melting point of the sample. In the former case, ice III (or V or VI) will revert to ice I, if pressure is released to 0.1 MPa at a temperature below 0°C. This change in ice form is accompanied by a brutal increase in sample volume that is detrimental for the sample (micro) structure.

**Pressure-shift freezing (Fig. 1c and Fig. 2a, b)**

A sample may be cooled well below 0°C under pressure (Fig. 2a, steps 1–2) then subjected to a sudden pressure release in few seconds (Fig. 2a, steps 3–4), thus inducing extensive undercooling (or supercooling) in the whole sample (Fig. 2a, step 3–4) and instant and uniform ice crystal nucleation throughout the sample depth (whatever the sample size and shape). Pressure-shift nucleation is uniform due to the isotropic character of pressure, and because temperature is uniform in the whole sample before pressure release and pressure equilibrium is established quasi-instantaneously throughout liquid or solid samples (isostatic principle). Nucleation is accompanied by the release of latent heat of ice I nucleation and by a temperature jump up to the sample freezing/

melting at close to atmospheric pressure (Fig. 2a, step 4). Only a partial liquid/solid transition is observed at this stage of the process, associated to nucleation and initial crystal growth. It is possible to calculate the amount of instantly frozen water from the recorded temperature jump (Fig. 2a, step 4). Processing liquid O/W model emulsions that contained (w/w) 0 to 25% fructose and correspondingly 82 to 57% water, we calculated that, respectively, 23% to 28% of the initial water in emulsion was instantly frozen by fast pressure release from 200 MPa/−18°C down to 0.1 MPa/−20°C (Thiebaud et al., 2002). Processing 'solid' smoked salmon mince (containing 64% water and 3% salt, w/w), we calculated that 30% of the initial water in samples was instantly frozen by fast pressure release from 207 MPa/−21°C down to 0.1 MPa/−26°C (Picart et al., 2005). These values agree with those calculated by Otero and Sanz (2000, 2006). The pressure-shift nucleation step may be followed by a complete freezing of the sample at 0.1 MPa (Ice I crystal growth). If the removal of latent heat of crystallization is fast enough during this latter step to limit the re-crystallization phenomena, the final ice crystals should be of small sizes and therefore less detrimental to sample microstructure than those resulting from most freezing conditions at atmospheric pressure (Martino et al., 1998; Lévy et al., 1999; Thiebaud et al., 2002; Sequeira-Muñoz et al., 2005; Dumay et al., 2006; Alizadeh et al., 2007). Selecting an appropriate negative temperature for the cooling medium will decrease the duration time of the freezing plateau at atmospheric pressure. Pressure release can be carried out rapidly or more slowly (Lévy et al., 1999; Thiebaud et al., 2002). When pressure is released slowly (Fig. 2b), the pressure-temperature coordinates partly follow the freezing/melting curve of the food sample (i.e., a curve corresponding to the freezing/melting temperature of pure water, with a shift towards lower temperatures equivalent to sample temperature depression at 0.1 MPa). Most freezing studies by pressure-shift have been carried out releasing the pressure close to the triple point (liquid water/ice I/ice III) of the phase diagram, i.e., from 207 MPa/−22°C. But the existence of a liquid metastable phase in the ice III domain in which an unfrozen sample may be kept in an undercooling state for a while, allows pressure release from higher pressure and lower temperature values than for the triple point (Schlüter et al., 2004; Otero et al., 2007).

**Pressure-induced freezing (Fig. 1d)**

A phase transition may be obtained by pressure increase. Ice I cannot be obtained through this process since the phase transition temperature decreases with pressure up to 207 MPa. This process is only applicable for

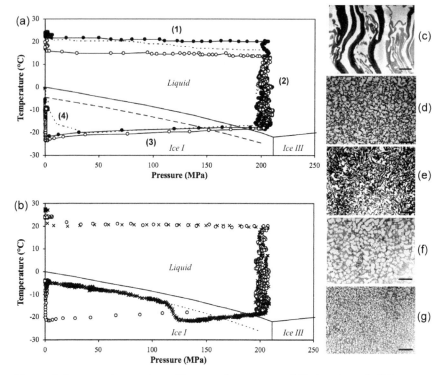

**Figure 2.** Pressure-temperature diagrams for pressure-shift freezing of oil-in-water emulsions (pH 7) from 207 MPa/–18°C (a, b) and micrographs of emulsions after freeze-substitution and toluidine blue staining using 5 µm slices (c–g). (a) P/T coordinates for emulsions prepared with (w/w) 3% sodium caseinate, 15% peanut oil plus 0% (---), 18% (●) or 25% fructose (○) and frozen by fast pressure release in 1–3s. (b) Comparison of fast (in 2–3 s) (○) and slow (in ~10 min) (×) pressure release for the emulsion containing 18% fructose. (c–e) Micrographs of emulsions without fructose and frozen by air-blast at –25°C (c), pressure-shift freezing with fast (d) or slow (e) pressure release. Emulsion without fructose and containing 0.25% sodium alginate frozen by fast pressure release (f). Emulsion containing 25% fructose and frozen by fast pressure release (g). Black and white areas correspond to the protein matrix and crystal imprints, respectively. Bar = 250 µm. For pressure shift-freezing experiments: data acquisition frequency was 10 Hz; temperature was measured at the sample center; temperature of the cooling medium was –29°C. (Lévy et al., 1999 and Thiebaud et al., 2002).

higher ice forms than ice I. As shown on Fig. 1d, freezing to ice V may be achieved by successive steps: a sample (e.g., at 10°C) is pressurized up to 220 MPa. A further cooling step down to –24°C is achieved at 220 MPa (where the sample is still unfrozen due to the metastable domain at 210–250 MPa and down to –24°C) (Schlüter et al., 2004). Then, the pressure is increased up to 500 MPa until crystallization of ice V occurs.

## Freezing followed by pressurization: pressure-assisted or pressure-induced thawing (Fig. 1e, f)

Independent of the previous cases, a sample may be frozen at atmospheric pressure and then subjected to pressurization. Partial melting (solid/liquid phase transition) or changes in ice form (solid/solid transition) may occur, depending on the pressure-temperature conditions. Ice I–ice III transition has been studied at low temperatures through pressure increase up to 320 MPa (Schlüter et al., 2004; Luscher et al., 2005; Urrutia Bennet et al., 2006). Pressure-assisted thawing (Fig. 1e) at constant pressure and/or pressure-induced thawing (Fig. 1f) through pressure increase may occur depending on the initial temperature of the previously frozen sample and PTM, and pressure level (Schlüter et al., 2004; Picart et al., 2004, 2005). In the case of pressure-assisted thawing, a frozen sample (Ice I, III or V) is thawed by increasing the temperature under a constant pressure level, the process driving force being the temperature gradient between the PTM and the sample. Pressure is then released down to 0.1 MPa at a temperature well above the sample freezing/melting point to avoid water re-crystallization during adiabatic expansion. In the case of pressure-induced thawing, a frozen sample (ice I) can be thawed at a subzero temperature by applying pressure at an appropriate pressure level, since the melting temperature of ice I is lowered under pressure. In this case, a partial melting of ice I takes place during the pressure build up (expressed by a sample temperature drop) and further thawing may follow under constant pressure (Picart et al., 2005). Due to the prolongation of the melting curve of ice I in the metastable domain of ice III, this process can be also obtained with pressure up to 210–250 MPa (Schlüter et al., 2004). Thawing assisted or induced by pressure may be envisaged for large food samples to reduce the thawing time (latent heat of melting ice I is lower than at 0.1 MPa). A food product could be thawed using pressure much more quickly than with conventional process, which might preserve its microbiological quality.

Different ice crystal size and shape can be achieved. The liquid/solid or solid/solid transitions that lead to ice crystal formation or changes in ice forms (with accompanying changes in phase density and sample volume) may influence the sample microstructure but also induce fatal structural or metabolic injury in microorganisms through tension effect on the cell envelope, changes in the melting temperature of membrane phospholipids, pressure-induced denaturation of membrane proteins or cell enzymes. Characteristics of ice I crystals induced by pressure-shift freezing depend on the processing parameters: pressure and temperature values at the onset point of pressure release, the related expansion and the subsequent undercooling inducing nucleation, the rate of pressure release, the PTM temperature for completion of freezing at atmospheric pressure, the food composition (e.g., fructose or polysaccharide content) (Lévy et al., 1999; Thiebaud et al., 2002; Dumay et al., 2006). Crystal sizes and crystal size distribution have been determined in O/W emulsion samples processed by fast or slow pressure release, and compared to those obtained by conventional freezing processes at atmospheric pressure or freezing under constant pressure (Lévy et al., 1999; Thiebaud et al., 2002). Some microphotographs are shown in Figs. 2c–g. Pressure-shift freezing by releasing pressure in few (1–3 sec) (or fast pressure release, Fig. 2a) initiates numerous ice nuclei uniformly distributed throughout the sample depth. When the latent heat is removed, nuclei grow to form crystal clusters of small sizes that displayed indented shapes without any specific orientation (Fig. 2d). By comparison air blast freezing at atmospheric pressure (Fig. 2c) or freezing under constant pressure (not shown) induced nucleation near the sample surface in direct contact with the cooling medium, then crystal growth from the sample surface to the center. Freezing by slow pressure release (i.e., in 10 min, Fig. 2b) changed ice crystal morphology leading to more jagged shapes (Fig. 2e). The higher the pressure and the lower the temperature at which expansion took place, the more instantaneous ice was formed, and also  the phase transition time  was shorter at a given cooling temperature (Barry et al., 1998; et al., 1999; Otero and Sanz, 2000). Pressure-shift freezing has also been applied to plant tissue or meat samples. The resulting ice crystals were small, granular in shape and dispersed throughout the sample, preserving sample microstructure (Koch et al., 1996; Fuchigami et al., 1996; Martino et al., 1998; Schlüter et al., 1998; Fuchigami et al., 2002; Fernandez et al., 2006). In contrast, freezing plant or fish muscle tissues at atmospheric pressure resulted in larger ice crystals at the center than the surface of samples, due to a marked temperature gradient through the sample depth (Martino et al., 1998; Chevalier et al., 2000). The sizes of ice crystals induced by fast pressure release appeared independent of their location and crystals were stable during 70 d storage in the frozen state (Chevalier et al., 2000). Only  a small size gradient  was noticed between the surface and the center of O/W emulsion (Levy et al., 1999; Thiebaud et al., 2002), turbot fillets (Chevalier et al., 2000) or porc muscle (Zhu et al., 2004). This was attributed to the slower removal of latent heat from the  central part of the sample compared  to the sample surface during ice crystals growth at atmospheric pressure (Lévy et al., 1999) and/or to the sample size and degree of undercooling (Zhu et al., 2004).

Freezing to ice I under constant pressure (or pressure-assisted freezing) resulted in larger needle-shaped ice crystals with a radial orientation and a marked size gradient from the sample surface to the center (Lévy et al., 1999; Fernandez et al., 2006). Luscher et al. (2005) compared different freezing pathways to form ice I, III or V, using samples of potato tissue. Freezing to ice I by fast pressure release gave the best results in terms of texture and non browning reactions after sample thawing compared to freezing at atmospheric pressure. By comparison, freezing to ice III or ice V did not prevent browning reactions after sample thawing in spite of lesser changes in sample volume and lesser membrane permeabilization induced by the process. This study points out the pressure-induced activation or inactivation of cellular enzymes such as polyphenol oxydase during and after processing, independently of structural modifications that HHP could induce.

Freezing by fast pressure release induced much less cell damage than conventional freezing at atmospheric pressure (i.e., cooled air blast, cooled solutions, liquid nitrogen). This beneficial effect clearly results from the formation of smaller ice crystals. But intracellular crystals might be detrimental for living cells (microorganisms or cell cultures). It is well known in biology that rapid crystallization is harmful to living cells (Taylor, 1960). Keeping cells alive using such HP-LT treatments is not a concern for food processing.

## HIGH PRESSURE PROCESSING AND INACTIVATION OF FOOD MICROORGANISMS

High pressure processing at 300–700 MPa is generally effective at reducing most vegetative bacteria, yeasts and molds. Microorganisms vary in their response to high pressure. The resistance of bacteria to pressure processing diminishes from spores to Gram-positive and Gram-negative vegetative cells. Gram-negative bacteria are more easily inactivated by high pressure than Gram-positive bacteria but an overlap exists between both groups (Smelt, 1998; Wuytack et al., 2002). Significant variation in pressure sensitivity between different strains of the same species has been reported and several studies agree that stationary growth phase cells (which have a more robust cytoplasmic membrane) are more pressure-resistant than exponentially phase cells for both Gram-positive and Gram-negative bacteria (Cheftel, 1995; Mackey et al., 1995; Simpson and Gilmour, 1997; Benito et al., 1999; McClements et al., 2001; Tay et al., 2003). Bacterial spores cannot be significantly inactivated by pressure alone at room or mild temperatures (Sale et al., 1970; Heinz and Knorr, 2001). Spores of yeast and molds are more easily inactivated at approximately 200 to 300 MPa and 4–45°C (Butz and Ludwig, 1991; Merkulow et al., 2000). High pressure induces changes in bacteria cell morphology, cell walls and membranes due to transition from a liquid-crystalline to a gel phase for membrane phospholipids (which causes irreversible changes in cell permeability), inactivation of enzymes, effects on biochemical reactions, cellular transcription and replication (Cheftel, 1995; Patterson et al., 1995; Smelt, 1998; Abee and Wouters, 1999) and leakage of ATP or UV-absorbing material from bacterial cells (Smelt et al., 1994; Benito et al., 1999; Ananta et al., 2004). The higher resistance to pressure of Gram-positive compared to Gram-negative bacteria may be linked to the rigidity of teichoic acids in the peptidoglycan layer of the Gram-positive cell wall. However, pressure inactivation seems to be related more to changes in membrane fluidity. Membranes rich in unsaturated fatty acids, poor in phosphatidylglycerol and lacking cholesterol have in general higher fluidity and are more pressure-resistant (Cheftel, 1995; Lado and Yousef, 2002; Russel 2002). Membrane fatty acyl components of barophilic organisms become more polyunsaturated with increase in growth pressure. Cell walls are less affected by high pressure than membranes but intracellular damage (bud scars, nodes and empty cavities between the cytoplasmic membrane and the cell wall) can also be observed using electron microscopy (Mackey et al., 1994; Tholozan et al., 2000; Ritz et al., 2002). Loss of membrane functionality has also been described by Wouters et al. (1998). In particular, enzymes and transport systems (cellular ATP-driven $K^+$ uptake system) are affected by pressure. Membrane-bound $F_0F_1$ ATPase may be dislocated or inactivated by pressure as well as other membrane ATPases responsible for ion motions (Smelt, 1998; Tholozan et al., 2000; Patterson, 2005). The pressure-resistance of bacterial enzymes varies greatly. Some enzymes withstand pressures above 500 MPa while others are inactivated from 200 MPa (Lado and Yousef, 2002; Patterson, 2005). Nucleic acids are relatively resistant to high pressure, but the enzyme-mediated steps involved in DNA replication and transcription are disturbed. High pressure causes condensation of nuclear material. At elevated pressures, DNA comes into contact with endonucleases that cleave it. Ribosome damage has also been suggested as an important determinant in pressure sensitivity (Earnshaw et al., 1995; Smelt, 1998; Wouters et al., 1998; Bozoglu et al., 2004; Patterson, 2005). Since ribosomes are implicated as temperature sensors, cold and heat shock proteins may also play a role in the stress response induced by high pressure processing (Russel, 2002).

Microbial inactivation depends on HP-treatment characteristics: pressure level and pressure holding time in combination with the processing temperature. However, inactivation is also greatly influenced by the food matrix in which microorganisms are present, the presence of antimicrobial substances (i.e., bacteriocins, lysozyme, lactoperoxydase system). Several studies using microorganisms inoculated in foods have proved the influence of the food matrix on microbial pressure sensitivity compared with buffer solutions. Food constituents such as proteins, carbohydrates and lipids, or ions such as $Ca^{+2}$ may have a protective effect with regards to pressure-induced inactivation (Simpson and Gilmour, 1997; Patterson and Kilkpatrick, 1998; Capellas et al., 2000; Gervilla et al., 2000). Validation studies need to be carried out with real foods and using the pathogenic/ surrogate cells which are commonly associated. Low water activities may protect microorganisms against pressure (Oxen and Knorr, 1993; Palou et al., 1997). A low food pH is a major advantage in high pressure inactivation procedures. It is not clear whether the drop in pH during high pressure processing affects microbial survival or not in addition to pressure effects (Roberts and Hoover, 1996).

A high-pressure treatment may not always completely inactivate microorganisms but may induce a substantial amount of sub-lethally injured cells (Wuytack et al., 2002; Yuste et al., 2003; Ananta et al., 2004; Bozoglu et al., 2004). Recovery of the injured cells will depend on the storage conditions after treatment. The degree of injury can be determined by comparing the cell growth using nutritive and selective media, nutritive media supplemented with NaCl or sodium dodecyl sulfate, low pH and over-layered media (Patterson et al., 1995; McClements et al., 2001; Chang et al., 2003; Picart et al., 2005). Flow cytometry techniques together with vital staining using fluorescent dyes showed the presence of viable sub-populations that do not form colonies and are therefore not detected on agar media (viable but not cultivable cells) (Ananta et al., 2004; Mathys et al., 2007; Volkert et al., 2008). Pressure-induced inactivation of microorganisms is complex, and plotting the $\log_{10}$ reduction against time does not always lead to a straight line relationship corresponding to a first-order kinetic (Fig. 3). Pronounced tailing is often observed after pressure processing which implies the use of non-linear kinetics consistent with Weibull, Log-logistic or $n^{th}$ order models, each of them with their particular underlying assumptions (Heinz and Knorr, 2001; Chen and Hoover, 2003; Tay et al., 2003; Patterson, 2005). Experimental data are regressively fitted to the selected model using an appropriate computing program, where goodness-of-fit is assessed by means of an adequate minimization of the discrepancy between predicted and observed data, usually the mean square error.

Better inactivation of vegetative cells is generally observed after pressure processing at temperatures above or below ambient temperature (20–25°C). Negative temperatures induce more effective microorganism inactivation.

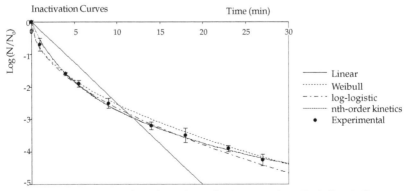

**Figure 3.** Inactivation curves for a hypothetical microorganism. Each line indicates a different kinetic model fit to corresponding experimental data (black circles).

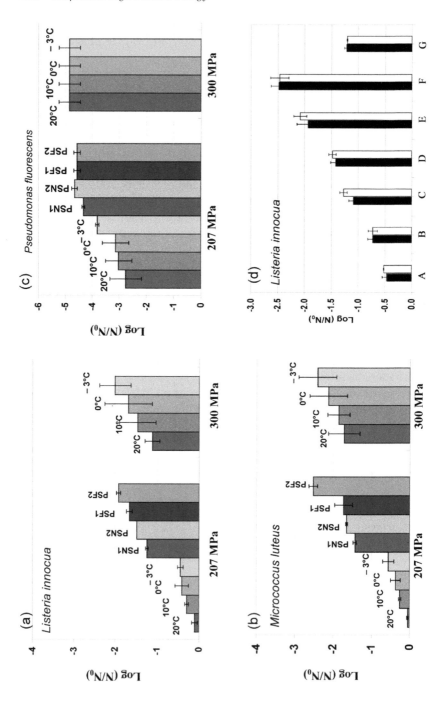

On the other hand, combined high-pressure-high-temperature treatments (HP-HT) have been proposed in order to achieve pressure sterilization. Preliminary studies reported a synergy between high pressure and low or negative temperatures to inactivate various microorganisms and yeasts in saline or buffered solutions (Takahashi, 1992; Hashizume et al., 1995; Hayakawa et al., 1998). Carlez et al. (1993) found higher inactivation ratios for *Pseudomonas fluorescens* and *Listeria innocua* inoculated in minced meat after pressurization at 5°C than 20°C (150 MPa or 300 MPa for 20 min). Gervilla et al. (1997) also observed a better inactivation of *Listeria innocua* inoculated in ewe's milk after pressurization at 5°C than 25°C (300 MPa, 5–15 min). Perrier-Cornet et al. (2005) show that decreasing the processing temperature down to –20°C without freezing could reduce the pressure level and/or pressure holding time necessary to get an expected inactivation ratio of *Saccharomyces cerevisiae* and *Lactobacillus plantarum*. Picart et al. (2004) confirmed the better efficiency of HP processing at low than ambient temperature at reducing microbial load for *Listeria innocua, Pseudomonas fluorescens* and *Micrococcus luteus* introduced in smoked salmon mince and evaluated after a resuscitation step to recover stressed cells (Figs. 4a–c). Low/sub-zero temperatures and pressure probably act synergistically in decreasing the cell membrane fluidity.

Pressure shift-nucleation (in 3 sec, from 207 MPa/–21°C) followed or not by freezing completion at atmospheric pressure was able to inactivate microorganisms more efficiency than pressurization at low temperature and constant pressure (207 MPa) without water-ice transition (Figs. 4a–d), probably due to the formation of both intra- and extracellular small size crystals inducing cell death. Pressure-shift freezing by slow pressure release (in 18 min from 207/21°C) was even more effective than by fast pressure

---

Figure 4. Inactivation of microorganisms inoculated in smoked salmon mince (pH 6.0) and pressurized at low or negative temperature. (a–c): samples inoculated with three strains together (*L. innocua, M. luteus* and *P. fluorescens*). Samples pressurized at 207 MPa or 300 MPa for 23 min and 20°C, 10°C, 0°C or –3°C. PSN1: cooling at 207 MPa for 23 min followed by pressure-shift nucleation (PSN) from 207 MPa/–21°C in 3 s. PSN2: cooling at 207 MPa for 30 min followed by PSN from 207 MPa/–25°C in 3 s. PSF1: same processing as for PSN1 followed by freezing completion at 0.1 MPa down to –25°C. PSF2: same processing as for PSN2 followed by freezing completion at 0.1 MPa down to –25°C. (d): *L. innocua* inactivation ratios after pressurization at 207 MPa and 4°C for 23 min (A) or 60 min (B); pressurization at 207 MPa and –21°C for 60 min without freezing (C); pressure-shift freezing form 207 MPa/–22°C using fast (in ~3 s; D) or slow (in ~18 min; E) pressure release. Pressure-induced thawing of a previously frozen sample (24 h at –28°C and 0.1 MPa) by pressurization at 207 MPa and sub-zero temperature (–29°C) followed by pressure release in ~3s (F); pressure-assisted thawing of a frozen sample (24 h at –28°C and 0.1 MPa) by pressurization at 207 MPa and low temperature (0–10°C) (G). Inactivation ratios were determined after a resuscitation step to allow recovery of stressed cells, then growth on selective media (a–c), or by numeration on non-selective (TSA-YE, dark bars) or selective (Oxford, white bars) medium (d). Means from two independent experiments plus deviation to the means (Picart et al., 2004 and 2005).

release probably due to the formation of ice I crystals of larger size and more irregular shapes, and tension effects on cells when the P/T coordinates follow the liquid water/ice I boarder line (as in Fig. 2b). On the contrary, freezing in still air at –15°C or – 40°C for up to 120 h (followed by thawing) did not inactivate microorganisms (Picart et al., 2004) but freezing at 0.1 MPa followed by pressurization could do (Fig. 4d). The best inactivation ratio was obtained after pressure-induced thawing followed by fast pressure release inducing freezing, i.e., a twin phase transition (solid/liquid/solid) (Fig. 4d). Ice I-III solid/solid transition also induced a more effective inactivation of *Listeria innocua* in phosfate buffer compared with HP-LT treatment in the liquid domain (Luscher et al., 2005). Shen et al. (2005) found similar results for *Bacillus subtilis* vegetative cells grown in TSB subjected to solid-solid phase transitions (0.1 MPa to 450 MPa at –25 or –45°C). Cellular damage was related to mechanical stress during phase transition phenomena.

The problem of pressure-resistant spores is still one of the challenges faced by high pressure technology. Spores have demonstrated resistance to more than 1000 MPa (Heinz and Knorr, 2001; Zhang and Mittal, 2008). Few data are available of the effect of water-ice transition on spore inactivation. Spores of *Clostridium* and *Bacillus* are the most relevant in food preservation. No relationship between pressure and heat resistance within the group of spore-formers has been found. The enhanced resistance of bacterial spores is linked to the protective effect of the membranes and coat layers surrounding the core, the low water activity in the spore core and the presence of dipicolinic acid (DPA) (Heredia et al., 1997; Palop et al., 1999). For HHP treatments applied at moderate temperatures (below 55°C), pressure between 50 and 300 MPa can trigger spore germination through activation of germinant receptors (similar to nutrient germination) which leads to pore formation in spore coats. Then the newly formed vegetative cells could die or could be killed using higher pressure levels in a further processing stage (Gould and Sale, 1970; Heinz and Knorr, 2001). This germination pathway proceeds faster at neutral pH (Raso et al., 1998). Pressures higher than 500–600 MPa and temperature up to 60°C would open $Ca^{+2}$- mediated DPA channels releasing DPA, leading to spore hydration and then inactivation (Paidhungat et al., 2002). Pressure cycling at at 200–400 MPa and temperatures up to 60°C has been proposed to inactivate bacterial spores showing efficient kinetics (Hayakawa et al., 1994; Furukawa et al., 2003). In this case, increase in DPA release and more numerous damaged spores were observed. Nevertheless, resistant sub-populations might persist.

High pressure combined with high temperature (HP-HT treatments) is envisaged as a fruitful strategy for spore inactivation in food systems (Toepfl et al., 2006; Wilson et al., 2008; Zhang and Mittal, 2008). Spore inactivation may be achieved by HP-HT processing following a mechanism that does not involve germination (Margosch et al., 2004). At 600–800 MPa and more than 60°C, DPA is released predominantly by a physicochemical rather than physiological mechanism indicating loss of core integrity and inactivation of cortex lytic enzymes. The DPA-free spores are inactivated by heat independent of the pressure level and pressure-assisted hydration. Numerous studies have been published on inactivation of *Bacillus* spp. or *Clostridium* spp. spores using HHP at mild or high temperature, or pulsed high pressure in combination with temperature (Heinz and Knorr, 2001; Black et al., 2007; Zhang and Mittal, 2008). Margosch et al. (2006) showed that increments in both pressure and temperature up to 1400 MPa and 110°C improved inactivation of *C. botulinum* spores up to 6.5 logarithmic cycles. However kinetics showed strong tailing and zones of pressure-temperature stabilization that should be avoided.

Indeed, taking advantage of the isostatic principle and of adiabatic heating during pressure build up to 600–1400 MPa, temperatures of 105–120°C can be reached in a few seconds starting from an initial product temperature of 70–90°C previously achieved by conventional heating process and chosen according to the product heat sensitivity (Meyer et al., 2000). After a few seconds (or minutes) holding time at high pressure, the pressure is released with a concomitant rapid cooling of the product down to the initial 60–90°C. Such high pressure-high temperature processing (HP-HT) (one or two pulses) may achieve food sterilization while limiting the product exposure to high temperature and allowing accurate control of treatment intensity required for spore inactivation. Such HP-HT treatments could be applied to solid products that cannot be treated by high-temperature-short-time processing actually used for liquid products (e.g., UHT processing using vapor infusion). Effects of such high pressure levels for short holding times in combination with high temperature remain to be seen in terms of food microstructure and texture, protein denaturation and preservation of small molecular weight molecules such as vitamins, aromas, pigments. An attempt to introduce this novel technology using a 35 liter HP vessel operating at 600 MPa has been made (Juliano et al., 2008). Such HP-HT processing requires sufficient temperature uniformity inside the HP vessel and heat transfer properties of the package material needs to be considered. Studies at varying pressure and temperature conditions should allow fitting kinetic inactivation models with a view to obtain a 12 $\log_{10}$ reduction for *C. botulinum* and assist processing validation and regulatory approval (Hartmann et al., 2003).

Inactivation of viruses is also a concern in the food area. As a result of virus diversity, pressure sensitivity is also heterogeneous. High pressure may act on viruses by altering envelop (if present), dissociating particles or changing their structure (Gaspar et al., 2002). Many enteroviruses such as hepatitis A virus, rotavirus, and norovirus surrogates (i.e., feline calicivirus) may be easily inactivated at pressures lower than 500 MPa (Grove et al., 2006). Notably, HHP treatments of 300 MPa and ambient temperatures was extremely efficient at reducing populations of halotolerant *Vibrio* spp. (*parahaemolyticus* and *vulnificus*) accumulated in oysters and shellfishs (Adams et al., 2002; Murchie et al., 2005). Other animal viruses related to food chain production such as avian influenza virus (subtype H7N7) and foot-and-mouth disease virus were inactivated combining pressure of 200-500 MPa and mild temperature (Oliveira et al., 1999; Isbarn et al., 2007). Temperatures below −10°C or above 50°C significantly increased pressure-induced virus inactivation (Chen et al., 2005). High pressure has also been described as a method for viral inactivation and vaccine development (Masson et al., 2001). Pressure inactivation preserves immunogenicity of the virus particles enabling stimulation of antibody responses. Processing at moderate pressure levels was also found useful for controlling infectivity of parasites such as *Trichinella spiralis* (200 MPa at 25°C; Ohnishi et al., 1992), development of fruit egg larvae (125 MPa at 25°C; Butz and Tauscher, 1995), *Anisakis simplex* larvae in fish (200–500 MPa; Molina-García and Sanz, 2002; Dong et al., 2003) and *Toxoplasma gondii* cyst in ground pork (more than 300 MPa is required; Lindsay et al., 2006).

## CONCLUSIONS AND PERSPECTIVES

HP processing at low or mild temperature appears to be well implemented in the food industry for 'niche' products. Crystallization (ice I) by fast pressure release (or pressure-shift freezing) has shown evidence of inducing much less cell and texture damage at the sample center (or surface) than conventional air blast freezing or even freezing in liquid nitrogen. Developments may be envisaged for some vegetables (not all) or for fish products (at pressure not exceeding 200 MPa to limit aggregation of myofibrillar proteins and loss of product quality after thawing). Inactivation of microorganisms (bacteria, parasites and viruses) induced by water/ice I phase transition or constant pressure and low temperature are beneficial. Thawing assisted/induced by pressure may be envisaged for large food samples to reduce the thawing time (latent heat of melting ice I is lower than at 0.1 MPa). In addition to faster thawing, absence of microbial growth or even microbial inactivation is expected. High pressure combined with high temperature is a process actually developed in view of spore inactivation. Processing optimization is needed to limit heating effects on food constituents. Beside food preservation aspects, HHP may also be used as a tool to structure and texture bio-materials.

# REFERENCES

Abee, T. and J.A. Wouters. 1999. Microbial stress response in minimal processing. Int. J. Food Microbiol. 50: 65–91.

Adams, H.H.R.M. and D.R. Farkas and M.T. Morrissey. 2002. Use of high-pressure processing for oyster shucking and shelf-life extension. J. Food Sci. 67: 640–645.

Ahn, J. and V.M.Balasubramaniam and A.E. Yousef. 2007. Inactivation kinetics of selected aerobic and anaerobic bacterial spores by pressure-assisted thermal processing. Intern. J. Food Microbiol. 113: 321–329.

Alizadeh, E. and N. Chapleau, M. de Lamballerie and A. le Bail. 2007. Effects of different processes on the microstructure of Atlantic salmon (*salmo salar*) fillets. Innov. Food Sci. Emerg. Technol. 8: 493–499.

Ananta, E. and V. Heinz. and D. Knorr. 2004. Assessment of high pressure induced damage on *Lactobacillus rhamnosus* GG by flow cytometry. Food Microbiol. 21: 567–577.

Barry, H. and E.M. Dumay and J.C. Cheftel. Influence of pressure-assisted freezing on the structure, hydration and mechanical properties of a protein gel. Pp. 343–353. *In*: N.S. Isaacs (ed.). 1998. High pressure Food Science, Bioscience and Chemistry. The Royal Society of Chemistry. Cambridge, UK.

Basset, J. and M.A. Macheboeuf. 1932. Etudes sur les effets biologiques des ultrapressions: résistance des bactéries, des diastases et des toxines aux pressions très élevées. Comp. Rend. Acad. Sci. 195: 1431–1433.

Basset, J. and M.A. Macheboeuf. 1933. Etudes sur les effets biologiques des ultrapressions. Etude sur l'immunité : Influence des pressions très élevées sur les antigènes et anticorps. Comp. Rend. Acad. Sci. 196: 67–69.

Basset, J. and M.A. Macheboeuf and G. Sandor. 1933. Etudes sur les effets biologiques des ultrapressions. Action des pressions très élevées sur les protéides. Comp. Rend. Acad. Sci. 197: 796–798.

Benito, A. and G. Ventoura, M. Casadei, T. Robinson and B. Mackey. 1999. Variation in resistance of natural isolates of *Escherichia coli* O157 to high hydrostatic pressure, mild heat, and other stresses. Appl. Environ. Microbiol. 65: 1564–1569.

Black, E.P. and P. Setlow, A.D. Hocking, C.M. Stewart, A.L. Kelly and D.G. Hoover. 2007. Response of spores to high-pressure processing. Compr. Rev. Food Sci. Food Saf. 6: 103–119.

Bozoglu, F. and H. Alpas and G. Kaletunç. 2004. Injury recovery of foodborne pathogens in high hydrostatic pressure milk during storage. FEMS Immunol. Med. Microbiol. 40: 243–247.

Bridgman, P.W. 1912. Water, in the liquid and five solid forms, under pressure. Proc. Am. Acad. Arts Sci. 47: 441–558.

Bridgman, P.W. 1914. The coagulation of albumen by pressure. J. Biol. Chem. 19: 511–512.

Buchheim, W. and A.M. Abou El Nour. 1992. Induction of milk fat crystallization in the emulsified state by high hydrostatic pressure. Fat Sci. Technol. 10: 369–373.

Butz, P. and H. Ludwig. 1991. Pressure inactivation of yeasts and molds. Pharmaz. Industrie 53: 584–586.

Butz, P. and B. Tauscher. 1995.Inactivation of fruit fly eggs by high pressure treatment. J. Food Proces. Preserv. 19: 161–164.

Butz, P. and R. Edenharder, A. Fernández García, H. Fister, C. Merkel and B. Tauscher. 2002. Changes in functional properties of vegetables induced by high pressure treatment. Food Res. Intern. 35: 295–300.

Butz, P. and Y. Serfert, A. Fernández Garcia, S. Dietrich, R. Lindauer and A. Bognar. 2004. Influence of high-pressure treatment at 25°C and 85°C on folates in orange juice and model media. J. Food Sci. 69: 117–121.

Butz, P. and A. Bognar, S. Dietrich and B. Tauscher. 2007. Effect of high pressure at elevated temperatures on thiamin and riboflavin in pork and model systems. J. Agric. Food Chem. 55: 1289–1294.

Capellas, M. and M. Mor-Mur, R. Gervilla, J. Yuste and B. Guamis. 2000. Effect of high pressure combined with mild heat or nisin on inoculated bacteria and mesophiles of goat´s milk fresh cheese. Food Microbiol. 17: 633–641.

Carlez, A. and J.P. Rosec, N. Richard and J.C. Cheftel. 1993. High pressure inactivation of *Citrobacter freundii, Pseudomonas fluorescens* and *Listeria innocua* in inoculated minced beef muscle. Food Sci. Technol. (LWT) 26: 357–363.

Carlez, A. and T. Veciana-Nogues and J.C. Cheftel. 1995. Changes in colour and myoglobin of minced beef meat due to high pressure processing. Food Sci. Technol. (LWT) 28: 528–538.

Chang, V.P. and E.W. Mills and C.N. Cutter. 2003. Comparison of recovery methods for freeze-injured *Listeria monocytogenes, Salmonella typhimurim* and *Campylobacter coli* in cell suspensions and associated with pork surfaces. J. Food Prot. 66: 798–803.

Cheftel, J.C. Effects of high hydrostatic pressure on food constituents: an overview. Pp. 195–209. *In*: C. Balny, R. Hayashi, K. Heremans and P. Masson (eds.). 1992. High Pressure and Biotechnology. Colloques INSERM, vol. 224. Jonh Libbey Eurotex Ltd., Montrouge, France.

Cheftel, J.C. 1995. Review: high-pressure, microbial inactivation and food preservation. Food Sci. Tech. Intern. 1: 75–90.

Cheftel, J.C. and J. Culioli. 1997. Effects of high pressure on meat: a review. Meat Sci. 46: 211–236.

Cheftel, J.C. and E. Dumay. Effects of high pressure on food biopolymers with special reference to β-lactoglobulin. Pp. 369–397. *In*: D.S. Reid (ed.). 1998. The Properties of Water in Foods—ISOPOW 6. Blackie Academic & Professional, London, UK.

Cheftel, J.C. and J. Lévy and E. Dumay. 2000. Pressure-assisted freezing and thawing: principles and potential applications. Food Rev. Int. 16: 453–483.

Cheftel, J.C. and M.Thiebaud and E. Dumay. 2002. Pressure-assisted freezing and thawing of foods: a review of recent studies. High Pressure Res. 22: 601–611.

Cheftel, J.C. and M. Thiebaud and E. Dumay. High pressure-low temperature processing of foods: a review. Pp. 327–340. *In*: R. Winter (ed.). 2003. Advances in High Pressure Bioscience and Biotechnology II. Springer Verlag, Berlin, Germany.

Chen, H. and D.G. Hoover. 2003. Modeling the combined effect of high hydrostatic pressure and mild heat on the inactivation kinetics of *Listeria monocytogenes* Scott A in whole milk. Innov. Food Sci. Emerg. Technol. 4: 25–34.

Chen, H. and D.G. Hoover and D.H. Kingsley. 2005. Temperature and treatment time influence high hydrostatic pressure inactivation of feline calicivirus, a norovirus surrogate. J. Food Prot. 68: 2389–2394.

Chevalier, D. and A. Sequeira-Muñoz, A. Le Bail, B. Simpson and M. Ghoul. 2000. Effect of freezing conditions and storage on ice crystal and drip volume in turbot (*Scophtalmus maximus*). Evaluation of pressure-shit freezing vs. air-blast freezing. Innov. Food Sci. Emerg. Technol. 1: 193–201.

Claeys, W.L. and Indrawati, A.M. Van Loey and M. Hendrickx. 2003. Review: are intrinsic TTIs for thermally processed milk applicable for high-pressure processing assessment? Innov. Food Sci. Emerg. Tech. 4: 1–14.

Considine T. and H.A. Patel, S.G. Anema, H. Singh and L.K. Creamer. 2007. Interactions of milk proteins during heat and high hydrostatic pressure treatments—A review. Innov. Food Sci. Emerg. Technol. 8: 1–23.

Cruess, W.V. Unfermented fruit beverages. Pp. 205–236. *In*: W.V. Cruess (ed.). 1924. Commercial Fruit and Vegetable Products. McGraw—Hill Book Company, New York, USA.

Czeslik, C. and O. Reis, R. Winter and G. Rapp. 1998. Effect of high pressure on the structure of dipalmitoylphosphatidylcholine bilayer membranes: a synchrotron-X-ray diffraction and FT-IR spectroscopy study using the diamond anvil technique. Chem. Phys. Lipids 91: 135–144.

Delbarre-Ladrat, C. and R. Cheret, R. Taylor and V. Verrez-Bagnis. 2006. Trends in postmortem aging in fish: understanding of proteolysis and disorganization of the myofibrillar structure. Crit. Rev. Food Sci. Nut. 46: 409–421.

Devine, C.E. and R.G. Bell, S. Lovatt, B.B. Chrystall and L.E. Jeremiah. Red meat. Pp. 51–84. *In*: L.E. Jeremiah (ed.). 1996. Marcel Dekker, Inc., New York, USA.

Dong, F.M. and A.R. Cook and R.P. Herwig. 2003. High hydrostatic pressure treatment of finfish to inactivate *Anisakis simplex*. J. Food Prot. 66: 1924–1926.

Dumay, E. and C. Lambert, S. Funtenberger and J.C. Cheftel. 1996. Effects of high pressure on the physicochemical characteristics of dairy creams and model emulsions. Food Sci. Technol. (LWT) 29: 606–625.

Dumay, E. and A. Laligant, D. Zasypkin and J.C. Cheftel. 1999. Pressure- and heat-induced gelation of mixed ß-lactoglobulin/polysaccharide solutions: scanning electron microscopy of gels. Food Hydrocolloids 13: 339–351.

Dumay, E. and L. Picart, S. Regnault and M. Thiebaud. 2006. High pressure-low temperature processing of food proteins. Biochim. Biophys. Acta 1764: 599–618.

Dykes, G.A. and S.M. Moorhead. 2001. Survival of three *Salmonella* serotypes on beef trimmings during simulated commercial freezing and frozen storage. J. Food Saf. 21: 87–96.

El Moustaine, D. and V. Perrier, L. Smeller and R. Lange. 2008. Full-length prion protein aggregates to amyloid fibrils and spherical particles by distinct pathways. FEBS J. 275: 2021–2031.

Earnshaw, R.G. and J. Appleyard and R.M. Hurst. 1995. Understanding physical inactivation processes: combined preservation opportunities using heat, ultrasound and pressure. Int. J. Food Microbiol. 28: 197–219.

Farkas, D.F. and D.G. Hoover. 2000. High pressure Processing. Kinetics of Microbial Inactivation for Alternative Food Processing Technologies, J. Food Sci. supplement 47–64.

Fernández, P.P. and L. Otero, B. Guignon and P.D. Sanz. 2006. High-pressure shift freezing versus high-pressure assisted freezing: effects on the microstructure of a food model. Food Hydrocolloids 20: 510–522.

Fernández García, A. and P. Butz and B. Tauscher. 2001. Effects of high-pressure processing on carotenoids extractability, antioxidant activity, glucose diffusion, and water binding of tomato puree (*Lycopersicon esculentum* Mill.). J. Food Sci. 66: 1033–1038.

Fuchigami, M. and K. Miyazaki, N. Kato and A. Teramoto. 1996. Histological changes in high pressure frozen carrots. J. Food Sci. 62: 809–812.

Fuchigami, M. and N. Ogawa and A. Teramoto. 2002. Trehalose and hydrostatic pressure effects on the structure and sensory properties of frozen tofu (soybean curd). Innov. Food Sci. Emerg. Technol. 3: 139–147.

Funtenberger, S. and E. Dumay and J.C. Cheftel. 1997. High pressure promotes β-lactoglobulin aggregation through SH/S-S interchange reactions. J. Agric. Food Chem. 45: 912–921.

Furukawa, S. and M. Shimoda and I. Hayakawa. 2003. Mechanism of the inactivation of bacterial spores by reciprocal pressurization treatment. J. Appl. Microbiol. 94: 836–841.

Gaspar, L.P. and A.C.B. Silva, A.M.O. Gomes, M.S. Freitas, B. Ano, W.D. Scharcz, J. Mestecky, M.J. Novak, D. Foguel and D. Silva. 2002. Hydrostatic pressure induces fusion-active state of enveloped viruses. J. Biol. Chem. 277: 8433–8439.

Gervilla, R.M. Capellas, V. Ferragut and B. Guamis. 1997. Effect of high hydrostatic pressure on *Listeria innocua* 910 CECT inoculated into Ewe's milk. J. Food Protec. 60: 33–37.

Gervilla, R. and V. Ferragut and B. Guamis. 2000. High pressure inactivation of microorganisms inoculated into ovine milk of different fat contents. J. Dairy Sci. 83: 674–682.

Grácia-Juliá, A. and M. René, M. Cortés-Muñoz, L. Picart, T. López-Pedemonte, D. Chevalier and E. Dumay. 2008. Effect of dynamic high pressure on whey protein aggregation: A comparison with the effect of continuous short-time thermal treatments. Food Hydrocolloids 22: 1014–1032.

Grove, S.F. and A. Lee, T. Lewis, C.M. Stewart, H. Chen and D.G. Hoover. 2006. Inactivation of foodborne viruses of significance by high pressure and other processes. J. Food Prot. 69: 957–968.

Gould, G.W. and A.J.H. Sale. 1970. Initiation of germination of bacterial spores by hydrostatic pressure. J. Gen. Microbiol. 60: 335–346.

Hartmann, C. and A. Delgado and J. Szymczyk. 2003. Convective and diffusive transport effects in a high pressure induced inactivation process of packed food. J. Food Engineer 59: 33–44.

Hashizume, C. and K. Kimura and R. Hayashi. 1995. Kinetic analysis of yeast inactivation by high pressure treatment at low temperatures. Biosci. Biotechnol. Biochem. 59: 1455–1458.

Hayakawa, I. and T. Kanno, Y.K. Yoshi and Y. Fujio. 1994. Oscillatory compared with continuous high pressure sterilisation on *Bacillus stearothermophilus* spores. J. Food Sci. 59: 164–167.

Hayakawa, K.Y. Ueno, S. Kawamura, S. Kato and R. Hayashi. 1998. Microorganism inactivation using high pressure generation in sealed vessels under sub-zero temperature. Appl. Microbiol. Biotechnol. 50: 415–418.

Hendrickx, M. and L. Ludikhuyze, I. Vanden Broeck and C. Weemaes. 1998. Effects of high pressure on enzymes related to food quality. Trends Food Sci. Technol. 9: 197–203.

Heinz, V. and D. Knorr. 2001. Effects of high pressure on spores. Pp. 77–113. *In*: M.E.G. Hendrickx and D. Knorr (eds.). 2001. Ultra High Pressure Treatments of Foods, Kluwer Academic/Plenum Publishers, New York, USA.

Heredia, N.L. and G.A. García, R. Luevanos, R.G. Labbe and. J.S. García-Alvarado. 1997. Elevation of the heat resistance of vegetative cells and spores of *Clostridium perfringens* type A by sub-lethal heat chock. J. Food Prot. 60: 998–1000.

Hite, B.H. 1899. The effects of pressure in the preservation of milk. Bull. West. Virg. Univ. Agric. Exp. Station 58: 15–35.

Hite, B.H. and N.J. Giddings and C.E. Weakley. 1914. The effect of pressure on certain microorganisms encountered in the preservation of fruits and vegetables. Bull. West. Virg. Univ. Agric. Exp. Station 146: 1–67.

Huppertz, T. and P.F. Fox and A.L. Kelly. 2004. Properties of casein micelles in high pressure-treated bovine milk. Food Chem. 87: 103–110.

Huppertz, T. and P.F. Fox, K.G. de Kruif and A.L. Kelly. 2006. High pressure-induced changes in bovine milk proteins: a review. Biochim. Biophys. Acta 1764: 593–598.

Isbarn, S. and R. Buckow, A. Himmelreich, A. Lehmacher and V. Heinz, 2007. Inactivation of avian influenza virus by heat and high hydrostatic pressure. J. Food Prot. 70: 667–673.

Juliano, P. and P.J. Fryer and C.Versteeg. 2008. *C. botulinum* inactivation kinetics implemented in a computational model of high-pressure sterilisation. Biotechnol. Progress. In press.

Kalichevsky, M.T. and D. Knorr and P.J. Lillford. 1995. Potential food applications of high-pressure effects on ice-water transitions. Trends food Sci. Technol. 6: 253–259.

Knorr, D. and O. Schlüter and V. Heinz. 1998. Impact of high hydrostatic pressure on phase transitions of foods. Food Technol. 52: 42–45.

Koch, H. and I. Seyderhelm, P. Wille, M.T. Kalichevsky and D. Knorr. 1996. Pressure-shift freezing and its influence on texture, color, microstructure and rehydration behavior of potato cubes. Nahrung. 40: 125–131.

Kolakowski, P. and E. Dumay and J.C. Cheftel. 2001. Effects of high pressure and low temperature on b-lactoglobulin unfolding and aggregation. Food Hydrocolloids 15: 215–232.

Krebbers, B. and A.M. Matser, M. Koets and R.W. Van den Berg. 2002. Quality and storage-stability of high-pressure preserved green beans. J. Food Eng. 54: 27–33.

Lado, B.H. and A.E. Yousef. 2002. Alternative food-preservation technologies: efficacy and mechanisms. Microb. Infect. 4: 433–440.

Le Bail, A. and D. Chevalier, D.M. Mussa and M. Ghoul. 2002. High pressure freezing and thawing of foods: a review. Int. J. Refrig. 25: 504–513.

Lévy, J. and E. Dumay, E. Kolodziejczyk and J.C. Chefel. 1999. Freezing kinetics of a model oil-in-water emulsion under high pressure or by pressure release. Impact on ice crystals and oil droplets. Food Sci. Technol. (LWT) 32: 396–405.

Lindsay, D. and M.V. Collins, D. Holliman, G.J. Flick and J.P. Dubey. 2006. Effects of high-pressure processing on *Toxoplasma gondii* tissue cysts in ground pork. J. Parasitol. 92: 195–196.

López-Pedemonte, T. and A.X. Roig-Sagués, S. De Lamo, R. Gervilla and B. Guamis. 2007a. High hydrostatic pressure treatment applied to model cheeses made from cow's milk inoculated with *Staphylococcus aureus*. Food Control 18: 441–447.

López-Pedemonte, T. and A. Roig-Sagués, S. De Lamo, M. Hernández-Herrero and B. Guamis. 2007b. Reduction of counts of *Listeria monocytogenes* in cheese by means of high hydrostatic pressure. Food Microbiol. 24: 59–66.

Ludikhuyze, L. and M. Hendrickx. Effects of high pressure on chemical reactions related to food quality. Pp. 167–188. *In*: M. Hendrickx and D. Knorr (eds.). 2001. Ultra High Pressure Treatments of Foods, Kluwer Academic/Plenum Publishers, New York, USA.

Ludikhuyze, L. and A. Van Loey Indrawati, S. Denys and M.E.G. Hendrickx. 2001. Effects of high pressure on enzymes related to food quality. Pp. 115–166. *In*: M.E.G. Hendrickx and D. Knorr. (eds.). 2001. Ultra High Pressure Treatments of Foods, Kluwer Academic/Plenum Publishers, New York, USA.

Luscher, C. and O. Schlüter and D. Knorr. 2005. High pressure-low temperature processing of foods: impact on cell membranes, texture, color and visual appearance of potato tissue. Innov. Food Sci. Emerg. Technol. 6: 59–71.

Luzzati, V. 1997. Biological significance of lipid polymorphism: the cubic phases. Curr. Opin. Struct. Biol. 7: 661–668.

Mackey, B.M. and K. Forestière and N. Isaacs. 1995. Factors affecting the resistance of *Listeria monocytogenes* to high hydrostatic pressure. Food Biotechnol. 9: 1–11.

Mackey, B.M. and K. Forestière, N.S. Isaacs, R. Stenning and B. Booker. 1994. The effect of high pressure on *Salmonella typhimurium* and *Listeria monocytogenes* examined by electron microscopy. Lett. Appl. Microbiol. 19: 429–432.

Makita, T. 1992. Application of high pressure and thermophysical properties of water to biotechnology. Fluid Phase Equilibria. 76: 87–95.

Margosch, D. and M.G. Gänzle, M.A. Ehrmann and R.F. Vogel. 2004. Pressure inactivation of Bacillus endospores. Appl. Environ. Microbiol. 70: 7321–7328.

Margosch, D. and M.A. Ehrmann, R. Buckow, V. Heinz, R.F. Vogel and M.G. Gänzle. 2006. High-pressure-mediated survival of *Clostridium botulinum* and *Bacillus amyloliquefaciens* endospores at high temperature. Appl. Environ. Microbiol. 72: 3476–3481.

Marigheto, N. and A. Vial, K. Wright and B. Hills. 2004. A combined NMR and microstructural study of the effect of high-pressure processing on strawberries. Applied Magnetic Resonance 26: 521–531.

Martino, M.N. and L. Otero, P.D. Sanz and N.E. Zaritzky. 1998. Size and location of ice crystals in pork frozen by high-pressure-assisted freezing as compared to classical methods. Meat Sci. 50: 303–313.

Masson, P. and C. Tonello and C. Balny. 2001. High pressure biotechnology in medicine and pharmaceutical science. J. Biomed. Biotech. 1: 85–88.

Matser, A.M. and B. Krebbers, R.W. van der Berg and P.V. Bartels. 2004. Advantages of high pressure sterilisation on quality of food products. Trends Food Sci. Technol. 15: 79–85.

Mathys, A. and B. Chapman, M. Bull, V. Heinz and D. Knorr. 2007. Flow cytometry assessment of Bacillus spore response to high pressure and heat. Innov. Food Sci. Emerg. Technol. 8: 519–527.

McClements, J.M.J. and M.F. Patterson and M. Linton. 2001. The effect of growth stage and growth temperature on high hydrostatic pressure inactivation of some psychrotrophic bacteria in milk. J. Food Prot. 64: 514–522.

Merkulow, N. and R. Eicher and H. Ludwig. 2000, Pressure inactivation of fungal spores in aqueous model solutions and in real food systems. High Press. Res. 19: 643–652.

Meyer, R.S. and K.L.Cooper, D. Knorr and H.L.M. Lelieveld. 2000. High-pressure sterilization of foods. Food Technol. 54: 67–72.

Molina-García, A.D. and P.D. Sanz. 2002. *Anisakis simplex* Larva killed by high-hydrostatic-pressure processing. J. Food Protec. 65: 383–388.

Molina-García, A.D. and L. Otero, M.N. Martino, N.E. Zaritzky, J. Arabas, J. Szczepek and P.D. Sanz. 2004. Ice VI freezing of meat: supercooling and ultrastructural studies. Meat Science 66: 709–718.

Mozhaev, V.V. and J. Heremans, J. Frank, P. Masson and C. Balny. 1996. High pressure effect on protein structure and function. Proteins Struc. Funct. Genet. 24: 81–91.

Murchie, L.W. and M. Cruz-Romero, J.P. Kerry, M. Linton, M.F. Patterson, M. Smiddy and A.L. Kelly. 2005. High pressure processing of shellfish: a review of microbiological and other quality aspects. Innov. Food Sci. Emerg. Technol. 6: 257–270.

Needs, E.C. and R.A. Stenning, A.L. Gill, V. Ferragut, and G.T. Rich. 2000. High-pressure treatment of milk : effects on casein micelle structure and on enzymatic coagulation. J. Dairy Res. 67: 31–42.

Novak, J.S. and V.K. Juneja. 2003. Effects of refrigeration or freezing on survival of *Listeria monocytogene*s Scott A in under-cooked ground beef. Food Control 14: 25–30.

Oey, I. and M. Lille, A. Van Loey and M. Hendrickx. 2008a. Effect of high-pressure processing on colour, texture and flavor of fruit-based food products: a review. Trends Food Sci. Technol. 10: 320–328.

Oey, I. and I. Van der Plancken, A. Van Loey and M. Hendrickz. 2008b. Does high pressure processing influence nutritional aspects of plant based food systems? Trends Food Sci. Technol. 10: 300–308.

Ohnishi, Y. and T. Ono, T. Shigehisa and T. Ohmori. 1992. Effect of high hydrostatic pressure on muscle larvaeof *Trichinella spiralis*. Japan. J. Parasitol. 41: 373–377.

Oliveira, A.C. and D. Ishimaru, R.B. Goncalves, P. Mason, D. Carvalho, T. Smith, and J.L. Silva. 1999. Low temperature and pressure stability of picornaviruses: implication for virus uncoating. Biophys. J. 76: 1270–1279.

Otero, L. and P.D. Sanz. 2000. High-pressure shift freezing. Part 1. Amount of ice instantaneously formed in the process. Biotechnol. Progr. 16: 1030–1036.

Otero, L. and P.D. Sanz. 2006. High pressure shift freezing: main factors implied in the phase transition time. J. Food Engineer 72: 354–363.

Otero, L. and G. OuseguiG. Urrutia Benet, C. de Elvira, M. Havet, A. Le Bail and P.D. Sanz. 2007. Modelling industrial scale high pressure low temperature processes. J. Food Engineer 83: 136–141.

Oxen, P. and D. Knorr. 1993. Baroprotective effects of high solute concentrations against inactivation of Rhodotorula rubra. Food Sci. Technol. (LWT ) 26: 220–223.

Paidhungat, M. and B. Setlow, B. Daniels, D. Hoover, E. Papafragkou and P. Setlow. 2002. Mechanisms of induction of germination of Bacillus subtilis spores by high pressure. Appl. Environ. Microbiol. 68: 3172–3175.

Palop, A. and P. Mañas and S. Condón. 1999. Sporulation temperature and heat resistance of Bacillus spores: a review. J. Food Saf. 19: 57–72.

Palou, E. and A. López-Malo, G.V. Barbosa-Canovas, J. Welti-Chanes and B.G. Swanson. 1997. Effect of water activity on high hydrostatic pressure inhibition of Zygosaccharomyces bailii. Letters Appl. Microbiol. 24: 417–420.

Patterson, M.F. 2005. Microbiology of pressure-treated foods. J. Appl. Microbiol. 98: 1400–1409.

Patterson, M.F. and D.J. Kilkpatrick. 1998. The combined effect of high hydrostatic pressure and mild heat on inactivation of pathogens in milk and poultry. J. Food Prot. 61: 432–436.

Patterson, M.F. and M. Quinn and A. Gilmour. 1995. The sensitivity of vegetative pathogens to high hydrostatic pressure treatment in buffered saline foods. J. Food Prot. 58: 524–529.

Payens, T.A.J. and K. Hereman. 1969. Effect of pressure on the temperature dependent association of b-casein. Biopolymers 8: 335–345.

Perrier-Cornet, J.M. and S. Tapin, S. Gaeta and P. Gervais. 2005. High-pressure inactivation of *Saccharomyces cerevisiae* and *Lactobacillus plantarum* at subzero temperatures. J. Biotechnol. 115: 405–412.

Picart, L. and E. Dumay, J.P. Guiraud and J.C. Cheftel. 2004. Microbial inactivation by pressure-shift freezing: effects on smoked salmon mince inoculated with *Pseudomonas fluorescens*, *Micrococcus luteus* and *Listeria innocua*. Food Sci. Technol. (LWT) 37: 227–238.

Picart, L. and E. Dumay, J.P. Guiraud and J.C. Cheftel. 2005. Combined high pressure–sub-zero temperature processing of smoked salmon mince: phase transition phenomena and inactivation of *Listeria innocua*. J. Food Engineer. 68: 43–56.

Qui, W. and H. Jiang, H. Wang and Y. Gao. 2006. Effect of high hydrostatic pressure on lycopene stability. Food Chem. 97: 516–523.

Rademacher, B. and J. Hinrinchs. Ultra high pressure technology for dairy products. Pp. 12–18. *In*: FIL-IDF (eds.). 2002. New Processing Technologies for the future. Bulletin of the IDF 374, Bruxelles, Belgium.

Randolph, W.T. and M. Seefeldt and J.F. Carpenter. 2002. High hydrostatic pressure as a tool to study protein aggregation and amyloidosis. Biochim. Biophys. Acta 1595: 224–234.

Raso, J. and M.M. Góngora-Nieto, G. Barbosa-Cánovas and B.G. Swanson. 1998. Influence of several environmental factors on the initiation and inactivation of Bacillus cereus by high hydrostatic pressure. Int. J. Food Microbiol. 44: 125–132.

Rastogi, N.K. and L.T. Nguyen and V.M. Balasubramaniam. 2008. Effect of pretreatments on carrot texture after thermal and pressure-assisted thermal processing. J. Food Eng. 88: 541–547.

Regnault, S. and M. Thiebaud, E. Dumay and J.C. Cheftel. 2004. Pressurisation of raw skim milk and of a suspension of phosphocaseinate at 9°C or 20°C. Effects on casein micelle size distribution. Intern. Dairy J. 23: 56–68.

Regulation 258/97/EC of the European Parliament and of the Council of 27 January 1997 concerning novel foods and novel food ingredients. European Commission, Brussel. Official Journal of the European Communities L 043, 14/02/1997, pp. 0001– 0006. Links: http://ec.europa.eu/food/food/biotechnology/novelfood/– http://europa.eu/scadplus/leg/en

Reid, D.S. Crystallisation phenomena in the frozen state. Pp. 313–326. *In*: M.A. Rao and R.W. Hartel (eds.). 1998. Phase/state Transitions in Foods. Chemical, Structural and Rheological Changes. Marcel Dekker, Inc., New York, USA.

Ritz, M. and J.L. Tholozan, M. Federighi and M.F. Pilet. 2002. Physiological damages of *Listeria monocytogenes* treated by high hydrostatic pressure. Int. J. Food Microbiol. 79: 47–53.

Roberts, C.M. and D.G. Hoover. 1996. Sensitivity of *Bacillus coagulans* spores to combinations of high hydrostatic pressure, heat, acidity and nisin. J. Appl. Bacteriol. 81: 363–368.

Rovere, P. and G. Carpi, S. Gola, G. Dall'Aglio and A. Maggi. HHP strawberry products: an example of processing in line. Pp. 445–450. *In*: R. Hayashi and C. Balny (eds.). 1996. High Pressure, Bioscience and Biotechnology, Elsevier Sciences, Amsterdam, The Netherlands.

Russel, N.J. 2002. Bacterial membranes: the effects of chill storage and food processing. An overview. Int. J. Food Microbiol. 79: 27–34.

Sale, A.J.H. and G.W. Gould and W.A. Hamilton. 1970. Inactivation of bacterial spores by hydrostatic pressure. J. Gen. Microbiol. 60: 323–334.

Sánchez-Moreno, C. and M.P. Cano, B. de Ancos, L. Plaza, B. Olmedilla, F. Granado and A. Martín. 2004a. Consumption of high-pressurized vegetable soup increases plasma vitamin C and decreases oxidative stress and inflammatory biomarkers in healthy humans. J. Nutr. 134: 3021–3025.

Sánchez-Moreno, C. and L. Plaza, B. de Ancos and L. Cano. 2004b. Effect of combined treatments of high-pressure and natural additives on carotenoid extractability and antioxidant activity of tomato puree (*Lycopersicum esculentum* Mill.) Eur. Food Res. Technol. 219: 151–160.

Schlüter, O. and V. Heinz and D. Knorr. Freezing of potato cylinders during high pressure treatment. pp. 317–324. *In*: N.S. Isaacs (ed.). 1998. High Pressure Food Science, Bioscience and Chemistry. The Royal Society of Chemistry. Cambridge, UK.

Schlüter, O. and G. Urrutia Benet, V. Heinz and D. Knorr. 2004. Metastable states of water and ice during pressure-supported freezing of potato tissue. Biotechnol. Progr. 20: 799–810.

Shen, T. and G. Urrutia Benet, S. Brul and D. Knorr. 2005. Influence of high-presure 6 low-temperature treatment on the inactivation of *Bacillus subtilis* cells. Innov. Food Sci. Emerg. Technol. 6: 271–278.

Sequeira-Muñoz, A. and D. Chevalier, B.K. Simpson, A. Le Bail and H.S. Ramaswamy. 2005. Effect of pressure-shift release versus air-blast freezing of carp (*Cyprinus carpio*) fillets: a storage study. J. Food Biochem. 29: 504–516.

Sila, D.N. and X. Yue, S. Van Buggenhout, C. Smout, A. Van Loey and M. Hendrickx. 2007. The relation between (bio-)chemical, morphological, and mechanical properties of thermally processed carrots as influenced by high-pressure pretreatment condition. Eur. Food Res. Technol. 226: 127–135.

Sila, D.N. and T. Duvetter, A. De Roeck, C. Smout, G.K. Moates, B.P. Hills, K.K. Waldron, M. Hendrickx and A. Van Loey. 2008. Texture changes of processed fruit and vegetables: potential use of high-pressure processing. Trends Food Sci. Technol. 10: 309–319.

Simpson, R.K. and A. Gilmour. 1997. The resistance of *Listeria monocytogenes* to high hydrostatic pressure in foods. Food Microbiol. 14: 567–573.

Smeller, L. 2002. Pressure-temperature diagrams of biomolecules. Biochim. Biophys. Acta 1595: 11–29.

Smelt, J.P.P.M. 1998. Recent advances in the microbiology of high pressure processing. Trends Food Sci. Technol. 9: 152–158.

Smelt, J.P.P.M. and A.G.F. Rijke and A. Hayhurst. 1994. Possible mechanism of high pressure inactivation of microorganisms. High Pressure Res. 12: 199–203.

Suzuki, K. 1960. Studies on the kinetics of protein denaturation under high pressure. Rev. Phys. Chem. Japan 29: 91–98.

Takahashi, K. Sterilization of microorganisms by hydrostatic pressure at low temperature. Pp. 303–307. *In:* C. Balny, R. Hayashi, K. Heremans and P. Masson (eds.). 1992. High Pressure and Biotechnology, Colloque INSERM, vol. 224. John Libbey Eurotext Ltd, Montrouge, France.

Talwar, G.P. and J. Basset and M. Macheboeuf. 1952. Influence des très hautes pressions sur l'évolution des solutions de gélatine en présence de trypsine. Comp. Rend. Acad. Sci. 235: 393–395.

Talwar, G.P. and J. Basset and M. Macheboeuf. 1953. Recherches sur la gélification des solutions de gélatine sous l'influence de pressions hydrostatiques élevées. Comp. Rend. Acad. Sci. 236: 2271–2272.

Tay, A. and T.H. Shellhammer, A.E. Yousef and G.W. Chism. 2003. Pressure death and tailing behaviour of *Listeria monocytogenes* strains having different barotolerances. J. Food Prot. 66: 2057–2061.

Taylor, A.C. 1960. The physical state transition in the freezing of living celles. Ann. N.Y. Acad. Sci. 85: 595–609.

Tewari, G. and D.S. Jayas and R.A. Holley. 1999. High pressure processing of foods: an overview. Sciences des Aliments 19: 619–662.

Thiebaud, M. and E. Dumay and J.C. Cheftel. 2002. Pressure-shift freezing of o/w emulsions: influence of fructose and sodium alginate on undercooling, nucleation, freezing kinetics and ice crystal size distribution. Food Hydrocolloids 16: 527–545.

Tholozan, J.L. and M. Ritz, F. Jugiau, M. Federighi and J.P. Tissier. 2000. Physiological effects of high hydrostatic pressure treatments on *Listeria monocytogenes* and *Salmonella typhimurium*. J. Appl. Microbiol. 88: 202–215.

Toepfl. S. and A. Mathys, V. Heinz and D. Knorr. 2006. Review: Potential applications of high hydrostatic pressure and pulsed electric field for energy efficient and environmentally friendly food processing. Food Rev. Intern. 22: 405–423.

Tonello, C. 2008. New trends in high pressure equipment for food industry. 46th EHPRG International Conferences, 7–11th September, Valancia, Spain.

Touakis, P.S. and P. Panagiotidis, N.G. Stoforos, P. Butz, H. Fister and B. Tauscher. Kinetics of vitamin C degradation under high pressure-moderate temperature processing in model systems and fruit juices. Pp. 310–316. *In:* N.S. Isaac (ed.). 1998. High Pressure Food Science, Bioscience and Chemistry. The Royal Society of Chemistry, Cambridge, UK.

Trujillo, A.J. and M. Capellas, J. Saldo, R. Gervilla and B. Guamis. 2002. Applications of high-hydrostatic pressure on milk and dairy products: a review. Innov. Food Sci. Emerg. Technol. 3: 295–307.

Urrutia Benet, G. and O. Schlüter and D. Knorr. 2004. High pressure-low temperature processing. Suggested definitions and terminology. Innov. Food Sci. Emerg. Technol. 5: 413–427.

Urrutia Benet, G. and N. Chapleau, M. Lille, A. Le Bail, K. Autio and D. Knorr. 2006. Quality related aspects of high pressure low temperature processed potatoes. Innov. Food Sci. Emerg. Technol. 7: 32–39.

Urrutia G. and J. Arabas, K. Autio, S. Brul, M. Hendrickx, A. Kakolewski, D. Knorr, A. Le Bail, M. Lille, A.D. Molina-Garcia, A. Ousegui, P.D. Sanz, T. Shen and S. Van Buggenhout. 2007. SAFE ICE: low temperature pressure processing of foods: safety and quality aspects, process parameters and consumer acceptance. J. Food Engineer 83: 293–315.

Van den Broeck, I. and L. Ludikhuyze, C. Weemaes, A. Van Loey and M. Hendrickx. 1998. Kinetics for isobaric-isothermal degradation of L-ascorbic acid. J. Agric. Food Chem. 46: 2001–2006.

Volkert, M. and E. Ananta, C.M. Luscher and D. Knorr. 2008. Effect of air-freezing, spray-freezing and pressure-shift freezing on membrane integrity and viability of *Lactobacillus rhamnosus* GC. J. Food Eng. 87: 532–540.

Wilson, D.R. and L. Dabrowski, S. Stringer, R. Moezelaar and T.F. Brocklehurst. 2008. High pressure in combination with elevated temperature as a method for the sterilization of food. Trends Food Sci. Technol. 19: 289–299.

Winter, R. 2002. Synchroton X-ray and neutron small-angle scattering of lyotropic lipid mesophases, model biomembranes and proteins in solution at high pressure. Biochim. Biophys. Acta 1595: 160–184.

Wouters, P.C. and E. Glaasker and J.P.P.M. Smelt. 1998. Effects of high pressure on inactivation kinetics and events related to proton efflux in *Lactobacillus plantarum*. Appl. Environ. Microbiol. 64: 509–515.

Wuytack, E.Y. and A.M.J. Diels and C.W. Michiels. 2002. Bacterial inactivation by high-pressure homogenisation and high hydrostatic pressure. Int. J. Food Microbiol. 77: 205–212.

Yasuda, A. and K. Mochizuki. The behaviour of triglycerides under high pressure: The high pressure can stably crystallize cocoa butter in chocolate. Pp. 255–259. *In*: C. Balny, R. Hayashi, K. Heremans and P. Masson. [eds.] 1992. High Pressure and Biotechnology. Colloque INSERM, vol. 224. John Libbey Eurotext, Montrouge, France.

Yuste, J. and M. Capellas, R. Pla, S. Llorens, D.Y.C. Fung and M. Mor-Mur. 2003. Use of conventional media and thin agar layer method for recovery of foodborne pathogens form pressure-treated poultry products. J. Food Sci. 68: 2321–2324.

Zhu, S. and A. Le Bail, H.S. Ramaswamy and N. Chapleau. 2004. Characterization of ice crystals in pork muscle formed by pressure-shift freezing as compared with classical freezing methods. J. Food Sci. 69: 190–197.

Zipp, A. and W. Kauzmann. 1973. Pressure denaturation of metmyoglobin. Biochemistry 12: 421–4228.

Zhang, H. and G.S. Mittal. 2008. Effects of High-pressure processing (HHP) on bacterial spores: An Overview. Food Rev. Int. 24: 330–351.

# PRESSURE AND CELL AND TISSUE FUNCTIONS

# 6

# Pressure Effects on Cells

*Stephen Daniels[1]* and *Yoram Grossman[2]*

## INTRODUCTION

Animals exhibit a variety of behavioural and physiological changes as a result of exposure to both constant raised pressure and to changes in pressure. Furthermore, having manifested changes at pressure animals frequently show some adaptation, either in terms of the frequency of signs of change or in the magnitude of the changes. In general, during compression above their normal ambient pressure both aquatic and terrestrial animals show increased locomotion, reduced co-ordination, impaired reflexes, tremor, seizures, paralysis, cardiac arrhythmia's, respiratory difficulty and ultimately death. Physiological changes that can be measured in the laboratory include changes in metabolism and thermo-regulation. In man more subtle physiological signs and symptoms are observed including nausea, sleep disruption, cognitive deficits and psychological responses (possibly related to stress). The changes in motor function and nervous system excitability apparently arise as a result of effects of pressure on the central nervous system (CNS) and have been termed collectively the High Pressure Neurological Syndrome (HPNS: Miller, 1974). This terms covers the complex set of neurological pathologies and although a single cause of HPNS is unlikely, if the major perturbations in CNS function are identified, then successful amelioration might provide substantial relief from HPNS.

Although bacteria and a few animals have been discovered living in the deep ocean (e.g. Mariana Trench 10.9 km deep, equivalent to a pressure of

[1] Wales Research and Diagnostic Positron Emission Tomography Imaging Centre, School of Medicine, Cardiff University, Cardiff CF14 4XN, UK.
E-mail: danielss@cf.ac.uk

[2] Department of physiology, Faculty of Health Sciences, Zlotowski Center for Neuroscience, Ben-Gurion University of the Negev, Beer-Sheva 84105, Israel.
E-mail: ramig@bgumail.bgu.ac.il

109 MPa) (Bruun, 1957) and there are interesting biochemical and chemical changes at pressures greater than 100 MPa (e.g. enzyme inhibition and protein) in general these pressures are incompatible with life. Accordingly only pressures in the range from atmospheric pressure to approximately 100 MPa will be considered.

In air-breathing animals raised partial pressures of respirable gases may lead to pulmonary and central nervous system oxygen toxicity (Clark, 1982), nitrogen narcosis (Bennett, 1982) and carbon dioxide dependent effects on respiratory responses, narcosis and hypoxia (Schaefer, 1969). However, in 1967 it was shown that helium acted essentially as a pure pressure transmitting fluid (Miller et al., 1967) and this opened the way for analytic experiments on the effects of pressure on mammals, including man.

## MANIFESTATIONS OF PRESSURE EFFECTS ON CELLS

There are many elements within cells that may exhibit sensitivity to changes in ambient pressure including the cell membrane, ion channels located in the plasma membrane (which mediate cellular excitability amongst other functions), neurotransmitter receptors (which mediate both fast and slow chemical neurotransmission), ion transporters and pumps (which also mediate cellular excitability and electrical homeostasis) and the cellular cytoskeleton (which has important functions in intracellular transport of materials and the regulation of cellular activity). These elements demonstrate differential sensitivity to pressure but the collective expression of these effects defines how various cell types will respond to pressure and hence how the organism as a whole will respond.

### Cell membranes

Phosphatidylcholines are the major membrane component in eukaryotic organisms and vital components of human lung surfactant and serum lipids. Accordingly, their phase-transition behaviour is of acute interest because of the impact that might have on a wide spectrum of biological processes, reviewed recently by Koynova and Caffrey (1998). Compression of dipalmitoyl-phosphatidylcholine (DPPC) bilayers is anisotropic; thus whilst pressure increases bilayer thickness, a net volume reduction occurs because of a simultaneous reduction in lateral spacing (Liu and Kay, 1977; Stamatoff et al., 1978).

The effect of pressure on the phase-transition temperatures is well documented for a variety of phosphatidylcholines (Koynova and Caffrey, 1998). The gel-phase to liquid crystalline transition temperature for a variety of saturated and unsaturated phosphatidylcholines increases with increasing pressure at approximately 0.24°C. MPa$^{-1}$ (Winter and Pilgrim,

1989). At pressures above 100 MPa additional gel phases have been reported (Prasad et al., 1987), which may be relevant to organisms adapted to living at very great depths. In the range of pressures that affect neurological function (2.5 MPa to 10 MPa) the transition temperature would change by less than 2°C. This is borne out by experiments on human erthyrocyte membranes (Finch and Kiesow, 1979). However, it is possible that for specialized regions of the lipid bilayer, with transition temperatures close to normal body temperature, relatively small increases in pressure might profoundly reduce membrane fluidity.

The role of phospholipids in cell signalling must also be considered where the association with the basal membrane, either plasma membrane or that of an intracellular organelle, may be influenced by pressure changes. The inositol phospholipids, a family of membrane lipids found in all mammalian cells, have important roles in cellular functions (Low, 1987; Higgins et al. 1989). It has been found that the strong hydrogen bonding with water molecules of the phosphate groups and a major portion of the carbonyl functions strengthens at pressure (up to 600 MPa). Pressure may also cause conformational changes that result in smaller area per molecule and an extension of the inositol moiety away from the phosphate group. These changes may be significant for physiological function, especially if they occur at lower pressures (Carrier and Wong, 1996).

## Cytoskeleton and motor proteins

Multisubunit or polymeric proteins are usually dissociated by high pressure, although the pressure sensitivity varies between different proteins and even, apparently, for the same protein in different species. Cytoskeletal proteins such as f-actin ('thin' stress fibers), tubulin ('thick' microtubules) and various 'intermediate filaments' stabilize the morphology of cells. Actin and tubulin also provide the structural basis for the action of motor proteins such as myosin, dynein and kinesin that generate muscle contraction, ciliary beating, and chromosome movement and intracellular transport, respectively.

### Actin

Filaments of actin, with other proteins, are arranged in bundles (cell microvilli), as a 'gel forming' mesh (beneath cell membranes), in short fragmented sections and as a double strand of the 'thin filament' of the muscle fibre. By interacting with the 'thick filaments' of myosin they produce muscle contraction. Similar but not identical assembly is found in specific regions of non-muscle cells that are capable of contracting (e.g. desmosomes, stress fibres). Self-assembly (polymerization) of the globular actin (g-actin) monomers to a single filament (f-actin) *in vitro* is associated with a large increase in volume ($\Delta V \sim 100$–$140$ cm$^3$.mole$^{-1}$) for one atmosphere-adapted animals ( for review, see Crenshaw et al., 1996). Pressure should therefore promote dissociation of the filamentous protein since that involves negative volume change. Bournes et al. (1988) have reported that actin stress fibres disassemble at 32 MPa and Crenshaw et al. (1996) reported similar findings in two other cell types: human *HeLa* fibroblasts and rat osteosarcoma cells. A study in guinea pig cerebellar brain slices (Yarom, Meiri and Grossman, personal communication) has revealed that Purkinje neurons in the tissue undergo dramatic morphological change when exposed to hydrostatic pressure above 60 MPa for 10–15 min. The dendritic tree transformed from normal smooth cylindrical segments covered with spines (small membranous protrusions) to abnormal 'pearl necklace'— like dendrites with very low spine density. The change was irreversible when pressure exceeded 80 MPa. It was suggested that this is the form a membrane cylinder will assume when it loses its cytoskeletal support and is unable to retract.

In skeletal muscle of fish adapted to life in the deep ocean, $\Delta V$ for actin polymerization is reduced from 100–140 cm$^3$.mole$^{-1}$ to 63 cm$^3$.mole$^{-1}$ for 3000 m dwelling fish and to only 9 cm$^3$.mole$^{-1}$ for 5000 m dwelling congeneric fish (Somero, 1990, 1992).

### Tubulin

A single tubulin protofilament is a polymer consisting of a chain of globular heterodimers (each consists of $\alpha$ and $\beta$ subunits). Thirteen such protofilaments in a concentric arrangement form the microtubule. This structure is dynamic; dimers may be added on one end and deleted on the other. Transport of molecules within the cell uses this system. The major component of the 'inner structure' of cilia and flagella, essential for beating mechanisms, are microtubules. They are also found in centrioles that generate simple microtubule spindles, formed during cell division and responsible for chromosome polar migration. Pressure (40 MPa) induces de-polymerization of spindle microtubules *in vivo* in various cells (Salmon, 1975a,b) and *in vitro* in spindle microtubules (Salmon et al. 1976) and extracts of tubulin from rabbit brain (Salmon 1975c). However, Bournes et al. (1988) and recently Crenshaw et al (1996) have proposed that microtubules are more resistant to pressure up to 40 MPa and that it is their normal assembly in cell regions, rather than simple de-polymerization that is impaired. Microtubule assembly was estimated to involve $\Delta V$ of 300–400 cm$^3$.mole$^{-1}$ in oocytes (Salmon 1975b), but lower estimates were reported for elongation of microtubules: 26 and 50 cm$^3$.mole$^{-1}$ for 35°C and 15°C respectively. Since pressures as low as 5–10 MPa were shown to be effective in slowing and modifying ciliary movement in *Paramecium* (Otter and Salmon, 1979, 1985) it was suggested that Ca$^{2+}$ signalling to the motility mechanism is compromised. However, a recent study from the same laboratory (Crenshaw and Salmon, 1996) revealed that pressure of up to 40 MPa did not change cytosolic [$Ca^{2+}$] of mouse fibroblasts. Therefore, Crenshaw et al. (1996) have proposed that in addition to a direct effect of pressure on de-polymerization of microtubules and certain filaments, pressure may impede some regulatory mechanism of cytoskeletal organization. A dual effect of pressure was also observed in two other ciliary systems. The beat frequency of water propelling sensory cilia in mollusc (*Hermissenda*) statocyst was gradually reduced by 30% when hydrostatically compressed to 40 MPa. At pressures higher than that, beating completely stopped (Salmon and Grossman personal communication). A similar response was observed for mucus transporting cilia in tissue cultures of frog (*Rana*) oesophageal epithelium (Waxman et al., 1991). Ciliary arrest was observed at pressures higher than 35 MPa. In both cases, complete and fast recovery was attained upon decompression. It seems that the first decrease in activity is due to some regulatory process, whereas complete ciliary arrest is due to the direct effect of pressure on microtubules assembly.

## Myosin

Myosin and actin are involved in contraction in most eukaryotic cells. The basic form of the myosin is a filamentous dimer, whose monomer is a polar molecule composed of 'tail' and 'head' regions. Under special conditions, myosin dimers aggregate through their tail regions to form bipolar 'thick filaments' of skeletal and cardiac muscle that contain hundreds of dimers. Much shorter and more labile filaments composed of only dozens of dimers, are found in smooth muscle fibres and non-muscle cells. In both types, force is generated when myosin heads interact with specific sites on the thin actin filaments, and contraction is associated with filaments sliding against each other. The process is triggered by increased cytosolic [$Ca^{2+}$] that induces either phosphorylation of myosin heads in smooth muscle and non-muscle cells, or enables hydrolysis of ATP by the myosin head (myosin ATPase) in skeletal and cardiac muscles. Surprisingly, myosin filaments remain stable even at pressures above 40 MPa in muscles and non-muscle types, probably due to the special tight assembly of the filament (Crenshaw et al., 1996). However, pressure does have a significant inotropic (positive and negative) effect on muscle contraction and its electrical activity. Since contraction is a complex process, other stages may be affected. $\Delta V$ for assembly of the actomyosin complex and the formation of high affinity cross-bridges is estimated to be approximately 100 $cm^3.mol^{-1}$ and would thus be impeded by pressure. In addition, myosin ATPase activity is inhibited by pressure, as is the response of other ATPases (for review see, Hogan and Besch, 1993). Additional processes that control free cytosolic [$Ca^{2+}$] such as the sarcoplasmic reticulum (SR) $Ca^{2+}$ ATPase pump and plasmalemal $Na^+$-$Ca^{2+}$ exchanger activity are impeded by pressure. Consequently, at pressure cytosolic $[Ca^{2+}]$ is expected to increase (due to the attenuated removal) and activate more rigorously muscle contraction (for review see, Hogan and Besch, 1993).

There is no information about pressure effects on the other known families of motor proteins, dynein and kynesin. However, since their structure and function are similar to that of myosin, it is reasonable to assume that they will respond to high pressure exposure in a similar manner.

### Intermediate filaments: Vinculin, Talin, Vimentin, Cytokeratin

These tough, stable filaments serve as mechanical support in all cell types. Their 8–10 μm diameter is larger than the thin actin filaments and smaller than the microtubules filaments. They form rope-like strands that include neurofilaments in neurons, cytokeratin filaments in epithelial cells, vimentin in fibroblast and glial cells and vinculin and talin in various types of cells. Under cellular conditions the assembly of intermediate filaments is irreversible and their disassembly requires destruction by specific proteases. However, Crenshaw et al. (1996) found that cytokeratin and vimentin filaments in *HeLa* cells exhibited a dramatic disruption at pressure above 20 MPa. In contrast, vimentin in rat osteosarcoma cells were not affected at pressures up to 40 MPa. Vinculin and talin filaments that are usually found in focal contact between the rat osteosarcoma cells, were absent in rounded cells at 40 MPa and were reduced in cells that remained flattened.

In summary, most cytoskeletal filaments are relatively stable at pressure. Usually pressures greater than 30 MPa and sometimes much greater are needed in order to directly disassemble the polymers. However lower pressures seem effective in impeding some regulatory mechanisms such as phosphorylation that determine cytoskeleton organization. Disruption of cytoskeletal elements occurs at pressure much higher than that reported to change ion channel function. However, if some of the cytoskeletal regulatory mechanisms are affected, reductions in filament cellular functions such as motility, contraction, adhesion, division, and morphology may be expected also at lower pressures.

### Cellular Proteins

Hydrostatic pressure effects on proteins in aqueous solution have been studied extensively (Heremans, 1982; Sibenaller and Somero, 1989; Somero, 1990). In general, water soluble proteins (and polypeptides) may undergo limited self-conformational changes when exposed to pressure. These processes involve small negative volume changes ($\Delta V$) and include transitions in the spatial coil structure (–1 to –3 $cm^3.mole^{-1}$), reversible denaturation (–20 to –75 $cm^3.mole^{-1}$) and some interactions with small molecules (–1 to –10 $cm^3 mole^{-1}$). Other self-organization processes, such as isomerization, exhibit positive $\Delta V$ (30–50 $cm^3.mole^{-1}$). Pressure will inhibit reactions when the overall change in reaction volume is positive and enhance reactions when the volume change is negative. Interactions between larger molecules are associated with larger positive volume changes; these include protein-protein association (25–390 $cm^3.mole^{-1}$), self-assembly of polymeric proteins (60–140 $cm^3.mole^{-1}$) and binding of cofactors or substrate molecules (30–70 $cm^3.mole^{-1}$). The volume change of these interactions depends on several, sometimes opposing, mechanisms such as the degree of ligand ionization, of protonation of protein groups and hydration of the protein surface.

## Enzymes

The effect of pressure on enzyme catalyzed reactions will be determined by the volume change associated with the critical, rate-determining step, frequently dissociation of the enzyme-substrate complex to yield the reaction products (Laidler, 1951). For a variety of enzymes (including myosin ATPase, trypsin, sucrase, myokinase) this volume change has been shown to be negative (Macdonald, 1982), although pressures greater than 40 MPa would be required to significantly increase the rate of reaction. In contrast, other enzymes (including creatine kinase, alkaline phosphatase and a number of ATPase's) have been shown to have a positive volume change on dissociation of enzyme and substrate (Macdonald, 1982) and pressure would therefore inhibit their action. Under physiological conditions the concentration of the substrate is not high when compared to the enzyme concentration and therefore association of enzyme with substrate or enzyme-substrate dissociation may be rate determining (Morild, 1977).

Although very high pressures are generally required to modify the activity of an enzyme in isolated conditions (Balny et al., 1997) much lower pressures (<10 MPa) can modify enzyme activity by altering the membrane environment of the enzyme (Sébert, 1997). Relatively low pressure (10 MPa) has been shown to modify glycolysis in Chinese freshwater crabs, increasing flux per unit time more than threefold (Sébert et al., 2000). Experiments on the inactivation of creatine kinase by high pressure indicate that low pressure inactivation arises from a change in environment that precedes dissociation of the enzyme dimer and unfolding of the hydrophobic core (Zhou et al., 2000).

It appears that under physiological conditions enzyme catalyzed reactions are not altered by pressure acting directly on the enzyme itself (until extreme pressures incompatible with biological processes are reached) but that change arises through modification of the enzyme's environment.

### Voltage-gated ion channels

Voltage-gated channels are proteins derived from a super-family of related genes which therefore exhibit remarkable homology in their structure and similarity in basic functions. They are usually activated by membrane depolarization and deactivate when the membrane potential returns to its resting level. Many channels also exhibit voltage-dependent channel inactivation, i.e. channel closure during maintained membrane depolarization. The conduction of electrical impulses is based on a delicate interplay between voltage-activated $Na^+$ and $K^+$ channels. Upon threshold depolarization from the resting potential, $Na^+$ channels open rapidly to cause fast depolarization followed by slower channel inactivation. In parallel, $K^+$ channels, opening at a slower rate, re-polarize the membrane. It is this voltage and time dependent channel interplay that accounts for the transient voltage 'spike' of the action potential.

The pore-forming antibiotic alamethicin, when incorporated into artificial lipid bilayer membranes, forms ion-conducting oligomers composed of a variable number of units in a voltage-dependent fashion (Bruner and Hall 1983). Exposure to pressure (100 MPa) causes both the voltage-dependent 'opening' and 'closing' reactions to become slower. Kinetic analysis of the effect of pressure indicates that the alamethicin channel has at least two closed states which are indistinguishable at normobaric pressure. The voltage-dependence of channel opening and the single-channel conductance are unaffected by pressure. These results imply that the formation and disassembly of a functional alamethicin channel are associated with molecular rearrangements that cause transient increases in free volume of the entire channel. In addition, the flow of ions through the channel does not require significant volume changes, very similar to free-ion diffusion in aqueous solutions.

*Na+ channels.* Early studies on $Na^+$ channel currents were carried out in squid giant axons (Henderson and Gilbert, 1975; Shrivastav et al. 1981; Conti et al. 1982a). Pressure (60 MPa) slightly depresses $Na^+$ current amplitude and slows the kinetics of activation and inactivation of the current. Inactivation seems more affected than activation and the voltage dependence of the steady-state inactivation is only  slightly affected. These data are consistent with $\Delta V$ in the range 30–70 $cm^3.mole^{-1}$. Conti et al. (1984) and Heinemann et al. (1987a) have reported considerable $\Delta V$ (17–28 $cm^3.mole^{-1}$) associated with channel gating currents, which precede channel opening. This represents the movement of charge as a result of conformational changes in the channel voltage sensor, rather than the flow of ions through the channel. Similar responses have been observed using other invertebrate preparations, such as snail (*Helix*) neurons (Harper et al., 1981) and lobster (*Panulirus*) axons (Grossman and Kendig, 1984); although current inactivation was not affected by pressure in *Helix* and with lobster axons transient increases in current were observed with fast pressurization (to 5–10 MPa), which reversed to leave a moderate decrease after 10–15 min of exposure.

The findings in vertebrate cells are more diverse. In single isolated myelinated axons of frogs (*Xenopus*) (Kendig, 1984) pressure (5–10 MPa) slows the rate of development of inactivation, i.e. channels stay open longer and more channels are available for opening. In another frog (*Rana*), action potential Na$^+$ currents recorded under the perineurial sheath of peripheral motor nerves (Grossman et al., 1991) were reduced by 25% at 10 MPa and the kinetics were slightly slower. Recent studies of Na$^+$ action potential currents in mammalian neurons revealed only a 10% decrease in amplitude and rise time at 10 MPa in single cerebellar Purkinje cells (Etzion and Grossman, 1999) and hippocampal pyramidal cells (Southan and Wann, 1996). A study on non-neuronal bovine adrenal chromaffin cells (Heinemann et al., 1987a) revealed similar pressure effects with considerable slowing of inactivation. Detailed statistical analysis of the steady-state current fluctuations in these cells (Heinemann and Conti, 1992) indicated that the single channel current amplitude was not affected, even at 45 MPa hydrostatic pressure, whereas the number of functional channels was reduced, by up to 50%.

In summary, in spite of species differences, membrane bound neuronal Na$^+$ channels seem quite resistant to compression. Pressure slightly reduces mean current amplitude by a minor slowing of the activation process or by decreasing the number of functional channels. In neurons in which inactivation is impeded more than activation, pressure may enhance or elongate, sometimes transiently, the Na$^+$ current.

*K$^+$ channels.* Pressure acts on non-inactivating 'delayed rectifier' K$^+$ currents in axons of invertebrate squid (Henderson and Gilbert, 1975; Shrivastav et al., 1981; Conti et al., 1982b), snail (Harper et al., 1981) and lobster (Grossman and Kendig, 1984) to slow the current rising phase after step depolarization, with an apparent $\Delta V$ of 25–60 cm$^3$.mole$^{-1}$, and increase the steady-state plateau current. In the lobster this increase developed with time after an initial decrease in the current (reminiscent of the Na$^+$ current response). Analysis of the firing pattern of Ca$^{2+}$ action potentials in guinea pig Purkinje cells indicated pressure enhancement of this current (Etzion and Grossman, 1999). Increased K$^+$ current was also proposed as the basis for slowing of the sinus-node pacemaker activity in various mammalian hearts, as a result of slower diastolic depolarization (Ornhagen and Hogan, 1977). Pressure depresses the activation of currents in snail ($I_{AK}$) (Harper et al., 1981) and in mouse *Shaker B* K$^+$ channels and mutants thereof, expressed in *Xenopus* oocytes (Schmalwasser et al., 1998; Meyer and Heinemann, 1997). However, pressure does not affect inactivation in the snail but in the mouse rapid N-type inactivation, mediated by the 'ball-and-chain' mechanism, is slowed at pressure but the slow C/P-type inactivation is accelerated.

K$^+$ channels activated by increased intracellular Ca$^{2+}$ ([Ca$^{2+}$]$_i$), rather than a change of membrane potential, are also modulated by pressure. Pressure decreases this channel conductance in rat hippocampal pyramidal neurons, as evident from the change in their action potential after-hyperpolarizing potential and the increase in firing frequency (Southan and Wann, 1996). Very similar results were recently obtained for pressure sensitive neurons in rat brain stem (solitary complex) compressed by helium to only 0.4 MPa (Dean and Mulkey, 2000). In contrast, high pressure increased the probability of openings of Ca$^{2+}$-activated BK-type channels in bovine chromaffin cells (Macdonald, 1997).

In summary, pressure slows the kinetics of activation and inactivation of membrane bound potassium channels as reported above for sodium channels. However, due to the larger impact on inactivation rather than activation and the existence of more than one molecular mechanism of inactivation, which may respond differently to pressure, the frequent overall effect of pressure is to increase the channel current.

*Ca$^{2+}$ channels.* Cytosolic Ca$^{2+}$ controls, modulates or triggers many fundamental cellular processes; e.g. muscle contraction, ciliary movement, synaptic transmission, egg fertilization and activation of various second messenger systems. In consequence several mechanisms such as membrane pumps, ion exchangers and buffering systems serve to keep the resting cytosolic [Ca$^{2+}$] as low as 0.1 μM. Ca$^{2+}$ may enter the cell from the external fluid via various types of channels or be released from internal stores via specific receptor channels.

Measurements of radiolabelled Ca$^{2+}$ influx into synaptosomes (sealed vesicles from broken nerve terminals, containing Ca$^{2+}$ channels and the synaptic release apparatus) revealed that pressure depresses the depolarization-dependent Ca$^{2+}$ influx (Gilman et al., 1986). However, in the same preparation, pressure slightly increased Ca$^{2+}$ influx through an artificially added Ca$^{2+}$ ionophore (Gilman et al., 1991). Otter and Salmon (1985) have provided indirect evidence for the effect of hydrostatic pressure on unclassified calcium channels in *Paramecium*. The brief reversal of swimming direction, which is Ca$^{2+}$ influx-dependent, normally occurred when the protozoan encountered the wall of the pressure chamber was inhibited by 10 MPa hydrostatic pressure. However, they failed to demonstrate an inhibition of Ca$^{2+}$ influx in a later study in mouse fibroblasts (Crenshaw and Salmon 1996).

Several high voltage-activated $Ca^{2+}$ channels have been described and termed N-, L-, P-, and Q- type, based on their kinetics, voltage range of activation and inactivation and pharmacological properties. An additional channel, which is resistant to all known channel-blocking drugs, was termed R-type. A low voltage-activated channel, which has been less well studied, is termed T-type. These channels differ in their $\alpha_1$ pore-containing protein subunit. Various $Ca^{2+}$ channels mediate synaptic transmission between CNS neurons in the hippocampus (Takahashi and Momiyama, 1993; Luebke et al., 1993; Wu et al., 1998) and cerebellum (Regehr and Mintz, 1994; Mintz et al., 1995). Specific GABA containing inhibitory interneurons in the hippocampus also utilize different $Ca^{2+}$ channels for their release processes (Poncer et al. 1997). Furthermore, colocalization of different $Ca^{2+}$ channels in a single motor nerve terminal of frog (Mallart, 1984, 1985), mouse (Penner and Dreyer, 1986) and CNS terminals (Lemos and Nowycky, 1989) has also been shown.

The effect of pressure on two types of co-localized $Ca^{2+}$ currents was first tested in the frog motor nerve (Grossman et al. 1991a,b; Kendig et al., 1993). The $Ca^{2+}$-dependent current is comprised of fast ($I_{CaF}$) and slow ($I_{CaS}$) components that reflect inward $Ca^{2+}$ current at the terminals (Mallart, 1984). Pharmacological analysis indicated that $I_{CaF}$ is carried by N-type channels and $I_{CaS}$ by L-type channels. Pressure (6.9 MPa) suppressed N-type channels by about 87%, whereas L-type channels were much less sensitive to pressure. Thus pressure may exert a differential effect on various types of $Ca^{2+}$ channels at nerve terminals. Heinemann et al. (1987a) reported that L-type $Ca^{2+}$ current was not sensitive to high pressure in mammalian chromaffin cells. Recently it has also been reported that similar results are obtained for P-type $Ca^{2+}$ action potentials in guinea pig cerebellar Purkinje cells (Etzion and Grossman, 1999). Finally, Etzion and Grossman (2000) have suggested that pressure blocks mainly the N-type $Ca^{2+}$ channel-dependent component of the synaptic response between the parallel fibres and Purkinje cells' synapse in guinea pig cerebellum.

Given the large diversity even within this single channel 'family' and the limited data available, pressure effects can not be generalized. A systematic study will be necessary in order to analyze the pressure response of the various $Ca^{2+}$ channels. It is conceivable that synapses utilizing N-type $Ca^{2+}$ channels will be more vulnerable to pressure exposure than others. The extent of the consequent behavioural modification will be dependent on the role of such a synapse is playing in the network activity.

### Chemically-activated receptor proteins

Receptors are proteins that bind effector substances and transduce that binding into a physiological effect. They are grouped, on the basis of structure and signal transduction mechanism, into four superfamilies; ionotropic, metabotropic, ligand-gated enzymes and protein-regulating receptors. Ionotropic receptors, also known as ligand-gated ion channels, are responsible for regulating fast synaptic neurotransmission. Activation of the receptor by binding a neurotransmitter causes a conformational change that allows ions to pass through a channel integral to the protein. Metabotropic receptors, also known as G-protein coupled receptors, bind neurotransmitters, hormones and neuropeptides that cause a conformational change in the receptor. In turn, this allows the receptor to bind to and activate a G-protein (guanyl nucleotide-binding protein). The activated G-protein then activates another protein or enzyme or ion channel, the 'so-called' second messenger, which effects the cellular response. These receptors primarily mediate slower processes than the ionotropic receptors. This is by far the largest family of receptors with approximately 1% of the human genome encoding the different proteins. The ligand-regulated enzymes (e.g. insulin receptor) are a smaller family of receptors responsible for the regulation of a variety of homeostatic systems. Lastly, the protein synthesis-regulating receptors are found in the cytosol and the nuclear membrane and, in response to binding hormones, steroids or retinoic acid, affect the transcription of DNA to RNA and hence the synthesis of proteins.

*Ionotropic Receptors.* The ionotropic receptor superfamily is composed of three families, categorized on the basis of amino acid homology in the primary structure; the Cys-loop receptors, the glutamate receptors and the P2X receptors. The Cys-loop family comprises the nicotinic acetylcholine receptor (nACh), the $\gamma$-aminobutyric acid-A receptor (GABA$_A$), the glycine receptor (Gly) and the third member of the 5-hydroxytryptamine family of receptors (5-HT$_3$). The nACh and 5-HT$_3$ receptors both gate a cation channel (permeable to Na$^+$, K$^+$ and Ca$^{2+}$) and are consequently responsible for mediating excitatory synaptic transmission. The GABA$_A$ and Gly receptors both gate a Cl$^-$ channel and are thus responsible for mediating inhibition. The glutamate receptor family, activated by glutamate and aspartate, comprises three distinct receptors known by the agonists that selectively activate them, the N-methyl-D-aspartate sensitive receptor (Glu$^{NMDA}$), the $\alpha$-amino-3-hydroxy-5-methyl-4-isoxazoleproprionic acid sensitive receptor (Glu$^{AMPA}$) and the kainic acid sensitive receptor (Glu$^{KA}$). These receptors all gate a cation channel, and are therefore excitatory, but Glu$^{NMDA}$ is far more permeable to Ca$^{2+}$ than the other two members of the family. Finally, the P2X familiy of receptors is activated by adenosine triphosphate (ATP). They gate a cation channel, therefore mediating fast excitatory synaptic transmission. At present nothing is known about the pressure sensitivity of the P2X family of receptors.

* Cys-loop Receptors. The nACh receptor exists in two major forms, the neuromuscular junction and neuronal types. The neuromuscular junction

form of the receptor is probably the most extensively studied of all receptor types and much of our knowledge of receptor structure and function has been derived from studies on this receptor ( for reviews, see Galzi et al., 1991; Unwin, 1993; Ortells and Lunt, 1995). The neuromuscular form of the nACh receptor is a pentameric structure composed of $(\alpha)_2\beta\varepsilon\delta$ subunits. Acetylcholine binds at the interface between each $\alpha$ and one other subunit. The effect of pressure on the neuromuscular form of the receptor has been investigated by Heinemann et al. (1987b). They calculated a volume of activation for the nACh receptor of some 48 $cm^3.mol^{-1}$ but found no evidence for any significant effect on volume of activation of the receptor by pressures less than 20 MPa. The neuronal form of the receptor whilst still pentameric only comprises two subunits $\alpha$ and $\beta$ but it has a far richer variety of isoforms since there are seven $\alpha$ and four $\beta$ subunits, which combine in unknown stoichiometries. It is possible that the neuronal form of the receptor is sensitive to pressure even though the neuromuscular form is not.

Glycine receptors show a marked reduction in sensitivity to glycine, with the $EC_{50}$ reduced by 60% at 10MPa (Shelton et al., 1993), whereas pressure has no effect on the $GABA_A$ receptor (Shelton et al., 1996). The effects of pressure on the glycine receptor become progressively greater at higher pressures. The volume change on activation of the glycine receptor was calculated to be 110 $cm^3.mol^{-1}$, considerably greater than for the nACh receptor. This would explain why the glycine receptor is more sensitive to pressure than the nACh receptor. A recombinant $\alpha1$ human glycine receptor was somewhat more sensitive to pressure, with a volume of activation of 149 $cm^3.mol^{-1}$ (Roberts et al., 1996). Since this homomeric receptor apparently binds at least three glycine molecules when activated, it lends support to the suggestion that the neuronal form of the nACh receptor may show more pressure sensitivity than the neuromuscular form.

The glycine dose-response data for the glycine receptor may be interpreted as though pressure is acting as a simple competitive antagonist. This action could be manifest either as a reduction in the affinity of glycine binding or in the transduction of binding to channel opening. A direct effect on channel opening or lifetime seems unlikely since the closely related $GABA_A$ receptor is unaffected. The apparent volume change on activation was large for simple chemical equilibria and indicates that considerable conformational changes occur when glycine binds. The effect of pressure may therefore be envisaged as opposing the conformational change that follows glycine binding thereby reducing the effective inhibitory transmission available.

* Glutamate Receptors. Native rat cerebella $Glu^{NMDA}$ receptors show a marked increase in sensitivity to glutamate (128%) at 10MPa (Williams et al., 1996; Daniels et al., 1998), implying that overall there is a negative $\Delta V$ associated with activation of this receptor. In contrast, $Glu^{KA}$ receptors, from

rat whole brain, are relatively unaffected (Daniels et al., 1991; Shelton et al., 1993). However, the response of $Glu^{NMDA}$ receptors is not straightforward. $Glu^{NMDA}$ receptors are composed of two subunits, R1 and one of R2A, R2B, R2C and R2D, in unknown stoichiometry. All $Glu^{NMDA}$ receptors contain at least one R1 subunit. The different R2 subunits control the conductance of the receptor, thus R1/R2A and R1/R2B are high conductance receptors (40–50 pS) whilst R1/R2C and R1/R2D are low conductance receptors (18–38 pS). These two receptor groups are differentially expressed in the brain (as revealed by the differential distribution of the R2 subunits: R1 is ubiquitous) and also have different sensitivity to voltage-dependent blockade by $Mg^{2+}$ and thus different activation properties (Monyer et al., 1992). In *Xenopus* oocytes the R1 subunit will express functional receptors without co-expression of one of the R2 subunits, probably by functional combination with an endogenous protein subunit. These receptors are insensitive to pressure (Daniels, unpublished observations), suggesting that the pressure sensitivity observed with $Glu^{NMDA}$ receptors expressed using a rat brain mRNA extract arose because one or more of the R2 subunits is pressure sensitive. Recombinant receptors R1/R2A, R1/R2B and R1/R2D expressed in *Xenopus* oocytes were not sensitive to pressure (Daniels, unpublished observations). To date the R1/R2C receptor has not been tested at pressure. However, given the localization of this receptor to the cerebellum and the well-known association of cerebella function and motor control it would be surprising if this particular receptor type were not pressure sensitive. This would also be in accord with the initial observation that $Glu^{NMDA}$ receptors from rat cerebellem were pressure sensitive.

The results with the $Glu^{NMDA}$ receptor demonstrate the need to carefully analyze the pressure sensitivity of all the different receptor subtypes before reaching an overall conclusion regarding the pressure sensitivity of the ionotropic receptors. All the ionotropic receptors share a rich subunit composition. Thus, unless some common structural motif can be identified that conveys pressure sensitivity, there is a major challenge to define the pressure sensitivities of this class of receptor.

*Metabotropic Receptors.* The metabotropic receptors possess 7 α helical transmembrane regions with a large intracellular loop joining transmembrane segments 5 and 6 and a long C-terminal tail from segment 7. The tail and loop contain binding sites for heterotrimeric guanine nucleotide binding proteins (G-proteins). Activation of G-proteins results in various biochemical and electrical effects that produce physiological effects. Three principal intracellular signalling pathways are activated by G-proteins. The first involves either the activation (Gs) or the inhibition (Gi) of adenylyl cyclase (AC). Stimulation of AC leads to the production of cyclic adenosine phosphate (cAMP) which in turn upregulates the formation of protein kinase A (PKA). The second pathway (Gq/G11) involves activation of phospholipase C (PLC), which leads to the formation of inositol 1, 4, 5-triphosphate ($IP_3$) the release of $Ca^{2+}$ from intracellular stores and the activation of calcium-calmodulin-dependent protein kinases. Activation of PLC also leads to the formation of diacylglycerol (DAG), which activates protein kinase C (PKC). The third pathway (Go) involves the activation of phospholipase $A_2$ ($PLA_2$), which produces arachidonic acid (AA) that is the precursor for the formation of various metabolites such as prostaglandins, thromboxanes and leukotrienes. The formation of DAG can also lead to AA formation.

Pharmacological experiments have indicated that pressure may affect both dopamine (D1, D2) and serotonin ($5\text{-}HT_{1b}$, $5\text{-}HT_{2c}$) receptors (Abraini and Rostain 1991; Abraini and Fechtali, 1992; Kriem et al., 1996, 1998). Dopamine D1 receptors couple through Gs proteins and hence activate AC, up-regulate cAMP and activate PKA. Dopamine D2 receptors couple through Gi/o proteins and can therefore inhibit AC and separately activate $PLA_2$ thus activating AA. Serotonin $5\text{-}HT_{1b}$ receptors also couple through Gi/o proteins. However $5\text{-}HT_{2c}$ receptors couple through Gq proteins that activate PLC leading to the release of intracellular calcium and PKC. These experiments, whilst indicating that the cellular processes mediated by these metabotropic receptors may be implicated in HPNS, cannot indicate whether pressure is modulating receptor function or steps in the intracellular signalling pathways activated by the receptors.

It has been found recently (Daniels, unpublished results) that the metabotropic receptor for the bacterial peptide N-formyl-L-methionyl-L-leucyl-L-phenylanaline, which couples through a Gq protein, when expressed in *Xenopus* oocytes is not sensitive to pressure. This suggests that none of the PLC pathway reactions result in a large volume change with which pressure could interact. In turn this suggests that it is likely that the modulation by pressure of the pharmacological response of $5\text{-}HT_{2c}$ receptors arises from a direct effect of pressure on the $5\text{-}HT_{2c}$ receptor and not on the intracellular pathway (PLC medicated) to which it couples. In separate experiments it was found that direct stimulation of a G protein endogenous to *Xenopus* oocytes by fluoride ions was potentiated by pressure (Daniels, unpublished results). Analysis indicated that the G protein stimulated by F⁻ was a member of the Gs family and therefore would positively couple to AC. Thus pressure appears to modulate either the up-regulation of AC, the production of cAMP or PKA (or a combination of these interactions). These experiments are supported by those of Siebenallar and Garrett (2002) who reported that AC activity and modulation in a variety of teleost fish species is affected by pressure. Pressure may therefore decrease the inhibitory actions of D2 receptors and stimulate the excitatory actions of D1 receptors. Furthermore, since both D1 and D2 receptors are predominantly expressed in the striatum and are known to be involved in motor control (cf Parkinson's disease; Gibb, 1992; Fearnley and Lees, 1991) this result does suggest that dopamine receptors may in part mediate the expression of acute HPNS.

### Transporters and pumps

Membrane transporters (e.g. Na/K pump, Na/K/2Cl cotransporter) have a crucial role in maintaining ionic and osmotic balance in a wide variety of cells. It is difficult to predict what effect pressure would have on these transport systems, just as it is difficult to predict the effect on enzyme catalyzed reaction rates. In human red cells $Na^+$ and $K^+$ are transported across the membrane by the sodium pump, chloride-dependent sodium-potassium co-transport and by a residual passive leak. It has been shown that over the pressure range atmospheric to 40 MPa the sodium pump is progressively inhibited, passive diffusion is markedly potentiated and the co-transport is initially inhibited but little further affected by pressures over 10 MPa (Hall et al., 1982). The authors suggested that the initial formation of the activated-$K^+$-membrane pore complex involving loss of water from the $K^+$ hydration shell was the rate-determining step in the sodium pump that was sensitive to pressure. No explanation was offered for the inhibition of the $Cl^-$-sensitive Na/K co-transporter. However, since the inhibition occurred during the initial pressurization from atmospheric to 10 MPa it would seem likely that this reflects a modification to the transporter protein-membrane environment and not a direct effect of pressure on the transporter protein itself. The increase in the passive flux by pressure has been ascribed to a pressure-dependent change in cell morphology (from biconcave disc to cup-shaped forms) leading to activation of a volume-sensitive KCl transport pathway (Hall and Ellory, 1986).

Glucose is transported across the red blood cell membrane by a transporter that alternatively presents binding sites to the exterior and interior of the cell (Thorne et al., 1992). The rates of re-orientation between inward- and outward-facing conformations are much higher for the glucose-bound protein than the unbound protein. Additionally, the rate constants for inward re-orientation exceed those for outward re-orientation. It was found that pressure specifically inhibits the translocation of glucose and not the binding of glucose to the carrier protein (Thorne et al., 1992).

Chondrocytes are the cells responsible for producing the extracellular matrix (cartilage) which protects joints. In part they must do this by sensing pressure changes and altering the composition of the matrix accordingly. It is possible that this is achieved by the effect of pressure on both specific and passive membrane transporters in these cells (Hall, 1999). In contrast to red blood cells, pressure on chondrocytes has been shown to inhibit the Na/K pump, the Na/K/2Cl cotransporter and the passive $K^+$ permeability (Hall, 1999). Increasing pressure therefore would reduce internal $K^+$ concentration

($[K^+]_i$) and increase $[Na^+]_i$ thereby inducing cell swelling. These changes are known to affect chondrocyte matrix synthesis.

In summary, alterations by pressure to transporter function are likely to be replicated across a wide spectrum of cell types and are probably not restricted simply to the transport of ionic species across membranes. Most neurotransmitters rely on specific transport systems to terminate their action at a synapse. Clearly inhibition of neurotransmitter re-uptake would have profound effects on neuronal function. As yet the effect of pressure on these systems has not been studied and it must be stressed that, as with enzyme function, the environment of the transporter may be critical to the manifestation of pressure effects.

## CONCLUSION

To understand how pressure affects biological processes is an enormous scientific challenge. Unlike a chemical, pressure acts instantaneously and ubiquitously. Although this means that the confounding problems of pharmacokinetics and metabolism are not an issue, in principle every chemical reaction contributing to the continued existence of the organism is a potential target.

It appears that the biological range of pressures, from atmospheric to 100 MPa, should be divided into two. A range (up to 20 MPa) that can be tolerated, to various degrees, by organisms that normally live at, or close to, atmospheric pressure and an upper range (from 20 to 100 MPa) inhabited by organisms adapted to life at pressure. Pressures above 40 MPa can have profound effects on the molecular structure and hence function of many proteins, including cytoskeletal elements. To understand the adaptations that enable bacteria, invertebrates and some vertebrates to exist at extreme pressures will require detailed analysis of the structural modifications to proteins that have occurred during the evolutionary process.

At the molecular level, pressures up to 20 MPa have little effect on enzyme function directly but may affect their phospholipid environment leading to indirect changes. Since many effects depend not only on the exposure pressure but also on the time of exposure, they may play a central role in the adaptation or acclimatization to pressure. Cytoskeletal filaments are relatively stabile at pressure but pressure can impede some regulatory mechanisms (e.g. phosphorylation) that determine cytoskeleton organization. Transporter proteins (for ionic species and amino acids) resemble enzymes in that their function is most likely to be modified by pressure affecting their environment. Voltage-regulated membrane bound ion channels are in the main quite resistant to pressure. Interesting exceptions are a few types of $K^+$ channels, where functions are enhanced and N-type $Ca^{2+}$ channels, which are depressed at pressure. Thus synapses using the latter channels may be more vulnerable to pressure exposure than others. Ionotropic receptor proteins that regulate fast chemical neurotransmission are obvious candidate targets for pressure effects. However, of the structurally related Cys-loop family of receptors only the glycine receptor, responsible for mediating inhibitory neurotransmission, is significantly affected (inhibited). The other principal family of receptors, the glutamate receptors, also show very selective pressure sensitivity. Only one of the four variants of the NMDA-sensitive glutamate receptor is sensitive to pressure (potentiated) and the non-NMDA glutamate receptors are not affected by pressure.

Much remains to be done before a comprehensive understanding of the effect of pressure on biological systems can be claimed. Indeed the existence of life at great ocean depths indicates the necessity for general adaptive mechanisms, especially of the nervous system. Finally, the prodigious depths achieved by the toothed whales and seals, which lie far beyond the abilities of other vertebrates, call for special attention (and are the subject of another chapter in this book). It suggests that a programme to reveal the genes responsible for selected key proteins, or more ambitiously to decode their genome and compare it with ours and that of other mammals, might be the way forward.

## REFERENCES

Abraini, J.H. and J.C. Rostain. 1991. Effects of the Administration of α-Methyl-Para-Tyrosine On the Striatal Dopamine Increase and the Behavioral Motor Disturbances in Rats Exposed to High-Pressure. Pharmacol. Biochem. Behav. 40: 305–310.

Abraini, J.H. and T.A. Fechtali. 1992. Hypothesis Regarding Possible Interactions Between the Pressure—Induced Disorders in Dopaminergic and Amino-Acidergic Transmission. Neurosc. Biobehavioral Rev. 16: 597–602.

Balny, C. and V.V. Mozhaev and R. Lange. 1997. Hydrostatic pressure and proteins: basic concepts and new data. Comp. Biochem. Physiol. A116: 299–304.

Bennett, P.B. Inert gas narcosis. Pp. 239–261. *In*: P.B. Bennett and D.H. Elliott (eds.). 1982. The Physiology and Medicine of Diving. 3rd Edit. Balliere Tindall. London.

Bournes, B. and S. Franklin, L. Cassimeris and E.D. Salmon. 1988. High hydrostatic pressure effects in vivo: changes in microtubule assembly, and actin organization. Cell Motil Cytoskeleton 10: 380–390.

Bruner, L.J. and J.E. Hall. 1983. Pressure effects on alamethicin conductance in bilayer membranes. Biophys J. 44: 39–47.

Bruun, A.F. 1957. Animals of the abyss. Scientific American 197: 50–57.

Carrier, D. and P.T.T. Wong. 1996. Effect of dehydration and hydrostatic pressure on phosphatidylinositol bilayers: An infrared spectroscopic study. Chem. Phys. Lipids. 83: 141–152.

Clark, J.M. Oxygen toxicity. Pp. 200–238. *In:* P.B. Bennett and D.H. Elliott (eds.). 1982. The Physiology and Medicine of Diving. 3rd Edit. Balliere Tindall. London.

Conti, F. and R. Fioravanti, J.R. Segal and W. Stuhmer. 1982a. Pressure dependence of the sodium currents of squid giant axon. J. Membr. Biol. 69: 23–34.

Conti, F. and R. Fioravanti, J.R. Segal and W. Stuhmer. 1982b. Pressure dependence of the potassium currents of squid giant axon. J. Membr. Biol. 69: 35–40.

Conti, F. and I. Inoue, W. Stuhmer and F. Kukita. 1984. Pressure dependence of sodium gating currents in the squid giant axon. Eur. Biophys. J. 11: 137–147.

Crenshaw, H.C. and E.D. Salmon. 1996. Hydrostatic pressure to 400 atm does not induce change of cytosolic concentration of $Ca^{2+}$ in mouse fibroblasts: measurements of fura-2 fluorescence. Exp. Cell Res. 227: 277–284.

Crenshaw, H.C. and J.A. Allen, V. Skeen, A. Harris and E.D. Salmon. 1996. Hydrostatic pressure has different effects on the assembly of tubulin, actin, myosin II, vinculin, talin, and cytokeratin in mammalian tissue cells. Exp. Cell Res. 227: 285–297.

Daniels, S. and D.M. Zhao, N. Inman, D.J. Price, C.J. Shelton and E.B. Smith. 1991. Effects of general anaesthetics and pressure on mammalian excitatory receptors expressed in *Xenopus* oocytes. Ann. NY Acad. Sci. 625: 108–115.

Daniels, S. and R. Roberts and N. Williams. Effects of high pressure on post-synaptic ionotropic receptors. Pp. 22–31. *In*: P.B. Bennett, I. Demchenko and R.E. Marquis (eds.). 1998. High Pressure Biology and Medicine. University of Rochester Press: Rochester, N.Y.

Dean, J.B. and D.K. Mulkey. 2000. Continuous intracellular recording from mammalian neurons exposed to hyperbaric helium, oxygen, or air. J. Appl. Phsiol. 89: 807–822.

Etzion, Y. and Y. Grossam. 1999. Spontaneous $Na^+$ and $Ca^{2+}$ spike firing of cerebellar Purkinje neurons at high pressure. Eur. J. Physiol. 437: 276–284.

Etzion, Y. and Y. Grossman. 2000. Pressure induced depression of synaptic transmission in the cerebellar parallel fibre synapse involves suppression of presynaptic N-type $Ca^{2+}$ channels. Eur. J. Neurosci. 12: 4007–4016.

Fearnley, J.M. and A.J. Lees. 1991. Aging and Parkinsons-disease—Substantia-nigra regional selectivity. Brain 114: 2283–2301.

Finch, E.D. and L.A. Kiesow. 1979. Pressure, anaesthetics and membrane structure: A spin probe study. Undersea Biomed. Res. 6: 41–45.

Galzi, J.L. and F. Revah, A. Bessis and J.P. Changeux. 1991. Functional architecture of the nicotinic acetylcholine-receptor—from electric organ to brain. Ann. Rev. Pharmacol. Toxicol. 31: 37–72.

Gibb, W.R.G. 1992. Neuropathology of Parkinsons-disease and related syndromes. Neurologic Clinics 10: 361–376.

Gilman, S.C. and K.K. Kumaroo and J.M. Hallenbeck. 1986. Effects of pressure on uptake and release of calcium by brain synaptosomes. J. Appl. Physiol. 60: 1446–1450.

Gilman, S.C. and J.S. Colton and Y. Grossman. 1991. A23187 stimulated calcium uptake and GABA release by cerebrocortical synaptosomes: effect of high pressure. J. Neural. Transm. 86: 1–9.

Grossman, Y. and J.J. Kendig, J.J. 1984. Pressure and temperature: Time dependent modulation of membrane properties in a bifurcating axon. J. Neurophysiol. 52: 692–708.

Grossman, Y. and J.C. Colton and S.C. Gilman. 1991a. Interaction of Ca channel blockers and high pressure at the crustacean neuromuscular junction. Neurosci. Lett. 125: 53–56.

Grossman, Y. and J.S. Colton. and S.C. Gilman. Reduced Ca currents in frog nerve terminals at high presure. Pp. 411–413. *In*: F.S. Elis, M.C. Nowycky and D.J. Triggle. (eds.). 1991b. Calcium Entry and Action at the Presynaptic Nerve Terminal. Anna NY Acad. Sci. 635.

Hall, A.C. 1999 Differential effects of hydrostatic pressure on cation transport pathways of isolated articular chondrocytes. J. Cell. Physiol. 178: 197–204.

Hall, A.C. and J.C. Ellory. 1986. Effects of high hydrostatic pressure on passive monovalent cation transport in human red cells. J. Membrane Biol. 94: 1–17.

Hall, A.C. and J.C. Ellory and R.A. Klein. 1982. Pressure and temperature effects on human red cell cation transport. J. Membrane Biol. 68: 47–56.

Harper, A.A. and A.G. Macdonald and K.T. Wann. 1981. The action of high hydro-static pressure on the membrane currents of helix neurones. J. Physiol. 311: 325–339.

Heinemann, S.H. and F. Conti. Non stationary noise analysis and its application to patch clamp recordings. *In*: B. Rudy and L.E. Iverson (eds.). 1992. Methods in Enzymology. Ion channels.Vol. 207. Academic Press. San Diego.

Heinemann, S.H. and F. Conti, W. Stuhmer and E. Neher. 1987a. Effects of hydrostatic pressure on membrane processes. Sodium channels, calcium channels, and exocytosis. J. Gen. Physiol. 90: 765–778.

Heinemann, S.H. and W. Stuhmer and F. Conti. 1987b. Single acetylcholine-receptor channel currents recorded at high hydrostatic pressures. Proc. Nat. Acad. Sci. USA, 84: 3229–3233.

Henderson, J.V. and D.L. Gilbert. 1975. Slowing of ionic currents in the voltage clamped squid axon by helium pressure. Nature 258: 351–352.

Heremans, K. 1982. High pressure effects on proteins and other biomolecules. Ann. Rev. Biophys. Bioengng. 11: 1–21.

Higgins, J.A. and B.W. Hitchin and M.G. Low. 1989. Phosphatidylinositol-specific phospholipase C of *Bacillus thurigensis* as a probe for the distribution of phosphatidylinositol in hepatocyte membranes. Biochem. J. 259: 913–916.

Hogan, P.M. and S.R. Besch. Vertebrate skeletal and cardiac muscle. Pp. 125–146. *In*: A.G. Macdonald (ed.). 1993. Advances in Comparative and Environmental Physiology, Effects of High Pressure on Biological Systems. Springer-Verlag. Berlin-Heidelberg.

Kendig, J.J. 1984. Ionic currents in vertebrate myelinated nerve at hyperbaric pressure. Am J Physiol 246: C84–C90.

Kendig, J.J. and Y. Grossman and S. Heinemann. Ion channels and nerve cell function. Pp. 88–124. *In*: A.G. Macdonald (ed.). 1993. Advances in Comparative and Environmental Physiology, Effects of High Pressure on Biological Systems. Springer-Verlag. Berlin-Heidelberg.

Koynova, R. and M. Caffrey. 1998. Phases and phase transitions of the phosphatidylcholines. Biochim. Biophys. Acta 1376: 91–145.

Kriem, B. and J.H. Abraini and J.C. Rostain. 1996. Role of 5-HT1b receptor in the pressure-induced behavioral and neurochemical disorders in rats. Pharmacol. Biochem. Behav. 53: 257–264.

Kriem, B. and J.C. Rostain and J.H. Abraini. 1998. Crucial role of the $5-HT_{2C}$ receptor, but not of the $5-HT_{2A}$ receptor, in the down regulation of stimulated dopamine release produced by pressure exposure in freely moving rats. Brain Res. 796: 143–149.

Laidler, K.T. 1951 Influence of pressure on the rates of biological reactions. Archs. Biochim. Biophys. 30: 226–236.

Lemos, J.R. and M.C. Nowycky. 1989. Two types of calcium channels coexist in peptide releasing vertebrate nerve terminals. Neuron 2: 1419–1426.

Liu, N.I. and R.L. Kay. 1977 Redetermination of the pressure dependence of the lipid bilayer phase transition. Biochem. 16: 3484–3486.

Low, M.G. 1987 Biochemistry of the glycosyl-phosphatidylinositol membrane protein anchors. Biochem. J. 244: 1–13.

Luebke, J.I. and K. Dunlap and T.J. Turner. 1993. Multiple calcium channel types control glutamatergic synaptic transmission in the hippocampus. Neuron 11: 895–902.

Macdonald, A.G. Hydrostatic pressure physiology. Pp. 157–188. *In*: P.B. Bennett and D.H Elliott (eds.). 1982. The Physiology and Medicine of Diving. 3rd Edit. Balliere Tindall. London.

Macdonald, A.G. 1997. Effect of high hydrostatic pressure on the BK channel in bovine chromaffin cells. Biophys. J. 73: 1866–1873.

Mallart, A. 1984. Presynaptic currents in frog motor endings. Pflugers Arch 400: 8–13.

Mallart, A. 1985. Electrical current flow inside perineurial sheaths of mouse motor nerves. J. Physiol. 368: 565–575.

Meyer, R. and S.H. Heinemann. 1997. Temperature and pressure dependence of shaker K channel N- and C-type inactivation. Eur. Biophys. J. 26: 433–445.

Miller, K.W. 1974. Inert Gas Narcosis, the High Pressure Neurological Syndrome, and the Critical Volume Hypothesis. Science 185: 867–869.

Miller, K.W. and W.D.M. Paton, W.B. Streett and E.B. Smith. 1967 Animals at very high pressures of helium and neon. Science 157: 97–98.

Mintz, I.M. and B.L. Sabatini and W.G. Regehr. 1995. Calcium control of transmitter release at a cerebellar synapse. Neuron 15: 675–688.

Monyer, H. and R. Sprengel, R. Shoepfer, A. Herb, M. Higuchi, H. Lomeli, N. Burnashev, B. Sakmann and P.H. Seeburg. 1992. Heteromeric NMDA receptors: molecular and functional distinction of subtypes. Science 256: 36–41.

Morild, E. 1977. Pressure neutralisation of substrate inhibition in the alcohol dehydrogenase reaction. J. Phys. Chem. 81: 1162–1166.

Ornhagen, H.C. and P.M. Hogan. 1977. Hydrostatic pressure and mammalian cardiac pacemaker function. Undersea Biomed. Res. 4: 347–358.

Ortells, M.O. and G.G. Lunt. 1995. Evolutionary history of the ligand-gated ion-channel superfamily of receptors. Trends Neurosci. 18: 121–127.

Otter, T. and E.D. Salmon. 1979. Hydrostatic pressure reversibly blocks membrane control of ciliary motility in *Paramecium*. Science 206: 358–361.

Otter, T. and E.D. Salmon. 1985. Pressure induced changes in $Ca^{2+}$ channel excitability in *Paramecium*. J. Exp. Biol. 117: 29–43.

Penner, R. and F. Dreyer. 1986. Two different presynaptic calcium currents in mouse motor nerve terminals. Pflugers Arch. 406: 190–197.

Poncer, J.C. and R.A. McKinney, B.H. Gahwiler. and S.M. Thompson. 1997. Either N- or P-type calcium channels mediate GABA release at distinct hippocampal inhibitory synapses. Neuron 18: 463–472.

Prasad, S. and R. Shashidhar, B. Gaber. and S. Chandrasekhar. 1987. Pressure studies on 2 hydrated phospholipids—1,2-dimyristoyl-phosphatidylcholine and 1, 2-dipalmitoyl-phosphatidylcholine. Chem. Phys. Lipids. 43: 227–235.

Regehr, W.G. and I.M. Mintz. 1994. Participation of multiple calcium channel types in transmission at single climbing fiber to Purkinje cell synapses. Neuron 12: 605–613.

Roberts, R.J. and C.J. Shelton, S. Daniels. and E.B. Smith. 1996. Glycine activation of human homomeric a1 glycine receptors is sensitive to pressure in the range of the high pressure nervous syndrome. Neurosci. Lett. 208: 125–128.

Salmon, E.D. 1975a. Pressure—induced depolymerization of spindle microtubule birefringence and spindle length. J. Cell. Biol. 65: 603–614

Salmon, E.D. 1975b. Pressure—induced depolymerization of spindle microtubules: Thermodynamics of in vivo spindle assembly. J. Cell. Biol. 66: 114–1127

Salmon, E.D. 1975c. Pressure—induced depolymerization of brain microtubules. Science 189: 884–886.

Salmon, E.D. and D. Goode, T.K. Maugel and D.B. Bonar. 1976. Pressure—induced depolymerization of spindle microtubules stability in Hela cells. J. Cell Bio. 69: 443–454.

Schaefer, K.E. Carbon dioxide effects under conditions of raised environmental pressure. Inert gas narcosis. Pp. 144–154. *In*: P.B. Bennett and D.H. Elliott (eds.). 1969. The Physiology and Medicine of Diving. 1st Edit. Balliere Tindall. London.

Schmalwasser, H. and A. Neef, A.A. Elliot and S.H. Heinemann. 1998. Two-electrode voltage clamp of *Xenopus* oocytes under high hydrostatic pressure. J. Neurosc. Methods 81: 1–7.

Sébert, P. Pressure effects on shallow water fish. Pp. 279–323. *In*: D.J. Randall and A.P. Farrell (eds.). 1997. Fish Physiology. Vol. 16. Academic Press. San Diego.

Sébert, P. and Y. Choquin and A. Péqueux. 2000. High pressure and glycolytic flux in the freshwater Chinese crab, *Eriocheir sinensis*. Comp. Biochem. Physiol. B126: 537–542.

Shelton, C.J. and M.G. Doyle, D.J. Price, S. Daniels and E.B. Smith. 1993. The effect of high pressure on glycine and kainate sensitive receptor channels expressed in *Xenopus* oocytes. Proc. Roy. Soc. Lond. B254: 131–137.

Shelton, C.J. and S. Daniels and E.B. Smith. 1996. Rat brain GABAA receptors expressed in *Xenopus* oocytes are insensitive to high pressure. Pharmaco. Comm. 7: 215–220.

Shrivastav, B.B. and J.L. Parmentier. and P.B. Bennett. A quantitative description of pressure induced alterations in ionic channels of the squid giant axon. Pp. 611–619. *In:* A.J. Bachrach and M.M. Matzen (eds.). 1981. Proceedings Sev-enth Symposium on Underwater Physiology. Undersea Society Inc.

Sibenaller, J.F. and G.N. Somero. 1989. Biochemical adaptation to the deep sea. Rev. Aquatic Sci. 1: 1–25.

Siebenaller, J.F. and D.J. Garrett. 2002. The effects of the deep-sea environment on transmembrane signalling. Comp. Biochem. Physiol. B131: 675–694.

Somero, G.N. 1990. Life at low volume change: hydrostatic pressure as a selective factor in the aquatic environment. Amer Zool. 30: 123–135.

Somero, G.N. 1992. Adaptation to highydrostatic pressure. Annu. Rev. Physiol 54: 557–577.

Southan, A.P. and K.T. Wann. 1996. Effects of high helium pressure on intracellular and field potential responses in the CA1 region of the in vitro rat hippocampus. Eur. J. Neurosci 8: 2571–2581.

Stamatoff, J. and D. Guillon, L. Powers, P. Cladis. and D. Aadsen. 1978. X-ray diffraction measurements of dipalmitoylphosphatidylcholine as a function of pressure. Biochem. Biophys. Res. Commun. 85: 724–728.

Takahashi, T. and A. Momiyama. 1993. Different type of calcium channels mediate central synaptic transmission. Nature 366: 156–158.

Thorne, S.D. and A.C. Hall and A.G. Lowe. 1992. Effects of pressure on glucose transport in human erthyrocytes. FEBS Letts. 301: 299–302.

Unwin, N. 1993. Nicotinic acetylcholine-receptor at 9-angstrom resolution. J. Mol. Biol. 229: 1101–1124.

Waxman, V. and A. Tarasiuk, Z. Priel. and Y. Grossman. High pressure suppresses motility of mucus transporting cilia. Pp. 135–141. *In:* N. Trikilias (ed.). 1991. Proceedings European Undersea Biomedical Society. Annual meeting, Heraklion-Crete, Greece.

Williams, N. and R.J. Roberts and S. Daniels. 1996. A study into pressure sensitivity of ionotropic receptors. Biophys. Mol. Biol. 65 (S1): 113.

Winter, R. and W. Pilgrim. 1989. A sans study of high-pressure phase-transitions in model biomembranes. Ber. Bunsenges. Phys. Chem. 93: 708–717.

Wu, L.G. and J.B. Borst and B. Sakmann. 1998. R-type Ca2+ currents evoke transmitter release at a rat central synapse. Proc. Natl. Acad. Sci. USA 95: 4720–4725.

Zhou, J.M. and L. Zhu. and C. Balny. 2000. Inactivation of creatine kinase by high pressure may precede dimer dissociation. Eur. J. Biochem. 267: 1247–1253.

# 7

# Pressure Sensing: Depth Sensors and Depth Usage

*Dr Peter J. Fraser*

## INTRODUCTION

Hydrostatic pressure affects all organisms as barometric pressure in air and hydrostatic pressure in water. It is normally close to 1 atmosphere (atm) (101.325 kPa) on land with a reduction for height e.g. around 50 kPa at a height of 5800m. As hydrostatic pressure in water, it is the natural proxy for depth and in simple terms changes by around 100 kPa or 1 atm for every 10m of descent down through the water. By convention the additional 1 atm provided by the weight of air above the water surface is taken as the starting value so that at a depth of 10m we consider the hydrostatic pressure to be 100kPa although the absolute value would be 100 kPa due to the water plus 100 kPa due to air pressure giving a combined value of 200 kPa (Hills, 1972). In addition fluid systems including hydraulic skeletons in many animals increase hydrostatic pressure above ambient for instance due to hearts pumping or in microenvironments including skeletal load bearing joints of vertebrates where transients occur. Measurements up to 10–20 MPa have been made from human hips (Macdonald, 1997). Hydrostatic pressure acts all around an organism and is in balance with no net force unlike a differential pressure. There has been considerable interest and uncertainty about how animals, especially those with no gas compartment, respond to hydrostatic pressure changes. A particular question is how do animals sense very small changes in pressure around 1 kPa corresponding to depths of 10 cm of water or less. The intention here is to present a selection of evidence to show that animals do sense hydrostatic pressure and make use of the

Institute of Biological and Environmental Sciences, Zoology Department, College of Life Sciences and Medicine, Aberdeen University, Tillydrone Avenue, Aberdeen, UK, AB24 2TZ. E-mail: p.fraser@abdn.ac.uk information as a proxy for depth in their normal lives. There is also now some limited information from known hydrostatic pressure sensors which are in fact also angular acceleration receptors in vestibular systems of crabs and sharks.

## DEPTH AND ITS USE BY ANIMALS

The marine environment has a considerable vertical component with average depths roughly 4000 m and maximum depths down to over 10000 m in the ocean trenches. Shallow seas and inshore areas have much smaller depth ranges down to several hundred meters. Given this vertical spatial dimension, it is not surprising that evidence has accumulated for the existence of an ability to sense depth in the animals which inhabit the sea . Certainly animals make use of this dimension. Diving animals have long been observed descending hundreds and even thousands of meters. Furthermore many animals are known to feed at these depths implying that a variety of animals are using this sort of range in their normal lives. Water depth over any point on the sea floor is further complicated by tidal changes which vary in a complicated way close to coastlines. As many as 390 harmonic constituents may account for tidal features at a particular spot with four of these components explaining 70% of the tidal period. These four are shown in Table 1. and illustrated in Figure 4. Although in open ocean, tidal amplitude is often small (around 0.5 m), near the shore or when constrained by features of the coastline, amplitudes can reach 10 m with around 5 m common. Clearly intertidal and shallow water species must organize their lives to deal with these changes which vary in a complex way with time and with locality on a scale relevant to individuals (see Thurman 2004 for a useful summary).

Tides are not the only factors influencing depth. Prevailing weather conditions can influence depth (by up to an estimated 6%) due to atmospheric pressure changes and wind, so in simple terms low pressure weather systems and onshore winds will increase surface height and hence depth, and vice versa for high pressure and offshore winds (Brown et al., 1991). Storm surges of greater than 30kPa have been recorded. Depth is also directly influenced by water flow as can be easily modelled by stirring liquid in a teacup, so tidal currents further lead to small changes in depth. For animals

**Table 1:** Tidal components and their period.

| Tidal Component | Period in hours |
|---|---|
| Principal Lunar | 12.42 |
| Principal Solar | 12.0 |
| Luni-solar Diurnal | 23.93 |
| Principle Lunar Diurnal | 25.82 |

near the bottom of the sea, these factors influence depth in particular often systematic and regular ways, but for all animals, their depth also varies by virtue of their own vertical motion. With accurate GPS based sea surface height measurements, systematic information on depth between surface and seabed is now available for many areas.

## DEPTH USAGE

Diving animals have long been observed descending hundreds and even thousands of meters. Furthermore many animals are known to feed at these depths implying that animals are routinely using this sort of range. Recently data storage tags have confirmed a Cuviers beaked whale, *Ziphius cavirostris*, down to below 1800 m depth for 85 min (Tyack et al., 2006). From the surface to around this depth range, the pressure spans around 20 MPa.

During a recent survey, a chance video observation from the ROV Isis, of krill which have long been considered important as surface layer animals, showed them ramming into the bottom to stir up the sediment and feeding on the detritus (Clark and Tyler, 2008). Video footage showed feeding krill near the bottom over a large range of depths down to 3500 m (35 MPa hydrostatic pressure), far from their previously considered lower limit of around 500 m. Clark and Tyler estimate that at a swimming speed of 20 cm s$^{-1}$ ascent to the surface from 3000 m would take around 4 h. Although there are issues with rapid changes of depth, many animals can survive 200 bar (20 MPa) corresponding to 2000 m depth, with extreme survival of 40 MPa. Schlieper (1972) looked at pressure tolerance of four different Baltic species and found considerable variation between 20 and 40 MPa in LD50s. Given the comfortable use of pressures up to 35 MPa by krill, including mature females in their natural environment, as seen by Clark and Tyler (2008), we may need to reconsider some of the data obtained in the laboratory from step changes of pressure.

Hydrostatic pressure, light levels and temperature all alter with depth. In the case of hydrostatic pressure, the relationship is very simple and largely linear apart from the fact that in deep water, high pressure induced increases in the density occur. These will slightly distort the relationship between pressure and depth at high hydrostatic pressures. Light levels vary in a complex way, with intensity and spectral composition greatly influenced by the amount of particulate matter and organic material in the water. Temperature relationships with depth tend to be complex with relatively stable thermoclines. Hydrostatic pressure is hence easily the best proxy for depth available.

In a study of different combinations of illumination, temperature and pressure on the vertical distribution of physostomous fish *Leuciscus leuciscus* and physoclistous fish *Perca fluviatilis*, all three factors influenced the distribution significantly with hydrostatic pressure considered the dominant and strongest factor (Pavlov et al., 2000).

Early experiments in the 1950s and 1960s soon established that a variety of planktonic animals showed behaviour best explained as part of depth regulation (Hardy and Bainbridge, 1951). Zooplankton subjected to an increase in hydrostatic pressure equivalent to depth changes of a few meters swam upwards so that they occupied the upper part of a pressurized chamber (Fig. 1). This behaviour was noted in most animals where it was sought over a wide taxonomic range including larval fish without swim bladders (Rice, 1964; Knight-Jones and Morgan, 1966 and Naylor, 1972). The regulatory mechanisms used by animals are likely to involve several other senses as well as pressure. In the simplest cases, animals such as larval squid (*Loligo forbesi*), ctenophores (*Pleurobrachius pileus*), crab megalopae (*Carcinus maenas*) and the coelenterate (*Aurelia aurita*) showed upward vertical displacement following increased pressure and *vice versa* for decreased pressure Rice, 1964).

The direction of light could influence pressure induced responses. With vertical light either upwelling or down-welling, an increase in pressure caused upward swimming i.e. opposing gravity. However, with horizontal light, increased pressure caused orientation towards the light and decreased pressure swimming away from the light. In other animals, light was important, but the responses were unaltered by the direction of the light. In all cases, an increase in pressure caused or enhanced movement directly towards a light source. In normal coastal shallow water, most light travels vertically down, so these responses are easily seen to be depth regulatory.

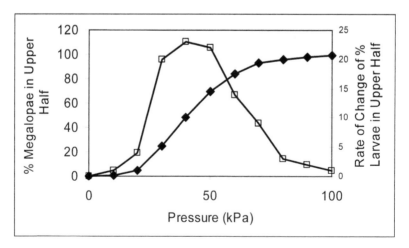

**Figure 1.** Influence of hydrostatic pressure on the distribution (■%) and rate of change of distribution (□%) of *Callinectes* megalopae larvae. Redrawn from Naylor, 2006.

Pressure decrease led to one of three different categories of response. Some animals ceased movement, and hence sank passively with no directed response relative to the light direction for example the hydromedusae, *Sarisa exima*, many crustacean zoea and the amphipod *Hyperia gaiba*. In one copepod, *Temora longiconrnis* and barnacle nauplii *Balanus balanoides*, a pressure decrease gave a photonegative response i.e. swimming away from the light direction. Finally in species such as *Calanus helgolandicus*, *Acartia clausi* and *Anomolocera patersoni*, a pressure decrease led to photonegative behaviour or decreased activity or a mixture of both behaviours.

These experiments indicated depth regulation with thresholds found to be in the order of 5–10 mBar ie 0.5–1 kPa corresponding to 5–10 cm depth. However Rice (1964) pointed out that the rates of change of pressure used in these experiments (up to 40 kPa per second) would be unlikely in the natural environment. Rates of change would be slower for example during vertical migration with gradual changes in pressure evoking compensatory swimming or passive sinking.

Some species tested failed to respond although it is uncertain whether this simply reflected an inactive period in the animal's natural activity cycle or the artificial setup in the laboratory. Enright (1962) found that short steps of pressure invoked transient randomly oriented increases in activity in the amphipod *Synchelidium* and Morgan (1969) found similar transient behaviour in *Nephtys* to an increase in pressure but found swimming inhibited by a decrease in pressure.

Ebb tide related pressure changes have been found to be particularly effective, increasing swimming activity in the amphipod *Caprella acanthifera* and the pycnogonid *Nyphon gracile*. This behaviour would help tidal transport seawards during the ebb and prevent stranding (Knight-Jones and Morgan, 1966). Tidal periodicities in locomotor behaviour have been linked to the ability of a variety of species to migrate up and down estuaries.

The ability of animals to move vertically to exploit on and off shore currents can account for horizontal transport. Warman et al. (1991) suggested that the activity patterns in the isopod *Euydice pulchra* would be sufficient for horizontal transport. Hill (1991) used mathematical models to show that net horizontal transport could be induced in any organism which migrates vertically. Forward et al. (1989) implied that rate of change of pressure rather than absolute pressure was responsible for the regulatory behaviour in many animals, although for settling blue crab larvae he found hydrostatic pressure less important than salinity (Forward et al., 1995). Larvae of the crab *Rhithropanopeus harrissii* show a graded series of increases in swimming speed to pressure increases of 0.3, 2.5 and 10.0 kPa respectively (Fig. 2).

Benthic animals commonly show endogenous free running tidal rhythms when taken from the wild into constant conditions and rhythms

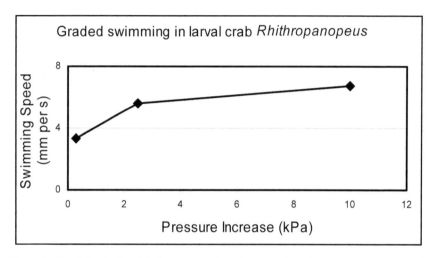

**Figure 2.** Graded relationship between swimming speed and pressure increase with increased sensitivity at low pressure increases necessary as part of a depth regulatory mechanism.

can be exogenously entrained by hydrostatic pressure as well as temperature, salinity and wave action (Warman and Naylor, 1995). Although there is plenty of evidence of entrainment, including using tide machines to entrain juvenile plaice in the laboratory (Gibson, 1973, 1982 and 1984), it is beyond the scope of this chapter to review this material other than pointing out that as seen in Table 1, the long periods involved mean that rates of change of pressure are extremely low and whatever is sensing pressure it is clearly cannot adapt at rates which would be normal for most sensory systems.

Recent work has shown the importance of depth regulation to zooplankton and solved a mystery regarding the identity of the fixed reference point which is necessary to allow plankton to assess the water currents they are carried in to optimize their feeding (Genin et al., 2005). Using a high resolution sonar system they measured the velocities in horizontal and vertical planes of planktonic organisms and current borne particles. Very different behaviour to horizontal and vertical currents was apparent. In the horizontal currents, planktonic organisms lived up to their name, wandering totally with the horizontal current. Hence graphs of plankton horizontal velocity against horizontal current show almost perfect following with a regression with a positive 45° slope (Fig. 3). In the vertical domain however the relationship of plankton velocity potted against vertical water current velocity shows perfect compensation with a negative 45° slope (Fig. 3a). Vertical currents are greatest as water shallows near the shore or at submerged features on the seabed which cause upwelling currents. Where

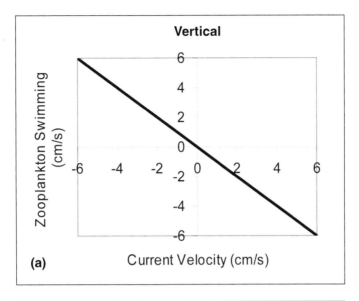

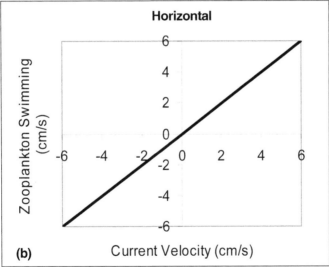

**Figure 3.** Zooplankton swimming velocity against current velocity **(a)** vertical components; **(b)** horizontal components. Redrawn from Genin et al. (2005).

these occur, animals tend to maintain depth. They use depth as a reference to allow them to swim optimally against the current at up to 10 body lengths per second hence maximizing their search for food. Genin et al. (2005) have modelled the effect of this behaviour given the known distribution of vertical currents in inshore bays and found that this behaviour is sufficient to account for the known spatial distribution of plankton in the bay. The aggregations of plankton are a very important process throughout the food chain and are now seen to be totally dependent on depth regulation. Many animals are known to regulate depth for considerable periods and very many have to synchronize their activity with complicated tidal cycles (Fig. 4) in a variety of ways.

Data storage tags are now showing depth usage in a variety of fish species ranging from plaice to tuna, Great white sharks, cod and juvenile Lemon sharks. From the depth records it is easy to work out when the plaice are on the bottom since the tidal cycles are clear. The plaice can also be seen to rise up from the bottom and swim in tidal streams until the next turn of the tide. By modifying depth the fish can use tidal transport for its migration to spawning grounds. Moreover the temperature and depth profiles can be used to identify the absolute position of the fish, so that their migration paths and use of tidal transport can be ascertained (Metcalfe et al., 2006). Many of the larger fish species are clearly bimodal in depth range, heating up near the surface and gaining enhanced sensory function over prey species in cold water at depths down to several hundred meters. Hence great white sharks during migration journeys spend time either in the top 5 m or at

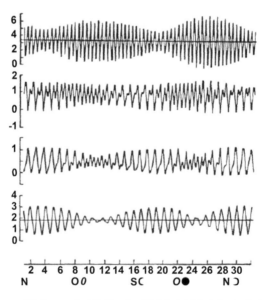

**Figure 4.** Tides over a 32 day period Left axis, tidal height (m).A single intertidal species may be spread over several tidal zones (modified from Thurman, 2004)
From Top to bottom. Semidiurnal Immingham, UK (North Sea).
Mixed, mainly semidiurnal Golden Gate, San Francisco Bay, USA. (W. Pacific).
Mixed, mainly diurnal Manila Bay, Philippines.
Diurnal Do-So'n, Vietnam (Gulf of Tonkin).

500m (Boustany et al., 2002). Tuna entering the Gulf of Mexico make deep dives to cold water to counteract the temperature rise when entering the gulf (Teo et al., 2007).

While observations of depth indicate regulation, more direct evidence of pressure sensitivity of fish has been determined by observing spontaneous locomotor behaviour after imposed pressure changes and by various types of conditioning experiments (see reviews by Blaxter, 1978 and 1980). Swimbladders in fish have a key role in depth usage. They alter buoyancy and are the most energy efficient way of maintaining buoyancy. The volume of gas is greatly influenced by the pressure as described by Boyle's law, so they impose the interesting property of allowing the fish to be at neutral buoyancy at one particular depth with deviations in either direction leading to passive upwards or downwards displacement. The presence of a swimbladder in fish correlates with increased hydrostatic pressure sensitivity (Fig. 5). Whereas animals with no gas compartments typically show thresholds in the range 0.5–2.0 kPa i.e. 5–20 cm of water pressure, fish with intact swimbladders have thresholds around 10 times lower in the range 0.04–0.5 kPa.

Blaxter 1980 argues that the swimbladder wall is the likely site for pressure operated stretch receptors, and it is known from his earlier work that deflation of the swimbladder in saithe (*Pollachius virens*) causes a loss of pressure sensitivity, although the restoration of sensitivity is greater than expected from a simple refilling model. The pro-otic bulla is a gas filled bony structure in clupeids and mormyrids. It is self contained, but has links to the swimbladder which allow the swimbladder to act as a gas reservoir following pressure changes in adult fish. Blaxter considers that the Bulla could act as a pressure sensor although it would adapt. In the case of the larval fish where the bulla is functional before the swimbladder, the bulla cannot adapt and hence could act as an absolute pressure sensor. Blaxter

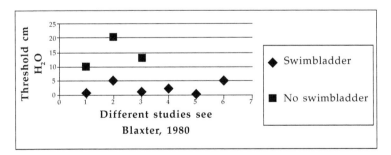

**Figure 5.** Thresholds for hydrostatic pressure sensitivity taken from Table 8 (Blaxter, 1980).

(1980) also points out that the sensitivity of fish to pressure change is high enough to allow them to sense waves passing overhead or meteorological pressure changes (20–30 mbar). The weather loach *Misgurnus fossilis* is said to indicate the weather by changes in its behaviour (Blaxter,1980).

## PRESSURE SENSORS

While stretch receptors in the swimbladder wall were obvious candidates as pressure receptors, it was clear that since many animals showed considerable pressure sensitivity in the absence of any gas compartment, or in the case of larval fish, pressure sensitivity preceded the filling of the swimbladder with gas, there had to be another sensor with the required sensitivity and lack of adaptation to be able to sense tidal cycles or act in depth regulation. Digby (1972) in particular proposed some ingenious ideas and experiments attempting to manipulate transient gas layers in order to make sense of what seemed a contradiction. Since compartments without gas were considered virtually incompressible, links to mechanoreceptors were ruled out. (see review in Macdonald and Fraser, 1999). Digby's results were inconclusive regarding a possible sensor and his ideas never got as far as identifying links into the nervous system. Nevertheless his experiments did support the existence of transient gas production at crustacean cuticle, and this may yet be seen to have a role in pressure transduction.

There had been early evidence for an involvement of the balancing organ in depth reception since animals with statocysts ablated, no longer showed depth compensatory responses. Accordingly recordings were made from receptors in a well characterized balancing system, the statocyst of crabs. These thread hair receptors in the statocyst of a crab, *Carcinus maenas* which were known as angular acceleration receptors also responded to steps and cycles of hydrostatic pressure (Fraser and Macdonald, 1994; Fraser et al., 1995), and a simple dilatometer based model showed how the limitation of low compressibilities for non gaseous materials could be overcome and allow links to mechanoreceptors.

Following this, using a comparative approach, Fraser and Shelmerdine found in the isolated vestibular system of the dogfish, *Scyliorhinus canicula* that hair cell afferents which respond to angular acceleration also responded to steps and cycles of hydrostatic pressure (Fig. 9). The dilatometer based model could not be applied to the vertebrate hair cells, so the mechanism involved remains unresolved (Fraser and Shelmerdine 2002; Fraser et al., 2003).

Since these findings, recordings have been made from statocyst neurones in a variety of crustacea caught in Bermuda, including deep scattering layer organisms *Systelasis debilis, Sergestes atlanticus* and *Sergia splendens* and a number of different crab species including the mud crab *Panopeus herbstii*, the swimming crabs *Callinectes ornatus, Necora puber, Macropipus depurator* and the edible crab *Cancer pagurus* as well as from littoral species such as the prawn, *Palaemon northopi*, and the spiny lobster *Panulirus argus* (Fraser et al., 2001; Fraser et al., 2008).

## PISTON MECHANISM

It is clear that the use of balancing system afferents as hydrostatic pressure receptors is a widespread phenomenon, but there are many unresolved questions regarding the mechanism and the mechaonoreceptor systems involved. Let us consider the receptor in the crab in more detail.

Thread hairs are found in a line inside the complicated statocyst in the crab and are bent by fluid displacement in horizontal or vertical canals depending on the axis of rotation of the crab. They hence normally act as angular acceleration receptors. Resting nerve activity recorded from thread hairs was shown to be altered by steps and cycles of hydrostatic pressure (Fraser and Macdonald 1994; Fraser et al., 1995)

The mechanism proposed involves differential compression of cuticular and tissue components of the hair and further assumes that the attachment of the linkage from the thread hairs to the sensory neurones could act like the piston of a syringe with the cuticular shaft acting as the barrel. An increase in pressure could hence lead to a decreased volume by moving the cuticular rod or chorda which in turn would push or pull on the sensory neurones (Fig. 6). It was calculated using the compressibility of seawater

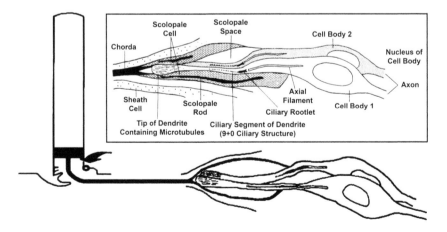

**Figure 6.** Diagrammatic view of thread hair showing fulcrum and the chorda linking the hair to the two dendrite tips, and piston like attachment of the chorda in the hair. See Macdonald and Fraser (1999) for detail of chordotonal organ (inset).

$(44 \times 10^{-6}$ per 100 kPa) and a simplifying assumption that the cuticle was non compressible or much less compressible than the fluid and tissue inside it, that a 100 kPa or 1 Bar pressure change would displace the chorda by around 17.6 nm. This was the same order of magnitude as that generated by a 1° bending of the 400 µm long by 2 µm diameter hair. Since it was non-controversial that bending the hair was the normal means of activation of the mechanoreceptors and a 1° bend was well above threshold, it was reasonable to accept that differential compression was sufficient to account for the responses. A variety of properties for this receptor are now known.

Resting activity in mixed units tracks slow cycles of hydrostatic pressure well (Fig. 7). Positive going and negative going relationships with pressure are indicated from the shape of the response shown in Fig. 8. Given that each thread hair is innervated by two neurones (Fig. 6), and these have different directional responses to bending the hair, we can explain the bimodal response to pressure and this is strong supporting evidence for the piston model.

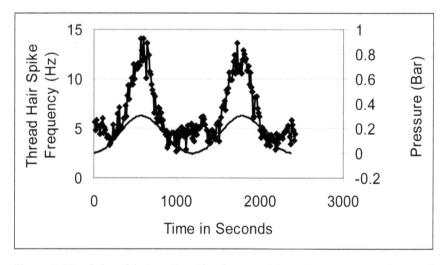

**Figure 7.** Mixed thread hair units with a large positive going response and a small negative going response to 20 minute period and 0.3 Bar (30 kPa) peak amplitude cycles of hydrostatic pressure.

## VERTEBRATE HYDROSTATIC PRESSURE SENSING

In the case of the shark, Fraser and Shelmerdine (2002) used an isolated vestibular system which was the original model preparation used by Lowenstein and co-workers to elucidate properties of semicircular canal angular acceleration receptors (Lowenstein and Sand, 1940; Lowenstein and Compton, 1978). They oscillated this in the plane of the horizontal

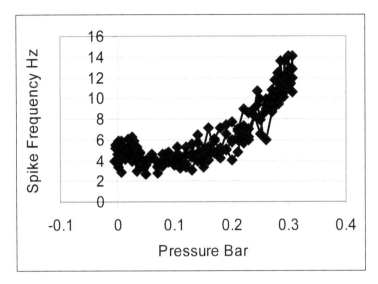

**Figure 8.** Relationship between thread hair spike frequency and pressure indicates a positive going and a negative going relationship.

canal for 64 cycles of oscillation at 1 Hz every 2 min while subjecting it to small steps and cycles of hydrostatic pressure. Resting and oscillation derived activity are both modulated by hydrostatic pressure as seen in Fig. 9. Fraser et al. (2008) estimated threshold responses to be around 4.3 mbar or 0.43 kPa. During each bout of oscillation under constant pressure there is adaptation (Fig. 10) but during imposed sinusoidal cycles with amplitude around 30 kPa, the normal adaptation is not present and the receptor firing frequency follows the sinusoidally modulated pressure well on ascending and descending phases (Fig. 11). There is little anatomically in common between the cuticular thread hair system in the crab and the vertebrate vestibular hair cells embedded in a gelatinous cupula, so the piston model cannot be applied.

Although there are ideas that hair cells in the macula neglecta of elasmobranchs might transduce hydrostatic pressure directly, these are highly speculative at present with hypothesized gas inclusions (Bell, 2007). Nevertheless, many of the ideas need to be followed up. What we do know from intracellular recordings together with canal indentation as a substitute for angular rotation is that adaptation is slow. Highstein et al. (2005) state that adaptation times for the hair cells fitted a log normal distribution with a mean of 112 sec. Afferents adapted much faster compared to hair cells with at least 2 orders of magnitude difference. Semicircular canal afferents are known to be extremely sensitive to transcupular endolymphatic pressure and dilational labyrinthine pressure with sensitivities of around 0.001 Pa and 0.05Pa (by 1 imp/s for 2Hz pressure stimuli (Kondrachuk and Boyle, 2009).

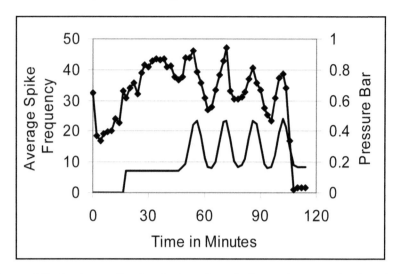

**Figure 9.** Mixed unit recording from horizontal canal afferents from an isolated vestibular system of the dogfish *Scyliorhinus canicula* responding to steps and cycles of hydrostatic pressure. Redrawn from Fraser et al. (2008).

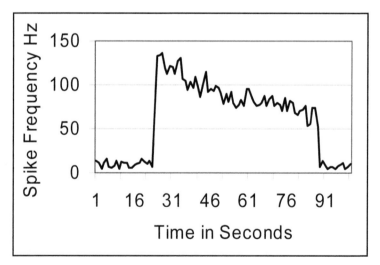

**Figure 10.** A single bout of 64 oscillations at constant pressure showing typical adaptation of the response with time.

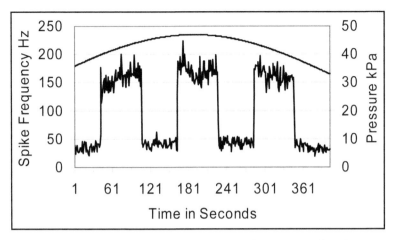

**Figure 11.** Three similar bouts of oscillation during the top part of a 15 min period sinusoidal pressure cycle.

## CONCLUSION

Although the piston mechanism seems to be a valid part of the crab hydrostatic pressure sensing system, using differential compression to produce nanometer level displacements, the transduction mechanism at the mechanoreceptor level is poorly understood and this can be said about all mechanoreceptor models although we have an intriguing list of components (Kung, 2005). We know from work on a chordotonal organ in *Drosophila* that the only two TRPV channels encoded in the genome NAN and IAV seem to form heteromeric channels that transduce vibrations into receptor potentials (Kim et al., 2003; Gong et al., 2004). They are found on the ciliary outer segments of the bipolar neurones but not elsewhere. It is not yet clear how they are involved in long time course displacements. A stiffening protein SPAM wraps round the chordotonal organ of *Drosophila* like hoops on a barrel (Cook et al., 2008), so the mechanics of the compartment may be more relevant to understanding pressure mechanisms than so far indicated by the described anatomy at the electron microscope level. Mechanoreception is just beginning to get the attention it deserves, and the mechanisms involved in hydrostatic pressure reception may well prove to be fundamental to what is taking place in a wide range of cells and tissues (see Macdonald and Fraser, 1999).

## REFERENCES

Bell, A. 2007. Detection without deflection? A hypothesis for direct sensing of sound pressure by hair cells. J. Biosci. 32: 385–404.

Blaxter, J.H.S. Baroreception. Pp. 375–409. *In*: M.A. Ali (ed.). 1978. Sensory Ecology. Plenum Press, New York, USA.

Blaxter, J.H.S. The effect of hydrostatic pressure on fishes. Pp 369–86. *In*: M.A. Ali (ed.). 1980. Environmental Physiology of Fishes Plenum Press, New York, USA.

Boustany, A.M. and S.F. Davis, P. Pyle, S.D. Anderson, B.J. Le Boeuf and B.A. Block. 2002. Satellite tagging: Expanded niche for white sharks. Nature 415: 35–36.

Brown, J. and A. Colling, D. Park, J. Phillips, D. Rothery and J. Wright. 1991. Ocean Circulation G. Bearman (ed.). Pergamon Press and the Open University.

Clarke, A and P. Tyler, 2008. Adult Antarctic Krill Feeding at Abyssal Depths. Current Biology 18: 282–285.

Cook, B. and R.W. Hardy, W.B. McConnaughey and C.S. Zuker. 2008. Preserving cell shape under environmental stress. Nature 452: 361–364.

Digby, P.S.B. 1972. Detection of small changes in hydrostatic pressure by Crustacea and its relation to electrode action in the cuticle. Symp. Soc. Exp. Biol. 26: 445–472.

Enright, J.T. 1962. Responses of an amphipod to pressure changes. Comp. Biochem. Physiol. 7: 131–45.

Forward, R.B. and C.A. Wellins. 1989. Behavioural responses of a larval crustacean to hydrostatic pressure: *Rhithropanopeus harrisii* (Brachyura: Xanthida). Marine Biology 101: 159–172.

Forward, R.B. and R.A. Tankersley, M.C. de Vries and D. Rittschof. 1995. Sensory physiology and behaviour of the blue crab (*Callinectes sapidus*) postlarvae during horizontal transport. Marine and Freshwater Behaviour and Physiology 26: 233–248.

Fraser, P.J. and A.G. Macdonald. 1994. Crab hydrostatic pressure sensors. Nature 371: 383–384.

Fraser, P.J. and R.L. Shelmerdine. 2002. Fish Physiology: Dogfish hair cells sense hydrostatic pressure. Nature 415: 495–496.

Fraser P.J. and A.G. Macdonald, and R.N. Gibson. Low pressure hydrostatic pressure receptors in the crab *Carcinus maenas* (L.). *In*: J.C. Rostain, A.G. Macdonaldand, R.E. Marquis. (eds.). 1995. Basic and Applied High Pressure Biology IV, France: Medsubhyp Int. 5: 59–68.

Fraser, P.J. and A.G. Macdonald and D. Steinberg. 2001. Hydrostatic pressure drives neural responses in shallow and deep crustacea. Proceedings of the Royal Institute of Navigation RIN 01: 17.1–7.

Fraser P.J. and S.F. Cruickshank and R.L. Shelmerdine. 2003. Hydrostatic pressure effects on vestibular hair cell afferents in fish and crustacea. Journal of Vestibular Research 13: 235–242.

Fraser, P.J. and S.F. Cruickshank, R.L. Shelmerdine and L.E. Smith. 2008. Hydrostatic pressure receptors and depth usage in crustacea and fish. Navigation: Journal of the Institute of Navigation Special Issue on Bio-Navigation 55: 159–161.

Genin, A. and J.S. Jaffe, R. Reef, C. Richter and P.J.S. Franks. 2005. Swimming Against the Flow: A Mechanism of Zooplankton Aggregation. Science 308: 860–862.

Gibson, R.N. 1973. Tidal and circadian activity rhythms in juvenile plaice *Pleuronectes platessa*. Mar. Biol. 22: 379–386.

Gibson, R.N. 1982. The effect of hydrostatic pressure cycles on the activity of young plaice *Pleuronectes platessa*. J. Mar. Biol. Assoc. UK 62: 621–635.

Gibson, R.N. 1984. Hydrostatic pressure and the rhythmic behaviour of intertidal marine fishes. Trans. Am. Fish Soc. 113: 479–483.

Gong, Z. and W. Son, Y.D. Chung, J. Kim, D.W. Shin and C.A. McClung 2004. Two interdependent TRPV channel subunits, Inactive and Nanchung mediate hearing in *Drosophila*. J. Neurosci. 24: 9059–9066.

Hardy, C. and R. Bainbridge. 1951. Effect of pressure on the behaviour of decapod larvae. Nature 168: 327–328.

Highstein, S.M. and R.D. Rabbitt, G.R. Holstein and R.D. Boyle. 2005. Determinants of spatial and temporal coding by semicircular canal afferents. J. Neurophysiol. 93: 2359–2370.

Hill, A.E. 1991. A mechanism for horizontal zooplankton transport by vertical migration in tidal currents. Marine Biology 111: 485–492.

Hills, G. 1972. The physics and chemistry of high pressures. Symp. Soc. Exp. Biol. 26: 1–26.

Kim, J. and Y.D. Chung, D.Y. Park, Y. Choi, S. Shin, D.W. Soh, H. Lee, H.W. Son, W. Yim, J. Park, C.S. Kernan and M.J. Kim C. 2003. A TRPV family ion channel required for hearing in *Drosophila*. Nature 424: 81–84.

Knight-Jones, E.W. and E. Morgan. 1966. Responses of marine animals to changes in hydrostatic pressure. Oceanogr. Mar. Biol. Ann. Rev. 4: 267–299.

Kondrachuk, A. and R. D. Boyle. 2009. Intralabyrinthine pressure and function of cupula-endolymphatic system. J. Gravitational Research. In Press.

Kung, C. 2005. A possible unifying principle for mechanosensation nature 436: 647–654.

Lowenstein, O. and A. Sand. 1940. The individual and integrated activity of the semicircular canals of the elasmobranch labyrinth. J. Physiol Lond. 99: 89–101.

Lowenstein, O. and G.J. Compton. 1978. A comparative study of the responses of isolated first-order semicircular canal afferents to angular and linear acceleration, analysed in the time and frequency domains. Proc. R. Soc. Lond. B202: 313–338.

Macdonald, A.G. 1997. Hydrostatic pressure as an environmental factor in life processes. Comp. Biochem. Physiol. 116A: 291–297.

Macdonald, A.G. and P.J. Fraser. 1999. The transduction of very small hydrostatic pressures. Comp. Biochem. Physiol. A122: 13–37.

Metcalfe, J.D. and E. Hunter and and A.A. Buckley. 2006. The migratory behaviour of North Sea plaice: Currents, clocks and clues. Marine and Freshwater Behaviour and Physiology 39(1): 25–36.

Morgan, E. 1969. The responses of *Nephtys* (Polychaeta: Annelida) to changes in hydrostatic pressure J. Exp. Biol. 50: 510–513.

Naylor, E. 2006. Orientation and navigation in coastal and estuarine zooplankton. Marine and Freshwater Behaviour and Physiology 39: 13–24.

Naylor, E. and R.J. Atkinson. 1972. Pressure and the rhythmic behaviour of inshore marine animals. Symp. Soc. Exp. Biol. 26: 395–415.

Pavlov, D.S. and R.V. Sadovskii, V.V. Kostin and A.L. Lupandin. 2000. Experimental study of young fish distribution and behaviour under combined influence of baro-, photo- and thermogradients. J. Fish Biol. 57: 69–81.

Rice, A.L. 1964. Observations on the effects of changes in hydrostatic pressure on the behaviour of some marine animals. J. Mar. Biol. Ass. UK 44: 163–175.

Schlieper, C. 1972. Comparative investigations on the pressure tolerance of marine invertebrates and fish. Symp Soc. Exp. Biol. 26: 197–207.

Teo, S.L.H. and A. Boustany, H. Dewar, M.J.W. Stokesbury, K.C. Weng, S. Beemer, A.C. Seitz, C.J. Farwell, E.D. Prince and B.A. Block. 2007. Annual migrations, diving behavior, and thermal biology of Atlantic bluefin tuna, *Thunnus thynnus*, on their Gulf of Mexico breeding grounds. Mar. Biol. 151: 1–18.

Thurman, C.L. 2004. Unravelling the ecological significance of endogenous rhythms in intertidal crabs. Biological Rhythm Research 35: 43–67.

Tyack, P.L and M. Johnson, N.A Soto, A. Sturlese and P.T. Madsen. 2006. Extreme diving of beaked whales. J. Exp. Biol. 209: 4238–4253.

Warman, C.G. and E. Naylor. 1995. Evidence for multiple cue-specific circatidal clocks in the shore crab *Carcinus maenas* Journal of Experimental marine Biology and Ecology 189: 93–101.

Warman, C.J. and T.J. O'Hare and E. Naylor. 1991. Vertical swimming in wave-induced currents as a control mechanism of intertidal migration by a sand-beach isopod. Marine Biology 111: 49–54.

# Pressure Effects on Mammalian Central Nervous System

*Yoram Grossman, Ben Aviner* and *Amir Mor*

## INTRODUCTION

Pressure represents a fundamental thermodynamic state variable and, as such, changes induced by pressure reflect alterations in reaction rates or equilibria. High pressure (HP) is expected to universally affect molecules such as membrane phospholipids and proteins such as enzymes, transporters, ionic channels, cytoskeleton, etc. (for review see Daniels and Grossman, 2003, and dedicated chapters in this book). Therefore, molecular 'resistance' to pressure is one of the most important factors determining animals' adaptation to deep-sea dwelling as well as to deep diving. Recent studies, however, suggest that pressure effects are not uniform and may be selective. Pressure is expected to slow reaction rates when increased volume is associated with the activated state of the molecule or the product of the reaction. It is expected to facilitate the process if the opposite is occurring. Thus, the consequent physiological effects are complex and dependent on many 'pressure sensitive' molecules in various cells. One of the most striking effects of pressure is on the nervous system. Exposure to HP is associated with the development of the high pressure neurological syndrome (HPNS) in humans and animals. HPNS signs and symptoms include malfunction of the autonomous nervous system, reduction of cognitive functions, decreased motor coordination, sleep disorders, and EEG changes. At greater pressures (deeper diving), serious signs such as tremors, convulsions and seizures leading to death may occur (see Chapter 19 in this book).

Therefore, in the present chapter, we shall review mainly HP effects in the physiological range of 0.1–10.1 MPa (1–100 ATA) on mammalian central

Department of physiology, Faculty of Health Sciences, Zlotowski Center for Neuroscience, Ben-Gurion University of the Negev, Beer-Sheva 84105, Israel.
E-mail: ramig@bgu.ac.il

nervous system (CNS). We shall attempt to reveal the associated cellular and circuit mechanisms in isolated CNS spinal cord and medulla, brain slices from various regions of the brain, and CNS synaptosomes. In some cases, we will refer to key data obtained from invertebrate preparations by using experimental methods that can not be implemented in CNS experiments at pressure.

## NEURONAL MEMBRANE PROPERTIES

The effects of pressure on individual ion channels (see Chapter 6 in this book) can account for many of the pressure effects on neuronal activity described in this section. Several earlier review articles deal with pressure effects on passive and active properties of excitable cells: Wann and Macdonald (1980); Macdonald (1982); Halsey (1982); Kendig et al. (1993); Hogan and Besch (1993); and Daniels and Grossman (2003).

In order to measure the electrical properties of neurons such as membrane resting potential ($V_{RP}$), membrane resistance ($R_m$) that can be calculated from the experimentally measured input resistance ($R_{in}$), and action potential (AP), intracellular recordings must be made. Due to the technical difficulties involved, these data are limited in invertebrate preparations and rare in mammalian CNS. Early reports indicated no change of $V_{RP}$ and small increase in $R_{in}$ in squid giant axon (Spyropoulos, 1957a), while lobster axon $V_{RP}$ was depolarized with some increase of $R_{in}$ (Grossman and Kendig, 1984), probably due to the pressure depression of electrogenic $Na^+–K^+$ ATPase activity. In molluscan neurons depolarization was also observed, but accompanied by a decrease in $R_{in}$ (Wann et al., 1979).

Only few studies have reported pressure effects on vertebrate neurons recorded intracellularly. There were no significant differences between mean values obtained for $V_{RP}$ and $R_{in}$ for two populations of rat CA1 hippocampal pyramidal neurons at atmospheric pressure and pressures up to 10 MPa (Southan and Wann, 1996). In contrast, Dean and Mulkey (2000) have reported that 38% of neurons sampled in the rat medulla solitary complex reversibly responded upon helium compression to as low as 0.3 MPa with 1–3 mV depolarization and 16% increase of $R_{in}$. This is probably an exceptionally pressure sensitive cell population, since the rest of the cells were not affected.

It is probable that pressure effect on mammalian CNS neurons electrical properties will differ from one cell to another cell due to differences in membrane channel populations, and the extent to which $V_{RP}$ depends on the activity of $Na^+–K^+$ electrogenic pump. A small depolarization may increase excitability by bringing a cell closer to the threshold for impulse initiation, but may partially inactivate sodium channels with its consequent effects on AP.

## ACTION POTENTIALS

### Single response

Alteration of channel kinetics by pressure will produce changes in a single AP shape, amplitude, duration, and conduction velocity. Remarkably, early investigation of isolated myelinated frog nerve (Grundfest and Cattell, 1935; Grundfest, 1936; Spyropoulos, 1957a, b) at hydrostatic pressure revealed slower conduction velocity, an increase in duration of the extracellularly recorded electrical signal, and a lengthening of the refractory period. Intracellular recordings (which are not feasible in CNS axons) in squid giant axon (Shrivastav et al., 1981), lobster axons (Grossman and Kendig, 1984, 1986) and molluscan cell bodies (Wann et al., 1979) revealed that various axon populations might respond differently to pressure. Yet, in most preparations, pressure (at steady state) slowed the kinetics, slightly reduced the amplitude and decreased the conduction velocity of the AP. These alterations mirror accompanying changes in peak sodium current amplitude. These studies also likened the effect of pressure to reduced temperature. The aforementioned findings were corroborated by later studies in several lobster neuromuscular synapses in which extracellular macropatch current recordings were performed from the presynaptic axon terminals (Grossman et al., 1991a; Golan et al., 1994, 1995, 1996). The small diameter terminals seem to be more sensitive to pressure since amplitude reduction in many of the synaptic sites could reach 30%.

For many years the information on neuronal activity of mammalian brain was obtained by extracellular field potentials recorded either *in vivo* from rodents' brain or *in vitro* from isolated rat brain structures or slices. This type of field potential recording reflects the activity of populations of neurons, axons, and synapses. Studies by Fagni et al. (1987a, b), Zinebi et al. (1988a, b), Talpalar and Grossman (1995, 2003, 2004), and Southan and Wann (1996) have demonstrated that orthodromic population spike that is evoked by synaptic mechanism may be reduced by various degrees. However, antidromic population spike that is induced by direct stimulation of the axons is usually only slightly changed. Etzion and Grossman (1999) succeeded for the first time in recording at pressure single Purkinge cell activity in guinea pig cerebellar brain slices. In these cells, the somatic $Na^+$ AP current was not affected by pressure, the dendritic $Ca^{2+}$ spike current amplitude was only slightly reduced, but its kinetics was slowed by 15%. Another study (Etzion and Grossman, 2000) has shown that the parallel fibers presynaptic volley, that reflects the AP (mostly $Na^+$ current) of a population of axonal terminals, is reduced by 25% at 10.1 MPa. In contrast, Mor and Grossman (2006) have reported only a tendency for a slight reduction in amplitude of the presynaptic volley of the Schaefer collaterals in the hippocampal CA1 area, although without statistical significance. Intracellular recordings from mammalian neurons revealed that neither low hyperbaric pressure (Dean and Mulkey, 2000) nor high pressure (Southan and Wann, 1996) significantly affect the kinetics of $Na^+$ AP. However, pressure did reduce the $K^+$-conductance dependent after hyperpolarization (AHP) of the spikes. This modulation should affect the firing pattern of the cells (see below).

Finally, few studies on nerve conduction in humans have been performed during simulated dives to 450m (Grapperon et al., 1988) and 360m (Todnem et al., 1989) in dry pressure chambers. Both studies reported reduced conduction velocity for fast sensory axons and increased delay at the distal part of the motor nerves.

## Patterns of action potential activity

Several patterns of AP activity have been tested under pressure conditions: branch point failure, spontaneous repetitive firing, and rhythmic (pacemaker) activity. The lobster axon used in the aforementioned studies has the interesting property of failing at its branch point during short trains of action potentials at physiological high frequencies (Grossman et al., 1979a, b). This low safety factor conduction is probably an important feature of distal dendrites and terminal brunching. Pressure at steady state impaired conduction through the branch point (Grossman and Kendig, 1986, 1987).

Mammalian CNS axons normally do not fire repetitively at pressure. However, a small portion of axons in crustacean (Kendig et al., 1978), squid (Spyropoulos, 1957a) and vertebrate peripheral nerve (Grundfest, 1936; Grundfest and Cattell, 1935) responded with spontaneous impulse generation or repetitive activity in response to single stimulus at 10–40 MPa pressures range.

Many CNS neurons exhibit variable oscillatory or pacemaker activity. Autorhythmic respiratory-drive frequency, recorded extracellularly in the cervical spinal ventral roots in isolated newborn rat brainstem-spinal cord preparation, was reduced at 10.1 MPa (Tarasiuk and Grossman, 1991). In contrast, in this preparation, pressure also induced an opposite response: increased spontaneous, stimulus-induced, or respiratory drive associated tremor-like activity in lumbar and cervical ventral roots motor neurons (Tarasiuk and Grossman, 1990). Moreover, Dean and Mulkey (2000) have shown that highly barosensitive neurons in the rat medulla increased their sustained firing rate at 0.3 MPa. In guinea pig cerebellar Purkinje cells (Etzion and Grossman, 1999) HP effects are complex. Pressure differentially altered the pattern of dendritic $Ca^{2+}$ APs activity, but did not affect the spontaneous somatic $Na^+$ spikes. Pressure prolonged the active period of spike firing by 52%, but had no effect on the quiescent period of the cycle.

Furthermore, within the active period, pressure reduced the mean firing rate by 65% and suppressed the development of typical firing of spike doublets. These changes are probably the result of pressure modulation of voltage and $[Ca^{2+}]_i$-dependent $K^+$ currents that underlie spike AHP and pacemakers activity. The changes indicate reduced excitability of the dendrites. In studies of rat hippocampal dentate gyrus axons (Talpalar and Grossman, 2003, 2004) HP did not affect the frequency response of the axons to direct stimulation. In contrast, the recent study of hippocampal CA1 pyramidal neurons (Mor and Grossman, 2007a) showed reduced capability of the neurons to produced $Na^+$ population spikes following repetitive synaptic activation under conditions of NMDAR activity alone.

In summary, pressure usually slightly increases $R_{in}$ and depolarizes neuronal membrane. Notable changes in $V_{RP}$ are observed in those neurons in which electrogenic $Na^+$-$K^+$ ATPase activity contributes considerably to the $V_{RP}$. Pressure slightly reduces the amplitude of both $Na^+$ and $Ca^{2+}$ APs in axons; nonetheless, the effect is quite prominent in axonal terminals. The most predominant pressure effect is the reduced kinetics of the AP which consequently decreases conduction velocity and frequency response. Pressure modulation of various $K^+$ channel types, that usually underlie spike AHP and pacemaker activity, probably determine the tendency to fire either spontaneous or evoked trains of APs at a particular frequency or specific pattern. These modifications, while not 'dramatic', will probably disturb normal CNS synaptic transmission (see below), impair time and event detection, and modify neuronal and circuit rhythmic activity.

## EXCITATORY SYNAPTIC ACTIVITY

It is conceivable that both HPNS motor hyperexcitability and cognitive impairments are the consequence of disturbance in network synaptic activity. We will describe only pressure effects on chemical synapses since electrical synapses have not been studied at pressure. The normal CNS activity is based on well-coordinated and well-balanced interactions between excitatory and inhibitory (see below) synaptic pathways. The most common excitatory transmitter in the CNS is glutamate. However, several different postsynaptic receptors are mediating glutamate transmission: α-amino-3-hydroxy-5-methyl-4-isoxazolepropionic acid (AMPA), kainate, quisqalate, and N-methyl-D-aspartate (NMDA) sensitive receptors. They all gate cation channel for $Na^+$, $K^+$, and $Ca^{2+}$, but NMDA receptor (NMDAR) has relatively much higher conductance of $Ca^{2+}$ (about 11% of the current is carried by $Ca^{2+}$).

In the 1980s, with the limited cellular methods for CNS studies, a large body of work was carried out on freely moving animals, testing their sensitivity to HP (threshold for tremor and seizure, HPNS) using pharmacological antagonists or sometimes agonists of the specific receptors.

Selective and non-competitive antagonists at non-NMDARs show both species and effect-specific actions. While some types have no effect against pressure in the rat (Wardley-Smith and Meldrum, 1984), others may be effective against tremor but not against seizures (Wardley-Smith et al., 1987), or effective against both symptoms (Pearce et al., 1994).

Initial experiments using the NMDAR antagonists (Meldrum et al., 1983) or competitive antagonists (Wardley-Smith and Meldrum ,1984; Wardley-Smith et al., 1987; Pearce et al., 1991, 1993) demonstrated various degrees of protection against hyperexcitability, tremors, and seizures caused by HP. Non-competitive NMDAR antagonists exhibited species and drug dependent effects. For example, ketamine is effective against all aspects of HPNS in the rat (Angel et al., 1984), while MK801 has no effect in the rat (Wardley-Smith and Wann, 1989), but gives some protection in the baboon (Pearce et al., 1990). It is conceivable that the differences between species, as well as in drug potency, may arise from different NMDAR subtypes existing in various animals (for details see Daniels and Grossman, 2003, and Chapter 6 in this book).

## Glutamatergic AMPA synapses

### Single response

All AMPA receptors (AMPAR) single field excitatory post synaptic potentials (fEPSP) tested so far in mammalian brain, as well as those at crustacean neuromuscular junction (see rev. Daniels and Grossman, 2003), are depressed and their kinetics is slowed at HP. This was demonstrated in brain slices preparations of rat hippocampal Shaefer collaterals synapse on CA1 pyramidal cells (Fagni et al., 1987a), guinea pig cerebellar parallel fibers (PF, Etzion and Grossman, 2000) and climbing fibers (CF, Etzion et al., 2008) synapses on Purkinje cells, and rat dentate gyrus medial perforant path synapse on granule cells (PP, Talpalar and Grossman, 2003, 2004). The most significant HP effects are the decrease of fEPSP amplitude, reduction of its initial slope, prolongation of decay time, and increase of synaptic delay. Although fEPSPs were depressed, the ability of the neurons to produce population spikes either remained stable (Talpalar and Grossman, 2004) or even enhanced (Fagni et al., 1987a). These indicate an increased gain of the system despite the severe 50% depression of the synaptic input to the cells. A similar effect was observed in isolated newborn rat medulla- spinal cord preparation. Synaptic responses were not studied directly in these models. However, in addition to increased excitability of the motorneurons (Angel and Halsey, 1987; Tarasiuk and Grossman, 1990), HP of 10.1 MPa slowed the kinetics but did not affect the amplitude of spinal monosynaptic reflex and even increased by 30% various polysynaptic reflexes (Tarasiuk and

Grossman, 1989; Tarasiuk et al., 1992). The reflexes' electrical responses recorded from cut ends of spinal ventral roots reflect a compound action potential (not synaptic potentials) elicited in the motorneurons pool and/ or population of medullar neurons as a result of reduced synaptic inputs. Thus, they also demonstrated increased gain of the system. Possible explanations for this unique phenomenon could be the involvement of NMDAR, reduced circuit inhibition, and possible boosting of the signal along the neuron dendrites by $Ca^{2+}$ channels activity (see below).

*Frequency modulation*

Synaptic plasticity as a function of activity is a fundamental feature of chemical synapses. HP alters the short term frequency response of the synapses. Pressure affects even the simple paired pulse modulation. HP reduces paired pulse depression (PPD) normally observed at short (< 20 ms) inter stimulus intervals (ISI), but increases paired pulse facilitation (PPF) normally observed at longer ISIs, in the dentate gyrus PP synapses (Talpalar and Grossman, 2003, 2004,). A qualitatively similar effect was observed in cerebellar PF synapses, but surprisingly, HP did not affect a general PPD in the cerebellar CF, a saturated-release type synapse (Etzion et al., 2008). In contrast to invertebrate synapses (see rev. Daniels and Grossman, 2003), frequency-dependent synaptic response for trains (5–10 stimuli) of AMPAR fEPSPs are bimodally modulated by HP. Normally, the second fEPSP is facilitated and the rest of the responses in the train are progressively depressed. HP enhanced this frequency dependent depression (FDD) at 50 Hz, while at 20 Hz the FDD was maintained. However, when the synaptic gain of the system (as above) was tested by evoking population spikes in the cells following a paired pulse protocol and short trains of stimuli, the picture changed. Despite the depression of single fEPSP, and in accordance with increased PPF, the gain of the system remained unchanged at HP for paired pulses. Population spike generation during 25 Hz train stimulation was conserved at pressure. However, at 50 Hz HP induced additional spikes per stimulus, indicating disruption of the regular low-pass filter properties of this dentate gyrus network (Talpalar and Grossman, 2004, 2005).

*Presynaptic mechanism*

HP suppression of evoked synaptic AMPA transmission is effected mainly by presynaptic mechanisms. Fagni and his colleagues (Fagni et al., 1987a, b; Zinebi et al., 1988a, 1990) examined in the hippocampal CA1 region the efficacy of various excitatory neurotransmitters that were perfused in the bath, in reducing fEPSP and population AP under normobaric and hyperbaric conditions. HP had no significant effect on the capacity of various agonists of the non-NMDA glutamate-like receptors such as of L-glutamate, quisqualate, L-aspartate, kainate, and D-homocysteate, in modulating the responses. Further support to the notion that AMPAR seems to be insensitive to pressure comes from direct tests of the receptors expressed in oocytes (see Chapter 6 in this book). The best way to demonstrate the physiological presynaptic effect is a detailed statistical quantal analysis of the release process. Such analysis at HP has been performed only in crustacean neuromuscular glutamatergic synapses (Golan et al., 1994, 1995, 1996). However, indirect studies suggest that the presynaptic mechanisms affected by HP in the CNS are similar to those of crustaceans. These include decreased quantal content (m), small variable change in probability of release (p), and reduction in the number of the release sites (n). Reduction in n is interpreted in the CNS synapses as a diminution of the readily releasable pool of vesicles (Talpalar and Grossman, unpublished data). All these parameter changes attest to presynaptic site of action. In contrast, HP has no effect on the quantal size (q), namely the unitary response to transmitter released from a single vesicle. This indicates lack of effect on the post synaptic receptor's response.

Previous experiments on synaptosomes prepared from rodents CNS tissue also support the notion of presynaptic effects of pressure. Synaptosome is a sealed broken nerve terminal, presumably containing the whole transmitter release apparatus, but without a postsynaptic component. The synaptosomes were preloaded with radioactive-labeled neurotransmitter, washed, and then the $Ca^{2+}$-dependent release caused by steady depolarization of the membrane (effected by increased $[K^+]_o$) was monitored. HP (6.9 MPa) suppressed release of the glutamate, but did not affect release of similar less abundant aspartate (Gilman et al., 1986a). The major pressure effect was slowing of the release during the first minute or two after stimulation, and in some cases, a moderate reduction in the maximal release. The technique used in these experiments can only measure transmitter release on a time scale of minutes or few seconds, which is considerably slower than the actual milliseconds time course of normal synaptic transmission; however, within this limitation, they demonstrate a large pre-synaptic effect of ambient pressure.

Exocytosis *per se*, i.e. the fusion of vesicles with the plasma membrane of the nerve terminals, could play a role in the pressure sensitivity of fast synaptic transmission. This question was addressed by Heinemann et al. (1987a) who studied degranulation (secretion) of bovine adrenal cells and peritoneal rat mast cells. Cells were stimulated by dialysis of 1 μM free calcium or 40 μM GTP γ-S (an intracellular second messenger) via a patch pipette, and the change in cells membrane capacity as an indicator for vesicles fusion was monitored by voltage clamp techniques. Even modest pressures caused an appreciable decrease in the rate of exocytosis in these non-neuronal cells. The sensitivity to pressure is 4–5 times greater than that of synaptic transmission (estimated by calculation of activation volumes). Since $Ca^{2+}$ influx is bypassed by this method, the results show that pressure interferes also with the exocytosis itself. However, the high sensitivity of the slow exocytosis may not directly reflect on the fast synaptic signaling, but since such a mechanism is ubiquitous in neurohormonal and neuromodulator secretion, it is postulated that pressure would have some long term modulatory effects on brain function. Yet some of the exocytosis intermediate steps may determine the frequency response of the fast synapses (as above).

## $Ca^{2+}$ and presynaptic release

Synaptic release is a multi-step mechanism. The first crucial stage is increased $Ca^{2+}$ influx and elevation of cytosolic $Ca^{2+}$ concentration at the nerve terminal following membrane depolarization by the invading AP. The subsequent step is fusion of docked vesicles with the terminal plasma membrane which ends in evoked release. In parallel, $Ca^{2+}$ entry activates mobilization processes that detach vesicles from their reserve pool, translocate them to the readily releasable pool and prime them at the releasing sites. Many previous experiments on invertebrate preparations, including use of theoretical models (Grossman and Kendig, 1990) and quantal analysis under reduced $[Ca^{2+}]_o$ conditions (Golan et al., 1994, 1995, 1996) revealed striking similarity to HP in parameters change. This strongly indicates that reduced $Ca^{2+}$ entry into the presynaptic terminals during the action potential is the major cause for depressed evoked synaptic transmission. While reduced $[Ca^{2+}]_o$ mimicked HP depression, increased $[Ca^{2+}]_o$ could antagonize HP depression in these experiments, both supporting this model. Indeed, HP also impeded depolarization–induced influx of $Ca^{2+}$ into CNS guinea pig cerebrocortical synaptosomes through unidentified channels (Gilman et al., 1986b). In contrast, HP did not change the $Ca^{2+}$ flux through artificial $Ca^{2+}$ ionophore (A 23187) inserted into these synaptosomes (Gilman et al., 1991). Furthermore, HP effects on single and paired responses in the CNS are consistent with the above mechanism. However, in vertebrate (Grossman et al., 1991b) and especially in mammalian CNS the release processes are more complicated, since the terminals usually contain more than one type of $Ca^{2+}$ channels (Takahashi and Momiyama, 1993) that may be differentially affected by HP (Aviner et al., 2008, 2009), and the frequency response may be nonlinear and biphasic (as mentioned above). HP affects the presynaptic terminals probably by impeding $Ca^{2+}$ influx through a combination of voltage gated $Ca^{2+}$ channels (Talpalar and Grossman, 2003). However, the most pressure sensitive $Ca^{2+}$ channel is the N-type, as demonstrated in frog neuromuscular terminals (Grossman et al.,

1991b), indicated for lobster neuromuscular synapse (Grossman et al., 1991a) and suggested for cerebellar CF synapse (Etzion and Grossman, 2000).

Experiments on modification of $[Ca^{2+}]_o$ in CNS synapses were similar to invertebrate results with respect to single responses, but varied by frequency response of different CNS synapses. In the dentate gyrus PP synapse, reduced $[Ca^{2+}]_o$ to half of its normal value (1mM) increased PPF to a greater extent than HP (10.1 MPa) at normal $[Ca^{2+}]_o$. This reduced $[Ca^{2+}]_o$, unlike pressure, turned FDD into a frequency dependent potentiation (Talpalar and Grossman 2003). Increased $[Ca^{2+}]_o$ (4–6 mM) at atmospheric pressure, saturated single fEPSP and increased FDD in response to short trains at 50 Hz. Increasing $[Ca^{2+}]_o$ to 4 mM at HP saturated synaptic release at a subnormal level (only 20% recovery of a single fEPSP), but enhanced FDD with an efficacy similar to that at atmospheric pressure. Analysis of the normalized fractions of release during the train and the consequent 'probability' of release revealed that HP may depress synaptic frequency response by reducing the size of the readily releasable pool of synaptic vesicles (Talpalar and Grossman unpublished data). In the cerebellar PF synapse, low $[Ca^{2+}]_o$, as well as N-type and N+P/Q –type $Ca^{2+}$ channel blockers that reduce presynaptic $Ca^{2+}$ entry, inhibit the single synaptic response to a similar extent as HP (Etzion and Grossman, 2000). These treatments significantly increased the PPF (118–125 %), but only at 20 ms ISI, whereas pressure effect was for a longer ISIs as well (Etzion et al., 2009). In the cerebellar CF synapse, low $[Ca^{2+}]_o$ (0.5 –1 mM) mimicked HP effect on the single response, but reduced also the PPD, while HP did not affect PPD. Doubling the level of $[Ca^{2+}]_o$ (4 mM) could not antagonize the depression of the response at HP. Further increase of $[Ca^{2+}]_o$ to 6 mM could slightly antagonize the effect of pressure on the amplitude of the first fEPSP, but it was accompanied by increased PPD.

In summary, HP depresses the response of all glutamatergic excitatory AMPAR-dependent synapses tested so far. The depression is primarily of presynaptic origin, i.e. reduced released quanta of transmitter and increased synaptic delay. Evidence on the role of $[Ca^{2+}]_o$ in single fEPSPs and very few $Ca^{2+}$ current measurements strongly indicate that pressure impedes $Ca^{2+}$ entry into the presynaptic terminals. Recent studies in the cerebellum suggest that pressure effects on various voltage gated $Ca^{2+}$ channels may be selective; the prevalence of N-type in the terminals seems to correlate with pressure vulnerability. However, during repetitive activation such as paired pulse and frequency modulations, HP probably interferes with additional mechanisms other than $Ca^{2+}$ entry that are involved with the cascade of synaptic release. Yet, frequency potentiation at pressure may partially compensate for the depression of single response. The extent at which HP will affect a particular synapse is dependent on its set-point release operation (with regard to saturation), the type of $Ca^{2+}$ channels associated with the release, and its frequency response characteristics.

Exocytosis itself may not play a major role in fast transmission (during which vesicles and terminal membranes fusion does occur), however it certainly is depressed in slow secretion of neuromodulators and neurohormons.

It is conceivable that the large depression in excitatory synaptic activity will cause functional difficulties for any neuronal circuit. However, it can not explain the hyperexcitability and epileptogenesis observed at HPNS.

## Glutamatergic NMDA synapse

In early studies in brain slices, Fagni and his colleagues (Fagni et al., 1987a, b; Zinebi et al., 1988a, 1990) examined in the hippocampal CA1 region the efficacy of various excitatory neurotransmitters and agonists that were perfused in the bath, in reducing fEPSP and population AP under at HP (as above). Whereas HP had no significant effect on the non-NMDA responses, it greatly increased NMDA and L-homocysteate efficacy. This indicates that pressure increases, rather than decreases, the activity of the latter two amino acids receptors. This intriguing result was confirmed by direct measurements of the current of NMDAR from rat cerebellum expressed in *Xenopus laevis* oocytes (see Chapter 6 in this book). Recently, Mor et al. (2008) have shown that this increase may be selective for specific combination of the NMDAR NR1-NR2 subunits. Mor and Grossman (2006, 2007a) have also succeeded in isolating the responses of NMDAR in Shaefer collaterals synapses onto the pyramidal cells in rat hippocampal CA1 brain slices by blocking AMPA and $GABA_A$ neurotransmission and removing $Mg^{2+}$ from the physiological solution (in order to enable receptor opening without prior depolarization of the AMPAR). These are the only available studies on NMDAR fEPSPs under HP conditions, the data of which are described below.

Since NMDARs are usually co-localized postsynaptically with AMPARs in neuronal synapses, they are exposed to glutamate released by the same presynaptic mechanism. Therefore a special paragraph on release is absent from this NMDAR description.

### Single response

The single fEPSP of NMDAR is enhanced by HP and its kinetics is greatly slowed down. HP (10.1 MPa) increased fEPSP delay by 16%, its maximal initial slope by 30%, elongated its decay time by 95%, increased its amplitude by 80%, and its time integral by 250%! This implies a great increase in the total charge influx which is carried by $Na^+$ and a significant portion of $Ca^{2+}$. Such abnormal elevation of postsynaptic (dendritic) influx of $Ca^{2+}$ may trigger known cytotoxic processes that could facilitate neuronal death (Lynch and Guttmann, 2002).

Additional findings demonstrate that HP greatly reduces the sensitivity of the fEPSP to $[Mg^{2+}]_o$, its natural ion-channel blocker, and to 2-amino-5-phosphonovaleric acid (AP-5), a pharmacological blocking agent (Mor and Grossman, 2007b). These important observations strongly indicate that HP causes conformational changes in the NMDAR.

### Frequency modulation

HP decreased PPF (slope measurements) by 29% and 36% at 10–20 ms ISIs (50–100 Hz) and 40 ms ISIs (25 Hz), respectively. However, the results were not statistically significant due to increased variability of the second responses for an unknown reason. Nevertheless, trend analysis revealed that PPF at high pressure was always smaller than the control values for each experiment. This is in accord with the increase in single response at high pressure. Frequency response was tested by trains of five fEPSPs at low frequency (25 Hz) and high frequency (50–100 Hz). At control pressure, for 25 Hz stimulation, the third, fourth, and fifth fEPSPs were depressed (FDD) following PPF of the second response. At HP, the first response was significantly augmented and PPF was reduced (as above), while the FDD was significantly enhanced. For 50–100 Hz trains, there was a similar behavior although HP did not affect the FDD. HP did not affect fEPSP train's time integral at 100 Hz, but it increased it by 96% at 25 Hz. This means again approximately two fold increase in $Ca^{2+}$ influx into the postsynaptic dendritic tree. It is important to note that in contrast to the PP intact glutamatergic synapses (as above), in which the synaptic efficacy in producing population spikes was increased at pressure (gain increase), the isolated NMDAR response was maintained. In fact not only the delay to the first spike is reduced, but also the number and frequency of population spikes induced in each fEPSP and their total number during a train were significantly reduced. In summary, increasing evidence indicate that HP potentiates NMDAR response. However, this increase alone has a complex effect on the capability of the synapse (deprived of its AMPA component) to generate population spikes. Despite the increase of a single fEPSP, consecutive fEPSPs just maintain synaptic efficacy with no gain increase of the system. Under these conditions there is no indication for major increase of excitability that is observed in intact hippocampal synapses (see above). It seems reasonable that the full effect of pressure modulation of the NMDAR on neuronal excitability would be attained only when the fast AMPAR component of the synapse is intact. While these findings may suggest partial explanation for HPNS, the increase of fEPSP time integral which is associated with excessive influx of $Ca^{2+}$, may also provide possible molecular basis for recent reports concerning long term deleterious health effects on professional divers associated with saturation dives.

## INHIBITORY SYNAPTIC ACTIVITY

The examination of synaptic inhibition at HP is of obvious importance in order to understand the underlying mechanisms associated with pressure induced hyperexcitability (i.e. HPNS). As described above, AMPAR excitatory synaptic input is depressed and despite some PPF can not contribute to the hyperexcitability. NMDAR response is increased despite the depression of presynaptic release of glutamate, and therefore may partially compensate for the impaired AMPAR excitation. Inhibition is an integral part of any neural circuit. Pressure-induced reduction in inhibitory processes, concomitantly with some compensation of the excitatory mechanisms will disrupt the delicate balance between excitation and inhibition in the CNS and may ultimately cause hyperexcitability. Since transmitter release from the presynaptic terminals is believed to be similar for excitatory and inhibitory synapses, it will not be surprising that HP similarly depresses inhibition and excitation in model synapses. However, if an additional effect on receptor activity is exhibited (e.g. NMDAR), pressure effects could be more complicated.

Several ionic mechanisms may be involved in two major types of known functional inhibition. In the most common 'postsynaptic inhibition', the excitatory and the inhibitory axons are both synapsing on the dendritic or somatic membrane of the target neuron, which in turn integrates their EPSPs and inhibitory postsynapic potentials (IPSPs). In the 'presynaptic inhibition', the inhibitory axons synapse on the excitatory axons prior to their terminals on the target neuron, and thus may selectively reduce their input by shunting the AP. γ-amino-butyric acid (GABA) in the brain and glycine in the spinal cord are the most common CNS inhibitory neurotransmitters. Here we will describe the HP effects on the GABA type A receptor ($GABA_A R$) and glycine receptor (GlyR), both ionotropic receptors that gate $Cl^-$ channel.

As mentioned above for the excitatory neurotransmission, in the 1980s an immense effort was invested in studying the role of the inhibitory neurotransmitters in HPNS as well by applying various related drugs (agonists and antagonists) to freely moving animals (usually rats). Competitive antagonists of $GABA_A R$ (e.g. bicuculline) and GlyR (e.g. strychnine), as well as drugs that block the chloride channel associated with the receptors (e.g. picrotoxin) are convulsant in action. They all seem to mimic pressure effects and thus are reminiscence of HPNS (Bowser-Riley et al., 1988). The $GABA_A R$ also possesses a recognition site for benzodiazepines (e.g. diazepam) which increase the receptor response and thus are powerful sedatives and anticonvulsants (i.e. antiepileptic). The benzodiazepines provide some protection against the effects of pressure (Halsey and Wardley-Smith, 1981), although at sedative doses. Prolonging the action of GABA by preventing its metabolism via inhibition of GABA transaminase (e.g. aminooxyacetic acid) or blocking its reuptake (e.g. 2, 4-diaminobutyric acid) provides some protection against pressure effects (Bichard and Little 1984). Increasing inhibition by administration of $GABA_AR$ agonist (e.g. muscimol) has no effect on the action of pressure on the whole animal. Aromatic propandiol compounds (e.g. mephenesin), which are known as muscle relaxants, are most effective against both pressure and strychnine induced convulsions, but not against GABA antagonists (Bowser-Riley, 1984; Bowser-Riley et al., 1989). In contrast, the aliphatic propandiol compounds (e.g. meprobamate) are effective anticonvulsants against GABA antagonists, but not against strychnine. It seems that the potency of these drugs against pressure effects was unrelated to their muscle relaxant efficacy. These partial lists of findings provide a compelling evidence to support a role for glycine and GABA mediated inhibitory neurotransmission in the etiology of HPNS.

### GABAergic synapses

Direct measurement of GABAergic IPSP at HP was carried out only in lobster neuromuscular synapses (Grossman and Kendig, 1988). The muscle is innervated by a single glutamatergic excitatory and a single GABAergic inhibitory motorneuron. Single IPSPs were depressed at HP (10.1 MPa), concomitantly with 60% reduction in the functional postsynaptic inhibition (evoked by short train of stimuli to the inhibitor) of the EPSP, which is the same extent as in the excitatory transmission itself. In a different crustacean neuromuscular model Golan et al. (1994) have demonstrated that GABAergic presynaptic inhibition of the excitatory synaptic input is similarly depressed at HP. Using spinal cord synaptosomes experiments (see above) Gilman et al. (1987, 1989a) have demonstrated that GABA release at pressure is reduced and its kinetics is slowed, similarly to the glutamate release, supporting again the notion that the major effect of pressure is presynaptic. This is also corroborated by later findings indicating that $GABA_AR$ is insensitive to pressure (Shelton et al., 1996).

An attempt to indirectly evaluate the degree of GABAergic inhibition involvement in CNS HP effects was carried out by Zinebi et al. (1988a, b) in rat hippocampal CA1 region. Antidromic stimulation of the pyramidal cells' axons evoked recurrent inhibition, which prevented the generation of single-unit action potential evoked by stimulation of the SC synaptic input to these neurons. This inhibition became ineffective in an all-or-none fashion at HP. Similarly GABAergic feed forward inhibition onto these cells was also decreased. Another attempt to evaluate the contribution of inhibition mediated by GABA$_A$R at HP was carried out recently in the PP synapse in rat dentate gyrus brain slices (Talpalar and Grossman, 2004). Partial blockade of the GABA$_A$R inhibition with low doze of bicuculline mimicked pressure effect, i.e. reduction of the frequency dependent attenuation of population spikes in the granular cells. Both HP and GABA$_A$R blockade increased the number of population spike generated during a short train of stimuli at 50 Hz. Furthermore, under effective block of the GABA$_A$R with higher doze of bicuculline, exposure to HP no longer increased excitability, but rather decreased it.

Functional inhibition of the MSR in a newborn rat's isolated spinal cord was also examined at HP (Schleifstein-Attias et al., 1994). MSR was recorded in hemisected spinal cord from severed end of lumbar ventral roots in response to stimulation of the corresponding dorsal root. Inhibition was evoked by prior stimulation of an adjacent dorsal root. By modulating the ISI between the two stimuli and the use of specific blockers, it was possible to demonstrate a complex time course of this inhibition. It is composed of an early phase (1–25 ms ISI) that is glycine dependent postsynaptic inhibition (see below), an intermediate phase (25–200 ms ISI) that is GABA$_A$R dependent postsynaptic (and possibly also presynaptic) inhibition, and a late phase (200–1000 ms ISI) that may represent non-synaptic 'primary afferent depolarization' inhibition (Rudomin, 1990; Hackman and Davidoff, 1991). HP completely abolished the glycinergic inhibition, reduced the GABAergic inhibition by 68 %, and did not affect the late phase. In a different experiment a strong inhibition of the MSR was induced following 1–15 conditioning stimuli applied to the adjacent ventral root (Schleifstein-Attias and Grossman unpublished data). Using various ISIs and specific receptor blockers (as above), it was demonstrated that stimulation of the motoneurons axons evoked a short term recurrent inhibition (5–300 ms ISI) that was GABA$_A$R, glycine, and acetylcholine receptor dependent (see below) and also a long lasting inhibition (0.3–100s) that was probably serotonin dependent (see below). HP reduced the short term inhibition but had no effect on the long lasting inhibition. It is important to note that the pressure-induced suppression of the inhibition could be mimicked under normobaric conditions by reducing $[Ca^{2+}]_o$ by 50 % of the normal concentration. Raising $[Ca^{2+}]_o$ by 50 % could compensate for HP (10.1 MPa) effects.

In summary, HP depresses the GABA$_A$R response of inhibitory synapses to the same extent as in the AMPAR-dependent synapses. Although strong direct evidence is missing, it seems conceivable that in the lack of HP effect on the GABA$_A$R, the depression is primarily of presynaptic origin, i.e. reduced released quanta of transmitter and increased synaptic delay. Only limited evidence suggests that reduced GABA release at HP in the CNS is caused by impaired Ca$^{2+}$ entry into the presynaptic terminals. The extent at which HP will affect a particular inhibitory synapse is probably similar to those associated with the AMPAR synapses (above).

### Glycinergic synapses

Spinal cord synaptosomes experiments have demonstrated (Gilman et al., 1987; Gilman et al., 1989a) that HP consistently reduced and slowed the kinetics of glycine release. Indeed, HP completely abolished the very effective (54%) glycine dependent postsynaptic inhibition of the MSR and greatly reduced the recurrent inhibition. Furthermore, the test MSR response was even slightly facilitated by the conditioning pulse at the adjacent dorsal root (Schleifstein-Attias et al., 1994). Such a complete 'shut down' is unusual compared to other neurotransmitters that usually exhibit 40–60% depression of the synaptic response at 10.1 MPa HP. A possible explanation could be a difference in the postsynaptic receptors behavior at HP (see Chapter 6 in this book). Rat GlyRs expressed in frog oocytes show no change in the maximal response, but unlike the $GABA_A R$ did demonstrate a significant increase in the doze-response curve ($EC_{50}$ increased by 60%) at 10.1 MPa (Shelton et al., 1993). Thus, postsynaptic reduction in receptor sensitivity will exacerbate HP suppression and would explain why the GlyR is more sensitive to pressure than the $GABA_A R$. A recombinant human GlyR has been also expressed in oocytes, exhibiting an even higher pressure sensitivity than the one in the rat  (Roberts et al., 1996). Since glycine is an abundant inhibitory neurotransmitter in the spinal cord, it is of no surprise that one of the first HPNS symptoms are reduced motor coordination and muscle tremor.

In summary, the complete blockade of glycine transmission both by pre- and post synaptic mechanism is responsible for the sensitivity of the spinal cord to HP induced tremor and convulsions. It is conceivable that the large depression in AMPAR excitatory synaptic activity is causing functional difficulties for any neuronal circuit. However, it is the great depression of the inhibitory synaptic transmission in addition to increased activity of the NMDAR that brings about the hyperexcitability and epileptogenesis observed at HPNS.

### MODULATORY SYNAPTIC ACTIVITY

Many of the CNS regulatory neurotransmitters are released by presynaptic mechanisms similar to those of the 'classical' excitatory and inhibitory neurotransmitters (as above), although the release could be slower. However, unlike the 'classical' neurotransmitters, they gate metabotropic receptors rather than ionotropic receptors. Thus, they do not directly change cellular ionic conductances or membrane potential. Instead, they bind to receptors that are coupled through various G-proteins, which activate intracellular signaling cascades altering cell function. Early Pharmacological experiments on whole animals have indicated that pressure may affect monoamines receptors of dopamine (D1, D2) and serotonin (5-HT1b, 5-HT2c) (Abraini and Rostain, 1991; Abraini and Fechtali, 1992; Kriem et al., 1996, 1998). Further experiments suggest (see Chapter 19 in this volume, Bennett and Rostain, 2003) that HP may decrease the inhibitory actions of D2 receptors and stimulate the excitatory actions of D1 receptors. Furthermore, since both D1 and D2 receptors are predominantly expressed in the striatum and are known to be involved in motor control (e.g. Parkinson's disease, Gibb, 1992; Fearnley and Lees, 1991), these results do suggest that dopamine receptors may in part mediate the expression of acute HPNS.

## Dopamine activity

As postulated, release of dopamine from synaptosomes preparations is depressed and slowed at HP (Gilman et al., 1988a; Gilman et al., 1989b). However, because of the great diversity of the signaling cascades, pressure effects are much more complicated. For example, D1 receptors stimulate adenylate cyclase (producing cAMP) and activate the cascade of protein kinase A. D2 receptors inhibit adenylate cyclase and separately activate arachidonic acid pathway (for details see Daniels and Grossman Chapter 6 in this book). Direct evidence regarding dopamine synaptic activity at HP in the CNS is not available. However, in recent years a major effort was carried out in measuring *in vivo* dopamine (and serotonin) levels in the sub-cortical brain structures using voltametry (Rostain and Forni, 1995) or microdialysis (Darbin et al., 1997) methods. Exposure to HP increased the extracellular concentration of dopamine in rat and monkey striatum in association with increased locomotion and motor activity (for review see, Bennett and Rostain, 2003). Similar increase of dopamine has also been described for the nucleus accumbens, which is associated with emotional and mental states. The exact mechanism of increased extracellular concentration of dopamine is not clear. Direct effect on synaptic release is less likely but can not be ruled out. The possibility of higher synthesis without any change in degradation or uptake is suggested to occur at pressure (Bennett and Rostain, 2003). Some of these effects may result from changes in the network behavior rather than local changes. The general reduction of tonic network inhibition on the nigrostiatal and mesolimbic pathways, a reasonable outcome of the pressure induced depression of inhibitory processes (mentioned above), could facilitate release. As suggested above, a different role of D1 and D2 receptors is also postulated.

## Serotonin activity

As postulated, release of serotonin (5-hidroxytriptamine, 5-HT) as well as norepinephrine from synaptosomes preparations is reduced and slowed under hyperbaric conditions (Gilman et al., 1988b; Gilman et al., 1989b). 5-HT1b receptors function through similar cascade as in D2 receptors. However, 5-HT2C receptors activate phospholipase C, leading to the release of intracellular $Ca^{2+}$ and protein kinase C activation. Further experiments in *Xenopus* oocytes (Daniels, unpublished data) suggested that pressure directly modulates the 5-HT2C receptors responses, whereas pressure effects on the 5-HT1b receptors appears to be mediated by up regulation of the cascade components.

In addition to the pharmacological studies, in recent years measurements of serotonin levels in the sub-cortical brain structures using voltametry and microdialysis were performed *in vivo* (as mentioned above for dopamine). Exposure to HP increased the extracellular concentration of serotonin in rat and monkey striatum and nucleus accumbens. It is suggested that pressure increases serotonin release by disinhibition of GABAergic post synaptic inhibition of the neurons via 5HT2A receptors. However, an opposite effect of reduction in release of serotonin may be mediated by another 5HT2C presynaptic autoreceptors (Bennett and Rostain, 2003, see also chapters 18 and 19 in this book).

In summary, dopaminergic and serotonergic systems are involved in pressure induced motor, locomotion, and possibly mental disorders (HPNS). The most prominent effect of pressure on the monoamines is elevation of their extracellular concentration in critical regions of the brain. However, the mechanism underlying this increase is unclear. Furthermore, this increase affects the receptors and their downstream signaling cascade in a very complicated and diverse mode. Thus, depending on the local pathways, pressure effects on any monoaminergic network could be differential, antagonistic, complementary, or even synergistic. The consequences of these changes are dependent on the network neurological task and the amount of control exerted by neighboring brain centers.

## Acetylcholine activity

Cholinergic processes have a well established modulatory role in arousal and motor activity (Jones, 2004) as well as in various cognitive processes (Perry et al., 1999; Saar et al., 2007). However, centrally acting antagonists at muscarinic acetylcholine receptors (mAChR) (e.g. atropine) or nicotinic acetylcholine receptors (nAChR) (e.g. mecamylamine) do not affect the development of hyperexcitability at pressure (Wardley-Smith et al., 1984). The nAChR (abundant in the vertebrates neuromuscular junction) is not sensitive to pressure (Heinemann et al., 1987b; for review see Chapter 6 in this book). However, studies in neuromuscular junction of frog (Ashford et al., 1982) and invertebrate central cholinergic synapses (Wann and Macdonald, 1980) indicate that HP slows the kinetics of the response and reduces spontaneous release of miniature EPSPs. Since the vertebrates (mammalian) motorneurons are cholinergic, acetylcholine is involved in various spinal inhibitory pathways that are activated by the motorneurons themselves. Pathways such as recurrent inhibition, Renshaw cells inhibition, and contralateral inhibition that involve activation of interneurons by the motorneurons are conceivably inhibited by HP, partially because of reduced release of acetylcholine (Schleifstein-Attias and Grossman unpublished data). For similar reasoning depressed acetylcholine release may underlie impairment of cognitive functions at pressure.

## SUMMARY

Presence of pressure sensitive molecules, especially ionic channels and receptors, in particular neurons and their localization in sub-cellular structures such as soma, axon or dendrites, will all concert the response of each cell to pressure exposure. Furthermore, on network levels, the incidence of affected neurons in the specific excitatory, inhibitory or modulatory network, the operating set-point of each synapse with regard to saturation that affect its frequency response, and the affected brain region(s), will all determine the final physiological response of the system to HP. More specifically, the mechanisms presently known to be involved in CNS response to HP exposure are summarized in Fig 1.

The less affected components of the system are the information lines of the presynaptic axons (Fig. 1, Input) and the postsynaptic axons (Fig. 1, Output). The action potentials amplitude and velocity are slightly reduced and their frequency response is not affected. However, the major characteristic of HPNS is the disruption of the delicate balance between excitation and inhibition. This could happen only if synaptic transmission (Fig. 1, Input component) and postsynaptic integration (Fig. 1, Integration) of the information are altered. Indeed, synaptic responses in both excitatory and inhibitory pathways (i.e. EPSPs and IPSPs) are suppressed at pressure. This general suppression is largely (but not exclusively) the consequence of reduced synaptic release due to impaired $Ca^{2+}$ entry into the presynaptic terminals. If both excitatory and inhibitory release mechanisms are similarly suppressed quantitatively, imbalance between the two systems may occur only if additional mechanisms take place. Partial compensation for the reduced single synaptic response is attained by increased facilitation and in some cases by additional frequency potentiation. In contrast, in several synapses the normal twin pulse depression is unaffected and the frequency

# High Pressure Modulation of CNS Synapses

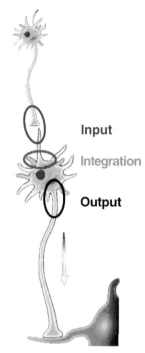

**Input**

- Action potential slighty reduced and slowed
- Presynaptic volley is decreased
- Synaptic response is delayed, slowed, decreased
- Synaptic release is suppressed
- Synaptic facilitation is increased
- Synaptic depression is unchanged/increased
- Synaptic frequency response is modulated
- Law $[Ca^{2+}]_0$ mimics, high $[Ca^{2+}]_0$ antagonizes pressure effects on single/twin responses
- Impaired presynaptic $Ca^{2+}$ influx; N-type $Ca^{2+}$ channel is most sensitive

**Integration**

**Output**

- Improved/maintained synapse - spike integration
- Increased probability for spikes in a train
- Possible boosting by dendritic/somatic electrogenesis
- Boosting by increased amplitude and duration of the NMDAR response; minor effects of AMPAR
- Depressed GlyR response; no effect on $GABA_A R$
- Selective effect on ionotropic receptor's isoforms
- Increased level of dopamine and serotonin
- Antagonistic effect on metabotropic receptor's signaling pathway
- Imbalance between excitation and inhibition

- Action potential is slighty reduced and slowed
- No change in frequency response

**Figure 1.** Summary of high pressure effects on synaptic transmission and network components.

depression is even increased. The role of synaptic frequency response in determining the type of 'behavioral' response is unclear. While the increase in synaptic efficacy in generating action potentials at 50 Hz may be related to seizure promotion, we may just speculate that the decline in the synaptic efficacy at < 25 Hz might correspond to minor reduction in motor performance (e.g. manual dexterity) or reduced cognitive capabilities. Another mechanism could be invoked due to differential pressure effect on the postsynaptic receptors. Boosting of excitatory response probably occurs when NMDAR (but not AMPAR) responses are increased, while strong disinhibition may be effected when the GlyR (but not $GABA_A R$) response is practically shut down. It is important to note that pressure effects may be selective for various isoforms of the receptors. Another mechanism that may bring about hyperexcitability could be additional activation of voltage dependent channels (such as specific $Ca^{2+}$ channels) along the dendrites or soma of the neurons. This may boost the reduced synaptic potentials and increase their efficacy in generating action potentials. Again, it is worth noting that pressure effects may be selective for various types of ionic channels. Finally, the excitatory and inhibitory systems are modulated by metabotropic receptors of monoamines such as serotonin and dopamine, while the monoaminergic neurons themselves can be inhibited by feedback circuits. Here the major effect of pressure is the increase of monoamines extracellular concentrations in the vicinity of their target neurons. However, this rise differentially affects various neurons (either in motor or cognitive system) depending on the type of receptor and its specific signaling cascade.

## CONCLUSION

To understand how pressure affects the CNS is an enormous scientific challenge. Many of the CNS functions and mechanisms are highly complex and remain a mystery for modern science. Apparently the task could be somewhat easier, since unlike a chemical, pressure acts 'instantaneously' and throughout the brain. This may mean that pharmacokinetics and metabolism are not confounding problems. However, this also indicates that the scope of research is wide open since, in principle, every molecule (protein) and any chemical reaction is a potential target. Furthermore, even when a family of a receptor, for example, is identified as pressure sensitive, pressure effects could be selective or differential for various isoforms of the receptor.

The common notion of short term pressure effects on the CNS are summarized above. Yet, our recent finding of the possibility of excessive $Ca^{2+}$ influx into the neurons during increased activation of the NMDAR suggests that HP may exert some long term effect on the cells' viability (neurotoxicity). Although the effect of single exposure can be minor, the long term effect following repetitive exposures during many years of deep diving may be significant.

There are many aspects of HPNS that are far less understood. Given the range of targets for modification by pressure and the complexity of the CNS, many aspects remain to be elucidated before a comprehensive understanding of the effect of pressure on CNS function could be claimed.

## ACKNOWLEDGEMENTS

This work was partially supported by a grant from the Office of Naval Research, USA to Yoram Grossman. Amir Mor is supported by the Eshkol Ph.D. Fellowship for Interdisciplinary Research of the Israeli Ministry of Science and Technology.

## REFERENCES

Abraini, J.H. and J.C. Rostain. 1991. Effects of the Administration of Methyl-Para-Tyrosine on the striatal dopamine increase and the behavioral motor disturbances in rats exposed to high-pressure. Pharmacol. Biochem. Behav. 40: 305–310.

Abraini, J.H. and T. Fechtali. 1992. A hypothesis regarding possible interactions between the pressure-induced disorders in dopaminergic and amino-acidergic transmission. Neurosci. Biobehav. Rev. 16: 597–602.

Angel, A. and M.J. Halsey. The effect of increased lumbosacral motoneuron excitability in the rat. Pp. 149–158. *In*: H. Jannasch, R. Marquis and A. Zimmerman. (eds.). 1987. Current perspectives in high pressure biology. Academic Press. London.

Angel, A. and M.J. Halsey, H. Little, B.S. Meldrum, J.A. Ross, J.C. Rostain and B. Wardley-Smith. 1984. Specific effects of drugs at pressure. Animal investigations. Philos Trans. R. Soc. Lond. B. Biol. Sci. 304: 85–94.

Ashford, M.L. and A.G. MacDonald and K.T. Wann. 1982. The effects of hydrostatic pressure on the spontaneous release of transmitter at the frog neuromuscular junction. J. Physiol. 333: 531–543.

Aviner, B. and M. Mor and Y. Grossman. 2008. Hyperbaric pressure facilitates $Ba^{2+}$ current in L-type $Ca^{2+}$ channel. Proc. 34th Eur. Undewater Baromed. Soc. Ann. Meeting: 30–33.

Aviner, B. and M. Mor and Y. Grossman. 2009. Pressure modulation of $Ca^{2+}$ channel activity may elucidate HPNS mechanisms. Proc. 35th Eur. Underwater Baromed. Soc. Ann. Meeting (in press).

Bennett, P.B. and J.C. Rostain. High pressure nervous syndrome. Pp. 323–357. *In*: A.O. Brubakk and T.S. Neuman, (eds.). 2003. Bennett and Elliott's Physiology and Medicine of Diving. W.B. Saunders. Edinburgh.

Bichard, A.R. and H.J. Little. 1984. Gamma-aminobutyric acid transmission and the high pressure neurological syndrome I. VIII Symp. Underwater Physiol.: 545–556.

Bowser-Riley, F. 1984. Mechanistic studies on the high pressure neurological syndrome. Philos. Trans. R. Soc. Lond. B. Biol. Sci. 304: 31–41.

Bowser-Riley, F. and S. Daniels and E.B. Smith. 1988. Investigations into the origin of the high pressure neurological syndrome: the interaction between pressure, strychnine and 1, 2-propandiols in the mouse. Br. J. Pharmacol. 94: 1069–1076.

Bowser-Riley, F. and S. Daniels, W.A. Hill and E.B. Smith. 1989. An evaluation of the structure-activity relationships of a series of analogues of mephenesin and strychnine on the response to pressure in mice. Br. J. Pharmacol. 96: 789–794.

Daniels, S. and Y. Grossman. Biological effects of pressure. Pp. 265–299. *In*: A.O. Brubakk and T.S. Neuman (eds.). 2003. Bennett and Elliott's Physiology and Medicine of Diving. W.B. Saunders. Edinburgh.

Darbin, O.J. and J. Risso and J.C. Rostain. 1997. Pressure induces striatal serotonin and dopamine increases: a simultaneous analysis in free-moving microdialysed rats. Neurosci Lett. 238: 69–72.

Dean, J.B. and D.K. Mulkey. 2000. Continuous intracellular recording from mammalian neurons exposed to hyperbaric helium, oxygen, or air. J. Appl. Physiol. 89: 807–822.

Etzion, Y. and Y. Grossman. 1999. Spontaneous Na+ and $Ca^{2+}$ spike firing of cerebellar Purkinje neurons at high pressure. Pflugers Arch. 437: 276–284.

Etzion, Y. and Y. Grossman. 2000. Pressure-induced depression of synaptic transmission in the cerebellar parallel fibre synapse involves suppression of presynaptic N-type $Ca^{2+}$ channels. Eur. J. Neurosci. 12: 4007–4016.

Etzion, Y. and A. Mor and Y. Grossman. 2009. Differential modulation of cerebellar climbing fiber and parallel fiber synaptic responses at high pressure. J. Appl. Physiol. 106: 729–736.

Fagni, L. and F. Zinebi and M. Hugon. 1987a. Evoked potential changes in rat hippocampal slices under helium pressure. Exp. Brain Res. 65: 513–509.

Fagni, L. and F. Zinebi and M. Hugon. 1987b. Helium pressure potentiates the N-methyl-D-aspartate- and D, L-homocysteate-induced decreases of field potentials in the rat hippocampal slice preparation. Neurosci Lett. 81: 285–290.

Fearnley, J.M. and A.J. Lees. 1991. Ageing and Parkinson's disease. substantia nigra regional selectivity. Brain. 114 (Pt 5): 2283–2301.

Gibb, W.R. 1992. Neuropathology of Parkinson's disease and related syndromes. Neurol Clin. 10: 361–376.

Gilman, S.C. and K.K. Kumaroo and J.M. Hallenbeck. 1986a. Effects of pressure on uptake and release of calcium by brain synaptosomes. J. Appl. Physiol. 60: 1446–1450.

Gilman, S.C. and K.K. Kumaroo and J.M. Hallenbeck. 1986b. Effects of pressure on uptake and release of calcium by brain synaptosomes. J. Appl. Physiol. 60: 1446–1450.

Gilman, S.C. and J.S. Colton and A.J. Dutka. 1987. Effect of pressure on the release of radioactive glycine and gamma-aminobutyric acid from spinal cord synaptosomes. J. Neurochem. 49: 1571–1578.

Gilman, S.C. and J.S. Colton and A.J. Dutka. 1988a. Release of dopamine from striatal synaptosomes. high pressure effects. Undersea Biomed. Res. 15: 13–18.

Gilman, S.C. and J.S. Colton and A.J. Dutka. 1988b. Pressure suppresses serotonin release by guinea pig striatal synaptosomes. Undersea Biomed. Res. 15: 69–77.

Gilman, S.C. and J.S. Colton and A.J. Dutka. 1989a. Pressure-dependent changes in the release of GABA by cerebrocortical synaptosomes. Undersea Biomed. Res. 16: 253–258.

Gilman, S.C. and J.S. Colton and A.J. Dutka. 1989b. Alterations in brain monoamine neurotransmitter release at high pressure. Exp. Brain Res. 78: 179–184.

Gilman, S.C. and J.S. Colton and Y. Grossman. 1991. A 23187-stimulated calcium uptake and GABA release by cerebrocortical synaptosomes. effects of high pressure. J. Neural Transm Gen. Sect. 86: 1–9.

Golan, H. and H.J. Moore and Y. Grossman. 1994. Quantal analysis of presynaptic inhibition, low ($Ca^{2+}$)0, and high pressure interactions at crustacean excitatory synapses. Synapse. 18: 328–336.

Golan, H. and J.S. Colton, H.J. Moore and Y. Grossman. 1995. Analysis of evoked and spontaneous quantal release at high pressure in crustacean excitatory synapses. Pflugers Arch. 430: 617–625.

Golan, H. and H.J. Moore and Y. Grossman. 1996. Pressure exposure unmasks differences in release properties between high and low yield excitatory synapses of a single crustacean axon. Neuropharmacology. 35: 187–193.

Grapperon, J. and A. Trousset, S. Bellard and L. Force. 1988. Effects of high pressure on nervous conduction velocity in man. Eur. J. Appl. Physiol. Occup Physiol. 58: 141–145.

Grossman, Y. and J.J. Kendig. 1984. Pressure and temperature: time-dependent modulation of membrane properties in a bifurcating axon. J. Neurophysiol. 52: 693–708.

Grossman, Y. and J.J. Kendig. 1986. Pressure and temperature modulation of conduction in a bifurcating axon. Undersea Biomed Res. 13: 45–61

Grossman, Y. and J. Kendig. 1987. Modulation of impulse conduction through axonal branchpoint by physiological, chemical and physical factors. Isr. J. Med. Sci. v. 23: 107–114.

Grossman, Y. and J.J. Kendig. 1988. Synaptic integrative properties at hyperbaric pressure. J. Neurophysiol. 60: 1497–1512.

Grossman, Y. and J.J. Kendig. 1990. Evidence for reduced presynaptic $Ca^{2+}$ entry in a lobster neuromuscular junction at high pressure. J. Physiol. 420: 355–364.

Grossman, Y. and J.S. Colton and S.C. Gilman. 1991a. Interaction of Ca-channel blockers and high pressure at the crustacean neuromuscular junction. Neurosci Lett. 125: 53–56.

Grossman, Y. and J.S. Colton and S.C. Gilman. 1991b. Reduced Ca currents in frog nerve terminals at high pressure. Ann. N.Y. Acad. Sci. 635: 411–412.

Grossman, Y. and I. Parnas, and M.E. Spira. 1979a. Differential conduction block in branches of a bifurcating axon. J. Physiol. 295: 283–305.

Grossman, Y. and I. Parnas and M. Spira. 1979b. Ionic mechanisms involved in differential conduction of action potentials at high frequency in a branching axon. J. Physiol. v. 295: 307–322

Grundfest, H. 1936. Effects of hydrostatic pressures upon the excitability, the recovery, and the potential sequence of frog nerve. Cold Spring Harbor Symposia on Quantitative Biology. Pp. 179–187.

Grundfest, H. and M. Cattell. 1935. Some effects of hydrostatic pressure on nerve action potentials Am. J. Physiol. 113: 56–57.

Hackman, J.C. and R.A. Davidoff. 1991. Dorsal root potentials in the isolated frog spinal cord: amino acid neurotransmitters and magnesium ions. Neuroscience 41: 61–69.

Halsey, M.J. 1982, Effects of high pressure on the central nervous system. Physiol Rev. 62: 1341–1377.

Halsey, M.J. and B. Wardley-Smith. 1981. The high pressure neurological syndrome. Do anticonvulsants prevent it? Br. J. Pharmacol. 72: 502–503.

Heinemann, S.H. and F. Conti, W. Stuhmer and E. Neher. 1987a. Effects of hydrostatic pressure on membrane processes. Sodium channels, calcium channels, and exocytosis. J. Gen. Physiol. 90: 765–778.

Heinemann, S.H. and W. Stuhmer and F. Conti. 1987b. Single acetylcholine receptor channel currents recorded at high hydrostatic pressures. Proc. Natl. Acad. Sci. USA. 84: 3229–3233.

Hogan and S. Besch. Vertebrate skeletal and cardiac muscle. Pp. 125–146. *In*: A. Macdonald (ed.). 1993. Advances in Comparative and Environmental Physiology, Effects of High Pressure on Biological Systems. Springer-Verlag. Berlin-Heidelberg,.

Jones, B.E. 2004. Activity, modulation and role of basal forebrain cholinergic neurons innervating the cerebral cortex. Prog. Brain Res. 145: 157–169.

Kendig, J.J. and T.M. Schneider and E.N. Cohen. 1978. Pressure, temperature and repetitive impulse generation in crustacean axons. J. Appl. Physiol. 45: 742–746.

Kendig, J. and Y. Grossman and S. Heinemann. Ion Channels and nerve cell function pp. 88–124. *In*: A. Macdonald (ed.). 1993. Advances in Comparative and Environmental Physiology, Effects of High Pressure on Biological Systems. Springer-Verlag.

Kriem, B. and J.H. Abraini and J.C. Rostain. 1996. Role of 5-HT1b receptor in the pressure-induced behavioral and neurochemical disorders in rats. Pharmacol Biochem. Behav. 53: 257–264.

Kriem, B. and J.C. Rostain and J.H. Abraini. 1998. Crucial role of the 5-HT2C receptor, but not of the 5-HT2A receptor, in the down regulation of stimulated dopamine release produced by pressure exposure in freely moving rats. Brain Res. 796: 143–149.

Lynch, D.R. and R.P. Guttmann. 2002. Excitotoxicity: perspectives based on N-methyl-D-aspartate receptor subtypes. J. Pharmacol Exp. Ther. 300: 717–723.

Macdonald, A. Hydrostatic pressure physiology. Pp. 157–188. *In*: P. Bennett and D. Elliott (eds.). 1982. Physiology and Medicine of Diving. Balliere Tindall. London.

Meldrum, B. and B. Wardley-Smit, M. Halsey and J.C. Rostain. 1983. 2-Amino-phosphonoheptanoic acid protects against the high pressure neurological syndrome. Eur. J. Pharmacol. 87: 501–502.

Mor, A. and Y. Grossman. 2006. Modulation of isolated N-methyl-d-aspartate receptor response under hyperbaric conditions. Eur. J. Neurosci. 24: 3453–3462.

Mor, A. and Y. Grossman. 2007a. High pressure modulation of NMDA receptor dependent excitability. Eur. J. Neurosci. 25: 2045–2052.

Mor, A. and Y. Grossman. 2007b. Magnesium and AP-5 blockade of NMDA receptor is inhibited by hyperbaric pressure. Proc. 33rd Eur. Undewater Baromed. Soc. Ann. Meeting: 31.

Mor, A. and S. Levy, M. Hollmann and Y. Grossman. 2008. Differential effect of high pressure on NMDA receptor currents in *Xenopus Laevis* oocytes: Diving and Hyperbaric Med. 38: 134–136.

Pearce, P.C. and C.J. Dore, M.J. Halsey, N.P. Luff, C.J. Maclean and B.S. Meldrum. 1990. The effects of MK801 on the high pressure neurological syndrome in the baboon (*Papio anubis*). Neuropharmacology 29: 931–941.

Pearce, P.C. and M.J. Halsey, C.J. MacLean, E.M. Ward, M.T. Webster, N.P. Luff, J. Pearson, A. Charlett and B.S. Meldrum. 1991. The effects of the competitive NMDA receptor antagonist CPP on the high pressure neurological syndrome in a primate model. Neuropharmacology 30: 787–796.

Pearce, P.C. and M.J. Halsey, C.J. Maclean, E.M. Ward, J. Pearson, M. Henley and B.S. Meldrum. 1993. The orally active NMDA receptor antagonist CGP 39551 ameliorates the high pressure neurological syndrome in *Papio anubis*. Brain Res. 622: 177–184.

Pearce, P.C. and C.J. Maclean, H.K. Shergill, E.M. Ward, M.J. Halsey, G. Tindley, J. Pearson and B.S. Meldrum. 1994. Protection from high pressure induced hyperexcitability by the AMPA/kainate receptor antagonists GYKI 52466 and LY 293558. Neuropharmacology 33: 605–612.

Perry, E. and M. Walker, J. Grace and R. Perry. 1999. Acetylcholine in mind: a neurotransmitter correlate of consciousness? Trends Neurosci. 22: 273–280.

Roberts, R.J. and C.J. Shelton, S. Daniels and E.B. Smith. 1996. Glycine activation of human homomeric alpha 1 glycine receptors is sensitive to pressure in the range of the high pressure nervous syndrome. Neurosci. Lett. 208: 125–128.

Rostain, J.C. and C. Forni. 1995. Effects of high pressures of various gas mixtures on rat striatal dopamine detected in vivo by voltammetry. J. Appl. Physiol. 78: 1179–1187.

Rudomin, P. 1990. Presynaptic inhibition of muscle spindle and tendon organ afferents in the mammalian spinal cord. Trends Neurosci. 13: 499–505.

Saar, D. and M. Dadon, M. Leibovich, H. Sharabani, Y. Grossman and E. Heldman. 2007. Opposing effects on muscarinic acetylcholine receptors in the piriform cortex of odor-trained rats. Learn Mem. 14: 224–228.

Schleifstein-Attias, D. and A. Tarasiuk and Y. Grossman. High pressure effects on modulation of mammalian spinal monosynaptic reflex. Pp. 193–199. *In*: P. Bennett, and R. Marquis (eds.). 1994. Basic and Applied High Pressure Biology. Rochester University Press. Rochester.

Shelton, C.J. and M.G. Doyle, D.J. Price, S. Daniels and E.B. Smith. 1993. The effect of high pressure on glycine- and kainate-sensitive receptor channels expressed in *Xenopus* oocytes. Proc. Biol. Sci. 254: 131–137.

Shelton, C.J. and S. Daniels and E.B. Smith. 1996. Rat brain GABA$_A$ receptors expressed in *Xenopus* oocytes are insensitive to high pressure. Pharmaco. Comm. 7: 215–220.

Shrivastav, B. and J. Parmentier and P. Bennett. 1981. A quantitative description of pressure induced alterations in ionic channels of the squid giant axon. Proc. Seventh Symp. on Underwater Physiology: 611–619.

Southan, A.P. and K.T. Wann. 1996. Effects of high helium pressure on intracellular and field potential responses in the CA1 region of the in vitro rat hippocampus. Eur. J. Neurosci. 8: 2571–2581.

Spyropoulos, C. 1957a. Responses of single nerve fibers at different hydrostatic pressures. Am. J. Physiol. 189: 214–218.

Spyropoulos, C. 1957b. The effects of hydrostatic pressure upon the normal and narcotized nerve fiber. J. Gen. Physiol. 40: 849–857.

Takahashi, T. and A. Momiyama. 1993. Different types of calcium channels mediate central synaptic transmission. Nature 366: 156–158.

Talpalar, A. and Y. Grossman. High pressure effects on the cortico-hippocampal connection pp. 45–49. *In*: J. Rostain, A. Macdonald and R. Marquis. (eds.). 1995. Basic and Applied High Pressure Biology v.

Talpalar, A.E. and Y. Grossman. 2003. Modulation of rat corticohippocampal synaptic activity by high pressure and extracellular calcium: single and frequency responses. J. Neurophysiol. 90: 2106–2114.

Talpalar, A.E. and Y. Grossman. 2004. Enhanced excitability compensates for high-pressure-induced depression of cortical inputs to the hippocampus. J. Neurophysiol. 92: 3309–3319.

Talpalar, A.E. and Y. Grossman. 2005. Sonar versus whales: noise may disrupt neural activity in deep-diving cetaceans. Undersea Hyperb. Med. 32: 135–139.

Tarasiuk, A. and Y. Grossman. 1989. Hyperbaric pressure depresses potentiation of polysynaptic medullospinal reflexes in newborn rats. Neurosci. Lett. 100: 175–180.

Tarasiuk, A. and Y. Grossman. 1990. Pressure-induced tremor-associated activity in ventral roots in isolated spinal cord of newborn rats. Undersea Biomed. Res. 17: 287–296.

Tarasiuk, A. and Y. Grossman. 1991. High pressure modifies respiratory activity in isolated rat brain stem-spinal cord. J. Appl. Physiol. 71: 537–545.

Tarasiuk, A. and D. Schleifstein-Attias and Y. Grossman. 1992. High pressure effects on reflexes in isolatedspinal cords of newborn rats. Undersea Biomed. Res. 19: 331–337.

Todnem, K. and G. Knudsen, T. Riise, H. Nyland and J. Aarli. 1989. Nerve conduction velocity in man during deep diving to 360 msm. Undersea Biomed. Res. 16: 31–40.

Wann, K. and A. Macdonald. 1980. The effects of pressure on excitable cells. Comp Biochem. Physiol. 66A: 1–12.

Wann, K. and A. Macdonald and A. Harper. 1979. The effects of high hydrostatic pressure on the electrical characteristics of helix neurons. Comp. Biochem. Physiol. 64A: 149–159.

Wardley-Smith, B. and B.S. Meldrum. 1984. Effect of excitatory amino acid antagonists on the high pressure neurological syndrome in rats. Eur. J. Pharmacol. 105: 351–354.

Wardley-Smith, B. and K.T. Wann. 1989. The effects of non-competitive NMDA receptor antagonists on rats exposed to hyperbaric pressure. Eur. J. Pharmacol. 165: 107–112.

Wardley-Smith, B. and B.S. Meldrum and M.J. Halsey. 1987. The effect of two novel dipeptide antagonists of excitatory amino acid neurotransmission on the high pressure neurological syndrome in the rat. Eur. J. Pharmacol. 138: 417–420.

Zinebi, F. and L. Fagni and M. Hugon. 1988a. Decrease of recurrent and feed-forward inhibitions under high pressure of helium in rat hippocampal slices. Eur. J. Pharmacol. 153: 191–199.

Zinebi, F. and L. Fagni and M. Hugon. 1988b. The influence of helium pressure on the reduction induced in field potentials by various amino acids and on the GABA-mediated inhibition in the CA1 region of hippocampal slices in the rat. Neuropharmacology. 27: 57–65.

Zinebi, F. and L. Fagni and M. Hugon. 1990. Excitatory and inhibitory amino-acidergic determinants of the pressure-induced neuronal hyperexcitability in rat hippocampal slices. Undersea Biomed. Res. 17: 487–493.

# Pressure and Osmoregulation

*A.J.R. Péqueux[1] and P. Sébert[2]*

## INTRODUCTION

Dealing with the effects of high pressure on osmoregulation should imply the study of the effects of high pressure on the various strategies developed by living species to control their body water in a wide variety of possible environments that they are able to invade. Potentially, an exhaustive investigation of all the aspects of this subject would be a hard task and result in an unbelievably large amount of information of limited interest for a clear and good synthetic approach of the question considered in a *'comparative high pressure biology'* context. As a matter of fact, from a general point of view, researchers have looked at the effects of pressure in two main ways. Some have used pressure as a tool to study the features of the structure and functioning of living organisms and their components. These living organisms may therefore even be terrestrial organisms which never experience high pressure during their normal life. The second category of researchers is more directly concerned with the role of pressure in the life of aquatic organisms and, in particular, with adaptations that enable them to live at depth in the ocean or in restricted fresh water (FW) deeper lakes. Probably due to the absence of photosynthetic plants in deep-waters, studies have essentially been conducted on animal species. It is therefore worth noticing that almost all of this work has involved studies on animals from shallow water that have been subjected to pressures of the orders of magnitude of these encountered in the deep-sea. If a few investigations have been conducted on animals collected at depth, it has been done mostly

[1] Ecophysiology Unit, University of Liège, 22, Quai Van Beneden, B-4020 Liège, Belgium.
E-mail: A.Pequeux@ulg.ac.be
[2] ORPHY-EA4324, Université Européenne de Bretagne-UBO, 6, Avenue Le Gorgeu, CS 93837, 29238 Brest Cedex3 France.
E-mail: Philippe.sebert@univ-brest.fr

on moribund organisms, and there is still even now an urgent need for the development of new equipment that will make it possible to approach the physiology of deep-sea animals under pressure.

At the beginning of the previous paragraph which dealt with the effects of high pressure on osmoregulation of animal species from aquatic environments should thus imply  a survey of whole-animal responses to combined actions of pressure and external salinities, whether fluctuating or not. In other words, it would imply detailed study of all the processes at work in the control of the thermodynamic activity of water in biological fluids, either intracellular either extracellular, of the animal species considered. This includes analyses of any kind of transport process and metabolic correlate involved in the adjustment and the control of the level of both inorganic and organic constituents.

That high hydrostatic pressure affects biological processes in diverse ways is clearly demonstrated and established in the various chapters of the present book. Potentially, the biological effects of hydrostatic pressure depend not only upon the organism and tissue being studied but also upon the environmental and evolutionary history of the animal considered. After a long period  when progress in this field  remained very slow, physiological and biochemical experiments have now been initiated, beyond descriptive reports of behavioural observations done on whole animals (Hochachka and Somero, 1973; Macdonald and Miller, 1976; Macdonald, 1981; Sébert, 1993; Siebenaller and Somero, 1989; Zimmerman and Zimmerman, 1976; for review, see also Péqueux, 2008). In  this view, it is quite reasonable to assume that one of the first processes to be affected by pressure is the cell membrane and its associated phenomena  of energy metabolism, a well studied topic in fish (Sébert and Macdonald, 1993; Sébert, 2003, 2008). Many pressure effects which can be considered as indirect evidences of membrane properties disturbance have been reported in the past but, by now, it is clearly demonstrated  that pressure directly affects the physico-chemical properties of the cell membrane of surface and deep-sea animals, and particularly their permeability. This is  covered in Chapter 4 of this book and  in several other review papers published previously (Cossins and Macdonald, 1989; Péqueux and Gilles, 1986; Somero et al., 1983). These disturbances can thus reasonably be thought to result in some impairment of membrane-related processes, among which processes involved in the maintenance of the hydromineral balance at the cellular level as well as at the blood level, in other words, osmoregulation.

The basic aim of this  chapter is to concentrate on a clear understanding of physiological and biochemical aspects of  the functioning under high hydrostatic pressure of tissues and mechanisms involved in osmotic control and regulation. One of its aims is also to try an integration of this with whole-animal responses.

## GENERAL CONSIDERATIONS ON OSMOREGULATION AND ADAPTATION STRATEGIES

### Coping with salinity

When dealing with salinity, one generally relates the osmolality of the external medium to that of the internal body fluids of multicellular organisms, i.e. blood or hemolymph. The relationship between both parameters is then usually expressed in terms of osmoconformity and osmoregulation. Osmoconformity means that their internal osmotic concentration remains close to that of the medium, hence their major problem is restricted to the maintenance of a cellular composition compatible with cellular life and activity. This is the basic problem encountered by monocellular organisms.

Salinity fluctuations induce aquatic organisms to exhibit two kinds of reactions. Stenohaline species are able to tolerate only a narrow range of external salinities, while euryhaline can withstand a wide range of external osmotic concentrations. These well-known and largely used terms may be difficult to understand as there does not exist any clear cut difference between both groups. This must therefore be used on a more relative than absolute way to characterize the animal's abilities to tolerate an osmotic stress. Such a situation may be considered as an illustration that salinity may affect aquatic organisms both directly and indirectly. If the meaning of a direct effect appears as evident, indirect effects include all the conditions where salinity acts together with other conditions which may themselves represent a stress, such as temperature, oxygen availability, the presence of toxic agents or, obviously, high hydrostatic pressure. Due to their unspecified nature and almost unknown possible interactions, these effects are not yet clearly understood nor even as yet identified.

In the present chapter centred on "high pressure and physiological functions among which osmoregulation", a clear cut distinction will have to be done between the effects of pressure on osmo-ionoregulation whichever animal species is considered (direct effects without real ecophysiological meaning, high pressure being used as a tool as mentioned in the introduction), and the situation resulting from the combined action of pressure, salinity, temperature, …in other words, how deep-sea animals have adapted to the combination of these parameters, identifying how osmoregulatory mechanisms have evolved to face pressure (a section of real ecophysiological meaning approaching the indirect effects quoted above).

### General considerations on osmoregulation

The major problem which animal species face is to maintain, or to restore if disturbed, a cellular volume and a basic pattern of intracellular solutes within some narrow range compatible with the different life-supporting cell activities. This function always results in a steady state situation where the animal's cells are in chemical disequilibrium with their surrounding medium. This medium may be external environment, as for monocellular organisms, but most of the time, it is an extracellular fluid or the hemolymph itself. Anyway, it is characterized by an amount of solutes whose concentration is always very different from that of the intracellular fluid. Such a situation generates a gradient of potential which favours the solutes to diffuse, provided the membrane is permeable. All animal species have therefore evolved basic molecular mechanisms to ensure a proper intracellular water and solutes balance. A first and basic way to effect osmotic regulation and which is of universal occurrence is thus to maintain the intracellular fluid isosmotic to the extracellular fluid, either by the external medium, or by the body fluid. In some cases the demand of the medium may be extremely rough which could explain that another strategy to achieve osmoregulation has been evolved, always working aside the mechanisms involved in the isosmotic regulation of the intracellular medium. This second way is to maintain the osmotic concentration of the extracellular fluids (ECF) more or less constantly regardless of the salinity of the surrounding medium. Both processes work together to control water movements at the cellular level. If the potential energy of water and solutes is similar in the ECF to that in the external medium, the composition of the ECF will remain constant without energy expenditure. But any difference resulting from a change of the external medium characteristics will create diffusive forces which will affect the ECF composition. The steady state will then only be maintained by generating a counter-flow identical to the leak but requiring energy expenditure. In that case, osmoregulation will appear as the balance between both processes. It thus appears that the maintenance of an equilibrium whatever the level considered, either intracellular or extracellular, will be achieved by two different categories of mechanisms relevant to the 'pumps and leaks' system: *'limiting processes'* acting on the permeability properties of the boundary structures in order to minimize the diffusive movements of osmotic effectors, and *'compensatory processes'* driving active movements of solutes to counterbalance the diffusive fluxes. Potentially, all these mechanisms are likely to be disturbed or affected by high pressure.

## HIGH PRESSURE EFFECTS ON ION MOVEMENTS AND OSMOREGULATION RELATED MECHANISMS

If one agrees with the idea that one of the first processes to be affected by pressure is the cell membrane and its associated phenomena, then this is now clearly supported by experimental evidences and if these disturbances can reasonably result in some impairment of membrane-related functions, among which is osmo-ionoregulation, there is no other option but to realize that the topic 'pressure on the osmoregulation function' has received only little, or scanty attention up to now. As a matter of fact, much of the research regarding the effects of pressure on membrane-related processes has focussed on excitable tissues, attempting to elucidate the etiology of the pressure-induced neurological disturbances.

## *In vitro* pressure experiments on extracellular body fluids

### *Long-term Pressure Exposure*

Yellow FW eels *Anguilla anguilla* survive moderate pressure applications and can acclimatize to pressure for more than one month without any apparent damage. The results of a study of a 30 d exposure to 101 ATA hydrostatic pressure of FW eels suggest that the hydromineral balance of the fish is significantly disturbed upon pressure treatment (Sébert et al., 1991, 2009; for review, see also Péqueux, 2008 ). In this study, hematocrit, osmolality, protein and ions ($Na^+$, $K^+$, $Cl^-$, $Ca^{2+}$, $Mg^{2+}$) content of the blood plasma were measured in eels submitted to pressure and compared with these from control animals kept at atmospheric pressure. The data comparison pointed to a slightly increased tendency of osmolality (9.5%) and of $Na^+$ and $Mg^{2+}$ contents in fish submitted to pressure. $Cl^-$ content also increased significantly, by 23%. These results suggest that the hydromineral balance of eel has been significantly disturbed upon pressure application as ions content and osmolality of plasma are still modified even after 1 mon under pressure. Shorter-term compression should assess this hypothesis but this work demonstrates that animals must undergo severe hydromineral balance disturbances. While it is not possible to identify from these results the exact nature of the changes probably occurring at the extracellular level, evidence suggests a redistribution of some inorganic ions in the course of the induction by pressure of a new steady state level of the hydromineral balance. These results demonstrate that animals can overcome these disturbances as they resume blood characteristics upon prolonged pressure exposure. It is interesting to note that the fish *A. anguilla* is known to accomplish a long reproduction migration to deep waters during its life cycle. It is therefore reasonable to consider it as, in some way, possibly 'pre-adapted' to high pressure (Dunel et al., 1996; Vettier et al., 2005). Further and more comprehensive knowledge of hydrostatic pressure effects *per se* should arise from comparative studies conducted on animals that normally do not encounter high levels of pressure during their life cycle, but nevertheless are aquatic ectotherms with comparable modes of life. Just like the eel, the Chinese crab *Eriocheir sinensis* spends the major part of its life in FW and migrates towards the sea for reproduction; moreover, both species are powerful hyper-regulators when in a dilute medium. However, in contrast with the eel, the Chinese crab remains confined to shallow waters.

From a series of experiments conducted on that species of crab, it surprisingly appears that the crab itself as well as the physiological processes involved in the control of its hydromineral balance, are outstandingly resistant to pressure when in FW. The animal withstands a 4 d pressure exposure at 101 ATA without any apparent damage (Sébert et al., 1997; Sébert and Péqueux, unpublished data). In agreement with the results previously reported on *Anguilla*, plasma osmolarity as well as $Na^+$ and $Cl^-$ contents of *Eriocheir* haemolymph exhibit a tendency to increase (15, 17 and 28% respectively), after 4 d of pressure exposure.

These results clearly establish that 101 ATA pressure exposure quite significantly affects the maintenance of the haemolymph hydromineral balance by disturbing the movements of ions involved in it. They also show that these effects are similar to whichever animals that were studied and must therefore be of general occurrence, reflecting disturbances of the same underlying mechanisms located at the cellular/molecular levels.

There is now a need to collect information on the nature of these mechanisms and on the way they are affected by pressure in order to better understand the origin of the disturbances reported above. Prior to this, it may be of interest to also consider the possible effects of shorter-term compression in order to discriminate whether these haemolymph impairments are due to a direct action of pressure *per se* on transport mechanisms themselves or are the consequence of a pressure-induced impairment of other mechanisms, for instance metabolic processes such as energy metabolism. A few experiments in this context have been conducted by Sébert et al. (1997) as well as scanty reports on other animals may be found in the literature.

### Short-term Pressure Exposure

Osmolality and inorganic ions ($Na^+$, $K^+$, $Cl^-$, $Ca^{2+}$, $Mg^{2+}$) content of the blood plasma were measured in FW *E. sinensis* following a short-term exposure (3 and 6 h) at 101 ATA (absolute hydrostatic pressure) (Sébert et al., 1997). In these conditions, the possibility exists of physiological processes that have not yet overcome the stress of the pressure change, while the authors confirmed the outstanding resistance of the crabs against hydrostatic pressure. It is reported that after 3 h pressure exposure, only plasma $Ca^{2+}$ shows a very slight increase tendency while all other ions remain unaffected. After 6 h at 101 ATA, the plasma $K^+$ content is decreased by 17 % while $Na^+$, $Cl^-$ and $Mg^{2+}$ remain unchanged; plasma $Ca^{2+}$ confirms an upward tendency. Besides the plasma $K^+$ decrease, an enhanced amount of $K^+$ may be measured in the surrounding FW, indicating a net $K^+$ loss from the animal towards the medium (Sébert et al. 1997). Concomitantly, a drastic enhancement of ammonia efflux occurs after 3 h pressure application (322 $\pm$ 27 $\mu$M ammonia.h$^{-1}$.Kg WW$^{-1}$ at 101 ATA, against 38 $\pm$ 13 $\mu$M.h$^{-1}$.Kg WW$^{-1}$ at 1 ATA, n = 4 animals). By now, these results support the idea that regulation facing pressure is much faster in the crab than in fish, even if the Chinese crab appeared by comparison with other crustaceans to be little affected by hydrostatic pressure.

For instance, the FW crayfish *Procambarus clarkii* appeared to be much more sensitive by even lower pressures. Roer and Shelton (1982) reported that exposure to 51 or 101 ATA immediately inhibit the Na$^+$ influx into the crayfish by approximately 80%. Even 16 or 26 ATA already have an inhibitory effect on Na$^+$ influx after 1 h. According to these authors, there was no effect of pressure on passive effluxes and these reductions in Na$^+$ influx were due to a severe inhibition by pressure of the active uptake of Na$^+$.

These conclusions corroborate other investigations conducted on FW gammarids (Brauer et al., 1980).

Another short-term pressure exposure investigation concerned 3 populations of green crabs *Carcinus maenas* acclimated to various salinities ( 100%, 25% and 10% sea-water SW). Following a 90 min exposure at 51 ATA, the authors reported a significant decrease exceeding 10% in the haemolymph Na$^+$ and Cl$^-$ concentrations of animals acclimated to and tested at 25% and 10% SW, while conversely, there was no change in the haemolymph Na$^+$ of full SW animals (Péqueux and Gilles, 1986; Roer and Péqueux, 1985).

These results clearly establish that short-term exposure to even relatively low pressure steps may affect quite significantly the maintenance of the haemolymph hydromineral balance of ionoregulators species of crustaceans. They support the idea that these impairments are due to a direct action of pressure *per se* on transport mechanisms which are likely to disturb the movements of ions involved in it. In the following section, the nature of these mechanisms and the way they are affected by pressure will be identified.

## *In vitro* pressure experiments on isolated organs, tissues, cells and on enzymes

### *Early experiments on isolated amphibian skin*

It is well known by osmoregulation and membrane physiologists that amphibian skin *in vitro* preparation has been frequently used to provide, in a rapid and rather easy way, a lot of information on the sensitivity of its permeability and transport properties to any variable. In the present case, it has been shown to be particularly convenient to investigate the effects of pressure on the transepithelial potential difference PD which is known to reflect the gross activity of ion transport mechanisms involved in the osmoregulation function.

In early experiments, Okada (1954) reported that the voltage of the isolated frog skin was considerably reduced under pressure. Later, Brouha and co-workers demonstrated that a pressure of 100 atmospheres (atm) induced a small and transient depolarization of the isolated abdominal skin epithelium from the frog *Rana temporaria* followed by a much larger and sustained hyperpolarization phase, while decompression was accompanied by a short hyperpolarization before the skin potential slowly resumed its initial level (Brouha et al., 1970).

These effects have been more systematically investigated subsequently over a wider range of pressure (Péqueux, 1976a). It was shown that application of pressure steps ranging from 25 up to 1000 atm induces PD changes of two types. First, there are short and transient PD variations essentially restricted to the application and to the release of pressure. The second type of PD modifications concerns more significant changes which persist as long as pressure is applied. According to the investigator, there is no evidence of any relation between PD variations of the first type and a specific effect of pressure on ionic permeabilities or active transport processes. They seem to be more likely related to transient alterations resulting from pressure-induced volume changes. Conversely, PD variations of the second type were reported to vary with the magnitude of the pressure applied. Pressure steps up to an order of magnitude of 500 atm induce a sustained hyperpolarization reaching a maximum by as much as 40 mV around 200–300 atm, while pressure applications in excess of 600 atm result in a prolonged depolarization which is irreversible when reaching the range 800–1000 atm. Relevant conclusions as to the origin of the sustained hyperpolarization were drawn from experiments where the ions composition of the salines bathing the skin was judiciously modified and where treatments such as pH changes or oxytocin were applied. All these experiments lead to the conclusion that an increase in $Na^+$ permeability across the cell membranes facing the outer face of the epithelium is the major source of the pressure induced hyperpolarization (Brouha et al., 1970; Péqueux, 1976a, b). $Na^+$ permeability appeared to be much more sensitive to pressure than others which furthermore suggests that permeability characteristics of the cellular membrane are disturbed selectively by pressure. In addition, evidence supports the idea that the skin adjusts the activity of its $Na^+$ intracellular concentration at a normal level as the intensity of the short-circuit current increases during pressure application, which reflects an increase in the transepithelial net influx of $Na^+$ known to be active. This also establishes that the integrity of the active transport machinery is maintained within this pressure range.

These conclusions are of importance as they emphasize the fact that, at least at the low pressures of 25–50 atm, the regulation machinery involved in the cell volume control remains quite functional and able to react immediately in a way to ensure salt balance.

Conversely, experimental evidence suggests that the sustained depolarization observed when larger pressure is applied is due to a direct action of pressure on the $Na^+$ pump itself. The PD fall is indeed of the same order of magnitude as the change resulting from an increase in the intracellular $Na^+$ content when the active output is inhibited at atmospheric pressure by specific drugs. As will be seen later in this chapter, the idea of a direct action of pressure on the $Na^+$ active transport mechanism is further substantiated by experimental evidence showing a severe pressure-induced inhibition of the pump-bound enzyme $(Na^+-K^+)ATPase$ itself.

The question which now arises is to know if these conclusions may be extended to other salt-transporting tissues, especially to those involved in the osmoregulation function.

### Other isolated organs, tissues and cells

*Isolated organs and tissues of fish.* Ionic and osmotic regulation under pressure has been relatively little investigated in fish (for review, see Péqueux, 2008). There are some scanty but interesting reports regarding the effects of pressure on these functions in non-excitable organs and tissues of animals which normally do not experience pressures higher than a few, even a few tens, atmospheres in their usual environment. In a context centred on osmoregulation in teleosts fish, it was interesting to investigate the action of pressure on the branchiae, as the organs are well known for being directly responsible for the control and the maintenance of the blood hydromineral balance. Isolated non-perfused preparations of gills from SW-acclimated eels *Anguilla anguilla* show a complex response to pressure in their passive and active transepithelial ionic movements relevant to the general physiological features which are responsible for the osmo- and ionoregulation in marine teleosts. Pressure *per se* has been reported to induce *in vitro* changes in $Na^+$, $K^+$ and $Cl^-$ contents of gill preparations, clearly due to a direct action on permeability and active transport mechanisms (Péqueux, 1972, 1979, 1981; Péqueux and Gilles, 1977, 1986). These effects appeared to be very complex and dependent on the pressure magnitude applied. As pressure is increased from 1 to 250 atm, a 36% decrease (by reference to the situation measured at atmospheric pressure) in passive $Na^+$ influx occurs, while there is a 150% $Na^+$ increase at 500 atm. It should be noted that all ionic species are not affected in the same way. Higher pressures are needed to increase $Cl^-$ content (350 atm and more to affect $Cl^-$ instead of 250 atm for $Na^+$), but then the magnitude of the $Cl^-$ increase becomes much larger than for Na$^+$ ions. This validates the point already expressed when dealing with amphibian skin that pressure acts selectively on the various transport mechanisms. Experimental evidence led also to the conclusion that, beside the increase of passive Na$^+$ influx, there is a concomitant inhibition of Na$^+$ active extrusion. The pressure-induced increase of gill tissue Na$^+$ content, therefore, appears as the combined result of an active transport inhibition and an increase of passive entry of the same ion along the transepithelial concentration gradient (Péqueux, 1981). In addition to both these effects, the exchange-diffusion process Na$^+$–K$^+$ is activated by pressure too, although it does not result in net variation of tissue Na$^+$, and there is a pressure-induced increase of the K$^+$ permeability (Péqueux, 1979a; Péqueux and Gilles, 1986). Clearly, these results demonstrate that hydrostatic pressure directly acts on and disturbs the permeability and ions transport mechanisms at work in the salt-transporting tissue covering the branchiae in fish and known to be involved in a vital way in the overall osmoregulation function at the body fluids level. This does not exclude an action at another level, more precisely the cell level itself, disturbed by the way processes involved in the osmo-ionoregulation control of the intracellular medium. For instance, in white muscle and gill tissues of the FW eels exposed for a long term (30 d) to 101 ATA hydrostatic pressure (results of plasma ions content of the same fish have already been discussed in a preceding section), only Na$^+$ and Cl$^-$ levels increase considerably ($\pm$ 30 and 50 % respectively) under pressure while K$^+$ and water remain unaffected (Sébert et al., 1991). It is clearly not possible to identify from these results the exact nature of the changes which possibly occur at the intracellular level. However, an effect on the volume of the extracellular space can reasonably be discarded by now as K$^+$ and water contents are not modified. As already pointed out, evidence suggests a redistribution of some inorganic ions in the course of the induction by pressure of a new steady state level of the hydromineral balance which the animal is able to overcome.

*Isolated organs and tissues of crab.* Almost similar results as those reported on the eel were obtained when submitting Chinese crabs *E. sinensis* for 4 d at 101 ATA (Sébert et al., 1997). In crab muscle and gill tissues, there is also a tendency for Na$^+$ and Cl$^-$ increase, which is quite substantial as far as muscle Na$^+$ content is concerned ($\pm$ 50%). At variance, K$^+$ and Mg$^{2+}$ contents do not appear to be modified by the pressure exposure. The rise in Ca$^{2+}$ content of anterior gills should be noted. These results further corroborate the general occurrence of hydromineral balance disturbance by hydrostatic pressure.

*In vitro experiments on mammal erythrocytes.* While the question of the nature of the pressure-induced K$^+$ movements reported above which occur in compressed isolated gills of SW eels is far from being clearly identified, more conclusive evidence concerning the pressure sensitivity of K$^+$ passive movements have been obtained with human erythrocytes much more accessible to *in vitro* experimentation under pressure. As a matter of fact, it is reported that the net $K^+$ efflux from red blood cells essentially considered as passive increases gently but almost linearly with pressure until 600–700 atmospheres. Above that pressure range, a much more pronounced increase in membrane $K^+$ permeability occurs (Péqueux et al., 1980; Zimmermann et al., 1980). It was further reported that pressure steps up to 400 ATA were also able to induce activation of $Cl^-$ dependent $K^+$ flux in the same material as well as in low $K^+$ sheep red cells which is not $Na^+$ dependent or inhibited by ouabain or bumetanide (Hall et al., 1982; Ellory et al., 1985). Considerable evidence supports the idea that this transport system is carrier-mediated. In apparent contradiction with these results are the data of Goldinger et al. (1981), who reported a 50% reduction of $K^+$ efflux from human erythrocytes upon a 150 ATA compression. However, the decrease turns to a dramatic increase in $K^+$ permeability when intracellular $Ca^{2+}$ concentration is raised by depletion of ATP (Knauf et al., 1975). These authors suggest that the mechanism of $K^+$ efflux enhancement is not the result of a direct effect of pressure on $K^+$ permeability, but rather the consequence of an increase in intracellular $Ca^{2+}$ either by increasing $Ca^{2+}$ permeability or reducing intracellular bound $Ca^{2+}$. Accordingly, increasing the internal $Ca^{2+}$ level also produces a dramatic increase in $K^+$ permeability at 1 ATA. No doubt, this conclusion emphasizes the uppermost difficulty to explain pressure effects and to discriminate between a direct or indirect action of the physical parameter. Other explanations for the mechanism of $K^+$ efflux increase observed under pressure brings up more direct possible effects on the membrane itself. One of them is based on the idea that increases in membrane permeability may be associated with a 'mechanical breakdown' of the cell membrane (Zimmermann, 1979; Zimmermann et al., 1977). Electrical breakdown of the cell membrane is known to result in a dramatic but reversible decrease in the membrane resistance and to be associated with a marked increase in membrane permeability. According to Zimmermann and co-workers, there are certain finite membrane areas for which the actual membrane thickness depends on the voltage across the membrane and the applied pressure. Their model predicts that the breakdown voltage of the cell membrane should be pressure-dependent and that a critical absolute hydrostatic pressure exists at which the intrinsic membrane potential is sufficiently high to induce a reversible 'mechanical breakdown'. This critical pressure is expected at about 650 atm, the huge but reversible increase in net $K^+$ efflux above 600 atm reported above is perfectly consistent with the predictions of the electro-mechanical model. This doesn't mean that other possible alternative explanations must then be *a priori* discarded for example, the possible occurrence of a pressure-induced phase transition in the lipid bilayer of the membrane. Furthermore, the different compressibilities of water, lipids and proteins could also produce tensions along the membrane surface, which could then result in the observed permeability increase. Mammal erythrocytes have also been used to further investigate the effects of hydrostatic pressure on $Na^+$ permeability and transport. The early work of Podolsky (1956) points out that a pressure of 80 atm reduces the $Na^+$ efflux from cat erythrocytes. In human erythrocytes, Goldinger and co-workers reported a 40–50 % inhibition of $Na^+$ efflux showing a roughly hyperbolic way, rising very steeply between 7 and 30 ATA and then reaching a plateau up to 400 ATA (Goldinger et al., 1980, 1981). According to these authors, the part of efflux corresponding to a $Na^+$ active transport process sensitive to ouabain was inhibited by 38% at 150 ATA, while diffusional efflux was inhibited by 30%. A somewhat different picture has been obtained by Péqueux and co-workers in the course of a more extensive study carried out on human and pig erythrocytes (Péqueux, 1979b, 1980; Péqueux and Parisis, 1980, Péqueux, unpublished data). These authors reported a 30–40 % inhibition at 100 atm of the overall efflux of $Na^+$ as well as of the 'ouabain-sensitive' $Na^+$ efflux which may be considered as a $Na^+$ pump component in human red blood cells. However, instead of keeping a plateau value, the inhibition becomes higher when the pressure is higher, being about 95% at 1000 atm and 37°C. If the discrepancy between the experimental data has not yet received conclusive explanation, it is important to point out that pressure is shown to inhibit active $Na^+$ transport in human erythrocytes, but also in pig and ox erythrocytes. In addition, in pig red cells, a $K^+$ 'ouabain-sensitive' influx was observed to be inhibited concomitantly, in support of the idea of a pressure effect on a $Na^+/K^+$ exchange process. Experimental evidence further established that the inhibition of active transport was not due to an effect of pressure on cell metabolism, hence to a reduction in the energy supply needed by the ion pumps, but rather to a direct effect on membrane-bound mechanisms themselves.

*Enzymes*

In a review dealing with iono- and osmo-regulation under pressure, the knowledge of how enzymes may be or are affected by the physical parameter is of significant importance. The actions of pressure on metabolic processes are among those which are the best documented and a large piece of information is to be found on the effects of pressure on their related enzymes. It is evident that any effect on enzymes involved in metabolic energy production will in turn implicate more or less rapidly an effect on transport mechanisms linked to energy supply, such as active transport mechanisms. In the previous sections, pressure-induced disturbance on ions 'pumps' have already been reported to occur. This kind of effect however remains essentially indirect and, when it occurs, always results in an inhibition by deficiency in energy provision. Therefore, the question does not exactly fit in with the subject of this chapter and interested readers could refer to the extensive specialized works and reviews. Much more evident into this framework is the interest to analyze the effects of hydrostatic pressure on the $(Na^++K^+)$ATPase activity of organs and tissues involved in salt transport. All the tissues 'pumping' $Na^+$, in the way they are involved in the iono-osmoregulation function, are known to have high activity of that enzyme system. This question has been investigated by several authors on the number of tissues implicated in osmoregulation in aquatic animals which may or may not be naturally submitted to hydrostatic pressure. As a general rule, most of the $(Na^++K^+)$ ATPases from surface animals remain little affected, if any, by moderate pressure steps of around 100–300 atm (Péqueux and Gilles, 1978, 1986). At higher pressure, enzyme activity is drastically inhibited reaching more than 80% at 1000 atm. Very similar results were obtained with mammal erythrocytes, exhibiting about 60% inhibition at 500 atm while remaining unaffected at 250 atm (Péqueux, 1979b). Qualitatively, pressure affects all tissues considered in the same way. Only quantitative differences occur, either in the percentage of inhibition or in the magnitude of pressure needed to induce inhibition. Slight activation of $(Na^++K^+)$ ATPase activity can even be observed at moderate pressure (Goldinger et al., 1981; Péqueux and Gilles, 1978). In human erythrocytes, Goldinger and co-workers (1981) reported up to 40% stimulation at 150 ATA. It must be kept in mind that basically the functioning of $(Na^++K^+)$ ATPases is impaired at high pressure. Such an effect is of great physiological significance as it might be directly related to the effects that pressure exerts on the active ionic movements across the epithelia and cell membranes reviewed above. A reasonably close parallelism can be established between the sensitivity to pressure of both processes in most of the tissues studied, either when considering the lack of direct effect at low pressure, or the progressive inhibition at higher pressure. In short, pressure induces inhibition of the $Na^+$ active transport by acting directly on the responsible enzyme. The situation is less clear at moderate pressures ranging up to 250–300 atm. Discrepancies sometimes exist between pressure effects on enzyme and $Na^+$ transport. This is particularly evident in the work of Goldinger and co-workers (1980, 1981) in which active $Na^+$ extrusion from human erythrocytes was inhibited by 40–50% at 50 atm while $(Na^++K^+)$ ATPase activity showed 40% activation. Some of these disparities may arise from differences in the methods used for the isolation and treatment of the enzyme extracts or from differences in their stability level. As a matter of fact, the $(Na^++K^+)$ ATPase activity is known to be very dependent on its lipid environment inside the cell membrane. In view of this, any difference in lipid composition could well lead to differences in pressure sensitivity, liable to explain the discrepancies reported so far. The results collected on the 'pressure-preadapted' fish *A. anguilla* exposed for 1 mon at 101 ATA (Sébert et al., 1991) are also interesting from which effects on the plasma, muscle and gills ions have already been reported and discussed in the above sections. Supporting the hypothesis of a $Na^+$ balance disturbance and/or impairment occurring under pressure at the cell level are the results of $(Na^++K^+)$ ATPase activity measured in the gill tissue after return at atmospheric pressure. As a matter of fact, the $(Na^++K^+)$ ATPase maximum activity exhibited a drop of 55% in tissues of pressure-exposed animals, suggesting an impaired functioning of the $Na^+$ active transport processes responsible at the gill level for both the maintenance of a low $Na^+$ level in the intracellular fluid and the control of the transepithelial $Na^+$ movements involved in the blood natremia regulation. The enzyme activity/external $Na^+$ concentration curve is drastically modified in compressed fish which suggests a significant change in $Na^+$ affinity of $(Na^++K^+)$ATPase in these animals. Whether these changes can be attributed to alterations of the kinetic properties of an existing enzyme protein or reflect the *de novo* synthesis of a new enzyme protein still remains a matter of speculation.

## CONCLUSIONS

As reported in the previous sections of this chapter, any kind of membrane-associated process, either passive diffusional movements, or energy mediated active transport mechanisms, is clearly subject to be significantly disturbed by hydrostatic pressure of even a few tens atmospheres, whether directly or indirectly. These data and these conclusions arise from studies conducted on a significant selection of animal species, organs and tissues originating essentially from the upper layers of the water column. They are actually of a general occurrence, able to affect any kind of surface animal species, organ and tissue. They also raise the question about the relationship between these alterations, more particularly those concerning energy-consuming mechanisms of ions transfers, and possible effects on energy metabolism itself. It has clearly been shown that hydrostatic pressure has specific effects, sometimes only transient, on energy metabolism (Simon et al., 1989; Sébert et al., 1997 for review). For instance, it is reported that the adenylate energy charge (EC) in the muscle of Chinese crabs is significantly lower after 3 h pressure exposure ($0.61 \pm 0.03$ against $0.86 \pm 0.02$) while it resumes its control value after 8 d under pressure (a complete recovery has even been observed after 6 h) (Sébert et al., 1997). From experiments conducted on FW eels kept at 101 ATA for 1 mon, it was concluded that, if pressure does not significantly affect the activity of metabolism-linked enzymes in liver, it induces activity changes in muscle which can be summarized as an improvement of the aerobic pathway in red muscle and of the anaerobic pathway in white muscle (Sébert et al., 1991). While these results could be disputed (Siebenaller and Somero, 1982), they support the idea that pressure has specific effects on the energy metabolism of the crustacean and fish species studied. They establish, moreover that ions balance impairment, as for instance $Na^+$ balance impairment, occurs at the tissue level at the same time when a new state of energetic metabolism results from adjustments of intertissue coupling of anaerobic and aerobic metabolisms induced by a long-term exposure to hydrostatic pressure (Sébert et al., 1991). It is also clear that (1) even when significant pressure-induced disturbances occur upon short-term exposure, they can be overcome quite easily upon long-term exposure and (2) the adaptive mechanisms triggered by long-term pressure exposure could persist for at least a few days after decompression (Simon et al., 1989). The ability of these animals to live for a prolonged time at pressure despite transient alteration of their energy metabolism, their plasma and cellular ionic balance results from a rapid compensation and subsequent acclimation to the disturbing effects of pressure. The nature of these mechanisms is far from being clearly identified and further investigation is obviously needed on whole animals to provide a clear picture of the pressure sensitivity and of the acclimation procedure of all the membrane-associated mechanisms at work in osmoregulating animals. One must assume that deep-waters animal species have developed specialized features in the course of their evolution history to successfully colonize deep environments. At least, the data reported in this chapter show the type of problems that hydrostatic pressure imposes on organisms and long-term pressure exposure experiments contribute to throw light on the adaptive processes that have been developed. They certainly allow identification of the various possible levels of organization, either cellular or molecular, which are sensitive to the physical parameter and point out the constraints that have to be overcome to cope with it. While the knowledge of these features is by no doubt of outmost interest, they remain outside the direct scope of a clear understanding of the problems encountered by surface diving species. As far as deep-waters species are concerned, it is evident that colonization of the deep water environment has been possible only after long-term ecological adaptation both at the physiological and the biochemical levels. Accordingly, deep-water organisms must have evolved special membrane properties to cope with the ambient pressure and with other ecological parameters specific to deep-water environments. More insight in the pressure biology of these species could therefore arise only from comparative studies conducted on abyssal organisms. However, information on osmoregulatory processes at work in deep-water animals, as well as on their pressure sensitivity, remains up to now very scanty or practically lacking, probably due to the fact that deep-water animals present the experimenter with the difficulty of collecting them in a reproducible, reasonably healthy condition. Collecting deep-water animals is usually carried out by trawling, use of baited hooks or sometimes by trapping. These animals almost certainly suffer from decompression during their retrieval which explains the difficulty to get healthy specimens able to survive at atmospheric pressure or even to be kept in a high pressure aquarium. Evidence suggests that membrane transport systems of deep-living animals are perturbed by reductions in pressure, much as the systems of shallow-living species are upset by increased pressure. Besides the usual problems of stress encountered when investigating surface animals, the risk of pathological disturbances is almost impossible to avoid. This means that samples collected on board from these animals as well as experiments conducted on their cells and tissues originate from moribund animals. Up to now, resuscitation attempts by recompression to the animal with customary pressure did not lead to satisfying results especially if the purpose is to experiment with osmoregulation. This explains why only a few experiments have been conducted in the laboratory on deep-water whole animals as well as *in vitro* on their tissues and cells and why most of the known investigations were biochemical experiments conducted on biological material collected on moribund animals and subsequently frozen. Clearly, investigations in this direction still requires the perfection of new techniques of capture and maintenance in the laboratory.

## OSMOREGULATION IN DEEP-SEA SPECIES

### Extracellular body fluids

In spite of the huge technical difficulties mentioned above, attempts to measure the osmolarity and the ionic composition of the plasma of deep-sea fish were carried out on board of ships during research cruises (Griffith 1981; Graham et al., 1985; Shelton et al., 1985). Accordingly, plasma inorganic ions and osmolarity, plasma proteins, urea and trimethylamine (TMAO) were measured and compared between several species of shallow-water marine teleosts, elasmobranches, as well as deep bentic and midwater teleosts. It is reported that elasmobranches have high osmolarity (1035 mOsm/l) because of high plasma urea (363 mM/l), TMAO (66 mM/l) and chloride (295 mM/l). Shallow water teleosts have low osmolarity (444 mOsm/l), chloride (176 mM/l), urea (4mM/l) and TMAO (14 mM/l). Deep benthic teleosts have higher osmolarities (576 mOsm/l) and chloride (242 mM/l) levels than shallow-water teleosts, as did midwater teleosts (561 mOsm/l, 267 mM/l). It is interesting to note that plasma TMAO is higher in benthic (51 mM/l) but not in midwater (12 mM/l) teleosts, and urea is low in midwater (1.0 mM/l) and benthic (1.5 mM/l) groups (Griffith, 1981). Basically, these results fit in with data of same parameters collected on nine species of deep-sea fish from depths of 900–4000 m (Shelton et al., 1985). Stress and morbidity are known to possibly raise osmolarity and chloride in marine teleosts. This could account for high values observed in midwater and benthic species which were sampled after considerable trauma as discussed in a previous section. By and large, these data suggest that deep-sea teleosts osmoregulate, at least from the point of view of their plasma patterns, as do shallow-water species. They do not support any idea that osmoregulatory specializations, such as ureosmotic regulation, evolve more rapidly in the deep-sea. Similar conclusions arise from more recent data collected on stenohaline osmoconformer crabs *Bythograea thermydron* captured at 2500 m in the deep-sea hydrothermal vent habitat (Martinez et al., 2001). This study investigating the salinity tolerance and the patterns of iono-osmoregulation of the deep-sea crab, demonstrates that the iso-osmotic regulation as well as iso-ionic regulation for $Na^+$ and $Cl^-$ are not affected by changes in hydrostatic pressure. In view of the ionic disturbances reported to occur in tissues and cells of surface animals when submitted or acclimated to hydrostatic pressure, it is clear that the physiology of osmoregulation of deep-water animals has adapted to cope with high pressure. It is then reasonable to predict that decompression under atmospheric pressure will damage the regulation processes in a way similar to disturbances observed upon compression of surface animals. This could be one of the problems explaining the poor physiological state of deep-sea animals brought up to the surface, making experimentation at atmospheric pressure almost impossible.

## Isolated organs, tissues cells and enzymes

### Tissues and cells

Till date and for the reasons pointed out above, the control of the hydromineral balance at the cellular level received even less attention than osmoregulation at the blood level. Information in this field is almost lacking. A few investigations were conducted on erythrocytes collected at sea on deep-sea fish such as *Coryphaenoïdes rupestris, C. armatus, Antimora rostrata, Synaphobranchus kaupi* and *Hoplostethus atlanticus*. Compared to the data from shallow-sea fish, their concentrations of $Na^+$ and $K^+$ are high and low respectively, both of them being erratic (Shelton et al., 1985). Surprisingly, recompressing the cells to their normal ambient pressure did not produce any particular effect. In agreement with the authors, it is difficult to avoid the conclusion that the cells were irreversibly damaged. In another study, Prosser et al. (1975) reported intramuscular $Na^+$ and $K^+$ concentrations to be unusual in *Antimora rostrata* and *Coryphaenoïdes* sp., suggesting the existence of leaky membranes. Here too, it is difficult to consider if these fish are in perfect physiological shape.

*Enzymes*

Deep-sea fish tissues among which gill tissue contain significant amounts of $Na^++K^+$-ATPase (Moon, 1985; Gibbs and Somero, 1989; Péqueux, unpublished results). In *Antimora rostrata* for instance, enzyme activity is within the range of those measured in shallow-water fish (Moon, 1985). However, unlike in shallow-water fish, enzymes in deep-sea fish are especially unstable. In addition to huge individual differences, enzyme activity drops down to insignificant values within hours when stored at 0° C and even –20°C (Moon, 1985; Péqueux, unpublished results). The sensitivity to pressure of the membrane-ATPase has been checked by some authors to seek the probable pressure-adaptive strategies in deep-living species. It is interesting to point out that $Na^++K^+$-ATPase is a membrane enzyme system for which any effects of pressure have been reported (Chong et al., 1985; Moon, 1985; Pfeiler, 1978; Gibbs and Somero, 1989). As a general rule, enzyme activity is inhibited by hydrostatic pressure, but inhibition is nonlinear. However, ATPases from deep-sea fish are less inhibited than those from shallow-water species. It was suggested that pressure greater than 200 atm was required for pressure adaptation (Gibbs and Somero, 1989). It is worth observing the results obtained using gill extracts from the abyssal fish *Antimora rostrata* captured at 1,500 meters (Moon, 1985). There is a $Na^++K^+$-ATPase activation up to 250 atm followed by a progressive inhibition of activity with further increase in pressure. Quite similar results were obtained by Pfeiler (1978) on the deep-sea (300–1500 m) fish *Anoplopoma fimbria*. Differences in ouabain sensitivity and stability further support the idea of distinctive catalytic properties of deep-sea enzymes. These effects may be of great physiological significance as they strongly suggest that, within the physiological range of pressures, deep-sea fish enzymes attain a more favourable conformation and are not inhibited. This means that active transport processes of inorganic ions are kept functional within the same pressure range too.

*Adaptation to the deep-sea life*

The low threshold pressure at which membrane-linked functions are disturbed in surface species, as well as the magnitudes of the pressure effects, indicate that life in the deep-waters has required significant biochemical adaptations in membrane systems. As all physiological processes depend on the integrity of these membrane structures composed of proteins and lipids, there is strong likelihood that structural changes induced by pressure at their level are at the origin of the necessary adaptations. Changes in protein-protein or protein-lipid interactions are possible candidates to induce conformational changes possibly sensitive to pressure. Experimental evidence has established the more direct involvement of lipids. As a matter of fact, pressure and temperature effects on lipid-dependent enzymes located in biological membranes (such as the $(Na^{+}+K^{+})$ ATPase) suggest that resistance adaptation to depth results from adjustments in membrane phospholipid composition to maintain proper membrane fluidity. In other words, the higher the percentage of phospholipids acyl chains bearing double bonds, the more fluid the membrane contains. Supporting this idea, Chong et al. (1985) have demonstrated that ATPase activity is reduced by pressure in a way which is closely matched by the reductions in the fluidity of the bilayer. Accordingly, there is evidence for homeoviscous adaptation in deep living animals where the decrease in fatty acid saturation with depth is statistically significant. This strongly suggests that conservation of membrane fluidity through the water column (Cossins and Macdonald, 1986), and, maybe for species inhabiting a wide depth range, the possibility to change membrane lipid composition as a function of environmental pressure (Gibbs and Somero, 1989). In addition to this, it has also been reported that many deep-sea proteins have evolved resistance to pressure. However, it is not the case of all proteins. In the recent past, it has been established that high level of protection can be afforded by a 'special' family of osmolytes, such as the protein-stabilizer trimethylamine N-oxide (TMAO) (Yancey et al., 2002; Yancey and Siebenaller, 1999). It has been demonstrated that TMAO, known to remain under 80 mmol/kg wet weight in most shallow teleosts, increases with depth in white muscle of several families of teleosts (up to 288 mmol/ kg at 2900 metres), as well as in red muscle (up to 93–118 mmol/kg, which is less than white muscle but more than shallow species) (Yancey et al., 2002). Similar results on several other species of teleost were already reported in a previous section. Moreover, the presence of TMAO does not appear to be restricted to fish. High levels of osmolyte were also reported in muscle tissue of FW amphipods collected at depth in the FW Baikal Lake (Zerbst-Boroffka et al., 2005). Moreover the authors reported that the elevated intracellular osmotic pressure was balanced by upregulating the extracellular heamolymph NaCl concentration. From a series of experiments carried out on several enzymes and pressure-sensitive proteins, Yancey has established that 200–250 mM TMAO protect ligand binding, protein stability and polymerization against pressure inhibition. Stabilizing osmolytes like TMAO are known to disfavour the formation of bound water (hydration water) around ligands and proteins, thus favouring protein folding and ligand binding (which reduce hydration water compared to unfolded and unbound states) (Yancey et al., 2002). In other words, the effects of pressure and TMAO would oppose each other in an additive manner. According to Yancey, other solutes can offset pressure effects too. This should be the case with trehalose, DMSO, D2O, polyols, ..., and still unidentified solutes (Yancey, 2005; Yancey et al., 2002, for review, see also Zerbst-Boroffka et al., 2005; Burg and Ferraris, 2008).

## GENERAL CONCLUSION

The data reviewed in this chapter emphasize that osmoregulation (in a broad sense) at elevated hydrostatic pressure is a fundamental problem in the physiology of deep-waters animals. Even for others, it is clear that pressure acts as a limiting factor in their vertical migrations. What also clearly appears in this chapter is that the field of pressure studies dealing with ion transport and osmoregulation is full of lacunae. This originates, no doubt, from the difficulty to get deep-water animals in a reasonably healthy condition. While some reports suggest that osmoregulation patterns of these animals do not look so different from those of surface animals, survival in and colonization of the very deep-waters is related only to long-term ecological adaptations both at the physiological and biochemical levels. Deep-waters animals have evolved special membrane properties such as 'homeoviscosity', to cope with the ambient pressure and with the other ecological parameters specific to deep environments. Enzymes have also been modified to operate at high pressure. In many cases, pressure sensitivity has been reduced by modification of bonds strength within molecular structures. As demonstrated more recently, high levels of TMAO have been selected to offset inhibitory effects of high hydrostatic pressure at the cell and the molecular levels. It is clear that more insight in the biology of deep-waters animals could arise only from studies conducted on healthy abyssal organisms. If captured carefully, some deep-living species (such as, vestimentiferan tubeworms and cephalopoda molluscs ) can presently be kept alive indefinitely at atmospheric pressure and methods are also available for successful maintenance under their respective habitat pressure in the laboratory. Accordingly, it is to be hoped that new technologies of capture and maintenance in the laboratory will be perfected for more invertebrates as well as for vertebrates too. Last but not the least, the interest with experiments conducted under pressure on surface animals must be pointed out as these investigations permit identification of the types of problems which hydrostatic pressure imposes on living organisms. They allow identification of the various possible levels of organization, either cellular or molecular, which are sensitive to the physical parameter and point out the constraints which have to be overcome to cope with it. Moreover, they might lead to conclusions of importance for a better and more complete understanding of many of the problems encountered by diving species.

# REFERENCES

Brauer, R.W. and M.Y. Bekman, J.B. Keyser, D.L. Nesbitt, S.G. Shvetzov, G.N. Sidelev and S.L. Wright. 1980. Comparative studies of Na+ transport and its relation to hydrostatic pressure in deep and shallow water gammarid crustaceans from Lake Baikal. Comp. Biochem. Physiol. 65A: 129–134.

Brouha, A. and A. Péqueux, E. Schoffeniels and A. Distèche. 1970. The effects of high hydrostatic pressure on the permeability characteristics of the isolated frog skin. Biochem. Biophys. Acta 219: 455–462.

Burg, M.B. and J.D. Ferraris. 2008. Intracellular organic osmolytes: function and regulation. J. Biol. Chem. 283: 7309–7313.

Chong, P.L. and P.A.G. Fortes and D.M. Jameson. 1985. Mechanisms of inhibition of (Na,K)—ATPase by hydrostatic pressure studied with fluorescent probes. J. Biol. Chem. 260: 14484–14490.

Cossins, A.R. and A.G. Macdonald. 1986. Homeoviscous theory under pressure. III. The fatty acid composition of liver mitochondrial phospholipids of deep sea fish. Biochem. Biophys. Acta 860: 325–335.

Cossins, A.R. and A.G. Macdonald. 1989. The adaptation of biological membranes to temperature and pressure: fish from the deep and cold. J. Bioen. Biomemb. 21: 115–135.

Dunel-Erb, S. and P. Sébert, C. Chevalier, B. Simon and L. Barthélémy. 1996. Morphological changes induced by acclimation to high pressure in the gill epithelium of yellow freshwater eel (Anguilla anguilla). J. Fish. Biol. 48: 1018–1022.

Ellory, J.C. and A.C. Hall and G.W. Stewart. 1985. Volume sensitive passive potassium fluxes in red cells. Proc. First Congress of Comparative Physiol. And Biochem. Liège. B43. Springer-Verlag, Berlin, Heidelberg, New York, Tokyo.

Gibbs, A. and G.N. Somero. 1989. Pressure adaptation of $Na^+/K^+$ ATPase in gills of marine teleosts. J. Exp. Biol. 143: 475–492.

Goldinger, J.M. and B.S. Kang, Y.E. Choo, C.V. Paganelli and S.K. Hong. 1980. The effect of hydrostatic pressure on ion transport and metabolism in human erythrocytes. J. Appl. Physiol. 49: 224–231.

Goldinger, J.M. and B.S. Kang, R.A. Moring, C.V. Paganelli and S.K. Hong. Effect of hydrostatic pressure on active transport, metabolism, and the Donnan equilibrium in human erythrocytes. Pp: 589–599. In: A.J. Bachrach and M.M. Matzen (eds.). 1981. Undersea Medical Society, Inc. Bethesda, Maryland.

Graham, M.S. and R.L. Haedriuch and G.L. Fletcher. 1985. Hematology of three deep-sea fishes: A reflection of low metabolic rates. Comp. Biochem. Physiol. 80A: 79–84.

Griffith, R.W. 1981. Composition of the blood serum of deep-sea fishes. Biol. Bull. 160: 250–264.

Hall, A.C. and J.C. Ellory and R.A. Klein. 1982. Pressure and temperature effects on human red cell cation transport. J. Membrane Biol. 68: 47–56.

Hochachka, P.W. and G.N. Somero. 1973. Strategies of biochemical adaptation. Philadelphia: Saunders, W.B. Company (ed.).

Knauf, P.A. and J.R. Riordan, B. Schuhmann, I. Wood-Guth and H. Passow. 1975. Calcium-potassium stimulated net potassium efflux from human erythrocyte ghosts. J. Membrane Biol. 25: 1–22.

Macdonald, A.G. Molecular and cellular effects of hydrostatic pressure: a physiologist's view. Pp: 567–575. In: A.J. Bachrach and M.M. Matzen. (eds.). 1981. Underwater Physiology VII. Undersea Medical Society, Inc. Bethesda, Maryland, USA.

Macdonald, A.G. and K.W. Miller. Biological membranes at high hydrostatic pressure. Pp: 117–147. In: D.C. Malins and J.R. Sargent (eds.). 1976. Biochemical and Biophysical Perspectives in Marine Biology, Vol. 3. Academic Press, London, New York.

Martinez, A.S. and J.Y. Toullec, B. Shillito, M. Charmantier-Daures and G. Charmantier. 2001. Hydromineral regulation in the hydrothermal vent crab *Bythograea thermydron*. Biol. Bull. 201: 167–174.

Moon, T.W. 1985. Effects of hydrostatic pressure on gill Na-K-ATPase in an abyssal and a surfacing-dwelling teleost. Comp. Biochem. Physiol. 52B: 59–65.

Okada, K. 1954. Effects of hydrostatic high pressure on the permeability of plasma membrane. V. On plasmolysis. Okayama Igakkai Zasshi. 66: 2095–2099.

Péqueux, A. Hydrostatic pressure and membrane permeability. Pp: 483–484. *In*: The Effects of Pressure on Organisms. 1972. Symposium XXVI. Soc. Exp. Biol. Cambridge University Press. Cambridge.

Péqueux, A. 1976a. Polarization variations induced by high hydrostatic pressures in the isolated frog skin as related to the effects on passive ionic permeability and active Na$^+$ transport. J. Exp. Biol. 64: 587–602.

Péqueux, A. 1976b. Effects of pH changes on the frog skin electrical potential difference and on the potential variations induced by high hydrostatic pressure. Comp. Biochem. Physiol. 55A: 103–108.

Péqueux, A. Effects of high hydrostatic pressures on ions permeability of isolated gills from sea water acclimated eels *Anguilla anguilla*. Pp: 127–128. *In*: R.Gilles (ed.). 1979a. Animals and Environmental Fitness. Vol. II. Abstracts. Pergamon Press, Oxford, New York.

Péqueux, A. Ionic transport changes induced by high hydrostatic pressures in mammalian red blood cells. Pp: 720–726. *In*: K.D. Timmerhaus and M.S. Barber (eds.). 1979b. High Pressure Science and Technology. Vol. I. Plenum Press, New York.

Péqueux, A. Osmoregulation and ion transport at high hydrostatic pressure. Pp: 405–425. *In*: R.Gilles (ed.) 1980. Animals and Environmental Fitness. Vol. I. Pergamon Press, Oxford, New York.

Péqueux, A. Effects of high hydrostatic pressures on Na$^+$ transport across isolated gill epithelium of sea-water acclimated eels *Anguilla anguilla*. Pp: 601–609. *In*: A.J. Bachrach and M.M. Matzen (eds.). 1981. Undersea Medical Society, Inc. Bethesda, Maryland.

Péqueux, A. Fish Osmoregulation in Special Environments. Pp: 1–39. *In*: P. Sébert, D.W. Onyango and B.G. Kapoor (eds.). 2008. Fish Life in Special Environments. Science Publishers, Enfield, NH, USA.

Péqueux, A. and R. Gilles. 1977. Effects of high hydrostatic pressures on the movements of Na$^+$, K$^+$ and Cl$^-$ in isolated eel gills. Experientia 33: 46–48.

Péqueux, A. and R. Gilles. 1978. Effects of high hydrostatic pressures on the activity of the membrane ATPases of some organs implicated in hydromineral regulation. Comp. Biochem. Physiol. 59B: 207–212.

Péqueux, A. and M. Parisis. 1980. Effects of high hydrostatic pressures on Na$^+$ and K$^+$ transports across membranes of human erythrocytes. Proceeding of the International Union of Physiological Sciences. Abstracts of the XXVIII International Congress, Budapest. Vol. XIV: 635.

Péqueux, A. and R. Gilles. Effects of hydrostatic pressure on ionic and osmotic regulation. Pp. 161–189. *In*: A.O. Brubakk, J.W. Kanwisher and G. Sundnes (eds.). 1986. Diving in Animals and Man. Tapir. Publishers. Trondheim, Norway.

Péqueux, A. and R. Gilles, G. Pilwat and U. Zimmermann. 1980. Pressure induced variations of K$^+$ permeability as related to a possible reversible mechanical breakdown in human erythrocytes. Experientia, 36: 565–566.

Pfeiler, E. 1978. Effects of hydrostatic pressure on (Na$^+$+K$^+$)ATPase and Mg$^{2+}$-ATPase in gills of marine teleost fish. J. Exp. Zool. 205: 393–402.

Podolsky, R.J. 1956. A mechanism for the effect of hydrostatic pressure on biological systems. J. Physiol. 132: 38–39.

Prosser, C.L. and W. Weems and R. Meiss. 1975. Physiological state, contractile properties of heart lateral muscles of fishes from different depths. Comp. Biochem. Physiol. 52: 127–131.

Roer, R.D. and M.G. Shelton. 1982. Effects of hydrostatic pressure on Na$^+$ transport in the freshwater crayfish *Procambarus clarkii*. Comp. Biochem. Physiol. 71A: 271–276.

Roer, R.D. and A. Péqueux. Effects of hydrostatic pressure on ionic and osmotic regulation. Pp. 31–49. *In*: A. Péqueux and R. Gilles. (eds.). 1985. High pressure effects on selected biological systems. Springer Verlag. Berlin, Heidelberg.

Sébert, P. 1993. Energy metabolism of fish under pressure: a review. Trends Comp. Biochem. Physiol. 1: 289–317.

Sébert, P. Fish adaptation to pressure. Pp. 73–95. *In*: A.L. Val and B.G. Kapoor (eds.). 2003. Fish Adaptation. Science Publisher, Enfield, USA.

Sébert, P. Fish muscle function and pressure. Pp. 233–255. *In*: P. Sébert, D.W. Onyango and B.G. Kapoor (eds.). 2008. Fish Life in Special Environments. Science Publishers, Enfield, NH, USA.

Sébert, P. and Macdonald A.G. Fish. Pp. 147–196. *In*: AG. Macdonald (ed.). 1993. Effects of High Pressure on Biological Systems. Advances in Comparative and Environmental Physiology. Springer Verlag, Berlin.

Sébert, P. and A. Péqueux, B. Simon and L. Barthélemy. 1991. Effects of long term exposure to 101 ATA hydrostatic pressure on blood, gill and muscle composition and on some enzyme activities of the FW eel (*Anguilla anguilla*). Comp. Biochem. Physiol. 98B: 573–577.

Sébert, P. and B. Simon and A. Péqueux. 1997. Effects of hydrostatic pressure on energy metabolism and osmoregulation in crab and fish. Comp. Biochem. Physiol. 116A: 281–290.

Sébert, P. and A. Vettier, A. Amérand and C. Moisan. High pressure resistance and adaptation of European eels. *In*: G. van den Thillart, S. Dufour and J.C. Rankin (eds.). 2009. Spawning migration of the European eel. Springer, Berlin (In Press).

Shelton, C. and A.G. Macdonald, A. Péqueux and I. Gilchrist. 1985. The ionic composition of the plasma and erythrocytes of deep sea fish. J. Comp. Physiol. B 155: 629–633.

Siebenaller, J.F. and G.N. Somero. 1982. The maintenance of different enzyme activity levels in congeneric fishes living at different depths. Physiol. Zool. 55: 171–179.

Siebenaller, J.F. and G.N. Somero. 1989. Biochemical adaptation to the deep sea. CRC Crit. Rev. Aquat. Sci. 1: 1–25.

Simon, B. and P. Sébert and L. Barthélemy. 1989. Effects of long-term exposure to hydrostatic pressure *per se* (101 ATA) on eel metabolism. Can. J. Physiol. Pharmacol. 67: 1247–1251.

Somero, G.N. and J.F. Siebenaller and P.W. Hochachka. Biochemical and physiological adaptations of deep-sea animals. Pp. 261–330. *In*: G.T. Rowe (ed.). 1983. Deep-sea Biology. J. Wiley and Sons, New York.

Vettier A. and A. Amérand, C. Moisan and P. Sébert 2005. Is the silvering process similar to the effects of pressure acclimatization on yellow eels? Resp. Physiol. Neurobiol. 145, 243–250.

Yancey, P.H. 2005. Organic osmolytes as compatible, metabolic and counteracting cytoprotectants in high osmolarity and other stresses. J. Exp. Biol. 208: 2819–2830.

Yancey, P.H. and J.F. Siebenaller 1999. Trimethylamine oxide stabilizes teleost and mammalian lactate dehydrogenases against inactivation by hydrostatic pressure and trypsinolysis. J. Exp. Biol. 202: 3597–3603.

Yancey, P.H. and W.R. Blake, J. Coley and R.H. Kelly. Nitrogenous solutes as protein-stabilizing osmolytes: counteracting the destabilizing effects of hydrostatic pressure in deep-sea fish. Pp: 13–23. *In*: P.A. Wright and D. MacKinlay (eds.). 2002. Nitrogen Excretion in Fish. Proceedings of the International Congress on the Biology of Fish. University of British Columbia, Vancouver, Canada. American Fisheries Society.

Zerbst-Boroffka, I. and R.M. Kamaltynow, S. Harjes, E. Kinne-Saffran and J. Gross. 2005. TMAO and other organic osmolytes in the muscles of amphipods (Crustacea) from shallow and deep water of Lake Baikal. Comp. Biochem. Physiol. 142A: 58–64.

Zimmerman, A.M. and S. Zimmerman. Influences of hydrostatic pressure on biological systems. Pp: 381–396. *In*: C.J. Lambertsen (ed.). 1976. Underwater Physiology V. FASEB, Bethesda, Maryland, USA.

Zimmermann, U. 1979. Physics of turgor- and osmoregulation. Ann. Rev. Plant. Physiol., 29: 121.

Zimmermann, U. and F. Beckers and H.G.L. Coster. 1977. The effect of pressure on the electrical breakdown in the membranes of *Valonia utricularis*. Biochem. Biophys. Acta. 464: 399.

Zimmermann, U. and G. Pilwat, A. Péqueux and R. Gilles. 1980. Electro-mechanical properties of human erythrocyte membranes: the pressure dependence of potassium permeability. J. Membrane Biol. 54: 103–113.

# Muscle Function and High Hydrostatic Pressures

## Cellular Mechanisms and Implications for Taxonomic Environmental Locomotion

*Oliver Friedrich*

*The brain might make the choice to approach or avoid environmental pressure*
*However, it is the muscle that carries us away*

## INTRODUCTION

Environmental variables are often used to investigate reaction rates and kinetics of cellular and subcellular processes in order to understand adaptational plasticity of organisms in terms of migratory activity and metabolism (e.g. Haman, 2006; Johnston, 2006; Guderley, 2004; Kawai, 2003). In fact, temperature is often considered a more ubiquitous parameter that affects cellular performance. As a matter of fact, life scientists and physiologists often ignore the impact of changes in ambient pressure, the second fundamental environmental parameter, to cellular physiology. This is surprising, as any vertical locomotion will give rise to pressure changes. These may be more subtle at high altitude, e.g. in flying birds (Scott and Milsom, 2006) or mountain dwellers (Marconi et al., 2005), as compared to vertical migrations in water, e.g. in fish (Tanaka et al., 2001), diving birds and mammals (Butler, 2006; Butler and Jones, 1997). In the first case, ambient pressure declines exponentially with altitude, being only 50% compared to

School of Biomedical Sciences, University of Queensland, Skerman Bldg, QLD 4072, St. Lucia, Brisbane, Australia.
E-mail: o.friedrich@uq.edu.au

Institute of Physiology & Pathophysiology, Department of Systems Physiology, University of Heidelberg, Im Neuenheimer Feld 326 69120 Heidelberg, Germany.
E-mail: oliver.friedrich@physiologie.uni-heidelberg.de

sea level at approx. 5,500 m. In water, hydrostatic pressure increases linearly with dive depth at approx. 0.1 MPa (1 atm) per 10 m of water column (depending on water salinity, Saunders and Fofonoff, 1976). More than 70% of the earth's surface is covered by oceans with depths up to almost 11,000 m at Mariana Trench and an overall average pressure of 38 MPa (Winter and Dzwolak, 2005). Given the fact that most of the earth's biomass is located in the oceans (Reynolds et al., 2005), with its most abundant representative group of mesopelagic (midwater) fishes showing pronounced diurnal vertical migration patterns (Salvanes and Kristoffersen, 2001), it is clear that pressure effects on organisms and cellular systems occur more as a general rule rather than being the exception. Yet, lessons from this comparative approach of animal responses to extreme conditions (Caputa, 2006) have stimulated high pressure bioscience in the last decades to gain insight into the cellular and molecular mechanisms that have evolved in high pressure residents to withstand hyperbaric environments (Morita, 2003, 2008; Altringham et al., 1982) or that allow 'high pressure visitors', e.g. human divers (Rostain, 1993), diving birds and mammals (Lafortuna et al., 2003; Hooker and Baird, 1999) to acclimatize to transient changes in hydrostatic pressures.

High pressure actions are mainly governed by the *Le Chatelier*'s principle, already detailed in various chapters in this book, simply stating that high hydrostatic pressures will inhibit any reaction that is associated with positive volume changes and favour its conversion rate to the lower volume state. Therefore, most cellular reactions that involve build-up of structural complexes (Meersman et al., 2006; Kharakoz, 2000) or enzymatic reactions are influenced by changes in ambient pressure (De Felice et al., 1999; Coelho-Sampaio et al., 1991). Whereas some molecules are stabilized by high pressures, i.e. stabilization of the DNA double helix through hydrogen bonds (Mentre and Hui Bon Hoa, 2001; Macgregor, 1998) or heat-shock protein mRNAs (Kaarniranta et al., 2003), probably the vast majority of biomolecular reactions is destabilized. This destabilization of protein or enzymes, for instance, is a function of both pressure level and exposure time: a slowed down reaction kinetics for lower to intermediate pressures (5 –70 MPa) and, finally, graded structural destabilization through depolymerization (~40 – 70 MPa), unfolding and dissociation for larger pressures, i.e. >100 MPa (Friedrich and Ludwig, 2007; Foguel and Silva, 2004; Smeller, 2002; Verjovski-Almeida et al., 1986; Swezey and Somero, 1985; Kutching, 1972). Recently, water has gained much attention in being primarily inflicted in these high pressure induced effects on the protein levels, because sterical protein structure was suggested to be disrupted by water penetrating protein cavities following weakening of hydrophobic protein interactions (Winter and Dzwolak, 2005; Collins et al., 2005; Foguel and Silva, 2004; Rashin et al., 1986).

Similar to proteins, biological membranes are very susceptible to changes in ambient pressure. Biomembrane compositions are very diverse among vertebrate species, even with a high intraindividual plasticity in response to nutritional and environmental impacts (Thomas and Rana, 2007; Switzer et al., 2004; Gibbs, 1998; Lee and Hirota, 1973). Much of our understanding of pressure-membrane interactions has come from studies on simplified model membranes of defined compositions (Nicolini et al., 2006; Winter and Dzwolak, 2005; Winter and Köhling, 2004). In contrast to dissociation seen in proteins, high hydrostatic pressures tend to order and compress membrane lipids due to conformational packing changes. Interestingly, a variety of saturated and unsaturated lipid membrane model systems show a very common slope of ~20°C/100 MPa at the boundary for a phase transition from the gel-like to the fluid state, in which acyl chains within the bilayer are conformationally disordered (Winter and Dzwolak, 2005). Using this slope, the *Clausius-Clapeyron* equation predicts an increase in the transition temperature with increasing pressure (Sébert, 2003). This has already important environmental implications for organisms living at high pressures in the deep sea. At depth, the additional lower temperature of around 4°C represents an additional stressor for the membranes, thus pushing them towards a gel-like equilibrium. Furthermore, high pressure has been shown to increase the bilayer thickness of different phosphatidyl-membranes and linearly decrease overall bilayer volume with pressure (Eisinger and Scarlata, 1987; Braganza and Worcester, 1986). This additional membrane stress would certainly have vast impacts not only on the bilayer itself, but also on embedded ion channels, e.g. mechanosensitive channels (Hamill, 2006; Macdonald and Martinac, 2005), thus altering downstream cell signalling (Ingber, 2006; Matthews et al., 2006). Although being a scalar, it is mostly the structure of protein pockets with water being driven into protein cavities that give high pressure an indirect vector-like mode of action through water pushing against peptide molecules in a directed fashion from non-cavity to cavity coordinates (Boonyaratanakornit et al., 2002). In addition to changes in protein inner energies, unfolding or changes in channel gating modes may occur (Heinemann et al., 1987; Conti et al., 1982a, b). In general, high hydrostatic pressures and low temperatures bear similar effects on chemical, biochemical and cellular reaction rates, i.e. slowing down reaction kinetics, at least for reactions having positive reaction volumes or transition kinetics with positive activation volumes (Sébert et al., 2004). Increasing pressure about 10 MPa has a roughly equivalent effect of slowing down reaction kinetics as decreasing temperature about 2°C (Macdonald, 1993; Brauer et al., 1985). However, it has to be kept in mind that the fundamental thermodynamic actions of temperature and pressure are quite different. Whereas an increase in temperature increases both inner energy and volume, pressure actions on volume changes depend on the sign of the reaction volume. Therefore, high pressure experiments have also become an elegant tool to separate changes in inner energy and mechanical energy arising from volume changes that would otherwise appear simultaneously in temperature experiments alone (Weber and Drickamer, 1983). Volume changes in the order of $Å^3$ can be resolved using high pressure experiments (Meyer and Heinemann, 1997).

From the above considerations, it is clear that normal physiological cellular and organ performance in an individual can be vastly modulated by environmental constraints. The rate limiting steps for effects of ambient pressure and temperature changes on biosystems are usually given by relocation velocities of the animals themselves, as changes in water salinity, temperature or climate can be neglected in that regard. For example, mesopelagic fish residing between 100 m and 1,000 m of depth perform slow diurnal vertical migrations covering several hundred metres of water column within hours (Salvanes and Kristoffersen, 2001). These animals balance the deteriorating effects of both increasing pressures and decreasing temperatures upon descend (or vice versa) on membrane and proteins directly on the subcellular level by a mechanism called *homeoviscous adaptation* (Macdonald, 1988; Macdonald and Cossins, 1985). Changes in membrane lipid compositions (Nevenzel, 1980) and protein sequence mutations (Morita, 2003) are most effective strategies and will be described later in this and other chapters of this book. Likewise, an increase in cellular osmotic pressure is often found in fish to balance hydrostatic pressure (Yancey, 2005). However, geographic long-term adaptations to temperature also exist (Johnson and Johnston, 1991). If one considers the very large vertical diving speeds of seals and mammals, i.e. 1–2 m/s, covering several hundreds to even ~1,500 m in 30–40 min (M. Castellini—this *book*, Hooker and Baird, 1999; Watkins et al., 1993), it is clear that reaction times would be far too short for conventional homeoviscous adaptation. Therefore, the question arises, "what are the principle pressure exposure limits for mammalian organisms and tissues?" This information, combined with the knowledge of cytoprotectant mechanisms in deep sea dwellers (Yancey, 2005) might be a valuable source for future interventions in biotechnology to extend pressure exposure limits in humans.

Skeletal muscle is the primary organ for voluntary locomotion in all vertebrates. In conjunction with environmental stressors, it is the vehicle that drives organisms into zones of higher pressures/lower temperatures, e.g. for predator/prey interactions. High pressure effects on skeletal muscle are a crucial determinant of locomotor performance that has to be taken into account when entering or leaving pressure zones. The complex function of intact skeletal muscle relies on a close interaction between membrane excitation and electro-chemico-mechanical coupling events that are being related to numerous endogenous muscle proteins and enzymes. Some of them can also be translated to similar proteins in other tissues. Therefore, skeletal muscle cells are an excellent cellular model to also investigate effects of prolonged high pressures, that mimic dive profiles of marine animals, or to define principle exposure limits for terrestrial animals. Muscle architecture and key proteins are very similar between taxonomic classes allowing extractions of results from one species to another. As stated by Medler (2002), "skeletal muscles are highly conserved in terms of the molecular mechanisms responsible for producing muscle contraction". However, this view might be too simple, because muscle is an organ of high plasticity that can substantially adapt to training or environmental conditions (Flück and Hoppeler, 2003; Somero, 1983).

The following chapter is designed to focus on the effects of high pressure on skeletal muscle performance; specifically, functional and structural changes involved with short- and long-term high pressure exposures and compensatory mechanisms. After a general introduction to muscle architecture, the steps transforming excitation through to contractile force output will be explained. Then, the specific effects of high pressure on amphibian, mammalian and fish muscles will be discussed and taxonomic implications being highlighted.

## SKELETAL MUSCLE PHYSIOLOGY: WHERE STRUCTURE MEETS FUNCTION

### Structure of the Sarcomere

Muscle contraction involves the coordinated activation of contractile proteins that are arranged in a highly linear array in skeletal muscle. A muscle cell—*muscle fibre*—consists of parallel bundles of myofibrils that are surrounded by mitochondria, cytoplasm and the sarcoplasmic reticulum (SR, equivalent to endoplasmic reticulum in other cells) which acts as an internal $Ca^{2+}$ store (Fryer and Stephenson, 1996). The elementary structural repetitive unit of a muscle fibre is the sarcomere (Fig. 1).

Recently, the functional elementary unit was found to be the half-sarcomere rather than the sarcomere (Telley et al., 2006). A sarcomere is composed of two half-sarcomeres of opposite orientation regarding the resulting pulling trajectory of the contractile proteins, actin and myosin. The regular striation pattern of single muscle fibres originates from the isotropic and anisotropic bands that appear in the light microscope and reflect the regular protein arrangement. One sarcomere is bordered by the Z-line where the actin filaments (thin filaments) are attached (see Table 1). In the middle of the sarcomere, the M-line builds the scaffold for anchoring the myosin filaments (thick filaments) that spread in opposite directions to each of the Z-lines (Fig. 1). The length of the thin filaments is regulated by an

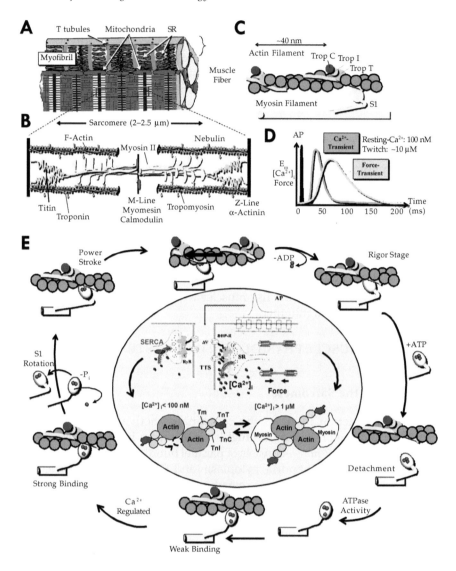

**Figure 1. Functional Micro-Anatomy of skeletal muscle. A,** overview of single muscle fibre architecture. Parallel myofibrils are wrapped by the sarcolemma with its t-tubular invaginations into the fibre. **B,** structure of the sarcomeres with the arrangement of the motorproteins, actin and myosin (microstructure in **C**), and supporting cytoskeleton proteins. **D,** temporal relationships between action potential (AP), Ca²⁺- and force-transient in skeletal muscle. **E,** cross-bridge cycle of actin-myosin interaction initiated by a rise in myoplasmic Ca²⁺ concentration during ec-coupling that opens actin's myosin binding site (inset). Ca²⁺ removal by SERCA terminates cycling. For details, see text.

**Table 1.** Important muscle proteins and their characteristics (modified from Friedrich, 2007).

| Protein | Properties and Function |
| --- | --- |
| actin | G-Protein: 42 kDa. Polymerization to F-actin. Main component of thin filaments, accounts for ~20 % of total muscle protein content. F-actin contains two coiled α-helices. ~1 μm long, ~8 nm thick. ~38 nm repeats. Filament length at Z-bands controlled by CapZ, towards M-lines by tropomodulin. |
| troponin (Tn) | Cooperative regulator-protein-complex of three subunits attached to thin filaments.<br>• *troponin T* → 30 kDa, anchors Tn-complex to actin-filament via binding to tropomyosin. Modulator of $Ca^{2+}$-sensitivity of the acto-myosin-ATPase.<br>• *troponin C* → 18 kDa, calmodulin-like protein, binds up to four $Ca^{2+}$ ions. '$Ca^{2+}$-switch', i.e. $Ca^{2+}$ binding changes its conformation to troponin I and tropomyosin and opens the myosin binding site.<br>• *troponin I* → 20 kDa, inhibitory subunit, inhibits myosin-ATPase-activity. |
| tropomyosin | ~37 kDa, two coiled α-Helices arranged along the groove of F-actin-filaments. At low myoplasmic $Ca^{2+}$, it blocks myosin binding site on actin filament. Translocation during $Ca^{2+}$-activation through the troponin complex opens up access to the myosin binding site. |
| tropomodulin | ~40 kDa, capping protein of actin filaments towards M-line. Regulates actin filament length. |
| α-actinin | ~100 kDa, main protein within the Z-line connecting the actin filaments. |
| myosin | Protein family of at least 18 members. Muscle myosin: type II myosin, 490 kDa. Consists of two heavy chains (myosin heavy chain MHC II: 205 kDa each), plus four light chains (myosin light chains MLC: ~20 kDa each). Molecular motor. ~40 % of total muscle protein content. Coiled-coil structure.<br>**MHC:** Consists of head region (S1: binding site for actin, ATP 2 MLC, '*lever arm region*'), neck region (S2) and tail region (meromyosin).<br>**MLC:** two pairs per myosin filament; one 'essential' and one 'regulatory' chain, each. |
| titin | Giant protein (3–3.8 MDa). Crosses entire half-sarcomere from Z- to M-line. Contains immunoglobulin- and fibronectin-like domains. Parts within the I-bands act as molecular spring and provide passive repulsive force during muscle lengthening. |
| nebulin | 600–900 kDa. Crosses the sarcomere from Z-line to the ends of the actin filaments ('ruler for actin filaments'). It is almost not distensible. |

ATP dependent polymerization at the barbed ends and is controlled by cap-proteins (Dos Remedios et al., 2003). Nebulin acts as a ruler along which the actin filaments can be linearly orientated to an average length of ~1 μm. Myosin filaments in both half-sarcomeres add up to a length of ~1.6 μm and their length is thought to be regulated by the giant protein titin that covers the whole half-sarcomere in mammalian muscle while their thickness is regulated by myosin binding protein C (for detailed reviews on muscle architecture see Au, 2004; Clark et al., 2002). Under resting conditions, the sarcomere length is about 2.2 µm. Active shortening as well as passive stretch are enabled by opposite gliding of thick and thin filaments against each other (Huxley and Hanson, 1954). The wide range of sarcomere lengths between ~1.8 µm and 4.0 µm results from different degrees of overlap between the acto-myosin complexes while the length of the motor filaments remains unchanged (Page, 1964). The giant filamentous protein titin (Table 1) limits successive stretching of the muscle through its elastic spring-like regions (Tskhovrebova et al., 1997).

## Motor Protein Interaction

The power producing interaction between actin and myosin involves different successive steps during which the energy rich triphosphate ATP is chemically split into ADP and inorganic phosphate through the ATPase activity of the myosin head (Fig. 1). Binding of ATP to the active pocket of myosin reduces the affinity of the myosin head to actin and both partners separate. After splitting ATP to ADP and $P_i$, both products still remain in the active myosin pocket which undergoes a conformational change in the neck region to grab for the next free myosin binding site along the actin filament. The affinity of myosin for actin increases and a weak binding state is being established. This binding state is usually predominant under conditions of unloaded shortening in muscle (Stehle and Brenner, 2000). Under physiological conditions, the release of the inorganic phosphate from the myosin induces the 'power stroke' from a high-affinity binding state that is associated with a conformational ~60° rotation of the myosin head and pulls the actin filament between 5 nm and ~12 nm towards the M-line (reviewed in Rüegg et al., 2002; Ruppel and Spudich, 1996). The strong binding state remains stable and the myosin attached to the actin filament unless ADP is being released from the pocket and a new ATP molecule binds to lower both partners' binding affinities with detachment of the myosin head. A new cycle can then be initiated (Fig. 1). If no ATP is present or ATP is being depleted, e.g. post-mortem, all cross-bridges are 'frozen' in the 'rigor state' (resulting in post-mortem muscle stiffness that also imposes problems to food processing of meat). Much of our understanding on motor-protein interaction has been obtained in recent years from '*in vitro*' motility studies where isolated myosin heads are coated on coverslips and movement of fluorescently labelled actin filaments can be directly observed under the microscope at different ionic and $Ca^{2+}$ conditions (Hook and Larsson, 2000; Kron and Spudich, 1986), as well as from studies involving optical tweezer traps that can resolve molecular force productions of single myosin heads (Kitamura et al., 1999; Molloy et al., 1995). It is worth noting that in the working muscle, cross-bridges are never synchronized in their power cycle, because this would prevent the development and maintenance of continuous isometric force if all myosin heads detached at the same time from actin. In fact, attachment and detachment are stochastic processes that are being regulated by myoplasmic $Ca^{2+}$ levels and the availability of energy substrates (ATP). Even in relaxed muscle, cross-bridges form and dissolve again, however, at a low probability, as compared to activated muscle. The sterical arrangement with one myosin filament surrounded by six actin filaments (Squire, 1974) ensures that the distance of adjacent binding sites on a single actin filament (~38.5 nm) does not have to be covered by one and the same myosin head previously attached to this actin filament. Instead, the myosin heads are being orientated in circular directions from the myosin axis so that cross-bridges with neighbouring actin filaments pull a new binding site towards the myosin head adjacent to one specific actin filament.

The cross-bridge formation and force production has to be finely regulated and tuned to the motorical demands (one should consider the very fine force tuning in small hand muscles). The most important regulator is myoplasmic $Ca^{2+}$, the level of which is controlled by internal stores, ion channels, pumps and ionic exchangers (see below). A set of very specialized proteins, the troponin-tropomyosin complex (Table 1) translates the $Ca^{2+}$ level to binding probability of myosin. Under low myoplasmic resting $Ca^{2+}$ conditions (~100 nM) the actin binding site for myosin is blocked by the filamentous protein tropomyosin that runs along the groove within the coiled actin helix. Every ~40 nm, three members of the troponin family (troponin-C, -T and -I) build a complex on the tropomyosin and the myosin binding site (Fig. 1). If the $Ca^{2+}$ levels increase due to muscle activation (~10–20 μM in a single twitch; Delbono and Stefani 1993), $Ca^{2+}$ binds with high affinity to troponin-C. The binding affinity is also determined by the regulatory troponin-T and -I subunits. The troponin complex acts as a 'Ca²⁺ switch', the activation of which results in a rotation of the connected tropomyosin away from the groove, now exposing the myosin binding site on the actin filament. Upon lowering of myoplasmic $Ca^{2+}$ at the end of muscle contractile activation, $Ca^{2+}$ unbinds from troponin-C and the tropomyosin is pushed back into the groove, thus blocking myosin's binding site. It should be noted that this $Ca^{2+}$ handling is governed by buffer properties of different $Ca^{2+}$ binding proteins with appropriate $\kappa_{on}$ and $\kappa_{off}$ rates and dissociation constants (Berchtold et al., 2000; Gillis et al., 1982).

The major determinant of force development kinetics is given by myosin-ATPase isoforms of muscle myosin II (Toniolo et al., 2007; Nyitrai et al., 2006). Slow (type I) and fast (type II) muscle myosin isoforms are distinguished with type I being predominantly present in non-fatiguing, oxidative slow-twitch muscle (e.g. posture muscles) and type II isoforms (that can be further divided into II A, II B, II X and some other less common ones) being predominantly expressed in fatigable, glycolytic fast-twitch muscle (e.g. many limb muscles). Apart from pure fibres (Friedrich et al.,

2008; Bottinelli et al., 1994), mixed distributions are present in many muscles (Schiaffino and Reggiani, 1994). The high plasticity of muscle is further reflected by adaptational transitions of myosin isoforms related to muscle exercise (Serrano et al., 2000) or as part of pathophysiology in muscle diseases (Hayes and Williams, 1998).

For further reading on muscle contraction regulation, $Ca^{2+}$ dynamics, muscle plasticity and signalling, the reader is referred to some recent reviews on the topics (Allen et al., 2008; Lynch and Ryall, 2008; Gunning et al., 2008; Rizzuto and Pozzan, 2008; Schiaffino et al., 2007).

## Membrane Excitability, Excitation-Contraction Coupling and $Ca^{2+}$ Cycling

Muscle contraction is governed by transient increases in intracellular $Ca^{2+}$ levels to activate the contractile machinery. One of the striking differences between skeletal muscle and cardiac or smooth muscle is that skeletal muscle is much less dependent on external $Ca^{2+}$ influx to feed the contractile system during contraction. This is mainly reflected by the very brief action potentials in skeletal muscle (60 to 100 times faster than in cardiac muscle). It should be noted, that recent evidence points towards some $Ca^{2+}$ entry during repeated contractions and store-depletion to replenish intracellular stores (Stiber et al., 2008; Launikonis and Rios, 2007; Cairns et al., 1998). Muscle excitation is initiated at the neuromuscular junction upon arrival of motor-nerve action potentials. In the nerve bouton, the excitatory transmitter acetylcholine (ACh) is released and binds to the nicotinic ACh-receptors in the post-synaptic muscle membrane (Lingle et al., 1992) where unspecific cation channels open and induce an excitatory postsynaptic potential (EPSP). This EPSP spreads along the neuromuscular junction and deep into adjacent transverse tubules (Dauber et al., 2000) where voltage-gated $Na^+$ channels are rapidly activated to trigger post-synaptic action potentials that spread along the sarcolemma and into the transverse tubular network. In skeletal muscle, the tubules are the exclusive localization of the L-type dihydropyridine (DHPR) $Ca^{2+}$ channels that mainly act as a voltage sensor under physiological conditions of brief depolarizations (Ahern et al., 2001; Berchtold et al., 2000; Rios and Brum, 1987) but also are conducting $Ca^{2+}$ channels under maintained depolarizations (Friedrich et al., 1999; Harasztosi et al., 1999; Donaldson and Beam, 1983). Deep in the extracellular invaginations of the t-tubules, the triads are the location of the excitation-contraction coupling (ec-coupling). Here, the t-tubules come into close contact with the intracellular membranes of the sarcoplasmic reticulum (SR; two SR contacts per t-tubule), the intracellular $Ca^{2+}$ storage organelle. The SR $Ca^{2+}$ concentration is with ~1 mM (Posterino and Lamb, 2003) close to the extracellular concentration and holds a similarly large driving force for $Ca^{2+}$ to be released into the cytoplasm as compared to extracellular $Ca^{2+}$ influx (resting myoplasmic $Ca^{2+}$ concentration ~100 nM). The DHPRs form tetrads with alternating juxtaposing contacts to a RYR1 receptor channel on the myoplasmic SR side (Fig. 2, Fill and Copello, 2002).

Each $\alpha$1-subunit of the DHPR makes a physical contact over the ~15 nm triadic cleft with the RYR1 via a protein loop at the cytoplasmic II-III

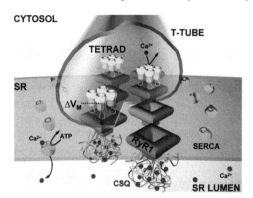

Figure 2. Geometry of the skeletal muscle triad junction and arrangement of the signalling proteins. In the t-tubules, tetrads of dihydropyridine receptor channels DHPRs are alternately facing RYR1 channels in the sarcoplasmic reticulum membrane. The junction is separated by a cleft of approximately 10–15 nm. The DHPRs have a physical protein coupling via the cytosolic II-III loop of the DHPR $\alpha$-subunit to the adjacent RYR1 ('the foot'). Sensing a tubular voltage change upon depolarization via the S4 segments of the DHPR results in a conformational rearrangement of charged residues ('charge movements') that is being transferred to the RYR1, thus opening the SR release channel. The physical coupling warrants a fast-coupling within ~10 ms. The luminal protein calsequestrin (CSQ) is thought to be involved in the control of the free luminal $Ca^{2+}$ concentration. The SR $Ca^{2+}$ ATPase (SERCA) refills the SR lumen with $Ca^{2+}$ via primary active transport. Taken from *Fill and Copello 2002 (with permission)*.

interlinker (Fig. 2). Thus, voltage changes that are sensed via the DHPR S4 transmembrane domain result in the movement of positively charged amino acid residues in the field. These 'charge movements' are associated with conformational coupling between DHPR and RYR1 to open the release channel (Papadopoulos et al. 2004). Within milliseconds, $Ca^{2+}$ is released into the myoplasm reaching concentrations of between ~5 μM and ~20 μM in response to a single action potential (Hollingworth et al., 1996; Delbono and Stefani, 1993). $Ca^{2+}$ then binds to the troponin C and initiates cross-bridge cycling (see above). With the end of the action potential, the SR membrane repolarizes, the RYR1 inactivates and $Ca^{2+}$ is pumped back into the SR via the sarco-endoplasmic reticulum pump (SERCA). The small $Ca^{2+}$ buffering protein parvalbumin acts as an intermediate buffer between the troponin-C and the SERCA and is particularly important for fast relaxation in small mammals.

The ec-coupling process shows some fundamental species differences in amphibians and mammals at the level of the triad geometry and protein interactions. In amphibian muscle, one tubule per sarcomere is present near the M-line, whereas in mammalian muscle, two tubules per sarcomere in the isotropic band near the Z-lines can be found (Gomez et al., 2006). Interestingly, fish muscles also show the mammalian distribution type with two triads per sarcomere, at least in those fish investigated so far (Korneliussen and Nicolaysen, 1975). Moreover, $Ca^{2+}$ release from RYR1 in amphibian muscle (RYRα) amplifies $Ca^{2+}$-release from parajunctional RYRβ in a process called $Ca^{2+}$ induced $Ca^{2+}$ release (CICR). This allows for sequential propagation of release sources in amphibian muscle, a pattern that has not been observed to date in mammalian muscle, probably due to postnatal loss of parajunctional RYR3 channels in mammals (Pouvreau et al., 2007). Therefore, CICR does probably not occur in mammalian muscle under physiological conditions (Zhou et al., 2005).

For further reading on skeletal muscle excitation-contraction coupling, microdomain $Ca^{2+}$ dynamics, and muscle membrane excitation, the reader is referred to some recent reviews (Bannister, 2007; Jurkat-Rott et al., 2006; Stephenson, 2006; Protasi, 2002).

## ACUTE HIGH PRESSURE EFFECTS ON SKELETAL MUSCLE FUNCTION AND STRUCTURE

### Neuromuscular Junction

The pressurization of whole animals (ranging from lower—frog, lizard—to higher vertebrates—mice, rats, hamster, monkeys) revealed different convulsion thresholds that correlated with brain complexity (Brauer et al., 1979). The higher the vertebrate, the lower was the threshold for convulsions. Although muscle activity was increased during these episodes, it was found that the observed effects of pressure on the effector organ muscle were probably secondary to neural hyperexcitability termed High Pressure Nervous Syndrome (HPNS). Apart from the central nervous system, presynaptic transmitter release at the neuromuscular junction itself is modulated by high pressure. However, in the range of high pressures where generalized convulsions occur (e.g. ~9.5 MPa in rats, ~14 MPa in frogs, ~20 MPa in newts; Brauer et al., 1979), neuromuscular transmission was depressed rather than augmented, e.g. at ~10 MPa in rats (Kendig and Cohen, 1976), ~16 MPa in frog (Ashford et al., 1982) or ~8 MPa in crustacean muscle (Campenot, 1975). In frog muscle, Ashford et al. recorded muscle membrane miniature-endplate-currents and-potentials right at the neuromuscular junction using extracellular focal glass electrodes (Ashford et al., 1982). Such spontaneous events represent quantal release of acetylcholine from presynaptic vesicles. Pressurization up to ~5 MPa exponentially decreased the frequency of transmitter release to the muscle, an effect that was completely reversible upon decompression (Ashford et al., 1982). It was concluded that presynaptic transmitter release was markedly depressed by high pressure. In this case, temperature decrease also mimicked these effects with an approximate 0.04°C/0.1 MPa relation (Ashford et al., 1982). Interestingly, the effects of pressure seem to be less prominent, apart from an increase in the decay time for post-synaptic currents that was suggested to be due to an increase in channel life time. This was confirmed in a patch-clamp study on acetylcholine receptor channels in membrane patches from cultured rat muscle cells that were acutely pressurized up to 60 MPa (Heinemann et al., 1987). The authors found an increase of both mean open and closed times at 40 MPa with apparent activation volumes of ~60 Å$^3$ and ~140 Å$^3$, respectively (Heinemann et al., 1987). Whether this also applies to the lower pressures associated with convulsion thresholds found in rats was, unfortunately, not assessed. However, although acetylcholine release for muscle activation might be already impaired at pressures that induce central nervous hyperexcitability, the neuromuscular transmission in the frog study did not fail even at ~21 MPa (Ashford et al., 1982). The robustness of postsynaptic muscle membrane parameters might account for this effect. For example, acetylcholine receptor single channel conductance was unchanged by pressures up to 60 MPa (Heinemann et al., 1987). This seems to be a very common feature of ion channels suggesting that the translocation process of the permeant ion through its channel pore is not associated with significant activation volumes (Macdonald, 2002, 1997; Heinemann et al., 1987).

## Effects of acute High Pressures (<15 MPa) on Contractile Function in activated Amphibian and Mammalian Muscle

Early studies on non-mammalian (mostly turtle and amphibian) muscle already found several differential effects of rapid (~min) high pressure applications to intact whole muscles (Cattell, 1935; Brown, 1934; Cattell and Edwards, 1930). It was commonly found that, in response to single electrical stimuli, isometric twitch tension (tension recorded with the muscle endings fixed to a force transducer) was almost instantaneously increased during high pressure applications, in particular for temperatures above 10°C (Cattell and Edwards, 1928), and completely returned to basal levels upon pressure release. This was consistent in different species. On the other hand, isometric tension during a tetanic stimulation was 'small and variable' (Cattell and Edwards, 1928) but turned into enhanced force depression when fatiguing the muscle under high pressures compared to atmospheric pressure (Cattell and Edwards, 1930). This reversal of augmenting single twitch force response by pressure was mimicked by cooling the muscle to 13°C at atmospheric pressure (Cattell and Edwards, 1930). Later studies on mammalian muscle largely confirmed the increase in twitch tension by pressures up to ~14 MPa (Ranatunga and Geeves, 1991; Kendig and Cohen, 1976). In rat diaphragm muscle, compression to 137 atm (13.7 MPa) markedly increased twitch tension but depressed simultaneously recorded EMG potential amplitudes in a $Ca^{2+}$ dependent manner (more depression at low 0.5 m M external $Ca^{2+}$). It was concluded that pressure-induced increases in twitch tension had to be due to pressure enhancement of ec-coupling steps or the contractile apparatus itself rather than acting on neuromuscular transmission that, in fact, inhibit synaptic transmission (Kendig and Cohen, 1976).

In the1990s, Ranatunga and colleagues conducted a series of careful studies to further elucidate the molecular basis underlying high pressure-induced alterations of skeletal muscle contractility during short-term pressurization up to 10 MPa. In both amphibian and mammalian intact muscle, the potentiation of twitch force by pressure was confirmed while tetanic stimulation clearly depressed tension in fast muscle (Vawda et al., 1996; Ranatunga and Geeves, 1991). Pressure dependent twitch tension enhancement was linearly related to pressure and reversible upon decompression (Fig. 3 A, B).

In fast rat EDL muscle studied at room temperatures, quick pressure jumps (completed within 1 min) up to 10 MPa increased peak twitch tension ~6 % $MPa^{-1}$ (Ranatunga and Geeves, 1991). Furthermore, kinetics of tension development was slowed down by pressure with an equal increase in both time-to-peak and half-relaxation by ~3 % $MPa^{-1}$ (Ranatunga and Geeves, 1991). When relating peak tension under pressure to time-to-peak, the authors calculated a relative rate of twitch tension rise with pressure of between two and $2.5 \times 10^{-2}$ $MPa^{-1}$. This explained the increase in twitch force solely by kinetics analysis, taking an increased rate of tension rise and a decreased rate of relaxation as a basis for first order transitions of the system (Ranatunga and Geeves, 1991). In single frog muscle fibres, a temperature-dependent study of twitch potentiation confirmed linear tension increases with pressure, however, the pressure sensitivity was largely shifted towards colder temperatures in amphibians compared to mammals (Vawda et al., 1996, 1995). In particular, the rate of twitch potentiation seen in rat fibres at room temperatures was already achieved in frog fibres at 4°C and was about fivefold larger in frog fibres than in rat fibres at room temperatures (Fig. 3). This is an interesting species difference probably reflecting adjustment of pressure sensitivity of contractile activation to different metabolic strategies (endo- vs. ectothermic).

Tetanic force tension seems to behave quite differently to twitch force during pressurization suggesting a different target for pressure. When

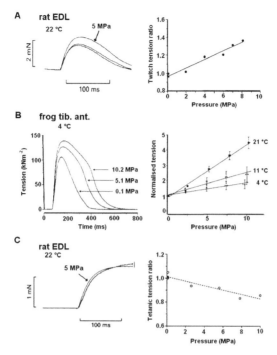

Figure 3. High pressure up to 10 MPa acutely increases twitch tension but depresses tetanic tension in amphibian and mammalian intact muscle. **A,** superimposed single twitch contractions during field stimulation of intact rat EDL muscle fibre bundles under atmospheric condition and following a fast (1 min) pressure jump to 5 MPa (arrow) and reversal after pressure release within 1 min. Twitch tension is linearly enhanced with pressure (right panel). Temperature 23°C. **B,** pressure-induced twitch enhancement is potentiated by higher temperatures in intact frog muscles (right panel). Example shown in the left panel at 23°C. **C,** tetanic tension during a train of stimuli applied at 100 Hz to intact rat muscle bundles is linearly decreased following fast jumps to various pressures but the kinetics seems to be accelerated. Taken with minor modifications from *Ranatunga and Geeves, 1991 (A, C) and Vawda et al., 1995; 1996, J. Muscle Res. Cell Motil. (B, original Fig.1; with kind permission from Springer Science and Business Media; A and C, original Figs. 2 and 4, with kind permission from Wiley-Blackwell Publishing).*

stimulating fast rat EDL fibre bundles at 100 Hz (Fig. 3 C), the fused tetanus tension plateau was typically depressed at high pressure, the amplitude and kinetics parameters being about 1% MPa$^{-1}$ for amplitude depression, 2% MPa$^{-1}$ decrease of tension rise half-time and ~2.5% MPa$^{-1}$ increase of exponential tension relaxation (Ranatunga and Geeves, 1991). Thus, compared to twitch kinetics, tetanic contractions under high pressure had a faster rising and a slower relaxation phase. Similar to the species differences for twitch tension characteristics, tetanic force was depressed in single intact frog fibres at lower temperatures between 4°C and 11°C with ~ 0.7% MPa$^{-1}$, while at 21°C, tetanic force was even potentiated by pressure (+0.4%

MPa⁻¹; Vawda et al., 1996). Moreover, the authors found that pressure effects on tetanic tension were even more complicated by muscle fatigue. In slightly fatigued muscle, pressure effects were similar to unfatigued muscle, i.e. pressure depressed tetanic tension. However, when muscle was severely fatigued, pressure induced an additional potentiation of tension compared to the reduced tension plateau in fatigued muscle under atmospheric conditions (Vawda et al., 1996).

What are the conclusions drawn from these experiments for the differential pressure action on skeletal muscle contractility? The major difference between single twitch and tetanic tension in intact skeletal muscle can be considered by the kinetics of $Ca^{2+}$ handling and levels of myoplasmic $[Ca^{2+}]$. Peak $[Ca^{2+}]$ levels in response to a single twitch stimulus have been ~20 µM in single mouse muscle fibres with a very fast and complete return to baseline levels (~100 nM) within ~15 ms (Hollingworth et al., 1996). During trains of stimuli, peak $[Ca^{2+}]$ is not much more elevated; however, there is a progressive slowing of the rate of $[Ca^{2+}]$ decay that results in a much longer elevated $[Ca^{2+}]$ level during tetanus (Hollingworth et al., 1996). Also, the rate of SR peak $Ca^{2+}$ release is vastly reduced during successive trains of action potentials. This is consistent with '$Ca^{2+}$ inactivation of $Ca^{2+}$ release' by the increased $Ca^{2+}$ in the triadic cleft binding to the SR $Ca^{2+}$ inactivation site of the RYR1, thus terminating release (Schneider and Simon, 1988; Baylor et al., 1983).

Pressure might increase twitch force in muscle by enhancing an early step in excitation-contraction coupling, mainly the SR $Ca^{2+}$ release (Vawda et al., 1996). In particular, pressure effects were similar to those where caffeine directly opened RYR1 when applied at atmospheric pressure (Vawda et al., 1996). Additionally, high pressures up to 10 MPa have been shown to increase the $Ca^{2+}$ sensitivity of thin filament activation (Fortune et al., 1994) and to impair the SR $Ca^{2+}$ ATPase activity for higher pressures (Hasselbach, 1988). Both would contribute to a positive inotropic effect of pressure. However, during tetanic stimulation, pressure effects on a maximally $Ca^{2+}$ activated fibre were suggested to be confined to the motor proteins and cross-bridges themselves (Fortune et al.,1989). In both resting and late fatigue state of muscle, pressure depressed tetanic tension but potentiated it during early fatigue. This bimodal action has not been yet finally explained but inorganic phosphate accumulation during muscle fatigue has been suggested to be involved (see Discussion in Vawda et al., 1996).

To further explore the pressure effects on the contractile apparatus under $Ca^{2+}$ saturating and lower $Ca^{2+}$ conditions, 'skinned' rabbit muscle fibres were used (Fortune et al., 1994). The advantage of the 'skinned fibre' preparation lies in the diffusional access to the myoplasm after either chemical permeabilization of the sarcolemma using saponin or glycerinated fibres or mechanical dissection of the plasma membrane (Launikonis and

Stephenson, 2002). Solutions controlling the free $[Ca^{2+}]$ can then be introduced and the mechanical apparatus gradually activated without the need for SR $Ca^{2+}$ release. Alternatively, if the SR is still intact, caffeine application enforces SR $Ca^{2+}$ release by directly opening the RYR1 and force transients can be measured.

When skinned rabbit psoas muscle fibre bundles were $Ca^{2+}$ activated at different pCa (from four to seven) at hydrostatic pressures (up to 10.1 MPa) at 12°C, tension amplitudes were potentiated by high pressures only at submaximal $Ca^{2+}$ concentrations of pCa > 6, whereas pressure depressed force levels at large $Ca^{2+}$ levels around pCa < 5 (Fortune et al., 1994). High pressure-induced tension depression was ~1% $MPa^{-1}$ at high $Ca^{2+}$, which is in perfect agreement with the results obtained from fused tetani in intact fibres (Ranatunga and Geeves, 1991). Although EGTA was used in the internal solutions to 'clamp' free $Ca^{2+}$, pressure has been shown to have only a minor influence on the $Ca^{2+}$ binding affinities of this buffer (< 0.3% at 10 MPa, Fortune et al., 1994) which also fits well with the functional data in intact vs. skinned fibres. The pressure depression of maximally activated acto-myosin complexes is likely to be due to inhibition of transitions from weak to strong-binding states and is compatible with a reported isomerization of actomyosin-S1 fragments by high pressures (Coates et al., 1985). It also shows that the force depressing effect of pressure must be well after the $Ca^{2+}$ activation of thin filaments, because in contrast, the normalized pCa-force curves under pressure were shifted somewhat towards larger pCa values (indicating even an increase in the $Ca^{2+}$ sensitivity of the contractile apparatus under pressure) and $Ca^{2+}$ binding to troponin-C remained unchanged (Fortune et al., 1994). So far, the depression of maximally $Ca^{2+}$ activated force by pressure seems to be restricted to cross-bridge transitions. For submaximal $Ca^{2+}$ activation in skinned fibres, the site of pressure action proposed earlier exclusively for early steps in ec-coupling had to be extended to the thin filaments. This was deduced from fast pressure relaxation experiments that suggested a thin filament-induced recruitment of cross-bridges with a ~2% $MPa^{-1}$ increase in force when $Ca^{2+}$ was low (Fortune et al., 1994). This shows that submaximally $Ca^{2+}$ activated muscle, i.e. during a twitch, is susceptible to high pressure at various steps of the ec-coupling cascade, all adding up to a positive inotropic effect, compared to maximum $Ca^{2+}$ activation where pressure effects at the cross-bridges prevail.

### Effects of acute High Pressures (10.1 MPa) on passive mechanical parameters in non-activated Mammalian Muscle

Besides the above mentioned differential effects of pressure on activated muscle, muscle is a complex elastic polymer that can be in two more important states. The resting state is reflected by low resting myoplasmic

$Ca^{2+}$ concentrations with mostly unattached cross-bridges. The other 'passive' state is given when cross-bridges are attached but do not cycle, as it is the case in ATP depleted muscle. In analogy to post-mortem rigor, this is called the rigor state of muscle. The elasticity of the latter is expected to be lower, i.e. the muscle is mostly stiff and presents a large resting tension. The elasticity changes of resting and rigor muscle were recorded and compared to other elastic non-muscle materials, ranging from glass to copper and rubber, during short-term exposures to pressures up to 10.1 MPa (Ranatunga et al., 1990). The amazing finding was that resting muscle under pressure behaved similar to rubber, exhibiting a net insensitivity of resting tension to pressure increases. Other amorphic materials, such as glass or collagen, showed substantial pressure sensitivity. Much of the elasticity in resting muscle can be attributed to titin and nebulin filaments. In contrast, muscle in the rigor state showed a large ~10% increase in resting tension during high pressure applications (Ranatunga et al., 1990). Therefore, fibre stiffness increases during high pressure. This adds up to the findings from $Ca^{2+}$ activated fibres (see above) suggesting that high pressure impairs the transition from weak to strong binding states but once in the strong binding state, it seems to stabilize this state in the rigor configuration. It is, therefore, tempting to speculate that under high pressures, detachment of cross-bridges by ATP binding is, to some extent, inhibited during the cross-bridge cycles. However, this still lacks experimental confirmation.

### Structural Changes of acute large High Pressure Application (up to 400 MPa) to Muscle

Structural changes of high pressure exposure to muscle are the domain of large pressures up to 400 MPa. Under such pressures, the functional integrity of a muscle cell is not given any more. In fact, dissociation, aggregation or gelation is expected to occur through disruption and rearrangement of noncovalent bonds. That is why most of such studies are performed on purified proteins or centrifuged cellular subfragments to avoid interference from effects to different proteins. The stability of a muscle cell is largely given by the integrity of the intracellular cytoskeleton. Pressure-induced depolymerization of F-actin into globular G-actin constituents has been widely studied (Ikeuchi et al., 2002; Garcia et al., 1992; Swezey and Somero, 1982). Early work by Ikkai and Ooi (1966) described irreversible denaturation of rabbit muscle actin following a 10 min exposure to between >150 MPa or >250 MPa, depending on whether ATP was present or not. The volume change for the F- to G-actin transformation was estimated to be ~80 ml per mol of monomer (Ikkai and Ooi, 1966). However, protein determinations were performed '*ex situ*' following pressure release. Using fluorescence spectroscopy during high pressure applications to purified rabbit F-actin,

Garcia et al. (1992) found that polymeric F-actin reversibly dissociates when exposed to pressures up to 240 MPa for 5 min. The reversibility was nicely visualized by comparing electron micrographs from F-actin filament preparations that were fixed at high pressure and stained in the post-decompression phase with ones that were pressurized and fixed and stained following decompression. The latter ones were identical to controls that were fixed without pressurization, whereas F-actin filaments were almost completely dissociated at the high pressure fixed probe (Garcia et al., 1992). The study also showed that actin subunits were packed differently in the filamentous polymer depending on the ionic strength and the cation present in the solution. F-actin in the presence of potassium had a higher dissociation tendency (half-maximal dissociation at ~50 MPa) than in the presence of divalent cations ($Ca^{2+}$, $Mg^{2+}$; half-maximal dissociation at ~150 MPa). Because in the intact muscle cell the predominant cation under resting conditions is indeed $K^+$, this would suggest a much larger pressure susceptibility for the less packed F-actin filaments of relatively large inter-subunit volumes ($\Delta V$~350 ml/mol, as compared to ~80 ml/mol in the presence of predominantly $Ca^{2+}$ or $Mg^{2+}$) in skeletal muscle *'in situ'* (Garcia et al., 1992). Such pressures can already affect animals living at abyssal depths, e.g. teleost fish *Coryphaenoides armatus* living at ~5,000 m, equivalent to ~50 MPa. Interestingly, there were no differences in the compressibility of G-actin and F-actin samples from this species (Swezey and Somero, 1985). Also, the reversibility of F-actin dissociation seen in short-term pressure exposures does not necessarily project to reversibility in long-term high pressure scenarios. But also species differences seem to be quite important. Mouse fibroblasts pressurized up to 40 MPa for 10 min showed irreversible loss of normal cell shape, i.e. rounding of cells at pressure >27.5 MPa, without any change in intracellular $Ca^{2+}$ concentrations (Crenshaw and Salmon 1996). For mammalian skeletal muscle (mouse), in contrast, no changes in cell morphology were observed during high pressure applications up to 30 MPa within ~20 min (Friedrich et al., 2006a). Even longer pressure exposures resulted in irreversible contractures that were related to a $Ca^{2+}$ deregulation of the contractile apparatus and mechanical activation rather than cytoskeleton dissociation (see below).

Different susceptibilities to pressure-induced destabilization of the cellular cytoskeleton may be finally related to the metabolic status the cell can keep during pressure challenges. It has been shown that actin becomes unstable if it loses bound nucleotides, e.g. during metabolic failure or mitochondrial dysfunction, and this results in irreversible denaturation of filamentous actin (Lewis et al., 1963). Both heavy meromyosin and ATP had a protective effect against pressure denaturation of F-actin for pressures larger than 150 MPa (Ikeuchi et al., 2002).

Although muscle myosin (myosin II) is also a polymer consisting of a total of six protein chains (Table 1), it seems to be more pressure resistant compared to F-actin, at least concerning structural transformation in pressure ranges affecting biological systems. When kept in 100 mM KCl and pH 6, myosin filaments gelatinized after 10 min exposure to 210 MPa (Yamamoto et al., 1990). Subsequent studies using ANS (8-anilino-1-naphthalene sulfonic acid) fluorescence spectroscopy to detect high pressure induced changes in the hydrophobic core of pressurized rabbit myosin subfragments revealed that fluorescence intensities originating from the rod fragments did not change at constant pressures up to 400 MPa. In contrast, the myosin head region (S1) was partially denatured between 300 MPa and 350 MPa for 10 min (Iwasaki and Yamamoto, 2003). Interestingly, avian flight muscle myosin seemed to be more pressure sensitive because myosin rod of chicken muscle was partially unfolded at 280 MPa (King et al., 1994). Also, global structural changes were found at the S1 fragment already from 150 MPa while local structural changes of the actin ATPase and binding sites were effected at higher pressures between 250 MPa and 300 MPa (Iwasaki and Yamamoto, 2002).

Pressure-induced degradation and inactivation of other skeletal muscle proteins studied in short-term exposures (~10 min) to between 200 MPa and 400 MPa include, for example, rabbit muscle proteasome (Yamamoto et al., 2005) or SR $Ca^{2+}$ ATPase (Okamoto et al., 1995).

## EFFECTS OF PROLONGED HIGH PRESSURE APPLICATIONS ON MAMMALIAN SKELETAL MUSCLE FUNCTION—LESSONS FROM TERRESTRIAL SPECIES FOR THE DEPTH LIMITS OF DIVING MAMMALS

So far, all examples given for high pressure effects on muscle function involved only use of short-term applications in the ~min range that mostly affected kinetics and amplitudes of physiological processes. This maybe reflects the major focus of those studies to use high pressure as a thermodynamic tool to evaluate volume changes and activation volumes of various steps during ec-coupling and contractile performance. However, such short-term applications rarely occur in nature and even in vertical migrating marine animals, pressure exposure is a matter of prolonged challenge to tissues and cells. From thermodynamic equilibrium considerations, short-term pressure applications can have more dramatic effects on protein stability and activities, whereas phase diagrams for such protein properties after a long-term exposure are closer to the equilibrium curve (Smeller, 2002). Long-term applications in conjunction with '*ex situ*' recordings, i.e. recordings of the parameter in question after the decompression in the post-decompression phase, have additional advantages. As both the prolonged high pressure exposure and the post-decompression phase can be considered as a steady-state, comparing protein kinetics and turnover amplitudes after a certain, prolonged p-T treatment with untreated controls returns information about the number of molecules that survived the treatment (Smeller, 2002). This means if pressure affected a cellular process in a fully reversible manner, e.g. reversible unfolding or dissociation of proteins, post-decompression values would be identical to controls. However, if a subfraction of proteins was irreversibly dissociated or degraded under prolonged pressure treatment, this protein fraction would be functionally 'turned off' and locked even beyond decompression into the post-decompression phase. Amplitude and kinetics analysis of a dynamic process measured in the post-decompression phase would then give insights into whether simply the number of molecules in the ensemble was reduced or whether the kinetics of transition steps was irreversibly altered. Examples from our research on ion channels will be presented below.

During the last years, our laboratory has focused on the effects of prolonged 3 hr high pressure applications on muscle contractility, excitability and $Ca^{2+}$ regulation of mouse muscle cells recorded in the post-decompression phase. Most of the *'ex situ'* experiments involved slow pressurization and decompression using two sequential ramp protocols of 0.2 MPa min$^{-1}$ and 0.2 MPa min$^{-1}$ steps to a maximum hydrostatic pressure between 5 MPa and 25 MPa that was kept for 3 hr at a temperature of 4°C. This temperature was chosen because of water's density maximum at this value. Empirically, fibre survival, as assessed by inspection and viability tests after decompression, was maximal at this temperature (Kress et al., 2001, 1999). The slow pressurization also ensured isothermic compression and decompression, avoiding adiabatic heating.

## Muscle Integrity and Contractile Function Following Prolonged High Pressure Applications

Following up to 20 MPa 3 hr pressure treatments, most muscle fibres were still viable, as judged from transparent appearance in the light microscope and their ability to twitch upon field stimulation. Following 22.5 MPa and 25 MPa treatments, most of the muscle fibres were contracted and irreversibly damaged with opalesque myoplasm and no visible striation patterns. Following 30 MPa treatments, the whole muscle was disintegrated and no more cells could be obtained for functional testing (Kress et al., 1999). Increasing external osmolarity by addition of 200 mM sucrose (~450 mosM total osmolarity) slightly increased the number of surviving fibres after 25 MPa (Kress et al., 1999). When assessing contractile function in the surviving, non-contracted, fibres, a marked decline of isometric maximum $Ca^{2+}$ activated (pCa = 4.5) force to 50% of control values was found following

25 MPa treatments, whereas up to 20 MPa, there was hardly any change (Kress et al., 2001). Likewise, the rate of force development was slowed down eightfold following 25 MPa applications but not following lower pressure applications. Stiffness of the fully $Ca^{2+}$ activated fibre bundles, however, was unaffected throughout, indicating that the strong binding state was not affected by pressure. The reduction in maximum force could be either due to a reduced number of cycling cross-bridges or a change in the $Ca^{2+}$sensitivity of the contractile apparatus. Assessing the pCa-force relations in skinned edl fibre bundles showed no change in the curves following pressure applications up to 20 MPa compared to controls but a significant shift towards smaller pCa values following larger 3 hr pressure exposures (i.e. from pCa ~5.8 for controls to pCa ~5.6; Kress et al,. 2001). This suggested a 70% decrease in the $Ca^{2+}$ sensitivity for the $Ca^{2+}$ induced activation of the myofilaments following 25 MPa long-term pressure treatments. Additionally, the stiffness of relaxed muscle increased biphasically in the pressure range tested. Up to 20 MPa, relaxed fibre stiffness increased roughly twofold, whereas for pressures >20 MPa, an additional fivefold increase occurred (Kress et al., 2001). Following 25 MPa, the relaxed fibre stiffness (~0.5 N/m) accounted for already 50% of the $Ca^{2+}$ activated stiffness (~1.1 N/m; Kress et al., 2001). The results were interpreted in terms of a model in which for increasing pressures, more and more cross-bridges might get irreversibly bound to the actin filaments even in the relaxed state and cannot be recruited for cycling and force development upon $Ca^{2+}$ activation. This decrease of cycling cross-bridges becomes a dominant problem for pressures >20 MPa when, taken together, resting muscle basically becomes 'stiff', slow in activation and less sensitive to $Ca^{2+}$ activation (Kress et al., 2001). It has to be kept in mind that these results could only be obtained from undamaged cells. So how could it then be that following pressures >20 MPa more and more muscle cells were found hypercontracted and irreversibly damaged when $Ca^{2+}$ activated force was impaired? When comparing SDS PAGE data from whole muscle homogenates following 20 MPa and 25 MPa with control muscles kept at atmospheric pressure, the band reflecting troponin-T was markedly reduced or almost absent following 3 hr 25 MPa treatments. So, although contractile performance declined with large pressures >20 MPa, the contractile apparatus is expected to become deregulated to some extent by degradation of thin filament regulatory proteins. This possibly induces irreversible contractures even at moderately increased intracellular $Ca^{2+}$ levels that would be otherwise necessary to activate the contractile apparatus at all for pressures larger than 20 MPa (in particular, 2.6 µM $Ca^{2+}$ after 25 MPa treatments would be needed to elicit 50% of isometric force from the shift in pCa-force curves compared to 1.6 µM in the controls, Kress et al., 2001).

## Muscle Membrane Integrity and Ion Channel Function following Prolonged High Pressure Applications

The contractile data following prolonged pressure applications suggest that contracted fibres must have experienced some cytosolic $Ca^{2+}$ overload in addition to the thin fibre filament deregulation. One possible source for this could be external $Ca^{2+}$ entry from the surface membrane. From recordings of creatine phosphokinase (CPK) activity in the medium following prolonged pressure exposure, no significant increases were found at 25 MPa compared with 20 MPa and controls ruling out a general breakdown of muscle membrane integrity that would be detected by leakage of intracellular enzymes. Also, resting membrane resistances recorded in the post-decompression phase were not decreased and even seemed larger following 20 MPa than after 10 MPa (Friedrich et al., 2002). Interestingly, resting membrane potentials were depolarized in a pressure-dependent manner that due to a persistent pressure-induced reduction of the selectivity of the resting membrane for $K^+$ over $Na^+$. A threefold increase of the $P_{Na}/P_K$ permeability ratio suggested an increase in intracellular $Na^+$ that was also confirmed by a shift of the reversal potential for $Na^+$ currents ($I_{Na}$) under voltage clamp conditions for pressures larger than 10 MPa (Friedrich et al., 2003, 2002). $I_{Na}$ potential amplitudes as well as $K^+$ currents ($I_K$) are irreversibly reduced after prolonged high pressure treatments. This already suggests a permanent impairment of membrane excitability in such high pressure treated muscle. $Ca^{2+}$ entry during tetanic excitation may also occur through the L-type $Ca^{2+}$ channel which usually does not become activated during single twitches. One way to test a putative increase of $Ca^{2+}$ flow through this channel was by recording voltage-dependent L-type currents ($I_{Ca}$) under maintained depolarization following high pressure treatments. Surprisingly, current amplitudes were persistently decreased similar to $I_{Na}$ amplitudes, thus ruling out $Ca^{2+}$ influx through this channel as a possible source for cytosolic $Ca^{2+}$ challenge (Friedrich et al., 2002). Because steady-state inactivation of channel entities was not irreversibly altered by high pressure (Friedrich et al., 2002), it was necessary to have a closer look at the current kinetics following different pressure treatments to elucidate the exact mechanisms for current reductions. It was found that for $I_{Na}$, pressures of 10 MPa and 20 MPa reduced amplitudes but did not significantly change activation or inactivation kinetics of the currents, indicating unaltered channel gating (Friedrich et al., 2006a). Therefore, the reduction in $I_{Na}$ amplitudes was compatible with an exclusively persistent reduction in functional ion channel density. In contrast, results for L-type channels were quite different: inactivation was slowed down fourfold while activation kinetics was slowed down twofold. Thus, one would have expected a net increase in $I_{Ca}$, rather than a decrease. This was interpreted as $Ca^{2+}$ channels being more susceptible to prolonged high pressure treatments in mammalian muscle, because not only the gating transitions of the channels were irreversibly impaired but also the channel density must have been decreased to a larger extent compared to $Na^+$ channels to explain the reductions in current densities seen (Friedrich et al., 2006a). This result showed that although both channels exhibit a high degree of homology of amino acid sequence (Jurkat-Rott and Lehmann-Horn, 2001), the three-dimensional sterical arrangement, which is crucial for proper channel function in muscle, seems to be differentially susceptible to prolonged high pressure exposures (Friedrich et al., 2006a). From the pressure dependence of current kinetics, the activation volumes $\Delta V^{\ddagger}$ for the irreversible, pressure-driven impairment of $I_{Na}$ and $I_{Ca}$ activation and inactivation could be determined in long-term pressurized muscle, for the first time. As expected, $\Delta V^{\ddagger}$ values for $Na^+$ channels were small ($< +30$ Å$^3$) compared to the very large values of $\Delta V^{\ddagger} = 510$ Å$^3$ for $Ca^{2+}$ channel activation and $\Delta V^{\ddagger} = 218$ Å$^3$ for inactivation, respectively, reflecting the larger pressure sensitivity of the latter.

## $Ca^{2+}$ Regulation during High Pressure Applications: Insights from novel 'in situ' High Pressure Microscopy Techniques

Taken together, the above mentioned mechanisms suggest a disturbance of intracellular $Ca^{2+}$ handling as the source for pressure-induced muscle cell damage to prolonged high pressures >20 MPa. An obvious source would be $Ca^{2+}$ leakage from the SR. Such a re-distribution of $Ca^{2+}$ from the SR to the myofibrillar space had been previously shown in ultrastructural EM studies from rabbit skeletal muscle treated with high pressures of between 100 MPa and 300 MPa for 5 min to study mechanisms of meat tenderization (Okamoto et al., 1995), but had never been attempted in living muscle cells. $Ca^{2+}$ release from the SR eventually occurs spontaneously as '*elementary $Ca^{2+}$ release events*' (ECRE) or '*$Ca^{2+}$ sparks*' in a concerted opening of ryanodine receptor clusters in mammalian muscle (Zhou et al., 2005). Following prolonged 3 hr pressure application, Schnee et al. (2008) recently studied ECRE kinetics, morphology and frequencies in mammalian mouse muscle fibres in the post-decompression phase: they found an exponential decline of ECRE frequencies from ~35 mm$^{-2}$s$^{-1}$ at atmospheric pressure to ~15 mm$^{-2}$s$^{-1}$ following 20 MPa and <10 mm$^{-2}$s$^{-1}$ following 30 MPa treatments. Although this suggests less frequent events, in the long run during a 3 hr high pressure treatment, the leak of $Ca^{2+}$ ions into the cytoplasm is substantial, because both the size of the $Ca^{2+}$ release source and the opening duration increased with pressure up to 20 MPa, but then decreased following 30 MPa applications (Schnee et al., 2008).

Clearly, $Ca^{2+}$ intracellular homeostasis in mammalian muscle seems impaired as muscle continues to be exposed to hydrostatic pressures.

However, a fixed time point of 3 hr applications, after which experiments would be performed in the post-decompression phase at atmospheric pressure, would not give further insights into the exact time course for which the impaired $Ca^{2+}$ regulation would render the muscle cell susceptible to irreversible mechanical activation and cell death in the contracted state. The results also show the principle limitations of '*ex situ*' approaches where everything happening during the pressure treatment would have to be considered a black box for which assumptions have to be deduced from the post-decompression recordings. For example, it would be suggestive that the 'pressure-level x exposure-time' product would be a major determinant for high pressure survival of cells, i.e. shorter toleration times at higher pressures. In order to overcome the principle constraints of '*ex situ*' studies, '*in situ*' high pressure experiments are desirable, because results are directly obtained under high (hydrostatic) pressure conditions. However, such approaches have been proven to be technically demanding, in particular in cases where electrophysiology rigs had to be (partially) incorporated into the pressure vessel (Macdonald and Martinac, 2005; Schmalwasser et al., 1998). Using an improved high pressure optical vessel with sapphire windows suitable for fluorescence microscopy (Hartmann et al., 2004), the morphology and integrity of single muscle fibres could be visualized during pressurizations up to 35 MPa. The fluorescent marker propidium iodide that only permeates into cells upon membrane damage was used. During the whole slow pressurization protocol, no membrane damage was observed. When holding the pressure at a maximum of 35 MPa, cell morphology and membrane integrity remained intact. After ~22 min, a contracture of the muscle fibre slowly developed within minutes. Amazingly, membrane integrity was still maintained during this contracture, supporting the postulated mechanism of activation by $Ca^{2+}$ release from internal stores (Friedrich et al., 2006a). Upon decompression, the fibre remained contracted and membrane integrity slowly broke down in the following one to two hours. This explains why following pressures >25 MPa, almost all muscle fibres were previously found contracted and irreversibly damaged after a 3 hr incubation (Friedrich et al., 2002; Kress et al., 2001).

To prove the proposed mechanism of altered $Ca^{2+}$ handling during prolonged pressure applications, we were able to adapt an optical pressure vessel to a confocal microscope to perform the first confocal $Ca^{2+}$ fluorescence experiments under high hydrostatic pressures in an excitable muscle cell. Intact cells were stained with the cytoplasmic $Ca^{2+}$ indicator Fluo-4 and the preferentially mitochondrial $Ca^{2+}$ indicator Rhod-2 and pressurized using a slow pressure ramp protocol. Figure 4 shows images from both $Ca^{2+}$ signals at atmospheric pressure and taken at 15 MPa during pressurization to 35 MPa.

From such images at all pressures, $Ca^{2+}$ fluorescence intensities within the fibre area were obtained. Their time courses are shown in the 3D plots. During pressurization, there was a marked drop in cytoplasmic $Ca^{2+}$ that was paralleled by an increase in mitochondrial $Ca^{2+}$ (Fig. 4). Qualitatively, this shows a redistribution of $Ca^{2+}$ between the two compartments. Quantitatively, it is, however, important to assess the pressure dependence of the dye-$Ca^{2+}$ binding affinities, as pressure is known to affect the affinity constants for several buffer systems, e.g. the ratiometric $Ca^{2+}$ dye Fura-2 (Crenshaw and Salmon, 1996). For Fluo-4, the '*in situ*' binding of $Ca^{2+}$ to the dye was assessed in the pressure range between 0.1 MPa and 250 MPa in a chemically permeabilized muscle fibre bathed in an internal solution with 100 nM free $Ca^{2+}$ using the high pressure confocal microscope (Friedrich et al., 2006b). Between 0.1 MPa and ~15 MPa, Fluo-4 fluorescence sharply dropped indicating a decrease in $Ca^{2+}$-Fluo-4 binding that was related to pressure 'per se' and not to changes in $Ca^{2+}$. However, between 15 MPa and 35 MPa, fluorescence had a plateau during continued compression indicating a stable affinity constant of $Ca^{2+}$ and Fluo-4 in this pressure range (Friedrich et al., 2006b). For the intact fibre experiment, this implicated that decreases in cytoplasmic $Ca^{2+}$ were smaller than anticipated from the initial pressurization plot. In fact, the exponential drop of Fluo-4 fluorescence in the intact pressurized fibre bathed in $Ca^{2+}$-free external solution occurred exponentially with a 'pressure constant' of ~4 MPa (Friedrich et al., 2006b).

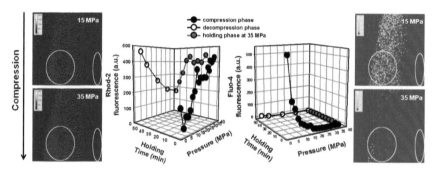

Figure 4. Confocal '*in situ*' high pressure $Ca^{2+}$ fluorescence microscopy in single intact mammalian muscle cells up to 30 MPa. Mitochondrial Rhod-2 (left) and cytoplasmic Fluo-4 (right) $Ca^{2+}$ fluorescence were monitored during slow pressurization (black circles), holding pressure at 35 MPa for ~50 min (red circles) and during decompression (white circles) in an intact skeletal mouse muscle fibre. During pressurization, Rhod-2 fluorescence increased while Fluo-4 fluorescence sharply decreased and remained stable from ~20 MPa while mitochondrial fluorescence kept increasing. This was compatible with $Ca^{2+}$ leaking from the SR into the cytoplasm being taken up and buffered by mitochondria. During the holding phase fluorescence remained stable until ~45 min, after which mitochondrial signals dropped sharply. To prevent a developing contracture, pressure release was immediately initiated, during which Rhod-2 fluorescence recovered markedly and Fluo-4 fluorescence only marginally (*according to data from Friedrich et al., 2006b*).

Now this means, that between 15 MPa and 35 MPa, cytoplasmic $Ca^{2+}$ hardly changed any more, yet, mitochondria were still taking up $Ca^{2+}$, as judged from the still markedly rising Rhod-2 fluorescence (pressure constant ~23 MPa). This is consistent with a pressure-induced $Ca^{2+}$ leak from the SR that would chronically overload cytoplasmic $Ca^{2+}$ in the pressurized muscle cell; however, this is prevented by mitochondrial $Ca^{2+}$ buffering (Friedrich et al. 2006b). It is important to note that SR $Ca^{2+}$ leak was confirmed in pressurized rabbit skeletal muscle (Okamoto et al., 1995). When pressure was held at a maximum of 35 MPa, the system was stable during a recording phase of ~45 min, after which mitochondrial $Ca^{2+}$ seemed to suddenly drop (Fig. 4) indicating $Ca^{2+}$ being re-poured into the cytoplasm, probably due to mitochondrial $Ca^{2+}$ overload, exhausted $Ca^{2+}$ buffering capacity and opening of giant mitochondrial transition pores (Bernardi et al., 1999). To prevent subsequent contracture in this experiment, pressure was released slowly at this point, which unexpectedly resulted in a recovery of mitochondrial $Ca^{2+}$

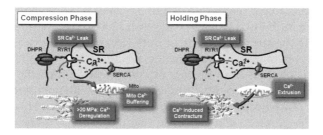

**Figure 5. Proposed model of high pressure acting specifically interacting with intracellular $Ca^{2+}$ regulation in mammalian muscle during slow pressurization holding the pressure.** During slow compression, $Ca^{2+}$ leaking from the SR is being taken up and buffered by mitochondria, thus preventing a rise in intracellular free $Ca^{2+}$ concentration. At some stage above 20 MPa, the contractile apparatus becomes deregulated by degradation of thin filament components. During the holding phase at maximum pressure, mitochondria keep buffering $Ca^{2+}$ until their buffering capacity is exhausted due to mitochondrial overload. $Ca^{2+}$ is being re-poured into the myoplasm, presumably by opening of giant transition pores and can now readily activate the deregulated motorproteins for an irreversible contracture.

uptake that even exceeded the initial atmospheric values threefold (Friedrich et al., 2006b). The reasons for this in mammalian fast-twitch muscle are not clear at this stage but from experiments in fish red muscle, mitochondrial function, oxidative phosphorylation and regulation of proton leak were markedly affected by pressures larger than ~3 MPa (Sébert et al., 2004).

A suggested model for pressure-intracellular $Ca^{2+}$-interaction is summarized in Fig. 5.

## High Pressure Effects on Mammalian Muscle—Implications for Diving Mammals?

The effects of high pressure on muscle performance and survival implicate a limited pressure-exposure time window for the reversibility of pressure-induced impairment of mammalian muscle function regarding contractility, excitability and $Ca^{2+}$ regulation. Ultimately spoken, a too extended pressure exposure will result in irreversible contracture of muscle by the mechanisms given above that freeze muscle in a non-functional state from which tissue death will occur.

Are such pressure limitations expected to affect any mammals?

Cetaceans are the prototype of deep-diving mammals that can dive beyond depths of 1,500 m (e.g. bottlenose whales, Hooker and Baird, 1999) and probably reach depths up to 3,000 m (sperm whales, Watkins et al., 1993). Therefore, cetaceans may well experience such prolonged pressure exposures as detailed above. A recent study on extreme diving behaviour of beaked whales found maximum depths of ~2,000 m and exposure times up to 90 min for individual whales (Tyack et al., 2006). The question, therefore, arises what the principle depth limits (i.e. pressure exposure limits, respectively) in mammalian organisms are and what organ limits further descent. It is obvious from taxonomic considerations that the brain (high pressure nervous syndrome), or lungs (air spaces) do not seem to prevent whales from such deep dives. The extreme profiles of whales fit surprisingly well into the pressure and exposure time frames that were assessed for terrestrial mammalian muscle before irreversible tissue damage occurred (Friedrich et al., 2006a, b). It is, therefore, tempting to speculate that skeletal muscle might be the limiting abyssal organ in mammals that prevents further pressure damage to the animal. For instance, on descent, muscle in deep diving cetaceans might provide similar dysregulations in $Ca^{2+}$ that induce contracture of some fibres at some stage during the pressure-exposure time product. As muscle is finely innervated with pain sensors, a beginning contracture should induce pain sensation which will be one of the peripheral signals to the brain to stop further descent and initiate ascent. The situation might be even more complex, because although fast-twitch fibres are present in diving mammals in abundance, the majority of muscle fibre types seem to be purely oxidative, at least in some species (Watson et al., 2003). Of course, further basic and taxonomic research will be needed to provide an unambiguous answer (Castellini et al., 2002).

More detailed information on Diving Mammals and Birds is found in the corresponding chapter in this book (M. Castellini).

## ADAPTATIONS OF SKELETAL MUSCLE FUNCTION TO PROLONGED HIGH PRESSURE EXPOSURES—IMPLICATIONS FOR DEEP SEA FISH

The deep sea realm ecosystems extend to the deepest marine layers known, exposing organisms to pressures as high as ~100 MPa. Moreover, pelagic animals, i.e. deep sea fish, show marked vertical migration patterns that make them even more susceptible to pressure fluctuations and temperature challenges besides the extreme long-term exposure at their dwell pressure and temperature (Somero, 1992). It is clear that evolutionary mechanisms must have evolved to overcome pressure constraints in tissues of these animals on a cellular level (Vettier et al., 2006; Sébert, 2003). Among different biochemical adaptations, e.g. low metabolic rates in locomotory muscle due to reduced ATP turnover capacities (Somero, 1992), the most fascinating general concepts to overcome increased ambient pressures are 'homeoviscous adaptation' of the membrane lipid composition (Macdonald and Cossins, 1985) and cellular production of organic osmolytes to counteract hydrostatic pressure effects on cell protein structures (Zerbst-Boroffka et al., 2005; Yancey and Siebenaller, 1999). For more detailed information, the reader is referred to Sébert (2008) and to the corresponding chapters in this book (R. Winter; A. Péqueux; A. Oliveria and P. Sébert).

The environmental plasticity of muscle can be impressively seen in deep sea fish or migrating fish that experience acclimatization to high pressures. In teleost fish, the environmental properties determine the rate of myogenesis, gene expression profiles, sub-cellular architecture as well as fibre types and sizes of swimming muscle (Johnston, 2006). In European eel (*Anguilla anguilla*, L.) that spawns from a non-migratory fresh-water yellow stage to a migratory silver stage during a ~6.000 km migration period from European rivers to the Sargasso Sea, acclimatization of slow twitch aerobic red muscle takes place in order to adapt to swimming depths between 200 m and 1,000 m (Rossignol et al., 2006). Specifically, 3 wk exposure to 10.1 MPa hydrostatic pressures resulted in a three-fold reduction in maximum isometric twitch and tetanic tension while activation and inactivation kinetics remained unchanged (Rossignol et al., 2006). As the recordings were conducted in the post-decompression phase after high-pressure acclimatization, they may be, at least to some degree, comparable to the prolonged high pressure treatments performed in mammalian muscle (Kress et al., 2001). It would be interesting to have more data on contractile performance in these fish during pressure acclimatization, i.e. performed at high hydrostatic pressures, to see whether the kinetics and amplitude behaviours would be similar to those ones obtained in frog and mammalian muscle during short-term pressure exposures (Fortune et al., 1996; Ranatunga and Geeves, 1991). Early studies in some mesopelagic fish (teleosts and elasmobranchs) fast and slow muscles that were also performed under atmospheric conditions showed isometric tension values comparable to those found in amphibian and mammalian muscles (Altringham and Johnston, 1982). One of the main problems faced with such experiments on deep sea fish muscle performance is the lack of preserved pressure environment once the fish are being trawled and brought to shore for experimentations. Decompression and warming upon surfacing already can induce substantial changes to membrane fluidity and kinetics of biochemical processes (Rossignol et al., 2006; Sébert, 2002). This becomes even more important since it has very recently been found from comparative sequence analysis of myosin heavy chain proteins of shallow-and deep-water living rattail fish that single amino acid substitutions in the latter rendered the myosin protein structure more compact and resistant to high pressures (Morita, 2008). Similar observations were made for actin where a few amino acid substitutions only were able to alter the protein's pressure sensitivity in terms of pressure adaptation (Morita, 2003).

These latter experiments show that the earlier statement that skeletal muscle proteins were highly conserved might not necessarily hold any more when pressure adaptations require genetic modulations (Morita, 2003).

## CONCLUDING REMARKS

High hydrostatic pressure effects on muscle are complex but very specific. Rather than unspecific damage, there is a defined sequence of alterations of single steps within the muscle activation cascade that critically depend on the pressure level, the rate of pressurization and the duration of pressure exposure. Within the mammalian realm, high pressure effects may be transferable and insights gained from terrestrial mammals may be of value to extrapolate to principle pressure limits in diving mammals. Deep sea fish also represent an excellent model to study prolonged high pressure applications and adaptational mechanisms that might need a time frame of pressure exposure not feasible to apply to mammalian samples. Nevertheless, experimental constraints like on site experimentation after trawling to minimize decay time of the preparation in fish muscle (v Wegner et al., 2008) or use of deep sea aquariums (Koyama, 2007) are only some of the technical difficulties that might have to be optimized for future studies of deep-sea fish muscle and other tissues. Taken together, there seem to be a lot of taxonomic similarities but also quite substantial differences of pressure affecting our primary locomotory organ. We might be still far away from an overall spanning theory of high pressure effects on muscle function but we are also a little bit closer by now.

## ACKNOWLEDGEMENTS

This work was supported by an Australian Research Council (ARC) International Fellowship to the author. I thank Dr. Bradley Launikonis, Joshua Edwards and Tanya Cully for careful reading of the manuscript and helpful suggestions.

## REFERENCES

Ahern, C.A. and J. Arikkath, P. Vallejo, C.A. Gurnett, P.A. Powers, K.P. Campbell and R. Coronado. 2001. Intramembrane charge movements and excitation-contraction coupling expressed by two-domain fragments of the $Ca^{2+}$ channel. Proc. Natl. Acad. Sci. USA 98(12): 6935–6940.

Allen, D.G. and G.D. Lamb and H. Westerblad. 2008. Skeletal muscle fatigue: cellular mechanisms. Physiol. Rev. 88(1): 287–332.

Altringham, J.D. and I.A. Johnston. 1982. The pCa-tension and force-velocity characteristics of skinned fibres isolated from fish fast and slow muscles. J. Physiol. 333: 421–449.

Altringham, J.D. and P.H. Yancey and I.A. Johnston. 1982. The effects of osmoregulatory solutes on tension generation by dogfish skinned muscle fibres. J. Exp. Biol. 96: 443–445.

Ashford, M.L.J. and A.G. Macdonald and K.T. Wann. 1982. The effects of hydrostatic pressure on the spontaneous release of transmitter at the frog neuromuscular junction. J. Physiol. 333: 531–543.

Au, Y. 2004. The muscle ultrastructure: a structural perspective of the sarcomere. Cell. Mol. Life Sci. 61: 3016–3033.

Bannister, R.A. 2007. Bridging the myoplasmic gap: recent developments in skeletal muscle excitation-contraction coupling. J. Muscle Res. Cell Motil. 28(4–5): 275–283.

Baylor, S.M. and W.K. Chandler and M.W. Marshall. 1983. Sarcoplasmic reticulum calcium release in intact frog skeletal muscle fibres estimated from Arsenazo III calcium transients. J. Physiol. 344: 625–666.

Berchtold, M.W. and H. Brinkmeier and M. Müntener. 2000. Calcium ion in skeletal muscle: its crucial role for muscle function, plasticity and disease. Physiol. Rev. 80(3): 1216–1265.

Bernardi, P. and L. Scorrano, R. Colonna, V. Petronilli and F. DiLisa. 1999. Mitochondria and cell death. Mechanistic aspects and methodological issues. Eur. J. Biochem. 264: 687–701.

Boonyaratanakornit, B.B. and C.B. Park and D.S. Clark. 2002. Pressure effects on intra- and intermolecular interactions with proteins. Biochim. Biophys. Acta 1595: 235–249.

Bottinelli, R. and R. Betto, S. Schiaffino and C. Reggiani. 1994. Unloaded shortening velocity and myosin heavy chain and light chain isoform composition in rat skeletal muscle fibres. J. Physiol. 478: 341–349.

Braganza, L.F. and D.L. Worcester. 1986. Structural changes in lipid bilayers and biological membranes caused by hydrostatic pressure. Biochemistry 25: 7484–7488.

Brauer, R.W. and R.W. Beaver, S. Lasher, R.D. McCall and R. Venters. 1979. Comparative physiology of the high pressure neurological syndrome—compression rate effects. J. Appl. Physiol. 46(1): 128–135.

Brauer, R.W. and M.R. Jordan, C.G. Miller, E.D. Johnson, J.A. Dutcher and M.E. Sheeman. Interaction of temperature and pressure in intact animals. Pp. 3–28. *In*: A.J.R.

Péqueux and R. Gilles (eds.). 1985. High Pressure Effects on Selected Biological Systems. Springer-Verlag, Berlin.

Brown, D.E.S. 1934. The effect of rapid changes in hydrostatic pressure upon the contraction of skeletal muscle. J. Cell. Comp. Physiol. 4: 257–281.

Butler, P.J. 2006. Aerobic dive limit. What is it and is it always used appropriately? Comp. Biochem. Physiol. A Integr. Physiol. 145(1): 1–6.

Butler, P.J. and D.R. Jones. 1997. Physiology of diving of birds and mammals. Physiol. Rev. 77(3): 837–899.

Cairns, S.P. and W.A. Hing, J.R. Slack, R.G. Mills and D.S. Loiselle. 1998. Role of extracellular $Ca^{2+}$ in fatigue of isolated mammalian skeletal muscle. J. Appl. Physiol. 84: 1395–1406.

Campenot, R.B. 1975. The effects of high hydrostatic pressure on transmission at the crustacean neuromuscular junction. Comp. Biochem. Physiol. 52B: 133–140.

Caputa, M. 2006. Animal responses to extreme conditions: a lesson to biomedical research. J. Physiol. Pharmacol. 57(Suppl. 8): 7–15.

Castellini, M.A. and P.M. Rivera and J.M. Castellini. 2002. Biochemical aspects of pressure tolerance in marine mammals. Comp. Biochem. Physiol. A 133: 893–899.

Cattell, M. 1935. Changes in the efficiency of muscular contraction under pressure. J. Cell Comp. Physiol. 6: 277–290.

Cattell, M. and D.J. Edwards. 1928. The energy changes of skeletal muscle accompanying contraction under high pressure. Am. J. Physiol. 86: 371–381.

Cattell, M. and D.J. Edwards. 1930. Reversal of the stimulating action of hydrostatic pressure on striated muscle. Science 71(1827): 17–18.

Clark K.A. and A.S. McElhinny, M.C. Beckerle and C.C. Gregorio. 2002. Striated muscle cytoarchitecture: an intricate web of form and function. Annu. Rev. Cell Dev. Biol. 18: 637–706.

Coates, J.H. and A.H. Criddle and M.A. Geeves. 1985. Pressure relaxation studies of pyrene-labelled actin and myosin subfragment-1 from rabbit skeletal muscle. Biochem. J. 232: 351–356.

Coelho-Sampaio, T. and S.T. Ferreira, G. Benaim and A. Vieyra. 1991. Dissociation of purified erythrocyte $Ca^{2+}$-ATPase by hydrostatic pressure. J. Biol. Chem. 266(33): 22266–22272.

Collins, M.D. and G. Hummer, M.L. Quillin, B.W. Matthews and S.M. Gruner. 2005. Cooperative water filling of a nonpolar protein cavity observed by high pressure crystallography and simulation. Proc. Natl. Acad. Sci. USA 102(46): 16668–16671.

Conti, F. and R. Fiovaranti, J.R. Segal and W. Stühmer. 1982a. Pressure dependence of the sodium currents of squid giant axons. J. Membr. Biol. 69: 23–34.

Conti, F. and R. Fiovaranti, J.R. Segal and W. Stühmer. 1982b. Pressure dependence of the potassium currents of squid giant axon. J. Membr. Biol. 69: 35–40.

Crenshaw, H.C. and E.D. Salmon. 1996. Hydrostatic pressure to 400 atm does not induce changes in the cytosolic concentration of $Ca^{2+}$ in mouse fibroblasts: measurements using fura-2 fluorescence. Exp. Cell Res. 227: 277–284.

Dauber, W. and T. Voigt, X. Härtel and J. Mayer. 2000. The t-tubular network and its triads in the sole plate sarcoplasm of the motor end-plate of mammals. J. Muscle Res. Cell. Motil. 21(5): 443–449.

De Felice, F.G. and V.C. Soares and S.G. Ferreira. 1999. Subunit dissociation and inactivation of pyruvate kinase by hydrostatic pressure. Oxidation of sulfhydryl groups and ligand effects on enzyme stability. FEBS Lett. 266: 163–169.

Delbono, O. and E. Stefani. 1993. Calcium transients in single mammalian skeletal muscle fibres. J. Physiol. 463: 689–707.

Donaldson, P.L. and K.G. Beam. 1983. Calcium currents in fast-twitch skeletal muscle of the rat. J. Gen. Physiol. 82(4): 449–468.

Dos Remedios, C.G. and D. Chhabra, M. Kekic, I.V. Dedova, M. Tsubakihara, A.D. Berry and N.J. Nosworthy. 2003. Actin binding proteins: regulation of cytoskeletal microfilaments. Physiol. Rev. 83(2): 433–473.

Eisinger, J. and S.F. Scarlata. 1987. The lateral fluidity of erythrocyte membranes. Temperature and pressure dependence. Biophys. Chem. 28: 273-281.

Fill, M. and J.A. Copello. 2002. Ryanodine receptor calcium release channels. Physiol. Rev. 82: 893–922.

Flück, M. and H. Hoppeler. 2003. Molecular basis of skeletal muscle plasticity—from gene to form and function. Rev. Physiol. Biochem. Pharmacol. 146: 159–216.

Foguel, D. and J.L. Silva. 2004. New insights into the mechanisms of protein misfolding and aggregation in amyloidogenic diseases derived from pressure studies. Biochemistry 43(36): 11361–11370.

Fortune, N.S. and M.A. Geeves and K.W. Ranatunga. 1989. Pressure sensitivity of active tension in glycerinated rabbit psoas muscle fibres: effects of ADP and phosphate. J. Muscle Res. Cell Motil. 9: 219–232.

Fortune, N.S. and M.A. Geeves and K.W. Ranatunga. 1994. Contractile activation and force generation in skinned rabbit muscle fibres: effects of hydrostatic pressure. J. Physiol. 474: 283–290.

Friedrich, O. 2007. Physiologie kompakt. Springer New York, Heidelberg, Tokyo. (in German).

Friedrich, O. and H. Ludwig. 2007. The high pressure scale for pressure-biosystems interactions: from sub-mega to giga-pascals. Pp. 96–103. In: F. Abe and A. Suzuki (eds.). Proceedings of the 4th International Conference on High Pressure Bioscience and Biotechnology, Tsukuba, Japan. J-stage. http://www.jstage.jst.go.jp/article/hpbb/1/1/96/ pdf.

Friedrich, O. and T. Ehmer and R.H.A. Fink. 1999. Calcium currents during contraction and shortening in enzymatically isolated murine skeletal muscle fibres. J. Physiol. 517: 757–770.

Friedrich, O. and K.R. Kress, H. Ludwig and R.H.A. Fink. 2002. Membrane ion conductances of mammalian skeletal muscle in the post-decompression state after high-pressure treatment. J. Membr. Biol. 188: 11–22.

Friedrich, O. and K.R. Kress, H. Ludwig and R.H.A. Fink. Activation of $Na^+$ and L-type $Ca^{2+}$ currents is more susceptible to prolonged high pressure treatments than voltage-dependent inactivation. Pp. 241–246. In: R. Winter (ed.). 2003. Advances in High Pressure Biosciences and Biotechnology II. Springer Heidelberg, Tokyo, New-York.

Friedrich, O. and K.R. Kress, M. Hartmann, B. Frey, K. Sommer, H. Ludwig and R.H.A. Fink. 2006a. Prolonged high-pressure treatments in mammalian skeletal muscle result in loss of functional sodium channels and altered calcium channel kinetics. Cell Biochem. Biophys. 45: 71–83.

Friedrich, O. and F.V. Wegner, M. Hartmann, B. Frey, K. Sommer, H. Ludwig and R.H.A. Fink. 2006b. 'In situ' high pressure confocal $Ca^{2+}$-fluorescence microscopy in skeletal muscle: a new method to study pressure limits in mammalian cells. Undersea Hyperb. Med. 33(3): 181–195.

Friedrich, O. and C. Weber, F. V. Wegner, J.S. Chamberlain and R.H.A. Fink. 2008. Unloaded speed of shortening in voltage-clamped intact skeletal muscle fibers from wt, mdx and transgenic minidystrophin mice using a novel high-speed acquisition system. Biophys. J. 94: 4751–4765.

Fryer, M.W. and D.G. Stephenson. 1996. Total and sarcoplasmic reticulum calcium content of skinned fibres from rat skeletal muscle. J. Physiol. 493(2): 357–370.

Garcia, C.R.S. and A. Amaral Jr., P. Abrahamsohn and S. Verjovski-Almeida. 1992. Dissociation of F-actin induced by hydrostatic pressure. Eur. J. Biochem. 209: 1005–1011.

Gibbs, A.G. 1998. The role of lipid physical properties in lipid barriers. Amer. Zool. 38: 268–279.

Gillis, J.M. and D. Thomason, J. Levefre and R.H. Kretsinger. 1982. Parvalbumin and muscle relaxation: a computer simulation study. J. Muscle Res. Cell Motil. 3(4): 377–398.

Gomez, J. and P. Neco, M. DiFranco and J.L Vergara. 2006. Calcium release domains in mammalian skeletal muscle studied with two-photon imaging and spot detection techniques. J. Gen. Physiol. 127(6): 623–637.

Guderley, H. 2004. Locomotor performance and muscle metabolic capacities: impact of temperature and energetic status. Comp. Biochem. Physiol. B. Biochem. Mol. Biol. 139(3): 371–382.

Gunning, P. and G. O'Neill and E. Hardeman. 2008. Tropomyosin-based regulation of the actin cytoskeleton in time and space. Physiol. Rev. 88(1): 1–35.

Haman, F. 2006. Shivering in the cold: from mechanisms of fuel selection to survival. J. Appl. Physiol. 100(5): 1702–1708.

Hamill, O.P. 2006. Twenty odd years of stretch-sensitive channels. Pflugers Arch. 453(3): 333-351.

Harasztosi, C.S. and I. Sipos, L. Kovacs and W. Melzer. 1999. Kinetics of inactivation and restoration from inactivation of the L-type calcium current in human myotubes. J. Physiol. 516: 129–138.

Hartmann, M. and M. Kreuss and K. Sommer. 2004. High pressure microscopy—a powerful tool for monitoring cells and macromolecules under high hydrostatic pressure. Cell. Mol. Biol. 50(4): 479–484.

Hasselbach, W. 1988. Presssure effects on the interactions of the sarcoplasmic reticulum calcium transport enzyme and dinitrophenyle phosphate. Z. Naturforsch. 43c: 929–937.

Hayes, A. and D.A. Williams. 1998. Contractile function and low-intensity exercise effects of old dystrophic (mdx) mice. Am. J. Physiol. Cell Physiol. 43: C1138–C1144.

Heinemann, S.H. and W. Stühmer and F. Conti. 1987. Single acetylcholine receptor channel currents recorded at high hydrostatic pressure. Proc. Natl. Acad. Sci. USA 84: 3229–3233.

Hollingworth, S. and M. Zhao and S.M. Baylor. 1996. The amplitude and time course of the myoplasmic free ($Ca^{2+}$) transient in fast-twitch fibers of mouse muscle. J. Gen. Physiol. 108: 455–469.

Hook, P. and L. Larsson. 2000. Actomyosin interactions in a novel single muscle fiber *in vitro* motility assay. J. Muscle Res. Cell. Motil. 21: 357–365.

Hooker, S.K. and R.W. Baird. 1999. Deep-diving behaviour of the northern bottlenose whale, *Hyperoodon ampullatus* (Cetacea: *Ziphiidae*). Proc. R. Soc. Lond. B266: 671–676.

Huxley, H.E. and J. Hanson. 1954. Changes in the cross-striations of muscle during contraction and stretch and their structural interpretation. Nature 173: 973–976.

Ikeuchi, Y. and A. Suzuki, T. Oota, K. Hagiwara, R. Tatsumi, T. Iti and C. Balny. 2002. Fluorescence study of the high pressure-induced denaturation of skeletal muscle actin. Eur. J. Biochem. 269: 364–371.

Ikkai, T. and T. Ooi. 1966. The effects of pressure on G-F transformation of actin. Biochemistry 5: 1551–1560.

Ingber, D.E. 2006. Cellular mechanotransduction: putting all the pieces together again. FASEB J. 20(7): 811–827.

Iwasaki, T. and K. Yamamoto. 2002. Effect of hydrostatic pressure on chicken myosin subfragment-1 Int. J. Biol. Macromol. 30: 2227–232

Iwasaki, T. and K. Yamamoto. 2003. Changes in rabbit skeletal muscle myosin and its subfragments under high hydrostatic pressure. Int. J. Biol. Macromol. 33: 215–220.

Johnson, T.P. and I.A. Johnston. 1991. Temperature adaptation and the contractile properties of live muscle fibres from teleost fish. J. Comp. Physiol. B161: 27–36.

Johnston, I.A. 2006. Environment and plasticity of myogenesis in teleost fish. J. Exp. Biol. 209(12): 2249–2264.

Jurkat-Rott, K. and F. Lehmann-Horn. 2001. Human muscle voltage-gated ion channels and hereditary disease. Curr. Opin. Pharmacol. 1: 280–287.

Jurkat-Rott, K. and M. Fauler and F. Lehmann-Horn. 2006. Ion channels and ion transporters in the transverse tubular system of skeletal muscle. J. Muscle Res. Cell Motil. 27(5–7): 275–290.

Kaarniranta, K. and M.A. Elo, R.K. Sironen, H.M. Karjalainen, H.J. Helminen and M.J. Lammi. 2003. Stress response of mammalian cells to high hydrostatic pressure. Biorheology 40(1–3): 87–92.

Kawai, M. 2003. What do we learn by studying the temperature effect on isometric tension and tension transients in mammalian striated muscle fibres? J. Muscle Res. Cell Motil. 24(2–3): 127–138.

Kendig, J.J. and E.W. Cohen. 1976. Neuromuscular function at hyperbaric pressures: pressure-anesthetic interactions. Am. J. Physiol. 230: 1244–1249.

Kharakoz, D.P. 2000. Protein compressibility, dynamics and pressure. Biophys. J. 79: 511–525.

King, L. and C.C. Liu and R.F. Lee. 1994. Pressure effects and thermal stability of myosin rods and rod minifilaments: fluorescence and circular dichroism studies. Biochemistry 33(18): 5570–5580.

Kitamura, K. and M. Tokunaga, A.H. Iwane and T. Yanagida. 1999. A single myosin head moves along an actin filament with regular steps of 5.3 nm. Nature 397: 129–134.

Korneliussen, H. and K. Nicolaysen. 1975. Distribution and dimension of the t-system in different muscle fiber types in the atlantic hagfish (*Myxine glutinosa*, L.). Cell Tiss. Res. 157: 1–16.

Koyama, S. 2007. Cell biology of deep-sea multicellular organisms. Cytotechnology 55: 125–133.

Kress, K.R. and O. Friedrich, H. Ludwig and R.H.A. Fink. Reversibility of high pressure effects in isolated adult murine skeletal muscle fibres tested at pressures up to 600bar and a temperature of +4°C. Pp. 551–554. *In*: H. Ludwig (ed.). 1999. Advances in High Pressure Biosciences and Biotechnology. Springer. Berlin, Heidelberg.

Kress, K.R. and O. Friedrich, H. Ludwig and R.H.A. Fink. 2001. Reversibility of high pressure effects on the contractility of skeletal muscle. J. Muscle Res. Cell Motil. 22(4): 379–389.

Kron, S.J. and J.A. Spudich. 1986. Fluorescent actin filaments move on myosin fixed to a glass surface. Proc. Natl. Acad. Sci. USA 83: 6272–6276.

Kutching, J.A. The effects of pressure on organisms: a summary of progress. Pp. 473–482. *In*: M.A. Sleigh and A. Macdonald (eds.). 1972. The effects of pressure on organisms. Cambridge University Press, Cambridge, UK.

Lafortuna, C.L. and M. Jahoda, A. Azzellino, F. Saibene and A. Colombini. 2003. Locomotor behaviours and respiratory patterns of the Mediterranean fin whale (*Balaenoptera physalus*). Eur. J. Appl. Physiol. 90: 387–395.

Launikonis, B.S. and G.D. Stephenson. 2002. Tubular system volume changes in twitch fibres from toad and rat skeletal muscle assessed by confocal microscopy. J. Physiol. 538(2): 607–618.

Launikonis, B.S. and E. Rios. 2007. Store-operated $Ca^{2+}$ entry during intracellular $Ca^{2+}$ release in mammalian skeletal muscle. J. Physiol. 583.1: 81–97.

Lee, R.F. and J. Hirota. 1973. Wax esters in tropical zooplankton and nekton and the geographical distribution of wax esters in marine copepods. Limnol. Oceanogr. 18: 227–239.

Lewis, M.S. and K. Maruyama, W.R. Carroll, D.R. Kominz and K. Laki. 1963. Physical properties and polymerization reactions of native and inactivated G-actin. Biochemistry 2: 34–39.

Lingle, C.J. and D. Maconochie and J.H. Steinbach. 1992. Activation of skeletal muscle nicotinic aceytlcholine receptors. J. Membr. Biol. 126(3): 195–217.

Lynch, G.S. and J.G. Ryall. 2008. Role of beta-adrenoceptor signalling in skeletal muscle: implications for muscle wasting and disease. Physiol. Rev. 88(2): 729–767.

Macdonald, A.G. 1988. Application of the theory of homeoviscous adaptation to excitable membranes: pre-synaptic processes. Biochem. J. 256(2): 313–327.

Macdonald, A.G. 1993. Effects of high pressure on biological systems. Advances in Comparative and Environmental Physiology, Vol. 17. Springer Verlag, Berlin.

Macdonald, A.G. 1997. Effect of high hydrostatic pressure on the BK channel in bovine chromaffine cells. Biophys. J. 73(4): 1866–1873.

Macdonald, A.G. 2002. Experiments on ion channels at high pressure. Biochim. Biophys. Acta 1595: 387–389.

Macdonald, A.G. and A.R. Cossins. 1985. The theory of homeoviscous adaptation of membranes applied to deep-sea animals. Symp. Soc. Exp. Biol. 39: 301–322.

Macdonald, A.G. and B. Martinac. 2005. Effect of high hydrostatic pressure on the bacterial mechanosensitive channel MscS. Eur. Biophys. J. 34: 434–441.

MacGregor, R.B. Jr. 1998. Effect of hydrostatic pressure on nucleic acids. Biopolymers 48(4): 253–263.

Marconi, C. and M. Marzorati, D. Sciuto, A. Ferro and P. Cerretelli. 2005. Economy of locomotion in high-altitude Tibetan migrants exposed to normoxia. J. Physiol. 569(2): 667–675.

Matthews, B.D. and D.R. Overby, R. Mannix and D.E. Ingber. 2006. Cellular adaptation to mechanical stress: role of integrins, Rho, cytoskeletal tension and mechanosensitive ion channels. J. Cell Sci. 119(3): 508–518.

Medler, S. 2002. Comparative trends in shortening velocity and force production in skeletal muscles. Am. J. Physiol. Regulatory Integrative Comp. Physiol. 283: R368–R378.

Meersman, F. and L. Smeller and K. Heremans. 2006. Protein stability and dynamics in the pressure-temperature plane. Biochim. Biophys. Acta 1764: 346–354.

Mentre, P. and G. Hui Bon Hoa. 2001. Effects of high hydrostatic pressures on living cells: a consequence of the properties of macromolecules and macromolecule-associated water. Int. Rev. Cytol. 20: 1–84.

Meyer, R. and S.H. Heinemann. 1997. Temperature and pressure dependence of Shaker K+ channel N- and C-type inactivation. Eur. Biophys. J. 26, 433–445.

Molloy, J.E. and J.E. Burns, J. Kendrick Jones, R.T. Treggear and D.C.S. White. 1995. Movement and force produced by a single myosin head. Nature 378: 209–212.

Morita, T. 2003. Structure-based analysis of high pressure adaptation of alpha-actin. J. Biol. Chem. 278(30): 28060–28066

Morita, T. 2008. Comparative sequence analysis of myosin heavy chain proteins from congeneric shallow- and deep-living rattail fish (genus *Coryphaenoides*). J. Exp. Biol. 211(9): 1362–1367.

Nevenzel, J.C. 1980. Lipids of midwater marine fish: family *Gonostomatidae*. Comp. Biochem. Physiol. 65B: 351–355.

Nicolini, C. and A. Celli, E. Gratton and R. Winter. 2006. Pressure tuning of the morphology of heterogeneous lipid vesicles: a two-photon excitation fluorescence microscopy study. Biophys. J. 91: 2936–2942.

Nyitrai, M. and R. Rossi, N. Adamek, M.A. Pellegrino, R. Bottinelli and M.A. Geeves. 2006. What limits the velocity of fast-skeletal muscle contraction in mammals? J. Mol. Biol. 355: 432–442.

Okamoto, A. and A. Suzuki, Y. Ikeuchi and M. Saito. 1995. Effects of high pressure treatment on $Ca^{2+}$ release and $Ca^{2+}$ uptake of sarcoplasmic reticulum. Biosci. Biotech. Biochem. 59: 266–270.

Page, S.G. 1964. Filament lengths in resting and excited muscle. Proc. R. Soc. Lond. B160: 460–466.

Papadopoulos, S. and V. Leuranquer, R.A. Bannister and K.G. Beam. 2004. Mapping sites of potential proximity between the dihydropyridine receptor and RYR1 in muscle using a cyan fluorescent protein-yellow fluorescent protein tandem as a fluorescence resonance energy transfer probe. J. Biol. Chem. 279: 44046–44056.

Posterino, G.S. and G.D. Lamb. 2003. Effect of sarcoplasmic reticulum $Ca^{2+}$ content on action potential-induced $Ca^{2+}$ release in rat skeletal muscle fibres. J. Physiol. 551: 219–237.

Pouvreau, S. and L. Royer, J. Yi, G. Brum, G. Meissner, E. Rios and J. Zhou. 2007. $Ca^{2+}$ sparks operated by membrane depolarization require isoform 3 ryanodine receptor channels in skeletal muscle. Proc. Natl. Acad. Sci. USA 104(12): 5235–5240.

Protasi, F. 2002. Structural interaction between RYRs and DHPRs in calcium release units of cardiac and skeletal muscle cells. Front. Biosci. 7: d650–d658.

Ranatunga, K.W. and M.A. Geeves. 1991. Changes produced by increased hydrostatic pressure in isometric contractions of rat fast muscle. J. Physiol. 441: 423–431.

Ranatunga, K.W. and N.S. Fortune and M.A. Geeves. 1990. Hydrostatic compression in glycerinated rabbit muscle fibers. Biophys. J. 58: 1401–1410.

Rashin, A.A. and M. Iofin and B. Honig. 1986. Internal cavities and buried waters in globular proteins. Biochemistry 25: 3619–3625.

Reynolds, J.D. and N.K. Dulvy, N.B. Goodwin and J.A. Hutchings. 2005. Biology of extinction risk in marine fishes. Proc. R. Soc. B272: 2337–2344.

Rios E. and G. Brum. 1987. Involvement of dihydropyridine receptors in excitation-contraction coupling in skeletal muscle. Nature 325: 717–720.

Rizzuto, R. and T. Pozzan. 2008. Microdomains of intracellular $Ca^{2+}$: molecular determinants and functional consequences. Physiol. Rev. 86(1): 369–408.

Rossignol, O. and P. Sébert and B. Simon. 2006. Effects of high pressure acclimatization on silver eel (*Anguilla anguilla*, L.) slow muscle contraction. Comp. Biochem. Physiol. A143: 234–238.

Rostain, J.-C. The Nervous System: Man and laboratory mammals. Pp. 198–238. *In*: A.G. Macdonald. (ed.). 1993. Effects of high pressure on biological Systems. Springer. New York, Berlin.

Rüegg, C. and C. Veigel, J.E. Molloy, S. Schmitz, J.C. Sparrow and R.H.A. Fink. 2002. Molecular motors: Force and movement generated by single myosin II molecules. News Physiol. Sci. 17: 213–218.

Ruppel, K.M. and J.A. Spudich. 1996. Structure-function analysis of the motor domain of myosin. Annu. Rev. Cell. Biol. 12: 543–573.

Salvanes, A.G.V. and J.B. Kristoffersen. Mesopelagic fishes. Pp. 1711–1717. *In*: J.H. Steele, S.A. Thorpe and K.K. Turekian. (eds.). 2001. Enyclopedia of Ocean Sciences. Vol.3. Elsevier Science Ltd. New York, USA.

Saunders, P.M. and N.P. Fofonoff. 1976. Conversion of pressure to depth in the ocean. Deep Sea Res. 23: 109–111.

Schiaffino, S. and C. Reggiani. 1994. Myosin isoforms in mammalian skeletal muscle. J. Appl. Physiol. 77(2): 493–501.

Schiaffino, S. and M. Sandri and M. Murgia. 2007. Activity-dependent signaling pathways controlling muscle diversity and plasticity. Physiology (Bethesda) 22: 269–278.

Schmalwasser, H. and A. Neef, A.E. Elliott and S.H. Heinemann. 1998. Two-electrode voltage clamp of Xenopus oocytes under high hydrostatic pressure. J. Neurosci. 81: 1–7.

Schnee, S. and F.V. Wegner, S. Schürmann, H. Ludwig, R.H.A. Fink and O. Friedrich. 2008. Microdomain $Ca^{2+}$ dynamics in mammalian muscle following prolonged high pressure treatments. J. Phys. Conf. Ser. 121: 112003.

Schneider, M.F. and B.J. Simon. 1988. Inactivation of calcium release from the sarcoplasmic reticulum in frog skeletal muscle. J. Physiol. 405: 727–745.

Scott, G.R. and W.K. Milsom. 2006. Flying high: a theoretical analysis of the factors limiting exercise performance in birds at altitude. Respir. Physiol. Neurobiol. 154(1–2): 284–301.

Sébert, P. 2002. Fish at high pressure: a hundred year history. Comp. Biochem. Physiol. A131: 575–585.

Sébert, P. Fish adaptations to pressure. Pp. 73–95. *In*: A.L. Val and B.G. Kapoor (eds.). 2003. Fish Adaptations. Science Publishers, Enfield, New Hampshire.

Sébert, P. Fish muscle function and pressure. Pp. 233–255. *In*: P. Sébert, D.W. Onyango and B.G. Kapoor. (eds.). 2008. Fish Life in Special Environments. Science Publishers, Enfield, USA.

Sébert, P. and M. Theron and A. Vettier. 2004. Pressure and temperature interactions on cellular respiration: a review. Cell. Mol. Biol. 50(4): 491–500.

Serrano, A.L. and E. Quiroz-Rothe and J.L. Rivero. 2000. Early and long-term changes of equine skeletal muscle in response to endurance training and detraining. Pflugers Arch. 441: 263–274.

Smeller, L. 2002. Pressure-temperature phase diagrams of biomolecules. Biochim. Biophys. Acta. 1595(1–2): 11–29.

Somero, G.N. 1983. Environmental adaptations of proteins: strategies for the conservation of critical functional and structural traits. Comp. Biochem. Physiol. A76(3): 621–633.

Somero, G.N. 1992. Biochemical ecology of deep-sea animals. Experentia 48(6): 537–543.

Squire, J.M. 1974. Symmetry and three-dimensional arrangement of filaments in vertebrate striated muscle. J. Mol. Biol. 90: 153–160.

Stehle, R. and B. Brenner. 2000. Cross-bridge attachment during high-speed active shortening of skinned fibers of the rabbit psoas muscle: Implications for cross-bridge action during maximum velocity of filament gliding. Biophys. J. 78: 1458–1473.

Stephenson, D.G. 2006. Tubular system excitability: an essential component of excitation-contraction coupling in fast-twitch fibres of vertebrate skeletal muscle. J. Muscle Res. Cell Motil. 27(5–7): 259–274.

Stiber, J. and A. Hawkins, Z.S. Zhang, S. Wang, J. Burch, V. Graham, C.C. Ward, M. Seth, E. Finch, N. Malouf, R.S. Williams, J.P. Eu and P. Rosenberg. 2008. STIM1 signaling controls store-operated calcium entry required for development and contractile function in skeletal muscle. Nat. Cell Biol. 10(6): 688–697.

Swezey, R.R. and G.N. Somero. 1982. Polymerization thermodynamics and structural stabilities of skeletal muscle actins from vertebrates adapted to different temperatures and hydrostatic pressures. Biochemistry 21(18): 4496–4503.

Swezey, R.R. and G.N. Somero. 1985. Pressure effects on actin self-assembly: interspecific differences in the equilibrium and kinetics of the G to F transformation. Biochemistry 24(4): 852–860.

Switzer, K.C. and D.N. McMurray and R.S. Chapkin. 2004. Effects of dietary n-3 polyunsaturated fatty acids on T-cell membrane composition and function. Lipids 39(12): 1163–1170.

Tanaka, H. and Y. Takagi and Y. Naito. 2001. Swimming speeds and buoyancy compensation of migrating adult chum salmon *Oncorhynchus keta* revealed by speed/depth/acceleration data logger. J. Exp. Biol. 204(22): 3895–3904.

Telley, I.A. and J. Denoth, E. Stüssi, G. Pfitzer and R. Stehle. 2006. Half-sarcomere dynamics in myofibrils during activation and relaxation studied by tracking fluorescent markers. Biophys. J. 90: 514–530.

Thomas, J.A. and F.R. Rana. 2007. The influence of environmental conditions, lipid composition and phase behaviour on the origin of cell membranes. Orig. Life Evol. Biosph. 37(3): 267–285.

Toniolo, L. and L. Maccatrozzo, M. Patruno, E. Pavan, F. Caliaro, R. Rossi, C. Rinaldi, M. Canepari, C. Reggiani and F. Mascarello. 2007. Fiber types in canine muscles: myosin isoform expression and functional characterization. Am. J. Physiol. Cell Physiol. 292: 1915–1926.

Tskhovrebova, L. and J. Trinick, J.A. Sleep and R.M. Simmons. 1997. Elasticity and unfolding of single molecules of the giant muscle protein titin. Nature 387: 308–312.

Tyack, P.L. and M. Johnson, N.A. Soto, A. Sturlese and P.T. Madsen. 2006. Extreme diving of beaked whales. J. Exp. Biol. 209: 4238–4253.

Vawda, F. and K.W. Ranatunga and M.A. Geeves. 1995. Pressure-induced changes in the isometric contractions of single intact frog muscle fibres at low temperatures. J. Muscle Res. Cell. Motil. 16: 412–419.

Vawda, F. and K.W. Ranatunga and M.A. Geeves. 1996. Effects of hydrostatic pressure on fatiguing frog muscle fibres. J. Muscle Res. Cell. Motil. 17(6): 631–636.

Verjovski-Almeida, S. and E. Kurtenbach, A.F. Amorim and G. Weber. 1986. Pressure-induced dissociation of solubilized sarcoplasmic reticulum ATPase. J. Biol. Chem. 261(21): 9872–9878.

Vettier, A. and C. Labbe, A. Amerand, G. Da Costa, E. Le Rumeur, C. Moisan and P. Sébert. 2006. Hydrostatic pressure effects on eel mitochondrial function and membrane fluidity. Undersea Hyperb. Med. 33: 149–156.

v Wegner, F. and S. Koyama, T. Miwa, and O. Friedrich. 2008. Resting membrane potentials recorded on-site in intact skeletal muscles from deep sea fish (*Sigmops gracile*) salvaged from depths up to 1.000 m. Mar. Biotechnol. 10: 478–486.

Watkins, W.A. and M.A. Daher, K.M. Fristrup and T.J. Howald. 1993. Sperm whales tagged with transponders and tracked underwater by sonar. Mar. Mamm. Sci. 9(1): 55–67.

Watson, R.R. and T.A. Miller and R.W. Davis. 2003. Immunohistochemical fibre typing of harbour seal skeletal muscle. J. Exp. Biol. 206: 4105–4111.

Weber, G. and H.G. Drickamer. 1983. The effect of high pressure upon proteins and other biomolecules. Q. Rev. Biophys. 16: 89–112.

Winter, R. and R. Köhling. 2004. Static and time-resolved synchrotron small-angle X-ray scattering studies of lyotropic lipid mesophases, model biomembranes and proteins in solution. J. Phys. Condens. Matter 16: S327–S352.

Winter, R. and W. Dzwolak. 2005. Exploring the temperature-pressure configurational landscape of biomolecules: from lipid membranes to proteins. Phil. Trans. R. Soc. A363: 537–563.

Yamamoto, K. and T. Miura and T. Yasui. 1990. Gelation of myosin filament under high hydrostatic pressure. Food Struct. 9: 269–277.

Yamamoto, S. and Y. Otsuka, G. Borjigin, K. Masuda, Y. Ikeuchi, T. Nishiumi and A. Suzuki. 2005. Effects of a high-pressure treatment on the activity and structure of rabbit muscle proteasome. Biosci. Biotechnol. Biochem. 69: 1239–1247.

Yancey, P.H. 2005. Organic osmolytes as compatible, metabolic and counteracting cytoprotectants in high osmolarity and other stresses. J. Exp. Biol. 208: 2819–2830.

Yancey, P.H. and J.F. Siebenaller. 1999. Trimethylamine oxide stabilizes teleost and mammalian lactate dehydrogenase against inactivation by hydrostatic pressure and trypsinolysis. J. Exp. Biol. 202: 3597–3603.

Zerbst-Boroffka, I. and R.M. Kamaltynow, S. Harjes, E. Kinne-Saffran and J. Gross. 2005. TMAO and other organic osmolytes in the muscles of amphipods (Crustacea) from shallow and deep water of Lake Baikal. Comp. Biochem. Physiol. A142: 58–64.

Zhou, J. and G. Brum, A. Gonzalez, B.S. Launikonis, M.D. Stern and E. Rios. 2005. Concerted vs. sequential. Two activation patterns of vast arrays of intracellular $Ca^{2+}$ channels in muscle. J. Gen. Physiol. 126(4): 301–309.

# Pressure and Reactive Oxygen Species

*C. Moisan,[1] A. Amérand,[1] Y. Jammes,[2] C. Lavoute[3] and J.J. Risso[3]*

## INTRODUCTION

All aerobic animal species living in terrestrial or marine habitats are concerned about reactive oxygen species (ROS). Indeed, reactive nitrogen species (RNS) such as ROS are produced as normal by-products of cellular metabolism. They have been implicated in numerous pathological conditions involving cardiovascular diseases, neurological diseases, diabetes, ageing, ischemia/reperfusion injury, hyperoxia and hypoxia, exposure to xenobiotics… Beside these well-known implications of the ROS/RNS, recent reviews about this topic have highlighted their importance in the maintenance of oxidant-antioxidant homeostasis known to be essential for cell survival and to function under normal conditions (Dröge, 2002; Valko et al., 2007).

This chapter will deal with the relationships between ROS/RNS and pressure not only in mammals through the hyperbaric oxygenation (HBO) and anoxia-reoxygenation in diving species but also in fish since some of them are confronted with high hydrostatic pressure.

[1] ORPHY EA4324, Université Européenne de Bretagne, Université de Brest, UFR Sciences et Techniques, France.
E-mail: christine.moisan@univ-brest.fr; aline.amerand@univ-brest.fr
[2] UMR MD2 P2COE, Faculté de Médecine, Université de la Méditerranée, France.
E-mail: jammes.y@jean-roche.univ-mrs.fr
[3] UMR MD2, Physiologie et Physiopathologie en Condition d'Oxygénation Extrême (P2COE), Université de la Méditerranée-Service de Santé des Armées, Institut de Médecine Navale du Service de Santé des Armées, BP 610 83800 Toulon Armées.
E-mail: j.j.risso@imnssa.net; cecile_lavoute@hotmail.com

This research field is complex, in particular in the hyperbaric context because of the existence of many and not fully elucidated paradoxes about ROS. The first of them is, of course, about oxygen itself: indeed, higher eukaryotic aerobic organisms cannot survive without this element but oxygen becomes indirectly toxic under certain conditions through the production of ROS. Exposure to hyperbaric oxygenation is worth investigating to gain more insight not only in the oxidative stress and its biological consequences (Pablos et al., 1997), but also for the treatment of various clinical conditions in relation with hypoxia (see Thom in this book).

Another paradox concerns mitochondria: essential organelles that provide cells with ATP which are both the main site of ROS production and a target for ROS. There is for example evidence that mitochondrial ROS production by attacking mitochondrial self-constituents is an important component of the aging process. The ROS could be not only by-products of aerobic metabolism and potentially toxic but also signalling agents (Dröge, 2002). For living animal species, the duality observed in the need for a highly-efficient oxidative phosphorylation and the necessity to decrease the damaging effects by ROS constitutes a true challenge. These aspects will be discussed in a metabolic and ecophysiological context through the physiological changes occurring in european eel during its migration.

Investigations about the effects of hyperoxia and hypoxia upon ROS production in tissues have highlighted an apparent paradox: as ROS production can be enhanced by both situations even though an increase during hypoxia remains difficult to explain (Turrens, 2003). In the field of high pressure and ROS, several situations or environmental conditions contribute to expose animals to oxidative stress or on the contrary tend to protect them against it. Diving mammals are ideal models to study antioxidant status because they must frequently cope with high rates of ROS production due to the alternation of apnea/reoxygenation involving ischemia/reperfusion processes (Filho et al., 2002).

Even though mammalian species are often used as models in investigations of oxidative stress or cellular physiological functions controlled, or induced, by oxidant-antioxidant status, fish are well-worth of interest. Indeed as in most of the ectotherm aquatic organisms, its metabolic rate is dependent on the physicochemical factors (oxygen concentration, temperature, pressure...) of its water environment. Under these conditions, coping with two fluctuating sources of ROS, exogenous (the seawater) and endogenous linked to metabolic processes is a must. Some specific habitats known as marine polar or deep sea are of particular interest to study ROS, the first considered as an oxidative habitat, the second as a refuge against oxidative stress (Camus and Gulliksen, 2004; Rees et al., 1998).

## GENERAL ASPECTS OF ROS AND RNS

### The oxygen paradox

Two to three million years ago, oxygen began to appear at high concentrations in the biosphere when photosynthetic organisms were able to photolyze water. Then, oxygen progressively became a major environmental element for mitochondrial energy production via the oxidative phosphorylation in all aerobic organisms. This energy process is associated with the reduction of oxygen to water by cytochrome oxidase and uses about 90% of the oxygen consumed by mitochondria. This aerobic life is paradoxical because of the 'oxygen paradox' as termed by Davies (1995), which means that oxygen is both required by organisms for their existence and inherently toxic to them because of its biotransformation into by- products such as ROS. So various ROS are produced as a consequence of normal oxygen utilization (Fridovich, 1995).

### ROS and RNS description

#### Chemistry of ROS and RNS

The term ROS refers not only to oxygen free radicals, but also to non-radical active species such as hydrogen peroxide ($H_2O_2$), which is also susceptible to generate free radicals through chemical reactions. A free radical is a chemical species (atom or group of atoms) with an impaired electron in the outer orbital. Atomic oxygen is itself a weak bi-radical because it has two electrons with anti-parallel spins, each occupying one orbital. So ROS are formed by incomplete reduction of oxygen and are usually short-lived and highly reactive; they include the superoxide anion ($O_2{}^{\bullet-}$), the hydroxyl radical ($HO^{\bullet}$) and the hydrogen peroxide ($H_2O_2$) (Halliwell and Gutteridge, 1999).

- Superoxide anion ($O_2{}^{\bullet-}$): this free radical results from the reduction of oxygen by one electron.

$$O_2 + e^- \rightarrow O_2^{\bullet-}$$

- Hydrogen peroxide ($H_2O_2$): this compound is produced by spontaneous dismutation of $O_2{}^{\bullet-}$ or through a reduction catalyzed by superoxide dismutases.

$$2O_2^{\bullet-} + 2H^+ \xrightarrow{\textit{Superoxyde dismutase}} H_2O_2 + O_2$$

- Hydroxyl radicals ($HO^{\bullet}$) are formed from either the superoxide anion according to the Haber-Weiss reaction and from $H_2O_2$ through a Fenton reaction catalyzed by a transitional metal such as iron $Fe^{2+}$ or copper ($Cu^{2+}$) (Haber and Weiss, 1934).

$$O_2^{\bullet-} + H_2O_2 \rightarrow O_2 + HO^- + HO^{\bullet} \text{ (Haber Weiss reaction)}$$

$$Fe^{2+} + H_2O_2 \rightarrow Fe^{3+} + HO^- + HO^{\bullet} \text{ (Fenton reaction)}$$

The hydroxyl free radical is likely the most powerful oxidant produced by biological systems and the most toxic of the ROS.

The term RNS includes radical species such as nitric oxide ($NO^{\bullet}$), nitrogen dioxide ($^{\bullet}NO_2$) as well as non-radical ones such as peroxynitrite ($ONOO^-$) and other. Briefly, among RNS, $NO^{\bullet}$ results from oxidation of L-arginine into L-citrulline. In the presence of NADPH and $O_2$, this reaction (see below) is catalyzed by a family of enzymes, the Nitric Oxide Synthase (NOS).

$$\text{L-Arginine} \xrightarrow[\text{NO synthase}]{\text{NADPH} + O_2} \text{L-Citrulline} + NO^{\bullet}$$

*The major sources of ROS and RNS*

Thus, the most important *RNS* ($NO^{\bullet}$) is formed by the Nitric Oxide Synthase (NOS) known as three isoforms, i.e neuronal NOS (nNOS), inducible (iNOS) and endothelial NOS (eNOS); a specific form of mitochondrial mtNOS has also been identified in some tissues (Jesek and Hlavata, 2005).These enzymes need co-factors such as $BH_4$ (tetrahydrobiopterin), FAD or FMN.

ROS can be generated by both exogenous and endogenous sources of free radicals or physical agents. UV-photolysis, radiation, pollution, xenobiotic agents and pesticides have been identified as exogenous sources (Rimbach et al., 1999). Concerning the endogenous sources, they are linked to several enzymatic and non enzymatic mechanisms, some important examples are presented here.

- Mitochondria are both considered as the main source of ATP synthesis and as the major sites of aerobic cellular ROS production (Boveris and Chance, 1973; Fig.1). Mitochondrial respiration accounts for about 90% of cellular oxygen uptake and about 1 to 2% of the consumed oxygen is converted to ROS. Within the mitochondria, reduced coenzymes such as NADH and $FADH_2$ are produced during the Krebs cycle and/or fatty acid oxidation. These coenzymes are oxidized and the liberated electrons are passed through the complex I (NADH-CoQ reductase) and II of the electron transport chain to the complexes III and IV (cytochrome C oxidase) to reduce oxygen. During this electron transport, protons are pumped across the mitochondrial inner membrane. A resulting proton motive force drives proton back to the mitochondrial matrix via the $F_0$-$F_1$ $ATP_{ase}$, thereby releasing the energy needed to phosphorylate ADP into

ATP (oxidative phosphorylation). During the respiratory chain activity, $O_2^{\bullet-}$ is generated as a natural by-product by electron leaks mainly at the levels of complexes I and III (Cadenas and Davies, 2000). This production may be increased especially when electron transport becomes slow like under non phosphorylating conditions (analogous to respiratory state 4 in the isolated mitochondria preparation) in comparison to phosphorylating conditions (state 3) and/or in all conditions which decrease the electron transfer rate. For example in state 4 conditions, the transport chain of electrons is highly reduced thus enhancing their reaction with oxygen (Bovaris and Chance, 1973; Skulachev, 1996; Korshunov et al., 1997).

- Enzymes such as NAD(P)H oxidase generate $O_2^{\bullet-}$. They are located on the cell membrane of macrophages and endothelial cells (Vignais, 2002) but also in cytochrome P450-dependent oxygenase cells. In the field of toxicology, cytochrome P450 has been thoroughly studied because of its implication in the metabolism of xenobiotics. (Fridovich, 1978). $O_2^{\bullet-}$ may have important signal transduction activity.
- Xanthine oxidase generates superoxide by converting hypoxanthine and xanthine into uric acid. This enzyme is an important source of ROS in many tissues under certain conditions such as ischemia-reperfusion or

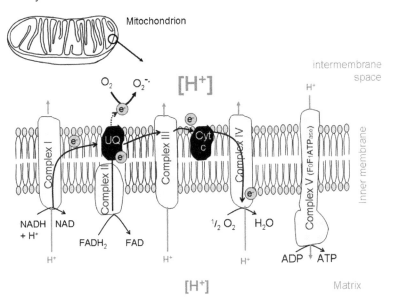

**Figure 1.** The mitochondrial electron transport chain is a site for free-radical production. $O_2^{\bullet-}$ is generated at the ubiquinone (UQ) level, more precisely at the levels of the two complexes I and III. Each of the respiratory chain complexes, except complex II, couples electron flow to proton pumping and the ensuing proton motive force is used by complex V to form ATP.

anoxia. On the other hand, under normal conditions xanthine oxidase accounts for only a minor proportion of total ROS production (Sies, 1991; Dröge, 2002)

- Auto-oxidation of catecholamines, or oxidation via the monoamine oxidase is a source of ROS in the central nervous system (Kil et al., 1996)
- NO• is an important chemical mediator produced by endothelial cells, neurons, macrophages and others (Dröge, 2002).

### Pro-oxidant and antioxidant balance at cellular level

ROS exist in all living tissues at low concentrations. Their steady state is determined by the balance between their production and their rates of removal by various antioxidants.

The regulation of this balance is necessary for the maintenance of redox homeostasis. In recent years the term redox state has been used to describe more generally the redox environment of a cell (Schafer and Buettner, 2001). Oxidative stress is defined as a disruption of this prooxidant—antioxidant balance which leads to potential damage (Sies, 1991). So a high ROS production and/or a decrease in antioxidant defence are largely involved in physiological mechanisms of numerous diseases (artherosclerosis, neurodegenerate diseases...).

### Antioxidant defences

Through evolution processes, aerobic organisms have developed antioxidant defence mechanisms. All organisms utilized antioxidant defences in an attempt to protect themselves against oxidant damage, repair or degrade oxidatively modified molecules (Halliwell and Gutteridge, 1999; Fig. 2).

*Antioxidant compounds:* Several natural compounds have the antioxidant ability of reacting with ROS. Non enzymatic defences include the fat-soluble vitamins alpha-tocopherol (vitamin E) and beta-carotene and several low molecular weight compounds e.g. ascorbic acid, uric acid, glutathione.

*Enzymatic antioxidant defences:* As they are considered as the most efficient system, they constitute the primary antioxidant protection.

- Superoxide dismutases (SOD): SOD are metalloenzymes, present in almost all living organisms. Three isoforms have been identified: mitochondrial Mn-SOD, cytosolic Cu/Zn-SOD and extracellular Cu/Zn-SOD. Through activation of SOD, $O_2^{•-}$ is converted to $H_2O_2$.

$$2O_2^{•-} + 2H^+ \xrightarrow{\text{Superoxyde dismutase}} H_2O_2 + O_2$$

## Source of production

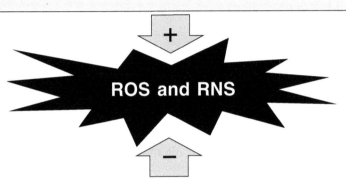

| **Endogenous** | **Exogenous** |
| --- | --- |
| - Mitochondria<br>- NADPH oxidase<br>- Cytochrome $P_{450}$<br>- Xanthine Oxidase<br>- NO synthase<br>- Auto-oxidation of<br>  catecholamines, hemoglobin,<br>  flavins | - UV radiation<br>- Xenobiotic exposure<br>- Oxygen concentration |

**+**

**ROS and RNS**

**−**

## Antioxidant systems

| **Enzymatic** | **Non enzymatic** |
| --- | --- |
| - Superoxide dismutase<br>- Catalase<br>- Glutathione peroxidase | - Vitamin E, C<br>- Beta-carotene<br>- Glutathione<br>- Uric acid |

**Figure 2.** The main sources of reactive oxygen species (ROS) and reactive nitrogen species (RNS) and antioxidant systems.

- Catalase (CAT): this enzyme catalyzes the conversion of $H_2O_2$ to water and oxygen. Catalase is mainly located in the peroxysomes and erythrocytes.
- Glutathion peroxidase (GPx): Glutathion peroxydase is located in the mitochondria matrix and the cytosol. This selenium-dependent enzyme reduces organic hydroperoxides (ROOH) and hydrogen peroxide in a reaction that involves glutathione.

Reduced glutathione (GSH) is oxidized to GSSG which is rapidly reduced back to GSH by glutathione reductase (GR) at the expense of NADPH forming a closed system.

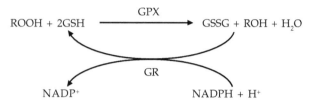

## ROS and RNS physiology

### Oxidative damage of ROS

At this stage it is worth briefly presenting the main deleterious ROS effects. When produced at high concentration, ROS may damage cellular components and tissues particularly targeting lipids, proteins, nucleic acids and glucides (Valko et al., 2007). Unsaturated fatty acids are particularly susceptible to be attacked by hydroxyl free radicals generating lipid peroxides. Peroxidation of unsaturated fatty acids make them more hydrophilic and consequently may alter the structure of the membrane and cause severe membrane disturbances such as decreased fluidity, inactivation of membrane bound receptors and increased ions permeability. For example, the brain is a vulnerable tissue to oxidative damage because it contains a high concentration of unsaturated fatty acids but also a large iron store and low antioxidant capacity (Halliwell, 1992).

Reactive oxygen species may oxidize amino acid residue side-chains into ketone or aldehyde derivatives. Histidine, arginine and lysine are the most susceptible amino acids for ROS-mediated protein carbonyl formation. Proteins may undergo deamination, decarboxylation and modifications in their aromatic rings which can lead to change in their three dimensional structure and activity. Protein carbonyl formation is a widely used marker for protein oxidation (Stadtman and Oliver, 1991).

Oxidative DNA damage can lead to DNA-protein irons linkage, strand breaks and base modifications. ROS induce hydroxylation of nucleic acid bases (formation of 8-hydroxy-2'deoxyguanosine, 8-OHdG). It is often used as a marker of DNA damage. Mitochondrial DNA is relatively more prone to damage as it is exposed to increased concentration of ROS (Evans et al., 2004).

### ROS and RNS in normal physiological functions

Like normal by-products of cellular metabolism, ROS participate in the maintenance of cellular 'redox homeostasis' known to be as crucial as pH or osmoregulation control. Indeed, according to recent studies, important physiological functions are controlled by intracellular redox sensitive signalling pathways (Dröge, 2002). This is illustrated by the following examples.

- The regulation of the tone of smooth muscle cells (Saran and Bors, 1990) is mediated by free radicals like NO•, which also acts as a signalling molecule in defence mechanisms or cardio-vascular regulation.
- The implication of ROS in inflammatory response is well documented. Inflammatory cells such as activated macrophages and neutrophils release various ROS against environmental pathogen. This response is called the oxidative burst and implies the phagocytic NADPH oxidase as cited above as a source of superoxide production (Shiloh et al., 1999).
- The generation of ROS by cells induces signal pathways also involved in cell growth and differentiation. Some transcription factors (nucleus factor-Kb and activator protein-1) have been clearly identified (Valko et al., 2007). ROS as $H_2O_2$ would be not only an effective agent in cell growth but also in cell adhesion , an important mechanism in embryogenesis.

### The special case of Anoxia-reoxygenation and oxidative stress

Diving animals, including mammals, reptiles and birds, have a high tolerance to very long apnea (Butler and Jones, 1997). Compared to non-diving animals, their oxygen stores are larger, blood buffers more efficient, the activity of glycolytic enzymes reduced, and the aerobic metabolism is increased in tissues poorly perfused as the liver, kidney and muscle (Behrisch and Elsner, 1984; Fuson et al., 2003) during the dive. Some elite breath-hold human divers also have the capacity to sustain apnea up to 7 min at rest (static apnea) inspite of their aerobic capacities not differing from those measured in untrained individuals.

Prolonged apnea duration leads to severe hypoxemia and cellular hypoxia. The succession of reduced oxygen supply to tissues and recovery on reoxygenation with the first post-apnea breath is responsible for an enhanced production of ROS (Dhaliwal et al., 1991). ROS exert cell damages, due to the oxidative stress, and also positive actions on potassium and calcium ionic channels (Reid, 1996) (Fig. 3). Post-apnea reoxygenation accompanies reperfusion succeeding reflex vasoconstriction in various organs (Craig, 1968; Sterba and Lundberg, 1988), responsible for a reduction of both glucose and oxygen supplies to tissues. This might explain the reduced lactate production measured in diving mammals (Butler and Jones, 1997) and trained breath-hold human divers (Joulia et al., 2002, 2003). Reintroduction of oxygenated blood raises the potential for ROS production which overwhelms the antioxidant defences.

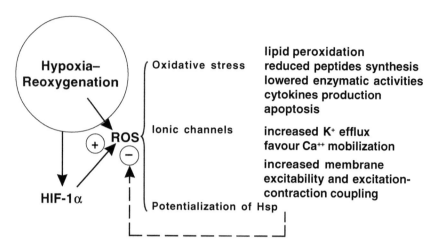

**Figure 3.** The main effects of reactive oxygen species (ROS) and interactions between ROS and signalling factors playing key roles in cell regulation like hypoxia inducible factor 1 (HIF-1α) and heat shock proteins (Hsp).

Ability to tolerate apnea for extended periods implies high tissue antioxidant defences to counteract the ROS generation associated with rapid transition from anoxia to reoxygenation when apnea has stopped. Diving animals (see chapter 15 in this book) have higher antioxidant status than non-diving ones in blood and tissues. In ringed seals (*Phoca hispida*), heart and kidney levels of superoxide dismutase (SOD) and heart levels of glutathione peroxydase (GPx) and glutathione-S-transferase (GST) are elevated compared to domestic pigs (Elsner et al., 1998; Vasquez-Medina et al., 2006). In parallel to their enhanced antioxidant capacities which lower ROS production, the muscle redox status greatly varies between species of seals (*Phoca vitulina, Arctocephalus gazilla, Halichoerus grypus*) and is related to their diving behaviour (Reed et al., 1994). This strongly suggests acclimatization to repeated dives. In swimming muscles and also in the heart, kidneys, and splanchnic organs of pinnipeds (seals and sea lions), the enhanced capacity to oppose ROS generation accompanies an elevation of mitochondrial volume density and activities of enzymes involved in aerobic pathways, favouring the aerobic fat-based metabolism under hypoxic conditions (Fuson et al., 2003). The increased mitochondrial activity is responsible for an enhanced ROS production (Hochachka et al., 1996). In diving reptiles, Pacific green turtles (Valvidia et al., 2007) and crocodiles (Furtado-Filho et al., 2007), the redox status in several organs differs from that measured in non diving reptiles in a direction of increased levels of antioxidants. The study by Furtado-Filho and coworkers also shows that seasonal changing wetland environment favouring the dives affects the level of oxidative stress markers in the brain, liver, kidneys and muscles.

Thus, adaptive processes to repeated events of anoxia-reoxygenation seem to also occur in diving reptiles.

In humans, static as well as dynamic apnea elicit an oxidative stress with increased plasma level of thiobarbituric acid reactive substances (TBARS) and consumption of reduced ascorbic acid and erythrocyte reduced glutathione (Joulia et al., 2002, 2003). Besides, elite breath-hold divers do not develop an oxidative stress after static or dynamic apnea, while their baseline levels of markers of oxidant-antioxidant status do not differ from 'naïve' individuals (Joulia et al., 2002). A recent study by Bulmer et al. (2008) reports the same differences between 'naïve' subjects and well trained free divers, showing that the blood level of antioxidants (SOD and TEAC, trolox equivalent antioxidant capacity) markedly increases after repeated apneas compared with controls. More interestingly, after a 3-mon training programme combining periods of breath-holding during steady state cycling exercise, 'naïve' subjects become non-respondent to static and dynamic apnea, i.e., that apnea-as well as exercise-induced oxidative stress is markedly reduced or absent (Joulia et al., 2003). Thus, the capacities to reduce the apnea-induced oxidative stress may be acquired. This corroborates animal observations in diving mammals and reptiles which suggest the development of adaptive metabolic response with their growth and diving practice.

The mechanism of reduced apnea- and exercise-induced oxidative stress in trained breath-hold human divers and diving animals is only suggested. Sjödin et al. (1990) were the first to hypothesize that a reduced oxygen supply to working muscle should decrease the ROS formation. Apnea is responsible for a reflex vasospasm in limb muscles (Craig, 1968; Sterba and Lundgren, 1988) which lowers the muscle oxygen supply during exercise. However, a reduced blood supply to muscles is probably not the sole factor explaining a reduced ROS production in response to static or dynamic apnea (Joulia et al., 2002 and 2003). Indeed, in humans acutely (Dousset et al., 2002) or chronically (Singh et al., 2001; Vij et al., 2005) exposed to hypoxia the blood oxidant-antioxidant status changes in the direction of an acclimatization to the oxidative stress. The kinetics of acclimatization processes has been approached by Vij et al (2005) who report no increase in antioxidant status after a 3-mon sojourn at 4,500 m and even an increased baseline oxidant activity (elevated level of thiobarbituric acid reactive substances, TBARS/reduced levels of ascorbic acid and SOD), whereas after 13 mon at 4,500 m the TBARS regresses back to pre-exposure level and blood concentration of all measured antioxidants is markedly elevated. Data by Vij et al. (2005) oppose other observations which report high level of oxidative stress in high altitude residents (Jefferson et al., 2004). It must be underlined that chronic hypoxemic patients suffering from respiratory insufficiency have elevated levels of plasma oxidative stress (Foschino

Barbaro et al., 2007; Jammes et al., 2008). However, chronic hypoxemia is not the sole hit factor in these patients who frequently undergo infections of the respiratory tract promoting oxidative and inflammatory reactions.

Similar increase in antioxidant defences are reported in animals which undergo long-term adaptations such as hibernation, estivation, dehydratation-rehydratation processes (Filho et al., 2002). This suggests a non-specific adaptive phenomenon and not an upregulation of heat shock proteins (Hsp) expression (see special chapter in this book). Numerous data suggest that Hsp would complement the existing endogenous antioxidants during and following the cellular oxidative stress, thereby protecting cells against the deleterious effects of ROS. Thus, in healthy subjects a close interrelationship exists between cellular Hsp expression and the redox status. Hsp expression reduces the ROS generation through the activation of antioxidants and, in turn, both the oxidant and antioxidant levels upregulate Hsp expression (Whitam and Fortes, 2008). Other data also suggest the responsibility of endogenous melatonin production in heart model of ischemia-reperfusion (Reiter and Tan, 2003; Tan et al., 2005). In diving reptiles, an upregulation of cognate and inducible Hsp (Hsp 72 and 73) is reported in the turtle's brain (Prentice et al., 2004; Scott et al., 2003), heart (Chang et al., 2000; Scott et al., 2003), liver, and skeletal muscle (Scott et al., 2003). After a forced dive, Hsp 72 expression increased two fold in the brain and threefold in the heart and skeletal muscle (Scott et al., 2003). However, in their study Ramaglia and Buck (2004) show that increased Hsp 72, 73 expression in heart and muscle of painted turtle only occurs if the dive duration exceeds 18 h and they conclude that "increased HSP expression is not critical in the early adaptation to anoxic survival and that short-term anoxia is probably not a stress for species adapted to survive long periods without oxygen". Thus, the role played by Hsp in promoting long-term anoxia tolerance is still debatable .

Opposing the protective role of Hsp on the oxidative stress, the hypoxia inducible factor (HIF-1$\alpha$) expressed in numerous organs (brain, kidneys, liver, heart and skeletal muscle) (Stroka et al., 2001) is activated during chronic (Bell et al., 2007) as well as intermittent hypoxia (sleep apnea syndrome) (Semenza and Prabhakar, 2007). The increased ROS production by hypoxia activates HIF-1$\alpha$ which in turn accentuates ROS generation and further amplifies HIF-1$\alpha$ activation. The presence of HIF-1$\alpha$ at high levels in seal tissues (*Phoca hispida*) is demonstrated (Johnson et al., 2005). There are no information on the net effect of hypoxia-induced Hsp upregulation and HIF-1$\alpha$ activation in the regulatory processes for ROS generation.

In conclusion, repeated exposure to anoxia elicits an adaptive response to the oxidative stress not only in diving animals but also in humans trained to breath-holding. The role played by an Hsp upregulation in this adaptive process is suspected but needs further investigation, namely in diving mammals and human free divers.

## ROS AND HYPERBARIC CONDITIONS

### Hyperbaric oxygen in mammals

This section describes the biochemical basis of hyperbaric oxygen (HBO) and subsequent oxygen toxicity especially at the Central Nervous System level (CNS). The brain is highly sensitive to oxygen toxicity which triggers alteration in neurotransmission and cerebral functions up to epileptic seizures.

At atmospheric pressure, the partial pressure of oxygen is equal to 0.21 ATA (21 kPa), which permits a normal diffusion of oxygen in the tissues. By increasing the ambient pressure, HBO exposure allows an elevation of the oxygen partial pressure and, therefore, facilitates the oxygen arterial diffusion. Thus, above 0.21 ATA, breathing hyperbaric oxygen induces an increase of arterial oxygen partial pressure and facilitates oxygen delivery to the tissues and to the cells. According to this property of oxygen diffusion, HBO therapy (from 1.8 to 2.8 ATA) is classically employed to treat several clinical disorders related to hypoxia, including decompression sickness, acute carbon monoxide intoxication, tissue infection, radiation necrosis and air embolism.

Oxidative damages appear in all tissues, but depend on the duration of exposure and the level of oxygen partial pressure. Brain oxidation triggers the most severe manifestations of oxygen toxicity, such as epileptic seizures. Indeed, above 2.5 ATA in men and 6 ATA in rats, hyperoxia-induced epileptic seizures lead to irreversible brain damages.

### *Symptoms of CNS oxygen toxicity*

Toxic effects arise upon enzyme activity, neurotransmission and cell viability of the Central Nervous System. Then, the normal brain function is disrupted as a consequence of intensive and synchronous burst activity from neuron assemblies, followed by convulsions and, destruction of neurons (Clark and Thom, 2003)

Lambertsen (1965) defined CNS toxicity in these terms: "The convulsion is usually preceded but not always by the occurrence of localized muscular twitching, especially about eyes, mouth and forehead. Small muscles of the hands may be involved and a lack of coordination of diaphragm activity in respiration may occur. Eventually an abrupt spread of excitation occurs and the rigid 'tonic' phase of the convulsions begins. Vigorous clonic contraction of the muscle groups of head and neck, trunk and limbs then occur becoming progressively less violent over 1 minute". These convulsions are generalized and of tonic-clonic type (grand mal seizures). On the electroencephalogram (EEG), seizures are composed of a first phase of spikes corresponding to the tonic period, followed by a phase of spike and waves (clonic period) and, finally, ending by an electrical silent (atonic period). After that, the EEG activities reappeared progressively (Harel et al., 1969; Rucci et al., 1967). According to the precursor studies of Gibbs et al. (1935) and of Cohn and Gersh (1945) there were no changes in EEG activities before the occurrence of epileptic seizure. However more recent investigations performed in rats with spectral analysis (Tobarti et al., 1981) have indicated some alterations up to 60 min before the seizures. In rats, seizures occur after breathing 3 ATA for 3–5h, 5 ATA for 5–90 min, and 6ATA for less than 20 min.

## *Hyperoxia and ROS/RNS production*

Hyperoxia is generally associated to photon emission and ROS production (Halliwell, 2006). A variety of biomolecules autooxidise when they are exposed to oxygen: for example dopamine, norepinephrine and epinephrine, DOPA, and thiols (cysteine). These compounds react with $O_2$ to produce superoxide, as well as others products. For example, the autoxidation of dopamine leads to the production of semiquinones which are themselves toxic and which may also lead to generation of ROS (Halliwell and Gutteridge, 1985). This mechanism is particularly present after Traumatic Brain Injury or blast induction and represents one of the most deleterious secondary insult. During the blast (120 and 140 kPa), overproduction of $NO^{\bullet}$ leads to ROS formation and over expression of heme oxygenase (HO-1). Nevertheless, paradoxically, synthesis of HO-1 has powerful antioxidant effects.

HO-1 is an inductible isoform of heme oxygenase, which catalyze the rate-limiting step in the oxidative degradation of heme to biliverdin. Using treatment of cell cultures with HBO as a model of oxidative stress, Rothfuss and Speit (2002a) put in evidence of an induction of HO-1 in isolated human lymphocytes, after a single HBO exposure, and protection of these cells against DNA damage by subsequent oxidative stress. Since the observed protective effects were abolished by co-treatment with the specific HO-1 inhibitor, tin-mesoporphyrin, their studies suggest that a low-level over expression of HO-1 provides protection against oxidative DNA damage induced by HBO.

Under hyperbaric conditions, overproduction of ROS, associated to decrease of scavenger efficiency, produces an oxidative stress leading to excess (160%) of $H_2O_2$ production *in vivo* (Yusa et al., 1987; Dirnagl et al., 1995). This effect is exacerbated by excitatory amino acids such as glutamate (Coyle and Puttfarcken, 1993).

*Physiopathology of HBO toxicity*

Brain tissue is especially sensitive to oxidative stress for several reasons. The brain shows a high consumption of $O_2$ (about 20% of basal consumption) to produce enough ATP to maintain neuronal intracellular ion homeostasis. Indeed, neurons show a high electric activity and dysfunction of mitochondria activity which may cause rapid damages.

In addition, subsequent to oxidative stress, cellular membranes lose their integrity, inducing receptor and channels dysfunction and release of excitatory amino acids as glutamate leading to alteration of the activity of adjacent neurons and to a strong increase of intracellular calcium. Then, intracellular activity is disrupted, leading to a cascade of deleterious events which trigger neuronal dysfunction and neurodegeneration.

One of the major hypothesis to explain the HBO toxicity, *in vivo*, is linked to the cerebral lipoperoxydation produced by an excess of ROS, associated with the notion of precritical lipoperoxydation threshold (Louge and Meliet, 2006).

Nitric oxide (NO•), from endothelial and neuronal origin, has also been considered as a major contributor to cerebral oxygen toxicity, notably by relaxing the hyperoxia- induced vasoconstriction (Courtière, 2006).

Demchenko et al. (2000, 2003) first verified the hypothesis of the interactions between ROS, NO• and vasomotricity in the apparition of the generalized hyperoxic crisis. They found that rats exposed to 400 and 500 kPa of pure oxygen developed an initial cerebral vasoconstriction, with a neuroprotective effect during 30 to 45 min, and followed with a vasodilatation phase that directly preceded the EEG neurotoxicity signs (spikes) .This vasodilatation phase is directly linked to NO• production. The mechanism by which NO• increases under hyperoxia is linked to an increased quantity of oxygen, as the principal substrate for NO• formation, and by an increased quantity of arginine, its other precursor (see generalities). In a second time, an induction of neuronal NO-synthase would occur, re-establishing the cerebral blood flow. The concomitant reaction of ROS and NO• would thus lead to the production of peroxynitrite (ONOO−), the neurotoxical action of which led to the convulsive hyperoxic crisis. NO• would also directly act on N-methyl-D-aspartate (NMDA) receptors, increasing the glutamatergic toxicity (Dawson et al., 1991; Globus et al., 1995).

*HBO and neurotransmitters*

Initially, one of the major hypotheses to explain the convulsive hyperoxic crisis was linked to an inhibition of gamma-aminobutyric acid (GABA) synthesis, the main inhibitor neurotransmitter. As a matter of fact, GABA concentration significantly decreases in rats exposed to 606 kPa $O_2$ (Wood and Watson, 1963). However, this decrease is reversible and disappears within 1 h after the end of HBO exposure. Although these results indicate alterations in GABA metabolism during hyperoxic exposure, they did not establish a direct cause and effect relationship between a decrease in brain GABA level and an induction of HBO convulsions (Clark and Thom, 2003). Moreover, changes in brain GABA levels are not always accompanied by appropriate changes in convulsion times. Hyperoxic effects on brain GABA metabolism may parallel the development of HBO seizures without a direct causal relationship or may be related to concurrent alterations in ionic gradients across cell membranes (Clark and Thom, 2003). Indeed HBO induced lipoperoxydation may directly act on excitable cell membranes, through ion channels or Na K ATPase, leading to an extracellular potassium accumulation and increasing neuronal excitability (Kovachich and Mishra, 1981).

CNS oxygen toxicity is known to affect most of neurotransmitter synthesis and degradation such as GABA, acetylcholine, glutamate, aspartate, dopamine and norepinephrine. In consequence, an imbalance between excitatory and inhibitory systems results in changes in spontaneous synaptic transmitter release (Colton and Colton, 1978) and pre- and post-synaptic response. Recent studies demonstrated strong alteration of dopaminegic neurotransmission under normobaric hyperoxia (1ATA of oxygen) resulting in a decrease of dopamine release (Adachi et al., 2001; Lavoute et al., 2008). As an explanation, an increase in catecholamine degradation enzymatic activity, the mitochondrial monoamine oxidase (MAO), has been suggested (Adachi et al., 2001). In cerebral tissue, lipid peroxydation has also been demonstrated to induce a decline in fusion between presynaptic membrane and synaptic vesicles (Omoi et al., 2006). In addition, a decreased activity of the glutamic acid decarboxilase (GAD), controlling the synthesis of GABA, has been observed during 10–15min at 6ATA, before convulsion appearance (Li et al., 2008). Moreover, exposures to 2, 3 or 4 ATA of $O_2$ enhance the hyperoxia-induced decrease of dopaminergic activity (Lavoute et al., 2008).

Inversely, epileptic seizures are associated with an increase in GAD activity, which could be indirectly associated to an elevation of excitatory activity (Li et al., 2008). In accordance with these results, hyperoxia-induced epileptic seizures at 5 or 6 ATA follows an over production of NO[.] (Demchenko et al., 2001; Thom et al., 2003), depending on an elevation of intracellular calcium through glutamate excitatory receptors.

### Hyperbaric oxygen-induced adaptative protection against oxidative stress

A major paradox concerning ROS is that they can have negative effects, being responsible of HBO toxicity or positive effects, by mediating $O_2$

toxicity (Clark and Thom, 2003). This phenomenon has been largely studied during the last decade.

It has been demonstrated that HBO (see chapter 20 in this book) leads to increased ROS production that can cause cellular damage with lipid, protein and DNA oxidation (Narkowicz et al., 1993; Benedetti et al., 2004); for this reason, HBO therapy may be considered as an excellent model system for the investigation of oxidative stress and its biological consequences (Benedetti et al., 2004).

Mori et al. (2007) showed that exposure of astrocytes to hyperbaric oxygen induces a time dependant apoptotic cell death, and that when astrocytes, damaged by oxidative stress, are co-cultured with normal neurons from the cerebrum of newborn rats, neuronal cell death is markedly induced. Their results imply that ROS induced by oxidative stress attack astrocytes to induce oxidatively denatured proteins in the cells that act as a neurotoxic factor, and that vitamin E protects neurons by inhibiting astrocyte apoptosis caused by oxidative stress. Vitamin C and E, riboflavin and selenium have been used successfully to scavenge the ROS produced by HBO, but Dunbar et al. (2005) have shown that melatonin exhibits protective effects against HBO-induced lipid peroxidation (300kPa hyperbaric oxygen, for 120 min).

Treatment of human subjects with HBO: 100% $O_2$ at a pressure of 2.5 ATA for a total period of three times 20 min causes DNA damage in lymphocytes: induction of this damage is found only after the first HBO exposure and or after further treatments of the same individuals (Speit et al., 2000). Furthermore, blood taken 24 h after HBO treatment is significantly protected against the *in vitro* induction of DNA damage by hydrogen peroxide, indicating that adaptation occurs due to induction of antioxidant defences. The authors put in evidence of increased levels of HO-1 in lymphocytes 24 h after HBO, while superoxide dismutase, catalase and the DNA repair enzymes are not enhanced in expression. The protection against DNA damage even occurs with a shortened HBO treatment, which did not induce DNA damage by itself, suggesting that the induction of DNA damage is not the trigger for adaptative protection. Increased sequestration of iron, as a consequence of induced HO-1 might be involved in the mechanism of adaptative protection after HBO treatment (Speit et al., 2000)

Preconditioning with 100 % $O_2$ for 24 h can induce ischemic tolerance via formation of $O_2$ free radicals, in transient focal cerebral ischemia (Zhang et al., 2004). An oxygen free radical scavenger, given during the preconditioning period, prevents the phenomenon. The results suggest that an enriched oxygen supply leads to an increased production of free radicals, which were the trigger for the subsequent ischemic tolerance process. Dirnagl et al. (2003) have demonstrated that triggers are detected by sensors which use transduction mechanisms such as nitric oxide or protein tyrosine phosphorylation, in order to impact effectors of preconditioning such as heat shock protein (HSP), free radical scavenger, GABA receptor upregulation and enhanced GABA release (Baker, 2004).

HBO preconditioning can induce ischemic tolerance in the brain (Miljkovic-Lolic et al., 2003) and in the spinal cord (Dong et al., 2002). Li et al. (2007) demonstrated that this tolerance against oxidative injury occurs via increased expression of heme oxygenase-1. This mechanism acts through the induction of ferritin synthesis, as a result of iron removal from the degradation of heme by HO-1.Ferritin may be released to restrict iron from participating in the Fenton reaction and, thus, may provide protection against oxidative damage by preventing the generation of the DNA-damaging hydroxyl radical (Rothfuss and Speit, 2002b).

The paradoxical effects played by the ROS and, especially, their protective effects can be summarized by the following message "what doesn't kill you makes you stronger" (Baker, 2004).

## ROS AND DEEP SEA ENVIRONMENT IN AQUATIC ECTOTHERMS

Seawater in relation to numerous parameters ($O_2$ content, temperature, UV radiation, biomass etc…) is an exogenous source of ROS for aquatic animals (Obermüller et al., 2005). Among ROS, anion superoxide and hydrogen peroxide are both abundant in the upper layers of the oceans and their concentration tend to decrease with increasing depth. ROS are also produced by living organisms themselves related to their metabolic processes. These interactions between environmental factors and ROS make fishes and marine invertebrates being used the most as models because of the environment-dependency of their metabolic rates, at least, for most of them (Filho, 2007). Changes in environmental parameters, e.g temperature, may require metabolic adaptations, which involve a number of physiological processes including structural and functional modifications of mitochondria (Guderley and St Pierre, 2002). These considerations explain why, in relation to these processes, changes in ROS and regulation of their production are both worth being studied. For example, antioxidant defence could protect fishes against the consequences of temperature variations (Filho, 1996).

The deep sea environment sometimes described as refuge against oxidative stress (Janssens et al., 2000) is characterized by some physicochemical parameters as low temperature, high pressure, decreased light… Pressure and temperature are two physical parameters with opposite effects in thermodynamic. So, studies about ROS performed for example in polar marine waters (low temperature) also contribute to a better understanding of the pressure effects on ROS metabolism.

Most marine animals are ectothermic that means their body temperature is in equilibrium with their environment. Consequently in many cases, species living at lower temperature have lower metabolic rates and inverse (Guderley and Kraffe, 2008). Theses processes concomitantly with ROS production have been studied in some specific habitats like polar marine waters. Because of the higher oxygen solubility and high UV radiation conditions in these cold waters, the species living in this habitat are potentially more vulnerable to oxidative stress (Abele and Puntarulo, 2004). The decline, at low temperature, of the metabolic performance by ectotherms submitted to the Arrhenius law explains the development, by polar ectotherm species, of strategies aimed at facing up this law.

In polar invertebrates and cold acclimated fish, mechanisms compensate for the slow diffusion rate of oxygen through tissues to the mitochondria at low temperature by increasing lipid content (Desaulniers et al., 1996). Another mechanism in Antarctic species consists in an elevation of their mitochondrial densities in the red muscle to compensate for reduced mitochondrial performance and low mitochondrial oxidative capacity (Johnston et al., 1998). The above two mechanisms facilitate the intracellular distribution of oxygen but have consequences such as an enhancement of mitochondrial ROS production and of their potential harmful effects. Indeed, in polar species the increase in lipid content and more precisely the high degree of unsaturation in membranes constitutes a potential target for ROS attack. In comparison to marine ectotherms from temperate environment, those from polar environments display an elevated level of antioxidant protection. Activities by tissues enzyme such as catalase and superoxide dismutase are higher in polar than those of temperate mollusc species submitted to the same assay temperature of 20°C (Abele and Puntarulo, 2004). Antioxidant activities by polar bivalves are higher than those by their Mediterranean homologs (Heise et al., 2003). These authors also found in parallel, conversions of oxygen to $H_2O_2$ of 3.7% (state 3) and 6.5% (state 4) at habitat temperature (1°C) and of 7% (state 3) and 7.6% (state 4) under experimental warming to 7°C. These percentages are high compared to the 1–3%.of oxygen used in conversion to ROS in mammalian isolated mitochondria, but because of the low metabolic observed in the  cold Antarctic waters, the absolute rates of mitochondrial ROS production are kept low.

Finally even though environmental factors, in polar habitats, contribute to exposure of animals to oxidative stress, the effects are counterbalanced by the low metabolic activity and elevated antioxidant levels. Among the antioxidant compounds that prevent cell constituents and membranes from oxidative damages, vitamin E plays a key role in vertebrates. The fishes from the Antarctic Ocean are known for the richness of their mitochondria in polyunsaturated fatty acids (PUFA) which makes them a potential target for ROS attack, but it is worth noting that these organisms contain more vitamin E than temperate species (Filho, 2007).

## ROS and Hydrostatic pressure: European eel example

While studies of ROS and temperature effects in fish are relatively numerous, only scarce data about the effects of high hydrostatic pressure are available and usually they refer to deep sea investigations. Some study have dealt with antioxidant systems in animals living in the deep sea waters: the lower activities by antioxidant enzymes (SOD and Glutathion peroxidase) in the muscles of deep sea fishes compared to those of shallow species were explained by a lower ROS production in relation with their low metabolic rate (Janssens et al., 2000). Similar results are reported in amphipods from different habitats (surface, sub- and deep- waters; Camus and Gulliksen, 2005). But these results remain difficult to interpret because of numerous factors liable to be implicated in the decline in metabolic activities of deep sea animals (Childress, 1995).

The European eel is often used as a model to gain more insight into the impact of hydrostatic pressure upon its metabolism. Among other fish species, the eel is characterized by a particular resistance to pressure (Sébert and Theron, 2001). In eel, the study of the effects of hydrostatic pressure on ROS regulation could be important in the context of its reproduction. To reproduce, the European eel (*Anguilla anguilla* L.) must migrate at depth over about 6,000 km from the European coast to the Sargasso Sea (its supposed breeding area). The period of time needed for this migration requires various adaptations to temperature, salinity, hydrostatic pressure, swimming… in order to cope with the high energy demand in an adverse environment and the fact that this trip is performed without feeding (Tesch, 2003; van Ginneken and Maes, 2005). As mentioned at the beginning of this chapter, ROS production is linked to the metabolism intensity. The important energy demand required to swim during approximately 5 mon at depth could increase ROS production. So the question is raised how do the European eels preserve themselves against the oxidative stress during their migration?

Before departure, eels metamorphose from yellow non-migratory and sexually immature stage to the silver migratory one, this transformation is called silvering process. Studies have been conducted in order to gain more insight into the role and/or the effect of this metamorphosis on the incredible capacity of the European eels, in particular in terms of energetic metabolism performance to resist to high hydrostatic pressure (Sébert and Theron, 2001). After acclimatization of yellow eels to hydrostatic pressure for, at least, 21 d, Simon et al. (1992) and then Theron et al. (2000) observed an enhancement of oxidative phosphorylation efficiency (significant increase of the P/O ratio, Table 1). In yellow eels a long-term exposure to pressure induced a readjustment of the enzyme activities with a decrease (about 50%) of the activities by the three primary complexes and an increase (about 30%) of the activity displayed by the last complex (cytochrome oxidase, dioxygen reducer) concomitant with a 30%-decrease of oxygen consumption in the red muscle (see table 1). These enzyme re-adjustments could allow a reduction of electron leak along the respiratory chain and, thus, of ROS production. But what happens in silver eels which have to cope naturally with hydrostatic pressure contrary to yellow eels?

The differences between both stages are not only morphological ( colour changes; expansion of the eyes and pectoral fins, Pankhurst, 1982a and Durif et al., 2000) but also physiological (rise of haemoglobin concentration, Johansson et al. 1974; increase of the red muscle volume, Pankhurst, 1982b; elevation of the red muscle lipid , Pankhurst, 1982a; Hoofd and Eggington, 1997). This transformation has often been described as a morphological and physiological preparation or pre-adaptation to the extreme migration conditions. For example, silver eel seems to be naturally pre-adapted to salinity exposure (Kirsch et al., 1975) or pressure exposure (Vettier et al., 2005). In fact, the silvering process performed at atmospheric pressure could permit mechanism establishment assuring hydrostatic pressure adaptation in terms of muscle energetic metabolism (Vettier et al., 2005; Amérand et al., 2006; Vettier et al., 2006). Indeed, yellow eels present significantly higher muscle oxygen consumption and ROS production than silver eels (Table 1) but after pressure acclimatization of yellow eels (at least 21 d at depth) no difference was observed between both stages. So, acclimatization of yellow eel to hydrostatic pressure could mimic the effects of the silvering process. On the other hand, silver eels appear to be unaffected by hydrostatic pressure

**Table 1.** Physiological data (mean ± SEM) of muscle from yellow and silver freshwater eels, Anguilla anguilla L., maintained at atmospheric pressure or after exposure to pressure.

| | Yellow eel | | Silver eel | | References |
|---|---|---|---|---|---|
| | 1 ATA | 101 ATA | 1 ATA | 101 ATA | |
| PUFA | 58.3 ± 2.4 | ND | 67.3 ± 0.5* | ND | Vettier et al., 2006 |
| Lipid (% of muscle) | 9.9 | ND | 42.5 | ND | Hoofd and Eggington, 1997 |
| P/O | 2.52 ± 0.04 | 2.87± 0.05 * | ND | ND | Theron et al., 2000 |
| Oxygen consumption nmol.min⁻¹.g FM⁻¹ | 501 ± 24 | 383 ± 28* | 388 ± 23* | 427±61 | Amérand, 2006 |
| HO• production ng.min⁻¹.g⁻¹ | 6.5 ± 0.9 | 5.5 ± 1.2 | 3.6 ± 0.6* | 4.5 ± 1.5 | " |
| Catalase activity µmol.min⁻¹.g | 3.1 ± 0.4 | 2.6 ± 0.3 | 3.5 ± 0.3 | 3.7 ± 0.2 | Amérand et al., 2006 |
| SOD activity U.g⁻¹ | 58 ± 10 | 40 ± 6 | 54 ± 5 | 50 ± 6 | " |
| Total MDA content nmol.g⁻¹ | 45.3 ± 7.2 | 51.7 ± 5.0 | 107.7 ± 14.5* | 104.1 ± 13.0 | " |

*The experimental pressure is 101 ATA (1000 m depth) and the rate of compression is 10atm.min⁻¹.*
*Tw = 15–17°C; ND: not determined. * significant difference from controls (1 ATA) yellow eels.*

in terms of red muscle oxygen consumption and ROS production (Table 1; Amérand et al., 2006); this indicates, at least at this level, that the migratory adaptation mechanisms established over the silvering process are not strengthened by silver eel acclimatization to hydrostatic pressure. The pre-adaptation to pressure resistance implies modifications of the membrane fluidity which are liable to affect metabolism intensity and ROS production by, for example, causing modifications of respiratory chain enzymes located within the inner mitochondrial membrane. This assumption is supported by the known effects of hydrostatic pressure as thermodynamic parameter on biological membranes (Macdonald, 1984; Somero, 1991). Because of an increase in the number of fatty acid unsaturations over the silvering process, the mitochondrial membrane fluidity of silver eels is higher than that of yellow eels (Vettier et al., 2006). So, the increase in lipid content and in unsaturated fatty acids is consistent with the increase in malondialdehyde content (MDA, result of the lipoperoxidation by ROS; Dröge, 2002) during the silvering process (Table 1). Therefore, despite the ROS production decrease and the lack of changes in antioxidant activity (superoxide dismutase or catalase, Table 1) the silvering process, by increasing the content in membrane poly-unsaturated fatty acids, increases the potential targets of ROS (Hulbert et al. 1989), which could eventually change the membrane fluidity (Carbonneau et al., 1991; Chen and Yu, 1994; Sébert et al., 1995).

The objective of the eel migration is reproduction supposedly to take place in the Sargasso Sea. Male and female silver eels present an important sexual dimorphism. Indeed males are twice and even thrice as smaller than females in terms of body-mass and -length (van Ginneken and Maes, 2005). Due to this difference of size it can be assumed that the swimming time and/or swimming speed to join the reproduction area are not same in both sexes. In females, the estimation of the average swimming speed has been experimentally to be about 0.5 BL.s$^{-1}$ (body length per second) for a migration in 4.5 mon (van den Thillart et al., 2004). Under natural conditions, males quit rivers 1 mon earlier than females but it is not enough to join the spawning area at the same time as females. It is thus hypothesized that males have to swim faster. Indeed males are able to swim faster because they exhibit both significant higher muscle oxygen consumption and maximal aerobic capacity than females (Amérand et al., 2009) with the potentially and deleterious increase in ROS production as shown in mammals during endurance exercise (Sachdev and Davies, 2008). After hydrostatic pressure exposure, differences between males and females are put forward in term of ROS metabolism. In fact, the positive relationship between ROS production and metabolism rate in fish muscle (Amérand, 2006; Mortelette et al., unpublished data) is maintained after pressure exposure only in females. On the other hand, after hydrostatic pressure exposure, males display a reversal effect on the above mentioned relationship which could tend to protect itself from the deleterious effects of an enhancement of metabolic rate. Though it would be interesting to complete this first approach, it can be speculated that migration for males could be facilitated under pressure because males swim faster without the negative oxidant effect of this swimming exercise. In a similar way, Scaion et al. (2008) showed through measurements of the oxygen consumed by a whole animal, that males seem to prefer a migration at depth (pressure, cold water and low UV radiation) whereas females swim more at the surface. Do males migrate at depth for better protection against oxidative stress? This behaviour could be necessary to compensate for the effects of high metabolic activity in males compared to females. But other investigations will be necessary to particularly analyze the antioxidant strategies like the enzymatic antioxidant and also the role of vitamin E for example, sexual hormones and or mitochondrial proton leak.

To conclude, European eels seem to use the environment factors (as hydrostatic pressure) to perform migration under the best physiological conditions while limiting oxidative stress.

## CONCLUSIONS

A brief summary of numerous characteristics of ROS and related molecules such as NO• in the hyperbaric context were described here. Their ubiquity appears yet more evident, particularly their roles in adaptive mechanisms as physiological signalling agents. In this chapter, the relationships between ROS and pressure were defined in different ways in terms of context and/ or problematic (apnea and anoxia-reoxygenation in diving animals, hyperbaric oxygen, hydrostatic pressure in fish), but some common aspects emerge.

ROS are not only potentially toxic species but they can also act as cellular signals at the origin of adaptive response to the oxidative stress. Hyperbaric oxygen treatment leads to an increased production of free radicals which would also be the triggers for the ischemic tolerance process in the brain. In diving animals exposed to repeated anoxia, an Hsp regulation would play a protective role on the oxidative stress and in turn, both the oxidant and antioxidant levels would upregulate Hsp expression. A permanent metabolic challenge exists for animals exposed to some specific conditions or environment. Diving animals submitted to anoxia/reoxygenation or ectotherms living in polar water increase their mitochondrial volume density to enhance their oxidative capacity while needing to elevate their antioxidant capacity. To prevent and face oxidative stress, several common strategies are developed in all species and tissues (antioxidant activities, vitamins C, E…). It is worth mentioning the higher specificity of some of them, e.g the mechanism established, after HBO exposure, by the induction of an isoform of heme oxygenase(HO-1) which increases iron sequestration and consequently decreases by ROS via the Fenton reaction.

Beside the endogen antioxidant defences commonly reported, some animals such as ectotherms seem to use their environment and to adapt their behaviour to preserve themselves against oxidative stress. In the context of eel migration at depth for example, the study of the effects of HP on the metabolic rate/ROS production relation leads one to suppose that eel males would be more prone to the development of an environmental strategy (migration at depth) while eel females would preferentially rely on the use of endogen antioxidant processes.

To get more knowledge of this vast wide topic (ROS and pressure), it is obvious that further investigations are needed for example, the study of the thermodynamic opposite effects of temperature and pressure on ROS metabolism with their interesting perspectives in physiology, ecophysiology and medicine.

## REFERENCES

Abele, D. and S. Puntarulo. 2004. Formation of reactive species and induction of antioxidant defence systems in polar and temperate marine invertebrates and fish. Comp. Biochem. Physiol. A Mol. Integr. Physiol. 138: 405–415.

Adachi, Y.U. and K. Watanabe, H. Higuchi, T. Satoh and E.S. Vizi. 2001. Oxygen inhalation enhances striatal dopamine metabolism and monoamineoxidase enzyme inhibition prevents it: a microdialysis study. Eur. J. Pharmacol. 422: 61–68.

Amérand, A. 2006. Approche *in vitro* des interactions metabolisme energétique/espèces réactives dérivées de l'oxygène dans le muscle de poissons; influence d'un facteur environnemental: la pression hydrostatique. Thesis. Université Européenne de Bretagne, Brest.

Amérand, A. and A. Vettier, P. Sébert and C. Cann-Moisan. 2006. Does hydrostatic pressure an effect on reactive oxygen species in eel? Undersea Hyperb. Med. 33: 157–160.

Amérand, A. and A. Vettier, C. Moisan, M. Belhomme and P. Sébert. 2009. Sex related differences in aerobic capacities and reactive oxygen species metabolism in the silver eel. Fish Physiol. Biochem. In press (DOI 10.1007/S10695-009-9348-0).

Baker, A. 2004. Ischemic preconditioning in the brain. Can. J. Anesth. 51: 201–205.

Behrisch, H.W. and R. Elsner. 1984. Enzymatic adaptations to asphyxia in the harbour seal and dog. Respir. Physiol. 55: 239–254.

Bell, E.L. and T.A. Klimova, J. Eisenbart, P.T. Schumacker and N.S. Chandel. 2007. Mitochondrial reactive oxygen species trigger hypoxia-inducible factor—dependent extension of the replicative life span during hypoxia. Mol. Cell. Biol. 27: 5737–5745.

Benedetti, S. and A. Lamorgese, M. Piersantelli, S. Pagliarani, F. Benvenuti and F. Canestrari. 2004. Oxidative stress and antioxidant status in patients undergoing prolonged exposure to hyperbaric oxygen. Clin. Biochem. 37: 312–317.

Boveris, A. and B. Chance. 1973. The mitochondrial generation of hydrogen peroxide. General properties and effect of hyperbaric oxygen. Biochem. J. 134: 707–716.

Bulmer, A.C. and J.S. Coombes, J.E. Sharman and I.B. Stewart. 2008. Effects of maximal static apnea on antioxidant defenses in trained free divers. Med. Sci. Sports Exerc. 2008, 40: 1307–1313.

Butler, P.J. and D.R. Jones. 1997. Physiology of diving of birds and mammals. Physiol. Rev. 77 : 837–899.

Cadenas, E. and K.J. Davies. 2000. Mitochondrial free radical generation, oxidative stress, and aging. Free Radic. Biol. Med. 29: 222–230.

Camus, L. and B. Gulliksen. 2004. Total oxyradical scavenging capacity of the deep-sea amphipod Eurythenes gryllus. Mar. Environ. Res. 58: 615–618.

Camus, L. and B. Gulliksen. 2005. Antioxidant defense properties of Arctic amphipods: comparison between deep-, sublittoral and surface-water species. Marine Biology 146: 355–362.

Carbonneau, M.A. and E. Peuchant, D. Sess, P. Canioni and M. Clerc. 1991. Free and bound malondialdehyde measured as thiobarbituric acid adduct by HPLC in serum and plasma. Clin. Chem. 37: 1423–1429.

Chang, J. and A.A. Knowlton and J.S. Wasser. 2000. Expression of heat shock proteins in turtle and mammal hearts: relationship to anoxia tolerance. Am. J. Physiol. Regul. Integr. Comp. Physiol. 278: R209–214.

Chen, J.J. and B.P. Yu. 1994. Alterations in mitochondrial membrane fluidity by lipid peroxidation products. Free Radic. Biol. Med. 17: 411–418.

Childress, J.J. 1995. Are there physiological and biochemical adptations of metabolism in deep-sea animals? Tree 10: 30–36.

Clark, J.M. and S.R. Thom. Oxygen under pressure. Pp. 358–418. *In*: A.O. Brubakk and T. S. Neuman. (eds.). 2003. Bennett and Elliott's Physiology and Medicine of Diving. Saunders, Elsevier, New-York, USA.

Colton, J.S. and C.A. Colton. 1978. Effect of oxygen at high pressure on spontaneous transmitter release. Am. J. Physiol. 235(5): 233–237.

Cohn, R. and I. Gersh. 1945. Changes in brain potentials during convulsions induced by oxygen under pressure. J. Neurophysiol. 8: 155–160.

Courtière, A. Toxicité de l'oxygène : biochimie et aspects physiopathologiques. Pp. 271–281. *In* : B. Broussolle and J.L. Meliet (eds.). 2006. Physiologie et Médecine de la plongée. Ellipse, Paris, France.

Coyle, J.T. and P. Puttfarcken. 1993. Oxidative stress, glutamate, and neurodegenerative disorders. Science 262: 689–695.

Craig, A.B. 1968. Depth limits of breath-hold diving (an example of Fennology). Respir. Physiol. 5: 14–22.

Davies, K.J. 1995. Oxidative stress: the paradox of aerobic life. Biochem. Soc. Symp. 61: 1–31.

Dawson, V.L. and T.M. Dawson, E.D. London, D.S. Bredt and S.H. Snyder. 1991. Nitric oxide mediates glutamate neurotoxicity in primary cortical cultures. Proc. Natl. Acad. Sci. USA 88(14): 6368–6371

Dhaliwal, H. and L.A. Kirshenbaum, A.K. Randhawa and P.K. Singal. 1991. Correlation between antioxidant changes during hypoxia and recovery on reoxygenation. Am. J. Physiol. 261: H632–H638.

Demchenko, I.T. and A.E. Boso, P.B. Bennett, A.R. Whorton and C.A. Piantadosi. 2000.Hyperbaric oxygen reduces cerebral blood flow by inactivating nitric oxide. Nitric Oxide 4(6): 597–608.

Demchenko, I.T. and A.E. Boso, A.R. Whorton and C.A. Piantadosi. 2001. Nitric oxide production is enhanced in rat brain before oxygen-induced convulsions. Brain Res. 917(2): 253–261.

Demchenko, I.T. and D.N. Atochin, A.E. Boso, J. Astern, P.L. Huang and C.A. Piantadosi. 2003. Oxygen seizure latency and peroxynitrite formation in mice lacking neuronal or endothelial nitric oxide synthases. Neurosci. Lett. 344(1): 53–56.

Desaulniers, N. and T.S. Moerland and B.D. Sidell. 1996. High lipid content enhances the rate of oxygen diffusion through fish skeletal muscle. Am. J. Physiol. 271: R42–47.

Dirnalg, U. and U. Lindauer, A. Them, S. Schreiber, H.W. Pfister, U. Koedel, R. Reszka, D. Freyer and A. Vilbringer. 1995. Global cerebral ischemia in the rat: online monitoring of oxygen free radical production using chemiluminescence *in vivo*. J. Cereb. Blood Flow Metab. 15(6): 929–940.

Dirnagl, U. and R.P. Simon and J.M. Hallenbeck. 2003. Ischemic tolerance and endogenous neuroprotection.Trends Neurosci. 13: 3–10.

Dong, H. and L. Xiong, Z. Zhu, S. Chen, L. Hou and Sakabe T. 2002. Preconditioning with hyperbaric oxygen and hyperoxia induces tolerance against spinal cord ischemia in rabbits. Anesthesiology 96(4): 907–912.

Dousset, E. and J.G. Steinberg, M. Faucher and Y. Jammes. 2002. Acute hypoxemia does not increase the oxidative stress in resting and contracting muscle in humans. Free Radical Res. 36: 701–704.

Dröge, W. 2002. Free radicals in the physiological control of cell function. Physiol. Rev. 82: 47–95.

Dunbar, K. and T. Topal, H. Ay, S. Oter and A. Korkmaz. 2005.Protective effects of exogenously administered or endogenously produced melatonin on hyperbaric oxygen-induced oxidative stress in the rat brain. Clin. Exp. Pharmacol. Physiol. 32(11): 926–30.

Durif, C. and P. Elie, S. Dufour, J. Marchelidon and B. Vidal. 2000. Analysis of morphological and physiological parameters during the silvering process of the European eel (*Anguilla anguilla*) in the lake of Grand-Lieu (France). Cybium 24: 63–74.

Elsner, R. and S. Oyasaeter, R. Almaas and E.D. Saugstad. 1998. Diving seals, ischemia-reperfusion and oxygen radicals. Comp. Biochem. Physiol. A119 : 975–980.

Evans, M.D. and M. Dizdaroglu and M.S. Cooke. 2004. Oxidative DNA damage and disease: induction, repair and significance. Mutat. Res. 567: 1–61.

Filho, D.W. 1996. Fish antioxudant defenses. A comparative approach. Brazilian journal of medical and biological research 29: 1735–1742.

Filho, D.W. 2007. Reactive oxygen species, antioxidants and fish mitochondria. Frontiers in Biosciences 12: 1229–1237.

Filho, D.W. and F. Sell, L. Ribeiro, M. Ghislandi, F. Carrasquedo, C.G. Fraga, J.P. Wallauer, P.C. Simoes-Lopes and M.M. Uhart. 2002. Comparison between the antioxidant status of terrestrial and diving mammals. Comp. Biochem. Physiol. A Mol. Integr. Physiol. 133: 885–892.

Foschino Barbaro, M.P. and G.E. Carpagnono, A. Spanevello, M.G. Cagnazzo and P.J. Barnes. 2007. Inflammation, oxidative stress and systemic effects in mild chronic obstructive pulmonary disease. Int. J. Immunopathol. Pharmacol. 20: 753–763.

Fridovich, I. 1978. The biology of oxygen radicals. Science 201: 875–880.

Fridovich, I. 1995. Superoxide radical and superoxide dismutases. Annu. Rev. Biochem. 64: 97–112.

Furtado-Filho, O.V. and C. Polcheira, D.P. Machado, G. Mourao and M. Hermes-Lima. 2007. Selected oxidative stress markers in a South American crocodilian species. Comp. Biochem. Physiol. C. Toxicol. Pharmacol. 146: 241–254.

Fuson, A.L. and D.F. Cowan, S.B. Kanatous, I.K. Polasek and R.W. Davis. 2003. Adaptations to diving hypoxia in the heart, kidneys and splanchnic organs of harbour seal (Phoca vitulina). J. Exp. Biol. 206: 4139–4154.

Gibbs, F.A. and H. Davis and W.G. Lennox. 1935. The electroencephalogram in epilepsy and in conditions of impaired consciousness. Arch. Neurol. Psychiatr. 34: 1133–1148.

Globus, M.Y. and R. Prado, J. Sanchez-Ramos, W. Zhao, W.D. Dietrich, R. Busto and M.D. Ginsberg. 1995. A dual role for nitric oxide in NMDA-mediated toxicity in vivo. J. Cereb. Blood Flow Metab. 15(6): 904–913.

Guderley, H. and J. St-Pierre. 2002. Going with the flow or life in the fast lane: contrasting mitochondrial responses to thermal change. J. Exp. Biol. 205: 2237–2249.

Guderley H. and E. Kraffe. Fish life in special environments: mitochondrial function in the cold. *In*: P. Sébert, D.W. Onyango and B.G. Kapoor (eds.). 2008. Fish life in special environments. Science Publishers, New Hampshire.

Haber, F. and J. Weiss. 1934. The catalytic decomposition of hydrogen peroxide by ion salts. Proc. R. Soc. London 147: 332–351.

Halliwell, B. 1992. Reactive oxygen species and the central nervous system. J. Neurochem. 59: 1609–1623.

Halliwell, B. 2006. Oxidative stress and neurodegeneration: where are we now? J.Neurochem. 97: 1634–1658.

Halliwell, B. and J.M. Gutteridge. 1985. The importance of free radicals and catalytic metal ions in human diseases. Mol. Aspects Med. 8(2): 89–193.

Halliwell, B. and J.M.C. Gutteridge. 1999. Free Radicals in Biology and Medicine. OUP, Oxford UK.

Harel, D. and J. Kerem and S. Lavy. 1969. The influence of high oxygen pressure on the electrical activity of the brain. Electroencephalogr. Clin. Neurophysiol. 27(2): 219.

Heise, K. and S. Puntarulo, H.O. Portner and D. Abele. 2003. Production of reactive oxygen species by isolated mitochondria of the Antarctic bivalve Laternula elliptica (King and Broderip) under heat stress. Comp. Biochem. Physiol. C Toxicol. Pharmacol. 134: 79–90.

Hochachka, P.W. and L.T. Buck, C.J. Doll, and C.S. Land. 1996. Unifying theory of hypoxia tolerance. Molecular/metabolic defense and rescue mechanisms for surviving oxygen lack. Proc. Natl. Acad. Sci. USA 03: 9493–9498.

Hoofd, L. and S. Egginton. 1997. The possible role of intracellular lipid in determining oxygen delivery to fish skeletal muscle. Respir. Physiol. 107: 191–202.

Hulbert, A.J. and P.L. Else. 1989. Evolution of mammalian endothermic metabolism: mitochondrial activity and cell composition. Am. J. Physiol. 256: R63–69.

Jammes, Y. and J.G. Steinberg, A. Ba, S. Delliaux and F. Brégeon. 2008. Enhanced exercise-induced plasma cytokine response and oxidative stress in COPD patients depend on blood oxygenation. Clin. Physiol. Funct. Imaging 28: 182–188.

Janssens, B.J. and J.J. Childress, F. Baguet and J.F. Rees. 2000. Reduced enzymatic antioxidative defense in deep-sea fish. J. Exp. Biol. 203: 3717–3725.

Jefferson, J.A. and J. Simoni, E. Escudero, M.E. Hurtado, E.R. Swenson, D.E. Wesson, G.F. Schreiner, R.B. Schoene, R.J. Johnson and A. Hurtado. 2004. Increased oxidative stress following acute and chronic high altitude exposure. High Alt. Med. Biol. 5: 61–69.

Jesek, P. and L. Hlavata. 2005. Mitochondria in homeostasis of reactive oxygen species in cell, tissue and organism. Int. J. Biochem. Cell. Biol. 37: 2478–2503.

Johansson, M.L. and G. Dave, A. Larsson, K. Lewander and U. Lidman. 1974. Metabolic and hematological studies on the yellow and silver phases of the European eel, Anguilla anguilla L. III. Hematology. Comp. Biochem. Physiol. B47: 593–599.

Johnson, P. and R. Elsner and T. Zenteno-Savin. 2005. Hypoxia-inducible factor 1 proteomics and diving adaptations in renged seals. Free Radic. Biol. Med. 39: 205–212.

Johnston, I.I. and J. Calvo and Y.H. Guderley. 1998. Latitudinal variation in the abundance and oxidative capacities of muscle mitochondria in perciform fishes. J. Exp. Biol. 201(Pt 1): 1–12.

Joulia, F. and J.G. Steinberg, F. Wolff, O. Gavarry and Y. Jammes. 2002. Reduced oxidative stress and blood lactic acidosis in trained breath-hold human divers. Respir. Physiol. Neurobiol. 133: 121–130.

Joulia, F. and J.G. Steinberg, M. Faucher, T. Jamin, C. Ulmer, N. Kipson and Y. Jammes. 2003. Breath-hold training of humans reduces oxidative stress and blood acidosis after static and dynamic apnea. Respir. Physiol. Neurobiol. 137:19–27.

Kil, H.Y. and J. Zhang and C.A. Piantadosi. 1996. Brain temperature alters hydroxyl radical production during cerebral ischemia/reperfusion in rats. J. Cereb. Blood Flow Metab. 16: 100–106.

Kirsch, R. and D. Guinier and R. Meens. 1975. [Water balance in the european eel (*Anguilla anguilla* L.). Role of the oesophagus in the utilisation of drinking water and a study of the osmotic permeability of the gills (author's transl)]. J. Physiol. (Paris) 70: 605–626.

Korshunov, S.S. and V.P. Skulachev and A.A. Starkov. 1997. High protonic potential actuates a mechanism of production of reactive oxygen species in mitochondria. FEBS Lett. 416: 15–18.

Kovachich, G.B. and O.P. Mishra. 1981. Partial inactivation of Na, K-ATPase in cortical brain slices incubated in normal Krebs-Ringer phosphate medium at 1 and at 10 atm oxygen pressures. J. Neurochem. 36 (1): 333–335.

Lambertsen, C.J. Efects of oxygen at high partial pressure. Pp. 1027–1046. *In*: W.O. Fenn and H. Rahn (eds.). 1965. Handbook of Physiology Vol 11. Section 3: Respiration. Washington, USA.

Lavoute, C. and M. Weiss, J.M. Sainty, J.J. Risso and J.C. Rostain. 2008. Post effect of repetitive exposures to pressure nitrogen-induced narcosis on the dopaminergic activity at atmospheric pressure. Exp. Neurol. 212(1): 63–70.

Li, Q. and J. Li, L. Zhang, B. Wang and L. Xiong. 2007. Preconditioning with hyperbaric oxygen induces tolerance against oxidative injury via increased expression of heme oxygenase-1 in primary cultured spinal cord neurons. Life Sci. 80: 1087–1093.

Li, Q. and M. Guo, X. Xu, X. Xiao, W. Xu, X. Sun, H. Tao and R. Li. 2008. Rapid decrease of GAD 67 content before the convulsion induced by hyperbaric oxygen exposure. Neurochem. Res. 33: 185–193.

Louge, P. and J.L. Meliet. Toxicité neurologique de l'oxygène. Pp. 271–281. *In*: B. Broussolle and J.L. Meliet (eds.). 2006. Physiologie et Médecine de la plongée. Ellipse, Paris, France.

Macdonald, A.G. 1984. The effects of pressure on the molecular structure and physiological functions of cell membrane. Phil. Trans. R. Soc. Lond. 304B: 47–68.

Miljkovic-Lolic, M. and R. Silbergleit, G. Fiskum and R.E. Rosenthal. 2003. Neuroprotective effects of hyperbaric oxygen treatment in experimental focal cerebral ischemia are associated with reduced brain leukocyte myeloperoxidase activity. Brain Res. 971(1): 90–94.

Mori, H. and M. Oikawa, M. Tamagami, T. Kumaki, H. Nakaune, R. Amaro, J. Akinaga, Y. Fukui, K. Abe and S. Urano. 2007. Oxidized proteins in astrocytes generated in a hyperbaric atmosphere induce neuronal apoptosis. J. Alzheimers Dis. 11(2): 165–174.

Narkowicz, C.K. and J.H. Vial and P.W. McCartney. 1993. Hyperbaric oxygen therapy increases free radical levels in the blood of humans. Free Radic. Res. Com. 19(2): 71–80.

Obermüller, B. and U. Karsten and D. Abele. 2005. Response of oxidative stress parameters and sunscreening compounds in Arctic amphipods during esperimental exposure to maximal natural UVB radiation. J. Exp. Mar. Biol. Ecol. 37: 100–117.

Omoi, N.O. and M. Arai, M. Saito, H. Takatsu, A. Shibata, K. Fukuzawa, K. Sato, K. Abe, K. Fukui and S. Urano. 2006. Influence of oxidative stress on fusion of pre-synaptic plasma membranes of the rat brain with phosphatidyl choline liposomes, and protective effect of vitamin E.J. Nutr. Sci. Vitaminol. (Tokyo) 52(4): 248–255.

Pablos, M.I. and R.J. Reiter, J.I. Chuang, G.G. Ortiz, J.M. Guerrero, E. Sewerynek, M.T. Agapito, D. Melchiorri, R. Lawrence and S.M. Deneke. 1997. Acutely administered melatonin reduces oxidative damage in lung and brain induced by hyperbaric oxygen. J. Appl. Physiol. 83: 354–358.

Pankhurst, N.W. 1982a. Changes in body musculature with sexual maturation in the European eel *Anguilla anguilla* (L.). J. Fish Biol. 21: 417–428.

Pankhurst, N.W. 1982b. Relation of visual changes to the onset of sexual maturation in the European eel *Anguilla anguilla* (L.). J. Fish Biol. 21: 127–140.

Prentice, H.M. and S.L. Milton, D. Scheurle and P.L. Lutz. 2004. The upregulation of cognate and inducible heat shock proteins in the anoxic turtle brain. J. Cereb. Blood Flow Metab. 24: 826–828.

Ramaglia, V. and L.T. Buck. 2004. Time-dependent expression of heat shock proteins 70 and 90 in tissues of the anoxic western painted turtle. J. Exp. Biol. 207: 3775–3784.

Reed, J.Z. and P.J. Butler and M.A. Fedak. 1994. The metabolic characteristics of the locomotory muscles of the grey seals (Halichoerus grypus), harbour seals (Phoca vitulina) and Antarctic fur seals (Arctocephalus gazella). J. Exp. Biol. 194: 33–46.

Rees, J.F. and B. de Wergifosse, O. Noiset, M. Dubuisson, B. Janssens and E.M. Thompson. 1998. The origins of marine bioluminescence: turning oxygen defence mechanisms into deep-sea communication tools. J. Exp. Biol. 201: 1211–1221.

Reid, M.B. 1996. Reactive oxygen and nitric oxide in skeletal muscle. News Physiol. Sci. 11: 114–119.

Reiter, R.J. and D.X. Tan. 2003. Melatonin: a novel protective agent against oxidative injury of the ischemic/reperfused heart. Cardiovasc. Res. 58: 10–19.

Rimbach, G. and D. Hohler, A. Fischer, S. Roy, F. Virgili, J. Pallauf and L. Packer 1999. Methods to assess free radicals and oxidative stress in biological systems. Arch Tierernahr 52: 203–222.

Rothfuss, A. and G. Speit. 2002a. Overexpression of heme oxygenase-1 (HO-1) in V79 cells results in increased resistance to hyperbaric oxygen (HBO)—induced DNA damage. Environ Mol. Mutagen. 40(4): 258–265.

Rothfussn, A. and G. Speit. 2002b. Investigations on the mechanism of hyperbaric oxygen (HBO)-induced adaptive protection against oxidative stress. Mutat. Res. 508: 157–165.

Rucci, F.S. and M.L. Giretti and M. La Rocca. 1967. Changes in electrical activity of the cerebral cortex and of some subcortical centers in hyperbaric oxygen. Electroencephalogr. Clin. Neurophysiol.22(3): 231–238.

Sachdev, S. and K.J. Davies. 2008. Production, detection, and adaptive responses to free radicals in exercise. Free Radic. Biol. Med. 44: 215–223.

Saran, M. and W. Bors. 1990. Radical reactions *in vivo*—an overview. Radiat. Environ. Biophys. 29: 249–262.

Scaion, D. and M. Belhomme and P. Sébert. 2008. Pressure and temperature interactions on aerobic metabolism of migrating European silver eel. Respir. Physiol. Neurobiol. 164: 319–322.

Schafer, F.Q. and G.R. Buettner. 2001. Redox environment of the cell as viewed through the redox state of the glutathione disulfide/glutathione couple. Free Radic. Biol. Med. 30: 1191–1212.

Scott, M.A. and M. Locke and L.T. Buck. 2003. Tissue-specific expression of inducible and constitutive Hsp70 isoforms in the western painted turtle. J. Exp. Biol. 206: 303–311.

Sébert, P. and M. Theron. 2001. Why can the eel, unlike the trout, migrate under pressure? Mitochondrion 1: 79–85.

Sébert, P. and J.F. Menez, B. Simon and L. Barthélémy. 1995. Effects of hydrostatic pressure on malondialdehyde brain contents in yellow freshwater eels. Redox Report 1: 379–382.

Semenza, G.L. and N.R. Prabhakar. 2007. HIF-1—dependent respiratory, cardiovascular, and redox responses to chronic intermittent hypoxia. Antioxid. Redox. Signal 9: 1391–1396.

Shiloh, M.U. and J.D. MacMicking, S. Nicholson, J.E. Brause, S. Potter, M. Marino, F. Fang, M. Dinauer and C. Nathan. 1999. Phenotype of mice and macrophages deficient in both phagocyte oxidase and inducible nitric oxide synthase. Immunity 10: 29–38.

Sies, H. Oxidative Stress: Introduction. *In*: H. Sies (ed.). 1991. Oxidative Stress: Oxidants and Antioxidants. Academic Press, London.

Simon, B. and P. Sébert, C. Cann-Moisan and L. Barthélémy. 1992. Muscle energetics in yellow fresh water eel (*Anguilla anguilla* L.) exposed to high hydrostatic pressure (101 ATA) for 30 days. Comp. Biochem. Physiol. 102B: 205–208.

Singh, S.N. and P. Vats, M.M.L. Kumria, S. Ranganathan, R. Shyam, M.P. Arora, C.L. Jain and K. Sridharan. 2001. Effect of high altitude (7,620 m) exposure on glutathione and related metabolism in rats. Eur. J. Appl. Physiol. 84: 233–237.

Sjödin, B. and Y.N. Westing and F. Apple. 1990. Biochemical mechanisms for oxygen free radical formation during exercise. Sports Med. 10: 236–254.

Skulachev, V.P. 1996. Role of uncoupled and non-coupled oxidations in maintenance of safely low levels of oxygen and its one-electron reductants. Q. Rev. Biophys. 29: 169–202.

Somero, G.N. 1991. Hydrostatic pressure and adaptation to the deep sea. Pp. 169–202. *In*: Prosser (ed.). Environmental and Metabolic Animal Physiology. Comparative animal physiology.

Stadtman, E.R. and C.N. Oliver. 1991. Metal-catalyzed oxidation of proteins. Physiological consequences. J. Biol. Chem. 266: 2005–2008.

Sterba, J.A. and C.E.G. Lundgren. 1988. Breath-hold duration in man and the diving response induced by face immersion. Undersea Biomed. Res. 15: 361–375.

Stroka, D.M. and T. Burkhardt, I. Desbaillets, R.H. Wenger, D.A. Neil, C. Bauer, M. Gassmann and D. Candinas. 2001. HIF-1 is expressed in normoxic tissue and displays an organ-specific regulation under systemic hypoxia. FASEB J. 15: 2445–2453.

Speit, G. and C. Dennog, U. Eichhorn, A. Rothfuss and B. Kaina. 2000.Induction of heme oxygenase-1 and adaptive protection against the induction of DNA damage after hyperbaric oxygen treatment. Carcinogenesis 21(10): 1795–1799

Tan, D.X. and L.C. Manchester, R.M. Sainz, J.C. Mayo, J. Leon and R.J. Reiter. 2005. Physiological ischemia/reperfusion phenomena and their relation to endogenous melatonin production: a hypothesis. Endocrine 27: 149–158.

Tesch, F.W. 2003. The Eel. 5th Edit. J.E. Thorpe Oxford.

Theron, M. and F. Guerrero and P. Sébert. 2000. Improvement in the efficiency of oxidative phosphorylation in the freshwater eel acclimated to 10.1 MPa hydrostatic pressure. J. Exp. Biol. 203 Pt 19: 3019–3023.

Thom, S.R. and D. Fisher, J. Zhang, V.M. Bhopale, S.T. Ohnishi, Y. Kotake, T. Ohnishi and D.G. Buerk. 2003. Stimulation of perivascular nitric oxide synthesis by oxygen. Am. J. Physiol. Heart Circ. Physiol. 284(4): 1230–1239.

Torbati, D. and A.J. Simon and A. Ranade. 1981. Frequency analysis of EEG in rats during the preconvulsive period of O2 poisoning. Aviat. Space Environ. Med. 52: 598–603.

Turrens, J.F. 2003. Mitochondrial formation of reactive oxygen species. J. Physiol. 552: 335–344.

Valko, M. and D. Leibfritz, J. Moncol, M.T. Cronin, M. Mazur and J. Telser, 2007. Free radicals and antioxidants in normal physiological functions and human disease. Int. J. Biochem. Cell Biol. 39: 44–84.

Valdivia, P.A. and T. Zenteno-Savin, S.C. Gardner and A.A. Aguirre. 2007. Basic oxidative stress metabolites in eastern Pacific green turtles (Chelonia mydas agassizzii). Comp. Biochem. Physiol. C. Toxicol. Pharmacol. 146: 111–117.

van den Thillart, G. and V. van Ginneken, F. Korner, R. Heijmans, R. Van der Linden and A. Gluvers. 2004. Endurance swimming of European eel. J. Fish Biol. 65: 312–318.

van Ginneken, V. and G.E. Maes. 2005. The European eel (*Anguilla anguilla*, Linnaeus), its lifecycle, evolution and reproduction: a literature review. Rev. Fish Biol. Fisheries 15: 367–398.

Vasquez-Medina, J.P. and T. Zenteno-Savin and R. Elsner. 2006. Antioxidant enzymes in ringed seal tissues: potential protection against dive-associated ischemia/reperfusion. Comp. Biochem. Physiol. C. Toxicol. Pharmacol. 142: 198–204.

Vettier, A. and A. Amerand, C. Cann-Moisan and P. Sebert. 2005. Is the silvering process similar to the effects of pressure acclimatization on yellow eels? Respir. Physiol. Neurobiol. 145: 243–250.

Vettier, A. and C. Labbé, A. Amérand, G. Da Silva, E. Le Rumeur, C. Cann-Moisan and P. Sébert. 2006. Effects of hydrostatic pressure on eel mitochondrial metabolism and membrane fluidity. Undersea Hyperb. Med. 33: 149–156.

Vignais, P.V. 2002. The superoxide-generating NADPH oxidase: structural aspects and activation mechanism. Cell. Mol. Life Sci. 59: 1428–1459.

Vij, A.G. and R. Dutta and N.K. Satija. 2005. Acclimatization to oxidative stress at high altitude. High Alt. Med. Biol. 6: 301–310.

Whitam, M. and M.B. Fortes. 2008. Heat shock protein 72: release and biological significance during exercise. Frontiers in Biosci. 13: 1328–1339.

Wood, J.D. and W.J. Watson. 1963. Gamma-aminobutyric acid levels in the brain of rats exposed to oxygen at high pressures. Can. J. Biochem. Physiol. 41: 1907–1913.

Yusa, T. and J.S. Beckman, J.D. Crapo and B.A. Freeman. 1987. Hyperoxia increases H2O2 production by brain in vivo. J Appl. Physiol. 63(1): 353–358.

Zhang, X. and L. Xiong, W. Hu, Y. Zheng, Z. Zhu, Y. Liu, S. Chen and X.Wang. 2004. Preconditioning with prolonged oxygen exposure induces ischemic tolerance in the brain via oxygen free radical formation. Can. J. Anaesth. 51(3): 201–205.

# PRESSURE AND
# LIVING ORGANISMS

# 12

# Piezophilic Prokaryotes

*D. Prieur,[1] D. Bartlett,[2] C. Kato,[3] Ph. Oger[4] and M. Jebbar[1]*

## INTRODUCTION AND A BRIEF HISTORY

Among the physical parameters to which living organisms are exposed, hydrostatic pressure is probably the least studied. However, most of the biosphere (in terms of volume) is exposed to elevated pressures that exist in the deep oceans, the deep aquifers and oil reservoirs or the deep subseafloor sediments. It has been recently estimated that the majority of Earth Prokaryotes are living in the deep surface (Whitman et al., 1998). In other words, the majority of Earth Prokaryotes live under elevated pressures, and this parameter cannot be ignored anymore by microbiologists.

Until recently, only the deep oceans were known to harbor piezotolerant or piezophilic prokaryotes. With an average depth of 3800m (maximum 10790 m in the Marianna Trench), and a coverage of Earth surface at 70%, it was estimated (Jannasch and Taylor, 1984) that the deep oceans represented about 62% of the biosphere.

[1] Laboratoire de Microbiologie des Environnements extrêmes, UMR 6197 (UBO, Ifremer, CNRS), Université de Brest, France.
E-mail: daniel.prieur@univ-brest.fr; mohamed.jebbar@univ-brest.fr

[2] 8750 Biological Grade, 4405 Hubbs Hall, Marine Biology Research Division, Center for Marine Biotechnology and Biomedicine, Scripps Institution of Oceanography, University of California, San Diego 92093-0202, La Jolla California, U.S.A.
E-mail: dbartlett@ucsd.edu

[3] Extremobiosphere Research Center, Japan Agency for Marine-Earth Science and Technology, 2-15 Natsushima-cho, Yokosuka 237-0061, Japan.
E-mail: kato_chi@jamstec.go.jp

[4] Laboratoire de Science de la Terre, UMR 5570 (CNRS-ENSL-UCBL), Université de Lyon, F-69364 Lyon, France.
E-mail: poger@ens-lyon.fr

The first records for deep-sea bacteria were given by Certes (1884) who cultivated micro-organisms from water samples collected at depths down to 5000 m during cruises of the ships « Le Travailleur » and « Talisman » in 1882–1883. Although the experiments were not performed under hydrostatic pressure, this report opened the way for deep-sea microbiologists. However this way was not so easy and depended on specific sophisticated tools for collecting samples at depth and processing them under pressure for bacteria cultivation.

The first questions addressed in the second half of the XXth century concerned the degradation of organic matter under low temperature and high pressure conditions. Zobell, Morita, Jannasch (see Prieur and Marteinsson, 1998 for review) and others tackled this problem and concluded after hundreds of experiments that heterotrophic activities of marine bacteria isolated from both surface and deep waters were considerably retarded under these conditions, and that probably no free-living bacteria adapted to deep-sea conditions existed. Schwarz, Colwell, Deming and especially Yayanos (see Prieur and Marteinsson, 1998 for review) demonstrated that deep-sea piezophilic bacteria did exist, but inhabited the digestive tracts of deep-sea invertebrates.

Thirty years ago, unexpected habitats were discovered in the deep-sea: hydrothermal vents, and the fascinating black smokers, venting fluids at temperatures up to 350–400°C. Many hyperthermophilic organisms (growing optimally at temperatures above 80°C) were isolated and some of them appeared to be piezotolerant, piezophiles and even obligate piezophiles. Still more recently, living bacteria were detected in a variety of subsurface environments, where elevated hydrostatic and lithostatic pressures are combined. This chapter will focus on recent data about this fascinating but still poorly explored world.

## PIEZOPHILIC PROKARYOTES FROM THE COLD DEEP OCEANS

Bacteria living in the deep-sea have several unusual features, which allow them to thrive in their extreme environments; ca. low temperature and high pressure. Several piezophilic and piezotolerant bacteria have been isolated and characterized from cold deep-sea sediments at depths ranging from 2,500 m to 11,000 m using sterilized sediment samplers by means of the submersibles *SHINKAI 6500* and *KAIKO* systems operated by the Japan Agency for Marine-Earth Science and Technology (Kato, 1999; Kato et al., 1995; 2004). Most isolated strains are not only piezophilic but also psychrophilic and cannot be cultured at temperatures higher than 20°C, known as 'Psychro-piezophiles'.

The isolated deep-sea piezophilic bacterial strains have been characterized in an effort to understand the interaction between the deep-sea environment and its microbial inhabitants (Kato et al., 1998; Margesin and Nogi, 2004; Yayanos et al., 1979). Thus far, all psychro-piezophilic bacterial isolates fall into the gamma-subgroup of the Proteobacteria according to phylogenetic classifications based on 5S and 16S ribosomal RNA gene sequence information (DeLong et al., 1997; Kato, 1999; Margesin and Nogi, 2004). DeLong et al. (1997) reported that the cultivated psychrophilic and piezophilic deep-sea bacteria were affiliated with one of five genera within the gamma-Proteobacteria subgroup: *Shewanella*, *Photobacterium*, *Colwellia*, *Moritella*, and an unidentified genus. The only deep-sea piezophilic bacterial species of these genera were called *Shewanella benthica* (Deming et al., 1984; MacDonell and Colwell, 1985) in the genus *Shewanella*, and *Colwellia hadaliensis* (Deming et al., 1988) in the genus *Colwellia*, at that time. Several novel piezophilic species have been identified within these genera based on the results of chromosomal DNA–DNA hybridization studies and several other taxonomic properties in the last decade. Both previously described and novel species of bacteria have been identified among the piezophilic bacterial isolates. Based upon these studies we have indicated that cultivated psychrophilic and piezophilic deep-sea bacteria could be affiliated with one of five genera within the gamma-Proteobacteria subgroup: *Shewanella*, *Photobacterium*, *Colwellia*, *Moritella*, and *Psychromonas* which was formerly classified as 'an unidentified genus' (Nogi et al., 2002). Figure 1 shows the phylogenetic relations between the taxonomically identified piezophilic species (shown in bold) and other bacteria within the gamma-Proteobacteria subgroup. The taxonomic features of the piezophilic genera were determined as described below.

**The genus *Shewanella***

Members of the genus *Shewanella* are not unique to marine environments of Gram-negative, aerobic and facultatively anaerobic gamma-Proteobacteria (MacDonell and Colwell, 1985). The type strain of this genus is *Shewanella putrefaciens*, which is a bacterium formerly known as *Pseudomonas putrefaciens* (MacDonell and Colwell, 1985; Owen et al., 1978). Recently, however, several novel marine *Shewanella* species have been isolated and described. These isolates are not piezophilic species and thus prior to the present report *S. benthica* and *S. violacea* were the only known members of the genus *Shewanella* showing piezophilic growth properties (Nogi et al., 1998b). *Shewanella* piezophilic strains, PT-99, DB5501, DB6101, DB6705, and DB6906, DB172F, DB172R, and DB21MT-2 were all identified as members of the same species, *S. benthica* (Nogi et al., 1998b; Kato and Nogi, 2001). The psychrophilic and piezophilic *Shewanella* strains, including *S. violacea* and *S. benthica*, produce eicosapentaenoic acid (EPA) and thus the production of such long-chain polyunsaturated fatty acid (PUFA) is a property shared by many deep-sea

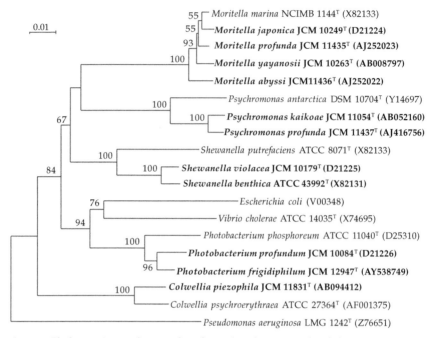

**Figure 1.** Phylogenetic tree showing the relationships between isolated deep-sea piezophilic bacteria (in bold) within the gamma-Proteobacteria subgroup determined by comparing 16S rRNA gene sequences using the neighbor-joining method (references for species description are indicated in the text). The scale represents the average number of nucleotide substitutions per site. Bootstrap values (%) are shown for frequencies above the threshold of 50%.

bacteria to maintain the cell-membrane fluidity under conditions of extreme cold and high hydrostatic pressure environments (Fang et al., 2003). *S. violacea* strain DSS12 has been studied extensively, particularly with respect to its molecular mechanisms of adaptation to high-pressure (Kato et al., 2000; Nakasone et al., 1998; 2002). This strain is moderately piezophilic, with a fairly constant doubling time at pressures between 0.1 MPa and 70 MPa, whereas the doubling times of most piezophilic *S. benthica* strains change substantially with increasing pressure. As there are few differences in the growth characteristics of strain DSS12 under different pressure conditions, this strain is a very convenient deep-sea bacterium for use in studies on the mechanisms of adaptation to high-pressure environments. Therefore, the genome analysis on strain DSS12 has been performed as a model deep-sea piezophilic bacterium (Nakasone et al., 2003).

## The genus *Photobacterium*

The genus *Photobacterium* was one of the earliest known bacterial taxa and was first proposed by Beijerinck in 1889. Phylogenetic analysis based on 16S rRNA gene sequences has shown that the genus *Photobacterium* falls within the gamma-Proteobacteria and, in particular, is closely related to the genus *Vibrio* (Nogi et al., 1998c). *Photobacterium profundum*, a novel species, was identified through studies of the moderately piezophilic strains DSJ4 and SS9 (Nogi et al., 1998c). *P. profundum* strain SS9 has been extensively studied with regard to the molecular mechanisms of pressure regulation (Bartlett, 1999) and subsequently genome sequencing and expression analysis (Vezzi et al., 2005). Recently, *P. frigidiphilum* was reported to be slightly piezophily: its optimal pressure for growth is 10 MPa (Seo et al., 2005). Thus, *P. profundum* and *P. frigidiphilum* are the only species within the genus *Photobacterium* known to display piezophily and the only two known to produce the long-chain polyunsaturated fatty acid (PUFA), eicosapentaenoic acid (EPA). No other known species of *Photobacterium* produces EPA (Nogi et al., 1998c).

## The genus *Colwellia*

Species of the genus *Colwellia* are defined as facultative anaerobic and psychrophilic bacteria (Deming et al., 1988), which belong to the gamma-Proteobacteria. In the genus *Colwellia*, the only deep-sea piezophilic species reported was *C. hadaliensis* strain BNL-1 (Deming et al., 1988), although no public culture collections maintain this species and/or its 16S rRNA gene sequence information. Bowman et al. (1998) reported that *Colwellia* species produce the long-chain PUFA, docosahexaenoic acid (DHA). We have recently isolated the obligately piezophilic strain Y223G$^T$ from the sediment at the bottom of the deep-sea fissure of the Japan Trench, which was identified as *C. piezophila* (Nogi et al., 2004). Regarding fatty acids, this strain did not produce EPA or DHA in the membrane layer, whereas high levels of unsaturated fatty acids (16:1 fatty acids) were produced. This observation suggested that the possession of long-chain PUFA should not be requirement for classification as a piezophilic bacterium; however, the production of unsaturated fatty acids could be a common property of piezophiles.

## The genus *Moritella*

The type strain of the genus *Moritella* is *Moritella marina*, previously known as *Vibrio marinus* (Colwell and Morita, 1964), and is one of the most common psychrophilic organisms isolated from marine environments. However, *V. marinus* has been reclassified as *M. marina* gen. nov. comb. nov. (Urakawa et al., 1998). *M. marina* is closely related to the genus *Shewanella* on the basis of 16S rRNA gene sequence data and is not a piezophilic bacterium. Strain DSK1, a moderately piezophilic bacterium isolated from the Japan Trench, was identified as *Moritella japonica* (Nogi et al., 1998a). This was the first piezophilic species identified in the genus *Moritella*. Production of the long-chain PUFA, DHA, is a characteristic property of the genus *Moritella*. The extremely piezophilic bacterial strain DB21MT-5 isolated from the world's deepest sea bottom, the Mariana Trench Challenger Deep at a depth of 10,898 m, was also identified as a *Moritella* species and designated *M. yayanosii* (Nogi and Kato, 1999). The optimal pressure for the growth of *M. yayanosii* strain DB21MT-5 is 80 MPa; this strain is unable to grow at pressures of less than 50 MPa but grows well at pressures as high as 100 MPa (Kato et al., 1998). The fatty acid composition of piezophilic strains changes as a function of pressure and in general greater amounts of PUFAs are synthesized at higher growth pressures. Approximately 70% of the membrane lipids in *M. yayanosii* are unsaturated fatty acids, which is a finding consistent with its adaptation to very high pressures (Nogi and Kato, 1999; Fang et al., 2000). Two other species of the genus *Moritella*, *M. abyssi* and *M. profunda*, were isolated from a depth of 2,815 m off the West African coast (Xu et al., 2003a); they are moderately piezophilic and the growth properties are similar to *M. japonica*.

### The genus *Psychromonas*

The genus *Psychromonas* described as psychrophilic bacterium, which also belongs to the gamma-Proteobacteria, is closely related to the genera *Shewanella* and *Moritella* on the basis of 16S rRNA gene sequence data. The type species of the genus *Psychromonas*, *Psychromonas antarctica*, was isolated as an aerotolerant anaerobic bacterium from a high-salinity pond on the McMurdo ice-shelf in Antarctica (Mountfort et al., 1998). This strain did not display piezophilic properties. *Psychromonas kaikoae*, isolated from sediment collected from the deepest cold-seep environment with chemosynthesis-based animal communities within the Japan Trench at a depth of 7,434 m, is a novel obligatory piezophilic bacterium (Nogi et al., 2002). The optimal temperature and pressure for growth of *P. kaikoae* are 10°C and 50 MPa, respectively, and both PUFAs, EPA and DHA, are produced in the membrane layer. *P. antarctica* does not produce either EPA or DHA in its membrane layer. DeLong and co-workers stated that strain CNPT-3 belonged to an unidentified genus of piezophiles (DeLong et al., 1997) and this strain proved to be closely related to *P. kaikoae*. Thus, the genus *Psychromonas* is the fifth genus reported to contain piezophilic species within the gamma-Proteobacteria. In addition, *P. profunda* is a moderately piezophilic bacterium isolated from deep Atlantic sediments at a depth of 2,770 m (Xu et al., 2003b).

This strain is similar to the piezo-sensitive strain *P. marina*, which also produces small amounts of DHA. Only *P. kaikoae* produces both EPA and DHA in the genus *Psychromonas*.

The piezophilic and psychrophilic *Shewanella* and *Photobacterium* strains produce EPA (Nogi et al., 1998b, 1998c), *Moritella* strains produce DHA (Nogi et al., 1998a; Nogi and Kato, 1999), and *Psychromonas kaikoae* produces both EPA and DHA (Nogi et al., 2002) but *Colwellia piezophila* does not produce such PUFAs (Nogi et al., 2004). The fatty acid composition of piezophilic strains changes as a function of pressure, and in general greater amounts of PUFAs are synthesized under higher-pressure conditions for their growth (DeLong and Yayanos, 1985, 1986). The fatty acid composition of piezophilic strains are distinct depending on their genus and usually high amounts of unsaturated fatty acids (49~71%) are involved in their membrane layer. The importance of unsaturated fatty acids in the growth of piezophiles is further described below in the section on membrane structure.

Most *Shewanella* sp. are isolated from ocean environments and some are psychrophilic or psychrotrophic bacteria. The piezophilic *Shewanella* species *S. benthica* and *S. violacea* are also categorized as psychrophilic at atmospheric pressure (Nogi et al., 1998b). *S. gelidimarina* and *S. frigidimarina* isolated from Antarctic ice (Bowman et al., 1997) and *S. hanedai* isolated from the Arctic Ocean (Jensen et al., 1980) are cold-adapted psychrotrophic bacteria that grow well at low temperature. A phylogenetic tree of these *Shewanella* species within the gamma-Proteobacteria subgroup constructed based on 16S rRNA gene sequences is shown in Fig. 2.

In this tree, two major branches are recognizable in the genus *Shewanella*, indicated by *Shewanella* group 1 and *Shewanella* group 2. Deep-sea *Shewanella* forming the *Shewanella* piezophilic branch are categorized as members of group 1. Interestingly, most *Shewanella* species shown to be psychrophilic or psychrotrophic also belong to group 1. The other species in group 1, *S. pealeana* and *S. woodyi*, isolated from ocean squid and detritus, respectively, grow optimally at 25°C (Leonardo et al., 1999; Makemson et al., 1997) and thus these strains might also be included in the group of cold-adapted bacteria. Most *Shewanella* species in group 2 are not cold-adapted bacteria. They grow well under mesophilic conditions at 25–35°C. *S. frigidimarina*, which can grow optimally below 25°C, is the only exception in this category, although this species belongs to group 2 (Kato and Nogi, 2001).

The growth of some of these *Shewanella* species under high-pressure conditions indicates that the members of *Shewanella* group 1 show piezophilic (*S. benthica* and *S. violacea*) or piezotolerant (*S. gelidimarina* and *S. hanedai*) growth properties, although the members of *Shewanella* group 2 generally show piezosensitive growth, i.e., no growth at a pressure of 50 MPa (Kato and Nogi, 2001). Only a limited number of experiments have been performed examining the growth of these bacteria under high-pressure conditions, but

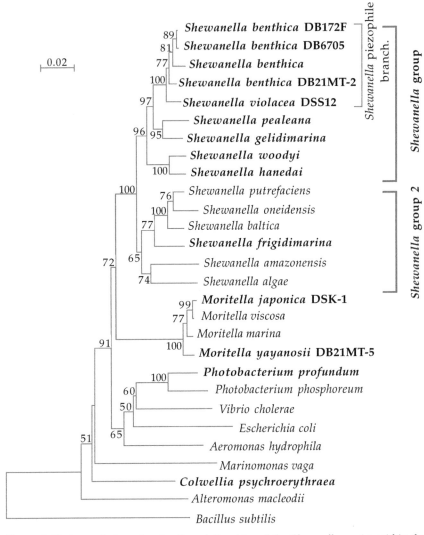

**Figure 2.** Phylogenetic tree showing the relationships of the *Shewanella* species within the gamma-Proteobacteria subgroup constructed based on 16S rRNA gene sequences with the neighbor-joining method. The scale represents the average number of nucleotide substitutions per site. Bootstrap values (%) were calculated from 1,000 trees. Psychrophilic and/or piezophilic bacteria are shown in bold.

generally members of *Shewanella* group 1 are characterized as cold adapted and pressure tolerant, whereas the members of *Shewanella* group 2 are mostly mesophilic and pressure sensitive. Some *Shewanella* species are known to produce PUFAs, particularly EPA. It is clear that the members of *Shewanella* group 1 produce substantial amounts of EPA (11–16% of total fatty acids), whereas members of *Shewanella* group 2 produce no EPA or only limited amounts. In terms of other fatty acids, the membrane lipid profiles of members of the genus *Shewanella* are basically similar. This observation also supports the view described above (Kato and Nogi, 2001).

On the basis of the properties of *Shewanella* species, we would like to propose that two major branches of the genus *Shewanella* should be recognized taxonomically, *Shewanella* group 1 and group 2 (Fig. 5). The two subgenus branches of *Shewanella* would be as follows: *Shewanella* group 1 is characterized as a group of high-pressure, cold-adapted species that produce substantial amounts of EPA and *Shewanella* group 2 is characterized as a group of mostly mesophilic and pressure-sensitive species.

The deep-sea bottom and other cold-temperature environments are probably similar in terms of microbial diversity. Members of *Shewanella* group 1 live in such environments, and most of them show piezophilic or piezotolerant growth properties. In this regard, it is interesting to consider the influence of the ocean circulation as deep ocean water is derived from polar ice (in the Arctic and/or Antarctic regions) that sinks to the deep-sea bottom (Schmitz, 1995), probably along with microbes. It was reported that *Psychrobacter pacificensis* isolated from seawater of the Japan Trench at a depth of 5,000–6,000 m was taxonomically similar to the Antarctic isolates *Psychrobacter immobilis*, *Psychrobacter gracincola*, and *Psychrobacter fridigicola* (Maruyama et al., 2000). The occurrence of *Psychrobacter* in cold seawater deep in the Japan Trench and at the surface of the Antarctic sea suggests that bacterial habitation of the deep-sea and their evolution have been influenced by the global deep ocean circulation linked to the sinking of cooled seawater in polar regions. Thus, it is possible that the ocean circulation may be one of the major factors influencing microbial diversity on our planet.

## ADAPTATIONS OF PIEZOPHILES DERIVED FROM GENOMICS AND GENETICS

The first deep-sea microbe adapted to low temperature and high pressure to be sequenced was *Photobacterium profundum* strain SS9, which was reported in 2005 . More recently, in 2008, genome analysis was completed on a deep-sea microbe which is tolerant to low temperature and high pressure, *Shewanella piezotolerans* strain WP3. Various degrees of whole genome sequence information are also available for psychrophilic or psychrotolerant piezophiles *Shewanella* sp. KT99, *Psychromonas* sp. CNPT3, *Moritella* sp. PE36 and *Carnobacterium* sp. AT7. The temperature and pressure growth optima of these bacteria along with the location and nature of their isolation is presented in Table 1.

Several general observations can now be made about the genome attributes of these organisms. The genome sizes of the two completely sequenced deep-sea bacteria indicates that their genome sizes are larger than those of other microbes within their genus affiliations. However, the remaining piezophile genome sequences suggest that this trend is not a universal property, and it may rather reflect the broad environmental adaptations of both *Photobacterium profundum* strain SS9 and of *Shewanella piezotolerans* WP3. All of the piezophiles thus far examined have large ratios of rRNA operons to genome size, with the most dramatic example being *Photobacterium profundum* strain SS9 which has the largest number of rRNA operons present in any member of the Bacteria or Archaea at 15. The piezotolerant deep-sea isolate *S. piezotolerans* WP-3 bucks this trend and only contains eight operons. Large numbers of rRNA operons are typically associated with responsiveness to nutrient shifts, a trait which could be adaptive in the deep sea where a large fraction of nutrients can be present in the form of sinking particulate organic carbon. An alternate possibility is that large numbers of rRNA operons and resulting increases in rRNA levels facilitate ribonucleoprotein assembly at high pressure. All of the sequenced piezophiles also have large average intergenic distances, on the order 150–200 bp. The reason for this is unknown, but could reflect modifications in gene regulation or chromosome organization.

**Table 1.** Low-temperature, high-pressure-adapted microbes for which genome sequence information is available.

| Strain | Phylogeny | Source | Pressure optimum | Temperature optimum |
|---|---|---|---|---|
| *Photobacterium profundum* SS9 | Gammaproteobacteria Vibrionales | Sulu Sea amphipod | 28 MPa | 15°C |
| *Shewanella piezotolerans* WP3 | Gammaproteobacteria Alteromonadales | Western Pacific Ocean sediment | 0.1–20 MPa | 15–20°C |
| *Shewanella benthica* KT99 | Gammaproteobacteria Alteromonadales | Kermadec Trench amphipod | ~98 MPa | >2°C |
| *Psychromonas* species strain CNPT3 | Gammaproteobacteria Alteromonadales | Central North Pacific Ocean amphipod | 52 MPa | 12°C |
| *Moritella* species strain PE36 | Gammaproteobacteria Alteromonadales | Patton Escarpment amphipod trap water | 41 MPa | 10°C |
| *Carnobacterium* species strain AT7 | Bacilli Lactobacillales | Aleution Trench water sample | 20 MPa | 20°C |

It is also of value to note a category of genes which are not present in deep-sea low-temperature and high pressure-adapted microbes, namely deoxyribodipyrimidine photolyase genes. These light-activated DNA repair enzymes are not present in autochthonous dark ocean residents. While this could seem like a trivial observation, it does have some relevance. Many surface-derived microbes are delivered into deep regions via downwelling and sinking matter, where they may survive for long periods of time. The presence or absence of photolyase genes can thus provide additional useful information to the actual depth preference (bathytype) of the microbe in question.

COG (Clusters of Orthologous Groups, ) analyses of shallow- and deep-sea bacteria indicate a tendency in the latter for fewer genes involved in transcription (category K), and more genes involved translation and ribosome structure (J), DNA replication and repair (category L), cell wall/membrane/envelope biogenesis (category M), motility (category N) and intracellular trafficking, secretion and vesicular transport (category U) (Fig. 3 ). In the case of the two moderately pressure-adapted microbes, SS9 and WP3, for which closed and finished sequence information is available, many duplicated genes exist. For SS9 this includes sequences associated with transposable elements, two complete operons for F1F0 ATP synthase and three complete sets of cytochrome cbb3 oxidase genes. WP3 contains an usually large number of decaheme cytochrome c reductase genes and those for outer membrane proteins likely to be involved in the reduction of insoluble substrates. It could be that versatility in electron acceptors is a feature of many deep-sea microbes. At the front end of metabolism many of the deep-sea microbes also seem to transport and catabolize a variety of substrates including a number of polymers such as chitin, cellulose and starch. At this time there is no evidence for a piezophile-specific gene.

Functional genomic studies have been undertaken for SS9. Microarray-based transcriptional profiling of gene expression at 0.1, 28 (optimum) and 45 MPa indicated a low pressure stress response in this organism as reflected by the upregulation of four different heat shock genes. In general genes whose products are associated with protein folding, glycolysis, amino acid or ion transport were upregulated at atmospheric pressure, whereas major shifts in certain fermentation and respiration genes were noted at higher pressures. Comparative genomics microarray hybridization experiments have also been performed for *P. profundum*, a species which includes both piezophilic and piezosensitive members. These experiments have indicated that piezosensitive *P. profundum* species lack a few genes which are present in the two piezophilic members examined, and these include a number of transporters. This transport connection is of particular interest because the membrane is the most compressible subcellular structure within a microbial cell, and because transport is the cellular process most influenced by pressure according to the SS9 transcriptional profiling data.

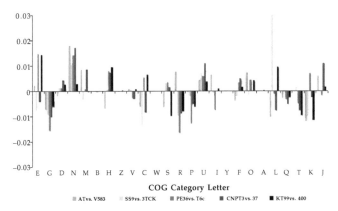

**Figure 3.** Biases in the distribution among major COG (Clusters of Orthologous Groups, (Tatusov et al., 2003) categories among low temperature-adapted, piezophilic bacteria. The plots present the normalized percent differences in COGs between *Carnobacterium* species strain AT7 and *Enterococcus faecalis* V583, *Photobacterium profundum* SS9 and *Photobacterium profundum* 3TCK, *Moritella* species strain PE36 and *Pseudoalteromonas atlantica* T6c, *Psychromonas* species strain CNPT3 and *Psychromonas ingrahamii* 37, and *Shewanella* species strain KT99 and *Shewanella figidimarina* NCIMB 400. The COG letters refer to the following: translation, ribosomal structure and biogenesis; J, RNA processing and modification; A, transcription; K, replication, recombination and repair; L, chromatin structure and dynamics; B, cell cycle control, cell division, chromosome partitioning; D, nuclear structure; Y, defense mechanisms; V, signal transduction mechanisms; T, cell wall/membrane/envelope biogenesis; M, cell motility; N, cytoskeleton; Z, extracellular structures; W, intracellular trafficking, secretion, and vesicular transport; U, posttranslational modification, protein turnover, chaperones; O, energy production and conversion; C, carbohydrate transport and metabolism; G, amino acid transport and metabolism; E, nucleotide transport and metabolism; F, coenzyme transport and metabolism; H, lipid transport and metabolism; I, inorganic ion transport and metabolism; P, secondary metabolites biosynthesis, transport and catabolism; Q, poorly characterized, general function prediction only; R, function unknown; S.

Transposon mutagenesis experiments have also been performed in SS9. 20,000 mutants were screened for cold and pressure sensitivity. The insertions within mutants possessing conditional phenotypes were mapped onto the existing genome sequence. Those genes whose disruptions were found to be associated with pressure-sensitive or pressure-enhanced growth are listed in Table 2. The functions affected were DNA replication, translation, membrane-associated processes, and metabolism. Remarkably, all of these are processes known to be pressure-sensitive from studies performed with mesophilic bacteria. The transposon mutagenesis data together and ongoing experimental work suggests that inhibitors of DNA replication inhibit growth at high pressure and activators of replication stimulate growth at high pressure (Gail Ferguson-Bradley et al., in preparation). Thus, DNA replication could be a key pressure point in SS9.

**Table 2.** *Photobacterium profundum* piezo-altered mutants.

| Structure or process affected | | Mutant phenotype |
| --- | --- | --- |
| **DNA** | | |
| *rctB* | chromsome II replication | Pressure-sensitive |
| *diaA* | chromosome replication | Pressure-sensitive |
| *seqA* | chromosome replication | Pressure-enhanced |
| *recD* | chromosome replication and repair | Pressure-sensitive |
| *hns* | nucleoid structure | Pressure-enhanced |
| **Ribosome** | | |
| *spoT* | stringent response | Pressure-sensitve |
| *suhB* | posttranscriptional gene control | Pressure-sensitive |
| **Membrane** | | |
| *rseC* | membrane protein assembly | Pressure-sensitive |
| *fabB* | unsaturated fatty acids | Pressure-sensitive |
| *fabF* | unsaturated fatty acids | Pressure-sensitive |
| **Metabolism** | | |
| Pyruvate kinase I | glycolysis | Pressure-sensitive |
| L-asparaginase I | a.a. metabolism | Pressure-sensitive |

## Membrane structure

A key distinction between the high temperature and the low temperature deep-sea microbes is that the latter experience greater ordering influences on their membranes. Increasing pressure raises the temperature of membrane phase transition from the gel state to the liquid crystalline state. The influence of temperature and pressure on this transition is such that a reduction of about 20°C is roughly equivalent to a pressure increase of 100 MPa.

Deep-sea bacteria and deep-sea animals are well known to modulate their fatty acid composition as a function of depth of isolation or pressure/temperature of cultivation. This is primarily achieved through changes in the degree of unsaturation. Other changes also occur, such as in lipid head groups, fatty acid lipid position and cis-trans isomerization. Which modifications lipids from archaeal species undergo in deep-sea environments is still largely unknown. Fatty acid unsaturation (the production of C-C double bonds) is likely to arise primarily during synthesis. But, most of the deep-sea bacterial genome sequences also indicate the presence of fatty acid desaturases, which could play a role in post synthesis unsaturation of fatty acids.

In addition to containing high levels of unsaturated fatty acids, most deep-sea bacterial membranes also contain long chain omega-3 polyunsaturated fatty acids (PUFAs) such as eicosapentaenoic acid (EPA, 20:5) or docosahexaenoic acid (DHA, 22:6). However, while most cold-adapted piezophiles produce omega-3-polyunsaturated fatty acids, all PUFA-producing microbes are not piezophiles. PUFAs are produced from biosynthetic pathways which are entirely distinct from those required for the production of saturated and monounsaturated fatty acids (MUFAs) and involves a polyketide biosynthetic process utilizing large multidomain proteins. Convincing evidence exists that MUFAs and PUFAs can be important for the growth of deep-sea microbes at high pressure. SS9 mutants defective in the production of MUFAs have been obtained following chemical or gene replacement mutagenesis. In the absence of an exogenous source of MUFAs these strains are highly pressure sensitive, depending on the amount of their reduction of monounsaturated fatty acids (Allen et al., 1999). EPA mutants in SS9 are not pressure sensitive, but EPA loss is accompanied by an increase in MUFAs in this species, and so it is possible that increased MUFAs can compensate for PUFA loss. In *Shewanella piezotolerans* WP3 an EPA mutant displays greatly reduced growth ability at both low temperature and high pressure .

Proteins within the membrane environment are also important for piezophiles. SS9 employs the inner membrane-spanning protein, ToxR as a pressure-responsive transcription factor. This protein is conserved among many bacteria within the *Vibrionaceae*. SS9 ToxR controls the expression of a number of genes, many of which themselves encode membrane proteins, including the *ompH* and *ompL* genes which encode outer membrane porins. The nature of the transmembrane domain segment of ToxR appears to determine its pressure-sensitivity in terms of transcriptional regulation (Rudi Vogel; personal communication). Another SS9 inner membrane protein, rseC, is required for growth at low temperature and high pressure. The RseC protein family has been proposed to promote the formation of membrane-associated complexes responsible for transfer of reducing power to key enzymes, such as those involved in nitrogen fixation in some microbes or for thiamine biosynthesis .

**Ribosome**

Ribosome stability and function are known to be pressure-sensitive in mesophilic bacteria. During the course of 16S rRNA gene sequence analyses as a part of phylogenetic analyses it was discovered that some piezophiles have small insertions of 9–12 bp which result in elongations to stems 10 and 11 (Escherichia coli positions 184–193 and 198–219, respectively). This includes rRNA genes derived from piezophilic members of the genera *Photobacterium, Shewanella* and *Colwellia*). Although these parts of the rRNA gene are hypervariable, elongations to stems 10 and 11 is an almost universal feature of piezophiles, at least among *Gammaproteobacteria*. Furthermore, semiquantitative PCR analyses indicate that the ratio of rRNA genes with longer to shorter stems roughly correlates with growth pressure optima. The basis of this correlation is unknown. Most microbiologists today appreciate the relevance of rRNA gene sequences in the context of phylogenetic investigations. However, small subunit rRNA possesses conserved and variable regions because of its critical roles in one of the universal processes of all of life; translation, including ribosome assembly, and translation initiation and elongation. The elongated stems interact with protein S20 of the small ribosomal subunit, a protein required for 70S ribosome assembly and translation initiation. In this regard it is interesting to note that piezophilic *Photobacterium* species also have small elongations in stem 49, another part of the 16S rRNA molecule that interacts with S20. The 16S rRNA molecule is also subject to base methylations and pseudouridylations. Deep-sea bacteria tend to have large numbers of pseudouridine synthase genes, but at present the most parsimonious explanation is that these are for ribosome assembly or function at low temperature, as many psychrotolerant and psychrophilic bacteria also have a large collection of these genes.

## Motility

Motility is one of the most pressure-sensitive activities known in mesophilic bacteria. Recent experiments employing a high pressure microscope to directly document bacterial swimming speeds as a function of pressure only serve to further reinforce this view (see also http://bartlettlab.ucsd.edu/ Motility%20at%20HP.html). In these experiments the mesophile *Escherichia coli* was only capable of swimming up to 50 MPa, but SS9 had a optimum pressure for swimming at 30 MPa and was capable of short-term swimming up to 150 MPa (above the know pressure limit for life). Remarkably, many of the sequenced moderately high pressure-adapted bacteria harbor two flagellar motility systems, one for swimming via a polar flagellum and another system for surface swarming through the use of lateral flagella (LAF). This includes *Photobacterium profundum* SS9, *Photobacterium profundum* DSK4, *Shewanella piezotolerans* WP3, and *Moritella* species strain PE36. The widespread use of LAF in deep-sea bacteria could reflect a greater reliance on surfaces. Many of the nutrients present at depth appear to come in the form of pulses of particulate organic carbon , and thus the ability to attach and migrate along particles may be particularly adaptive in this setting. Experiments with WP3 indicate that LAF in this organism are both induced at low temperature as well as needed for growth at low temperature. In the case of SS9 LAF transcriptional induction requires a functional polar flagellum, high pressure and high viscosity (a condition that mimics the physical influence of a surface on a bacterial cell). Why WP3 and SS9 appear to differentially regulate their LAF systems is unknown.

## Pressure sensing

Just as there appears to be no universal piezophile gene, there appears to be no universal piezophile pressure sensor. In SS9 the membrane-localized oligomeric transcription factor ToxR, together with its associated protein ToxS functions as a pressure sensor. FabF appears to directly or indirectly respond to pressure to adaptively control monounsaturated fatty acid biosynthesis, and even the polar flagellum could function in some way as a pressure-sensor for LAF regulation in this organism. The piezophile *Shewanella violaceae* DSS12 regulates the expression of the sigma 54-controlled gene *glnA* in response to pressure and this appears to be tied in to regulation of NtrC levels. WP3 induces genes for its polar flagellum at high pressure via a signal transduction process as yet unknown. It thus appears likely that deep-sea microbes have evolved many modifications of existing regulatory systems to better coordinate gene expression, protein abundance and enzyme activity with pressure. How microorganisms travel the kilometers of vertical distance required to experience substantial pressure differences is a matter of conjecture.

## Environmental genomics

Genome analyses of low temperature, deep-sea microbes has not completely relied on material obtained from pure cultures. Environmental (metagenomic) studies have been undertaken from a 3 km depth within the eastern Mediterranean and from a depth of 4 km within the North Pacific Subtropical Gyre. These and other culture-independent studies have highlighted how little is still known about microbial communites present at depth. Often large fractions of the planktonic community are dominated by members of the Rhizobiales within the *Alphaprotebacteria*, Planctomycetes, Acidobacteria, Chloroflexi, and Marine Group I Crenarchaeota, groups not present in any piezophile culture collections. Consistent with the above description genome analyses of various piezophilic *Gammaproteobacteria*, the environmental genomic projects uncovered disproportionately large numbers of genes associated with surface-associated growth such as those for polysaccharide and secondary metabolite biosynthesis, large numbers of transposase genes and large numbers of genes encoding enzymes involved in complex biopolymer and xenobiotic degradation, but in contrast to the pure culture genomics fewer motility genes were identified. The cultured piezophile representative also lack TRAP transporters, which utilize electrochemical gradients rather than ATP hydrolysis for substrate transport. This class of transporters is well represented at depth in both the deep Pacific and Mediterranean samples.

## THERMO-PIEZOPHILIC PROKARYOTES FROM DEEP-SEA HYDROTHERMAL VENTS

Deep-sea hydrothermal vent black smokers, combining elevated temperatures and pressures were discovered only in 1979 (Corliss et al., 1979). Most of the hyperthermophilic organisms from these environments were obtained in pure culture in the laboratory after enrichments performed at high temperature in liquid media that were slightly pressurized to avoid boiling when temperatures reached 100°C. Some of these organisms, all belonging to the Thermococcales, (strains GB-D, GB4, *Pyrococcus abyssi* GE5, ES1, ES4) were then exposed to elevated hydrostatic pressures and appeared to be at least barotolerant (growth was not affected by elevated pressures) but some were considered as piezophiles, since their growth rates were enhanced when grown under elevated pressures (Deming and Baross, 1993). Some could even grow at temperatures beyond (a few degrees) their optimal temperatures for growth at atmospheric pressure, only under pressurized conditions. It was noted in a review by Deming and Baross (1993) that for piezophilic strains, optimum pressure for growth was above pressure existing at locations where samples were collected.

Similar experiments were carried out with the deep-sea methanogen *Methanococcus jannaschii*, that grows on a $H_2/CO_2$ gaz mixture. Again, elevated pressures enhanced the growth rate, but also the methanogenic activity. For the highest temperatures tested, growth and methanogenesis were uncoupled, and only the latter was enhanced by elevated pressures (Miller et al., 1988).

A significant progress was accomplished by performing enrichment cultures and following subcultures under both elevated pressures and temperatures. From a sample collected at a depth of 3.5 km, on the Mid-Atlantic-Ridge, and preserved under pressure until laboratory microbiological work was performed, Marteinsson et al. (1997, 1999) isolated the hyperthermophilic archaeon *Thermococcus barophilus*. This anaerobic organism grows by fermentation of organic compounds in a temperature range of 75–95°C, with an optimum of 85°C, when cultivated under atmospheric pressure. Cultivation at elevated temperatures under hydrostatic pressure resulted in a shift of the maximum temperature for growth (95 to 100°C), and a significant increase of the growth rate. Optimum pressure for growth was determined at 40 MPa, which is above the pressure existing at the location of sample collection (34 MPa). Total cell proteins were extracted after cultures under atmospheric and hydrostatic pressures and their profiles determined by SDS-PAGE. While an undetermined protein was only expressed under hydrostatic pressure conditions, a 60kD protein was expressed under atmospheric pressure, that was related to stress proteins already reported for hyperthermophilic Archaea. This result clearly confirmed that *Thermococcus barophilus* was a piezophilic organism.

A similar approach was followed by Alain et al. (2002) who processed a sample collected at 2600m depth on the East Pacific Rise, and again preserved in a pressurized container. They isolated a thermophilic organism from the *Bacteria* domain, described as a novel species *Marinitoga piezophila*. This organism grows in a temperature range of 45–70°C, with an optimum of 65°C. Optimum pressure for growth was determined at 40 MPa, which is above pressure existing at the sampling site. Again, although *M. piezophila* could grow under atmospheric pressure, the growth rate was significantly enhanced when hydrostatic pressure was applied. Interestingly, morphology of cells was considerably affected by pressure conditions. When grown under hydrostatic pressure, cells of *M. piezophila* appeared as regularly shaped short rods. Decrease of pressure down to atmospheric pressure resulted in elongation of cells until only irregular filaments were observed. Similar changes in cell morphologies as a function of pressure had been reported by Ishii et al. (2004 ), while growing *Escherichia coli* under different pressure conditions. However, in this particular case, filamentous cells were more and more abundant when hydrostatic pressure applied had increased. Inhibition of the septum formation during cell division, and particularly inhibition of the *ftsZ* gene by the elevated pressure was suggested. An opposite response could explain the changes observed for *M. piezophila*.

After studying psychrophilic piezophiles for many years, Yayanos (1986) concluded that piezophily was a feature of psychrophilic bacteria coming from depths below 2000m, and that piezophily increased with depth. The obligate piezophiles were obtained from samples collected  at great depths, below 6000–7000 m, and essentially from the deepest trenches of the oceans such as the Marianna Trench or the Okinawa Trench. Until last year, the deepest hydrothermal vents known were located on the mid-Atlantic ridge at depths of 3600m. However, a deeper active vent field named « Ashadze » was discovered and explored in 2007, at 14°N on the mid-Atlantic Ridge during the French-Russian cruise « Serpentine ». Black smoker samples collected during this cruise were preserved in pressurized containers, and brought to the laboratory for enrichment cultures under elevated pressures and temperatures. Back in the laboratory, pieces of black smokers were inoculated in various culture media and incubated at elevated temperature and hydrostatic pressure. An anaerobic enrichment culture carried out at 95°C and 42 MPa led to isolating of strain CH1 (Zeng et al., 2009). CH1 is highly motile, coccoid-shaped heterotrophic and strictly anaerobic. Rapid growth was observed with peptone, tryptone, and yeast extract, but poor growth occurred on casaminoacids. No growth was observed on carbohydrates, alcohols or single amino acids. CH1 is an Archaeon whose 16S rDNA almost complete sequence (1424 bp) indicated it belongs to the genus *Pyrococcus*, close to *P. furiosus*. CH1 grows in a temperature range from 85 to 105°C, with an optimum at 98°C and a doubling time of 51 min. No growth was observed for hydrostatic pressures below 20 MPa. Optimal growth pressure was recorded at 52 MPa, and growth still occurred at 110 MPa. Under similar culture conditions, CH1 relatives (*P. furiosus*, *P. horikoshi*, *P. glycovorans*, and *P. abyssi*) did grow at atmospheric pressure, but never grew for pressures above 50 MPa. Obligate psychro-piezophiles are known for many years, but are originated from greater depths. Isolated from a sample collected at 4200 m, with an optimum pressure of 52 MPa, CH1 is the first obligate piezophilic hyperthermophilic Archaeon isolated so far. These data should encourage further investigations on pressure loving micro-organisms, since it was estimated that the majority of prokaryotic organisms on Earth inhabit the deep subsurface environments, where elevated pressure is a major parameter.

## Genomic approach

In addition to their ecological importance, the piezophilic hyperthermophilic prokaryotes have fascinated researchers' interests in both fundamental and applied research. To date the complete genome sequences of two piezophilic hyperthermophilic archaea, *Methanocaldococcus jannaschii* and *Pyrococcus abyssi* have been published (Cohen et al., 2003; Graham et al., 2001). A transcriptional study was performed in *M. jannaschii* under shifting extremes of temperature and hyperbaric pressure (Boonyaratanakornkit et al., 2007). Transcriptional profiles for cells grown at 88°C and 50 MPa, heat-shocked at 50 MPa, and pressure-shocked to 50 MPa, shared a subset of genes whose differential expression was attributed to elevated pressure. The core pressure response was limited and consisted of differential expression of four genes among them two encoding hypothetical proteins with no homologs in any other species, one encoding a conserved hypothetical protein that contains characteristic of a glutamine amidotransferase class-II, and the last codes for a replication protein A-related protein that is involved in chromosomal replication and DNA recombination and repair (Kelly et al., 1998).

*In silico* comparison of 141 orthologous proteins from genomes of a piezosensistive archaeon *Pyrococcus furiosus* and a piezophile archaeon *P. abyssi* (Di Giulio, 2005) was made. The pattern of asymmetries in the amino acid substitution process identifies the amino acids arginine, serine, glycine, valine and aspartic acid as those having the most piezophilic behavior, and tyrosine and glutamine as the least piezophilic (Di Giulio, 2005). By correlating hydrostatic pressure with amino acids polarity showed that the more polar amino acids are more piezophilic. In another predictive study, a comparison of six deep sea genomes (among them *P. abyssi* and *M. jannaschii*), based on the percentage of predicted highly expressed genes was made (Xu and Ma, 2007). The predicted highly expressed genes are those genes that have codon frequencies similar to those of the ribosomal protein, principal transcription/translation factors genes, and the major chaperone/degradation genes, but deviates strongly from the average genes of the genome (Karlin and Mrazek, 2000). This study showed that the percentage of predicted highly expressed genes out of a whole genome positively correlates to optimal growth temperature, whereas such positive correlation seems not to exist between environmental hydrostatic pressure and the percentage of predicted highly expressed genes, indicating that the gene expression levels of the used deep sea prokaryotes are highly individual (Xu and Ma, 2007).

Only very few studies  focussed on genomic studies of piezophilic hyperthermophilc prokaryotes in comparison to piezo-psychrophiles, the recent sequencing of the whole genomes of two true piezophilc hyperthermophilic prokaryotes *T. barophilus* and *M. piezophila* will increase the number of available genomes which will encourage developing genomic, transcriptomic and proteomic studies to characterize the piezophily features in piezophilic hyper/thermophilic prokaryotes.

## CULTIVATION OF PIEZOPHILIC ORGANISMS

Cultivation of piezophilic organisms requires specific equipment. At the laboratory scale, batch cultures are carried out in syringes that are pressurized using stainless steel bioreactors. For hyperthermophiles utilization of glass syringes is recommended (Martiensson et al., 1997). However, these devices do not allow for production of large amount of biomass, or sub sampling for time series experiments. Complex devices such as the 'Deepbath' have been designed for these purposes.

For handling piezophiles for further study, JAMSTEC designed a 'deep-sea baro(piezophile) and thermophile isolation and cultivation system', referred to as the 'DEEPBATH' system. The DEEPBATH system consists of four separate devices: 1) a pressure-retaining sampling device, 2) dilution device under pressure conditions, 3) isolation device, and 4) cultivation device (Kato, 2006). The system is controlled by central regulation systems and the pressure and temperature ranges of the devices are from 0.1 to 65 MPa and from 0 to 150°C, respectively. The capacity of the cultivation devices (2 sets) is 1.5 liters each and, therefore, cultures of up to 3 liters can be obtained. The construction of the system and the sample stream are shown in Fig. 4.

The DEEPBATH system can be used at temperatures up to 150°C, allowing high-pressure cultivation studies using hyperthermophilic microorganisms, which can grow at around 100°C. Erauso et al. (1993) reported the cultivation of the hyperthermophilic archaeon *Pyrococcus abyssi*,

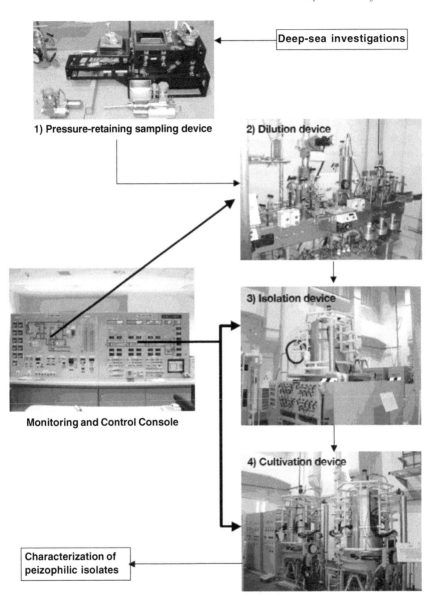

**Figure 4.** The DEEPBATH system. The system is composed of four devices: 1) pressure-retaining sampling device, 2) dilution device under pressure conditions, 3) isolation device, and 4) cultivation device. The system is controlled by Monitoring and Control Console.

isolated from a deep-sea hydrothermal vent, under elevated pressure conditions. They indicated that there were problems associated with the effects of repeated decompression and repressurization processes, which affected the quality of the growth results. DEEPBATH is currently the only system available which permits sampling without any change in pressure and temperature. Using this system, we have performed high-pressure and high-temperature (HP-HT) cultivation studies of two novel hyperthermophilic Archaea isolated from submarine hydrothermal vents, *Thermococcus peptonophilus* (isolated from the Izu-Bonin Trough at a depth of 1,400 m; Gonzalez et al., 1995) and *Pyrococcus horikoshii* (isolated from the Okinawa Trough at a depth of 1,300 m; Gonzalez et al., 1998). The optimal growth temperatures of *T. peptonophilus* and *P. horikoshii* are 85°C and 95°C, respectively. The results of HP-HT growth studies are shown in Fig. 5. It was interesting that the temperature for growth of both hyperthermophilic Archaea could be increased under higher-pressure conditions, and their growth profiles appeared piezophilic. In particular, the growth profile of *P. horikoshii* at 103°C was the same as that of an obligatory piezophilic microorganism. Such piezophilic growth profiles in deep-sea hyperthermophiles are common, as indicated by Deming and Baross (1993), who suggested that these microbes might originate from deep subsurface environments. It is interesting to consider that the origin of life might have come from the subsurface. The DEEPBATH system should be useful in the field of subsurface microbiology in future investigations.

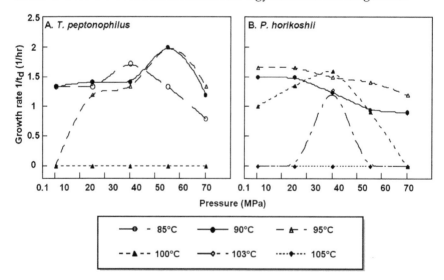

**Figure 5.** Effects of pressure on the growth rate of *Thermococcus peptonophilus* (A) and *Pyrococcus horikoshii* (B) at various temperatures. The growth rate is shown as $1/t_d$, where $t_d$ represents the doubling time in hours.

## NOVEL TECHNIQUES TO MONITOR PIEZOPHILIC MICROORGANISMS

Up to recently, monitoring of cells and cellular activities at high pressure was mainly limited to observations before and after pressure treatment. The samples were analyzed, subcultured or imaged *ex situ* prior to and following the depressurization step to evaluate the effect of the pressure treatment on cellular features (Abe, 2004; Bartlett, 2002; Giovanelli et al., 2004; Sato and Kobori, 1995; Sebert et al., 2004; Shimada et al., 1993). This procedure has two major disadvantages : a) the depressurization step has the potential to lead to artefactual measurements; and b) it does not allow for kinetic measurements because the pressurization/depressurization steps required for analysis interfere with the reactions under scrutiny. Indeed, if one were to monitor the activity of an enzyme, that is severely but reversibly inhibited by pressure, the activity measured under inhibitory pressure would not be null but would correspond to the activity of the enzyme during the pressurization and depressurization steps; Thus, it is crucial to be able to access and monitor the cells and cellular processes under pressure *in situ* to determine the real impact of high pressure on living systems. Recent advances in high pressure incubator design have opened new perspectives in high pressure science. The following pages will concentrate on advances in *in situ* imaging and cell activity monitored by *in situ* spectroscopy which offer novel ways to image and analyze cells.

### *In situ* high resolution, high pressure/high temperature imaging

Proper imaging of cells is a prerequesite of *in situ* analysis to confirm the integrity of morphological alterations of the cells under scrutiny. A few of the high pressure apparatus available have been designed with observation windows of several materials including glass, quartz, sapphire or diamond (Adams et al., 1973; Bassett et al., 1993; Chervin et al., 1995; Grasset, 2001; Gregg et al., 1994; Hirsch and Holzapfel, 1981). Up to recently it was not possible to obtain high quality microscopic images with a high resolution of microscopic objects in these conventional high pressure cells (Besch and Hogan, 1996, Deguchi and Tsujii, 2002; Hogan et al., 1981; Pagliaro et al., 1995; Perrier-Cornet et al., 1995, 1999; Reck et al., 1998; Salmon and Ellis, 1975; Sharma et al., 2002). One solution to lengthen the working distance of the objectives has been to adapt biconvex lenses between the window and the objectives. However, this further degraded the microscopic picture resolution, reducing picture quality (Butz, 1987). Another option has been the use of the Diamond anvil cell (DAC), a common tool used in high pressure physics (see below). Finally, some research groups have undertaken the task of adapting high pressure vessels to fit standard microscopic stages:

The following paragraphs present the three high-pressure cells offering sub-micro resolution.

## Design of a high pressure flow cell for microscopic observations

The Koyama High pressure cell (Fig. 6, Panel A) is a flow cell designed to fit a standard inverted microscope with off the shelf long distance microscope objectives (Koyama et al., 2001). Pressurization, up to 100 MPa, is obtained by two high pressure pump which can recirculate the fluid inside the cell at a variable rate. The cell is thermalized by circulation of water. Observations are performed through a 2-mm-thick glass window with a minimum working distance of 3.8 mm, and an estimated resolution ca. 500 nm. Using this chamber system, Koyama and colleagues were able to monitor the aberrant morphologies of growing *E. coli* cells under pressure, as well as quantifying morphological parameters of several eucaryotic cell types (Koyama et al., 2001, 2003, 2005).

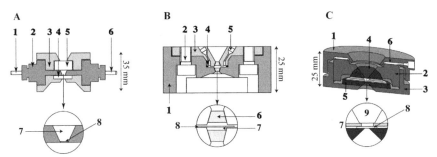

**Figure 6.** Sectional drawings of the High Pressure Cells for microscopic imaging. Each cell has been position with the observation side at the bottom. **Panel A:** The Koyama Flow Through cell : 1) medium in; 2) connection; 3) cell body; 4) bottom glass window; 5) top glass window; 6) medium out; 7) sample chamber; 8) plastic coverslip. (adapted and simplified from Koyama et al., 2001). **Panel B:** The HPDS high pressure cell. 1) main body; 2) screw system; 3) lid; 4) o-ring; 5) lockable connections; 6) sapphire windows; 7) sample chamber; 8) transparent plastic plate (adapted from Hartmann et al., 2004). **Panel C:** The low Pressure DAC. 1) lid with membrane chamber; 2) piston; 3) body; 4 and 5) hemispheric and flat tungsten carbide seats; 6) diamond anvil; 7) metal gasket; 8) diamond window; 9) diamond window (adapted from Oger et al., 2006).

## Design of a high pressure sapphire cell for microscopic observations

The HPDS High pressure cell design (Fig. 6, Panel B) is based on the design of flow through sapphire cells (Hartmann et al., 2004). It has two main modifications: The first being a reduction in thickness to allow the cell to fit inverted microscope stages. The second is the use of a small plastic coverslip to isolate the sample volume from the pressurizing liquid. Pressure range is 0.1 to 300 MPa. In combination with an inverted microscope and an analysis system it allows brilliant microscopic color pictures with an optical resolution better than 0.56 μm (Frey et al., 2006; Friedrich et al., 2006a, 2006b; Hartmann et al., 2004).

### Design of a biology decidated low-pressure DAC (lpDAC)

The Zebda new DAC design (lpDAC, Figure 6, Panel C) is based on that of the DAC optimized for IR-spectroscopy (Chervin et al., 1995). Such a DAC is made up of two diamonds, flat at the tip, and a metal disc with a small hole. The sample is placed within the hole, and pressure is applied on it by applying a force on the two diamonds. Diamond anvils are thick, i.e. usually 1.5 to 2.5 mm in height (Adams et al., 1973; Bassett et al., 1993; Chervin et al., 1995; Hirsch and Holzapfel, 1981). This has some major drawbacks: Diamond has a higher refractive index (2.4) than glass (1.5), for which common objectives are corrected. In addition, imaging is performed through 2 mm of diamond in contrast to 0.23 mm of glass for most coverslips. This results in many optical aberrations, leading to poor quality images (Oger et al., 2006; Sharma et al., 2002). In the new design, one of the anvils is replaced by a diamond window (250 to 600 μm thick) (Daniel et al. unpublished data, Oger et al., 2006; Picard et al., 2007). This is the side chosen for observations and measurements. The pressure is remotely controlled by inflating an internal membrane ram. This allows a fine control and an excellent stability of the pressure applied to the sample. The 24 mm-high cell fits most microscope stages, and can be used with commercially available objectives. A novel pressure calibrant used in the 0.1–2GPa range has been designed specifically for this application (Picard et al., 2006) So far, the present DAC has reached 1.4GPa. The optical resolution in the lpDAC has been estimated at ca. 150nm using confocal microscopy and image interpolation (Oger et al., 2006; Picard et al., 2007).

### Monitoring microbial activities *in situ* under high pressure and/or temperature

One of the main drawbacks encountered while performing experiments under high pressure is the difficulty to analyze the biological sample during the course of the experiment and analyze it without a depressurization step. Thus, doubts will always remain that part, or all, of the measured effect is due to activities occurring during the compression or decompression steps. Many parameters such as pH, temperature, pressure, and a few molecules can be monitored *in situ* provided the proper sensor or probe is used. Thus far, very few instrumented high pressure incubators have been constructed. These can usually monitor one or a few parameters, e.g. temperature, pressure, pH, or gaz content. Several spectroscopic techniques have long been used in chemistry and biochemistry to monitor chemical or enzymatic reactions (Clegg et al., 2001; Fletcher et al., 2003; Mulvaney and Keating, 2000; Pallikari et al., 2001; Shi et al., 1995; Skoulika et al., 2000). In conjunction with spectroscopic techniques, Diamond anvil cells (DACs) are probably the most powerful and versatile high pressure devices, which allow simultaneous *in situ* observation and measurements. Due to the transparency of diamonds, the sample may be probed using an incident radiation comprised between IR and energetic Xray radiations, except in the deep UV range between ca. 0.1 and 250 nm while being pressurized. This approach has long been used to study the behavior of materials under high pressure in physics (Dubessy et al., 1989; Kuba and Knözinger, 2002), Earth sciences (Guillaume et al., 2003; Martinez et al., 2004) and biophysics (De Klerk et al., 2001; Lange and Balny, 2002; Schweitzer-Stenner, 2001), but has never been applied to the monitoring of cellular processes.

In addition to its imaging capabilities, the lpDAC has been designed to reduce spectroscopic signal absorption, and thus allow *in situ* spectroscopies as novel routine tools to monitor microbial cell actvities *in situ* under HP and HT. The following two paragraphs illustrate two such applications in the monitoring of carbon and metal metabolism.

### In situ *Raman Spectroscopy*

The Raman effect which was discovered ca. 70 years ago is due to the inelastic interaction of light with interatomic vibrations (Nakamoto and Ferraro, 2002). When light interacts with an atom, most of it is reflected, the rest being elastically scattered (Rayleigh diffusion) or inelastically scattered (Raman diffusion). Since it probes interatomic vibrations, Raman spectroscopy has a great potential for the detection of polyatomic molecules. Hence, Raman spectroscopy is a routine tool to detect and monitor the behavior of minerals under pressure or to monitor chemical and enzymatic reactions (Aarnoutse and Westerhuis, 2005). Raman spectroscopy is quantitative, since the intensity of the Raman signal is directly linked to the concentration of the analyte. Thus the concentration of the analyte can be determined from the intensity of its Raman spectrum. The absorption and diffusion of light in the experimental setup make this property difficult to exploit directly (Pelletier, 2003). However, the concentration of analyte can be calculated by using an appropriate internal standard and calibration curves (Aarnoutse and Westerhuis, 2005).

Semi-quantitative Raman spectroscopy was first used to monitor microbial metabolism by Picard et al. (2007). The metabolic behavior of baker's yeast *Saccharomyces cerevisiae*, a commonly used piezotolerant model, was monitored to determine the non-permissive pressure for glycolysis. Under anoxic conditions, yeast utilizes one glucose molecule to produce energy leaving two molecules of ethanol as the final product. Glycolysis was monitored through the quantification of the appearance of ethanol, by measuring the symmetric C–C stretching mode of ethanol at 883 cm$^{-1}$, and using the sulfate S–O stretching mode at 980 cm$^{-1}$ as the internal calibrant (Fig. 7). Experiments performed under high hydrostatic pressures (HHP) up to 100 MPa showed yeast fermentation to stop at 87±7 MPa. This is ca. 37 MPa higher than what was predicted by Abe and Horikoshi (1998) from indirect evidence based on the acidification of the yeast vacuole during fermentation under HHP. Interestingly, *in situ* Raman monitoring showed an unexpected pressure-dependent speed of reaction with two separate domains. Below 10 MPa, the reaction rate is increased up to three fold and ethanol production is enhanced up to 5%. Above 20 Mpa, the yield of the reaction decreases linearly as a function of pressure (Picard et al., 2007).

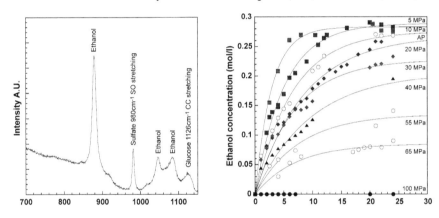

**Figure 7.** *In situ* monitoring of ethanol production in yeast under high hydrostatic pressure. Left: Detail of the Raman spectrum of the culture medium inside the DAC after 24 h of incubation. The symmetric C–C stretching mode of ethanol at 883cm$^{-1}$ is clearly visible. Right : Kinetics of ethanol production by yeast from ambient pressure (AP) to 100 MPa.

*Limits of in situ Raman spectroscopy.* Raman spectroscopy is a non-intrusive, non-destructive spectroscopy that does not require sample preparation. As exemplified above, Raman spectroscopy allows one to measure the appearance or disappearance of a specific metabolite *in situ* under HHP and/or HT with a precision comparable to *ex situ* techniques. The range of molecules that can be quantified by this technique encompasses carbohydrates with carbon chains of less than 6 units, with a higher sensitivity towards shorter and more symmetrical molecules. Thus many common metabolic products such as, $CH_4$, methanol, lactate, acetate, pyruvate, etc, as well as $H_2S$. can be measured and monitored by Raman spectroscopy.Therfore, semi-quantitative Raman spectroscopy has a great potential for the monitoring of microbial metabolism. Its main limitation is that the analyte needs to have a Raman spectrum with at least one peak unmasked by the Raman spectrum of the growth medium (Aarnoutse and Westerhuis, 2005; Pelletier, 2003).

## Monitoring microbial activities by in situ X-ray spectroscopy

Many ecologically relevant microbial metabolisms cannot be investigated by IR or Raman spectroscopies in the DAC. This for example is the case for the dissimilative reduction of metals (DMR) which fuels the sediment and hydrothermal vents ecosystems. However, DMR leads to a change in valence of the metal used as an energy source. Thus, X-ray-based element spectroscopies such as XANES (X-ray Absorption Near Edge Structure) or EXAFS (Extended X-ray Absorption Fine Structure) which are sensitive to the redox environment of elements can potentially be used to monitor these reactions.

Similar to Raman spectroscopy, X-ray spectroscopies are non intrusive quantitative spectroscopies, although X-ray may induce great damage to the cells under scrutiny (Battista et al., 1999; Daly et al., 2004; Ghosal et al., 2005; Mattimore and Battista, 1996). In fact, under the conditions used to acquire a single X-ray spectrum, none of the known microbial isolate survived, even the most irradiation tolerant strain of *Deinococcus radiodurans* (Oger et al., 2008). In the case of live cells, radiation damage to cells cannot be reduced, but the impact of these damages to the culture and the cellular activities can be limited, by controlling the volume of culture in contact with the beam. This implies using the brightest known X-ray sources as well as focussing the X-ray beam to a few microns on the sample.

X-ray spectroscopies have been used previously to determine the speciation of microbiologically produced minerals (Gnida et al., 2007; Harada et al., 2006; Harris et al., 2003; Lemelle et al., 2003; Oger et al., 2006; Oger et al., 2008; Popescu et al., 2007; Sarret et al., 2006; Sors et al., 2005; Thavarajah et al., 2007; Villalobos et al., 2006). They were first applied by Oger et al. (2008) for monitoring selenite reduction by a living bacterium. The *A. tumefaciens* strain C58 used in this study reduces selenite ($Na_2SeO_3$, IV) and selenate ($Na_2SeO_4$, VI) into elemental selenium (Se, 0) and selenides such as dimethyl-selenide ($CH_3$-Se-$CH_3$, II) (Garbisu et al., 1996; Mougel et al., 2001). The kinetics of selenite reduction by *A. tumefaciens* was determined by a combination of µXANES (X-ray Absorption Near Edge Structures) and µXRF (X-ray fluorescence) (Oger et al., 2004). Spectra acquired during a time series clearly show a gradual replacement of selenite in solution, with the apparition of a shoulder on the µXANES spectra. After 24 h, the reduction is complete, and the µXANES spectrum no longer shows a contribution of selenite. In contrast to Raman or IR spectroscopies, the concentration of species present in the solution can be quantified from linear combinations of known standards. In the *A. tumefaciens* example, out of seven different redox species of selenium three were shown to explain the experimental data (Fig. 8). Hence, three selenites are reduced to one metallic selenium and two selenides (Oger et al., 2004). Experiments performed under hydrostatic pressures, in the lpDAC up to 100 MPa showed selenite reduction to stop at 60±1 MPa (P.O. unpublished data). Similar to glycolysis in yeast, we observed a two-domain pressure-dependent reaction rate for selenite reduction. Selenite reduction proceeds faster than at ambient pressure and yield is increased below 5 MPa. Above 10 Mpa, the yield of the reaction decreases linearly as a function of pressure until it stops.

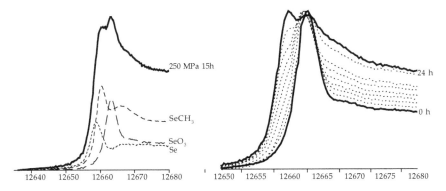

**Figure 8.** Selenium metabolism by *Agrobacterium tumefaciens* Selenium speciation was monitored inside the pressured vessel by μXANES for 24 h. (Left) Deconvolution of a μXANES Spectrum (15 h of incubation at 30°C). Each of the standard required to reproduce the experimental spectrum is drawn to scale. $SeO_3$ : selenite; SeCH3 : di-methyl selenide; Se : elemental selenium. (Right) Reduction of selenite by A. tumefaciens at 25 MPa and 30°C in LB medium over 24 h.

*Limits of in situ X-ray spectroscopy.* X-ray spectroscopy is a quantitative non-intrusive spectroscopy that does not require sample preparation. As exemplified above, it allows one to measure the change in valence of a specific metabolite *in situ* under HHP. The panel of elements that can be monitored by X-ray spectroscopy depends on the beamline and the high pressure incubator used. When using the lpDAC, only elements with atomic numbers greater than iron can be visualized, due to the absorption of diamond. Practically, X-ray spectroscopy is the method of choice for monitoring the reduction of oxide in the environment. Thus, quantitative X-ray spectroscopy has a great potential to study microbial oxide metabolism under pressure. The main limitation of X-ray spectroscopy is that unlike Raman or IR spectroscopy, it is destructive, since no cell can survive the acquisition of a single spectrum. This requires working with a beam to cell culture volume ratio lower than 1/500. In these conditions, more than 75% of the cells survive the multiple spectral acquisitions of a 24 h kinetic.

## WHAT IS THE FUTURE FOR *IN SITU* MONITORING?

The above examples clearly demonstrate the great potential of *in situ* spectroscopic analyses to obtain meaningful and novel quantitative measurements of metabolism under high pressure for piezophilic and non-piezophilic microorganisms. Spectroscopic methods offer a novel alternative to obtain metabolic data for microorganisms under HHP. Under no circumstances will these methods supplant the traditional methods based on culturing and molecular analyses. However, these constitute a perfect complement to all the HHP techniques developed to date. In addition, these recent techniques may open new areas in high pressure biology, two of which are briefly described below.

### On site instrumentation

One the main problems in HHP biology, which is mirrored in other areas of environmental microbiology, is the inability to cultivate all the organisms from an ecosystem in the laboratory. Thus, one is unable to monitor this ecosystem in HHP incubators. Raman and IR microspectrometry may offer an alternative for on site instrumentation to monitor organic molecules and possibly microbial metabolic reactions. Instruments as small as a cubic foot have been designed and sent to space, or Mars (Rull Perez and Martinez-Frias, 2006), making the instrumentation of a seep or a hydrothermal vent a likely possibility for the near future (Brewer et al., 2004).

### High pressure incubators with ported spectroscopic ability

Besides X-rays, visible, IR or UV light can be transferred from one point to the other by means of the appropriate optic fiber. Thus, the collection head and analytic spectrometer can be physically separated, even in separate rooms. This is achieved at a small cost, a reduction in signal intensity, which might be compensated by an increase of either the acquisition time or the excitation power. This allows one to engineer incubators with spectroscopic optical windows, which would give access to the cultures, which have much larger volumes and shapes than the three HHP cells presented above.

### CONCLUSIONS

In these last 20 years, significant progress was achieved in the field of microbiology of natural environments exposed to elevated pressures. The estimates showing that most of the prokaryotic biomass on Earth was located in the subsurface, and consequently exposed to elevated pressure has certainly boosted high pressure microbiology of natural environments. Many questions are still unanswered: what is the limit of the biosphere at depth? What are the carbon and energy sources, the electron acceptors used by these organisms and at which rate? Do they significantly influence the geopshere dynamics?

To answer these questions and many others, there is an urgent need for more exploration campaigns, for more advanced technologies for explorations at depth, and also for laboratory experiments. New technology for laboratory experiments should include a combination of devices simulating physical and chemical conditions at depths and non destructive investigation techniques as well. Techniques to be developed for *in situ* exploration of remote planets, or samples collected on these planets (Mars, for instance) will certainly contribute to improve our knowledge on the deep Earth biosphere.

## REFERENCES

Aarnoutse, P.A. and J.A. Westerhuis. 2005. Quantitative Raman reaction monitoring using the solvent as internal standard. Anal. Chem. 77: 1228–1236.

Abe, F. 2004. Piezophysiology of yeast: Occurrence and significance. Cell Mol. Biol. 50: 437–445.

Abe, F. and K. Horikoshi. 1998. Analysis of intracellular pH in the yeast *Saccharomyces cerevisiae* under elevated hydrostatic pressure: a study in baro- (piezo-) physiology. Extremophiles. 2: 223–228.

Adams, D.M. and S.J. Payne and K. Martin. 1973. Fluorescence of diamond and Raman spectroscopy at high-pressures using a new design of diamond anvil cell. Appl. Spectr. 27: 377–381.

Alain, K. and V.T. Marteinsson, M.L. Miroshnichenko, E.A. Bonch-Osmolovskaya, D.Prieur and J.L. Birrien. 2002. *Marinitoga piezophila* sp. nov., a rod-shaped, thermo-piezophilic bacterium isolated under high hydrostatic pressure from a deep-sea hydrothermal vent. Int. J. Syst. Evol. Microbiol. 52: 1331–1339.

Allen, E.E. and D. Facciotti and D.H. Bartlett. 1999. Monounsaturated but not polyunsaturated fatty acids are required for growth of the deep-sea bacterium *Photobacterium profundum* SS9 at high pressure and low temperature. Appl. Environ. Microbiol. 65: 1710–1720.

Bartlett, D.H. 1999. Microbial adaptations to the psychrosphere/piezosphere. J. Molec. Microbiol. Biotechnol. 1: 93–100.

Bartlett, D.H. 2002. Pressure effects on *in vivo* microbial processes. BBA Prot. Struct. Mol. Enzymol. 1595: 367–381.

Bassett, W.A. and A.H. Shen, M. Bucknum and I.M. Chou. 1993. A new diamond-anvil cell for hydrothermal studies to 2.5 GPa and from −190°C to 1200°C. Rev. Sci. Instru. 64: 2340–2345.

Battista, J.R. and E.M. Earl and M.-J. Park. 1999. Why is *Deinococcus radiodurans* so resistant to ionizing radiation? Trends Microbiol. 7: 362–365.

Beijerinck, M.W. 1889. Le *Photobacterium luminosum,* Bactérie luminosum de la Mer Nord. Arch. Néerl. Sci. 23: 401–427. (in French).

Besch, S. and P. Hogan. 1996. A small chamber for making optical measurements on single living cells at elevated hydrostatic pressure. Undersea Hyperbar Med. 23: 175–184.

Boonyaratanakornkit, B.B. and L.Y. Miao and D.S Clark. 2007. Transcriptional responses of the deep-sea hyperthermophile *Methanocaldococcus jannaschii* under shifting extremes of temperature and pressure 11: 495–503

Bowman, J.P. and J.J. Gosink, S.A. McCammon, T.E. Lewis, D.S. Nichols, P.D. Nichols, J.H. Skerratt, J.T. Staley and T.A. McMeekin. 1998. *Colwellia demingiae* sp. nov., *Colwellia hornerae* sp. nov., *Colwellia rossensis* sp. nov. and *Colwellia psychrotropica* sp. nov.: psychrophilic Antarctic species with the ability to synthesize docosahexaenoic acid (22:6?3). Int. J. Syst. Bacteriol. 48: 1171–1180.

Bowmam J.P. and S.A. McCammon, D.S. Nichols, J.H. Skerratt, S.M. Rea, P.D. Nichols and T.A. McMeekin. 1997. *Shewanella gelidimarina* sp. nov. and *Shewanella frigidimarina* sp. nov., novel Antarctic species with the ability to produce eicosapentaenoic acid (20:5?3) and grow anaerobically by dissimilatory Fe(III) reduction. Int. J. Syst. Bacteriol. 47: 1040–1047.

Brewer, P. and G. Malby, J. Pasteris, S. White, E. Peltzer, B. Wopenka, J. Freeman and M. Brown. 2004. Development of a laser Raman spectrometer for deep-ocean science. Deep-Sea Res. 51: 739–753.

Butz, P. 1987. Biochemische Systeme unter hohem hydrostatischen Druck, thesis, Heidelberg.

Chervin, J.C. and B. Canny, J.M. Besson and P. Pruzan. 1995. A diamond-anvil cell for IR microspectroscopy. Rev. Sci. Instru. 66: 2595–2598.

Certes, A. 1884. Sur la culture, à l'abri des germes atmosphériques, des eaux et des sediments rapports par les expeditions du Travailleur et du Talisman, 1882–1883. C.R. Acad. Sci. Paris 98: 690–693.

Clegg, I. and N. Everall, B. King, H. Melvin and C. Norton. 2001. On-line analysis using Raman spectroscopy for process control during the manufacture of titanium dioxide. Appl. Spectr. 55: 1138–1150.

Cohen, G.N. and V. Barbe, D. Flament, M. Galperin, R. Heilig, O. Lecompte, O. Poch, D. Prieur, J. Quérellou, R. Ripp, J.C. Thierry, J. Van der Oost, J. Weissenbach, Y. Zivanovic and P. Forterre. 2003. An integrated analysis of the genome of the hyperthermophilic archaeon *Pyrococcus abyssi*. Mol Microbiol. 47: 1495–1512

Colwell, R.R., and R.Y. Morita. 1964. Reisolation and emendation of description of *Vibrio marinus* (Russell) Ford. J. Bacteriol. 88: 831–837.

DeLong, E.F. and D.G. Franks and A.A. Yayanos. 1997. Evolutionary relationship of cultivated psychrophilic and barophilic deep-sea bacteria. Appl. Environ. Microbiol. 63: 2105–2108.

Corliss, J.B. and J. Dymond, L.I. Gordon, J.M. Edmond, R.P. Von Herzen, R.D. Ballard, K. Green, D. Williams, A. Bainbridge, K. Crane and T.H. van Andel 1979. Submarine thermal springs on the Galapagos Rift. Science 203 : 1073–1083.

Daly, M.J. and E.K. Gaidamakova, V.Y. Matrosova, A. Vasilenko, M. Zhai, A. Venkateswaran, M. Hess, M.V. Omelchenko, H.M. Kostandarithes, K.S. Makarova, L.P. Wackett, J.K. Fredrickson and D. Ghosal. 2004. Accumulation of Mn(II) in *Deinococcus radiodurans* facilitates gamma-radiation resistance. Science 306: 1025–1028.

Daniel, I. and P.M. Oger, A. Picard, H. Cardon and J.-C. Chervin. A Diamond Anvil Cell adapted for low-pressure high-resolution investigations. Unpublished data.

De Klerk, B. and T. Smellinckx, C. Hugelier, N. Maes and F. Verpoort. 2001. Monitoring of the polymerization of norbornene by in-line fiber-optic near IR-FT Raman. Appl. Spectr. 55: 1564–1567.

Deguchi, S. and K. Tsujii. 2002. Flow cell for *in situ* optical microscopy in water at high temperatures and pressures up to supercritical state. Rev. Sci. Instru. 73: 3938–3941.

DeLong, E.F. and A.A. Yayanos. 1985. Adaptation of the membrane lipids of a deep-sea bacterium to changes in hydrostatic pressure. Science 228: 1101–1103.

DeLong, E.F. and A.A. Yayanos. 1986. Biochemical function and ecological significance of novel bacterial lipids in deep-sea prokaryotes. Appl. Environ. Microbiol. 51: 730–737.

Deming, J.W. and J.A. Baross. 1993. Deep-sea smokers: Windows to a subsurface biosphere? Geochim. Cosmochim. Acta 57: 3219–3230.

Deming, J.W. and H. Hada, R.R. Colwell, K.R. Luehrsen and G.E. Fox. 1984. The nucleotide sequence of 5S rRNA from two strains of deep-sea barophilic bacteria. J. Gen. Microbiol. 130: 1911–1920.

Deming, J.W. and L.K. Somers, W.L. Straube, D.G. Swartz and M.T. Macdonell. 1988. Isolation of an obligately barophilic bacterium and description of a new genus, *Colwellia* gen. nov. System. Appl. Microbiol. 10: 152–160.

Di Giulio, M. 2005. A comparison of proteins from *Pyrococcus furiosus* and *Pyrococcus abyssi*: barophily in the physicochemical properties of amino acids and in the genetic code. Gene 346: 1–6

Dubessy, J. and B. Poty and C. Ramboz. 1989. Advances in C-O-H-N-S fluid geochemistry based on micro-Raman spectrometric analysis of fluid inclusions. Eur. J. Mineral. 1: 517–534.

Erauso, G. and A.-L. Reysenbach, A. Godfroy, J.R. Meunier, B. Crump, F. Partensky, J.A. Baross, V. Marteinsson, G. Barbier, N.R. Pace and D. Prieur. 1993. *Pyrococcus abyssi* sp. nov., a new hyperthermophilic archaeon isolated from a deep-sea hydrothermal vent. Arch. Microbiol. 160: 338–349.

Fang, J. and M.J. Barcelona, Y. Nogi and C. Kato. 2000. Biochemocal function and geochemical significance of novel phospholipids of the extremely barophilic bacteria from the Mariana Trench at 11,000 meters. Deep-sea Res. 147: 1173–1182.

Fang, J. and O. Chan, C. Kato, T. Sato, T. Peeples and K. Niggemeyer. 2003 Phospholipid FA of piezophilic bacteria from the deep sea. Lipids 38: 885–887.

Fletcher, P. and S. Haswell and X. Zhang. 2003. Monitoring of chemical reactions within microreactors using an inverted Raman microscopic spectrometer. Electrophoresis. 24: 3239–3245.

Frey, B. and M. Hartmann, M. Herrmann, R. Meyer-Pittroff, K. Sommer and G. Bluemelhuber. 2006. Microscopy under pressure—An optical chamber system for fluorescence microscopic analysis of living cells under high hydrostatic pressure. Microsc. Res. Tech. 69: 65–72.

Friedrich, O. and K. Kress, M. Hartmann, B. Frey, K. Sommer, H. Ludwig and R. Fink. 2006a. Prolonged High-Pressure treatments in mammalian skeletal muscle result in loss of functional sodium channels and altered calcium channel kinetics. Cell Biochem. Biophys. 45: 71–83.

Friedrich, O. and F. Wegner, M. Hartmann, B. Frey, K. Sommer, H. Ludwig and R. Fink. 2006b. *In situ* high pressure confocal $Ca^{2+}$- fluorescence microscopy in skeletal muscle: a new method to study pressure limits in mammalian cells. Undersea Hyperbar Med. 33: 181–195.

Garbisu, C. and T. Ishii, T. Leighton and Buchanan. 1996. Bacterial reduction of selenite to elemental selenium. Chem. Geol. 132: 199–204.

Ghosal, D. and M.V. Omelchenko, E.K. Gaidamakova, V.Y. Matrosova, A. Vasilenko, A. Venkateswaran, M. Zhai, H.M. Kostandarithes, H. Brim, K.S. Makarova, L.P. Wackett, J.K. Fredrickson and M.J. Daly. 2005. How radiation kills cells: Survival of *Deinococcus radiodurans* and *Shewanella oneidensis* under oxidative stress. FEMS Microbiol. Rev. 29: 361–375.

Giovanelli, D. and N.S. Lawrence and R.G. Compton. 2004. Electrochemistry at high pressures: A review. Electroanalysis. 16: 789–810.

Gnida, M.G. and E.Y. Sneeden, J.C. Whitin, I.J. Pickering and G.N. George. 2007. Sulfur X-ray absorption spectroscopy of living mammalian cells: taurine uptake/release and oxidative stress in renal epithelial cell cultures. Biophys. J. 336A–336A.

Gonzalez, J.M. and C. Kato and K. Horikoshi, 1995. *Thermococcus peptonophilus* sp. nov., a fast-growing, extremely thermophilic archaebacterium isolated from deep-sea hydrothermal vents. Arch. Microbiol. 164: 159–164.

Gonzalez, J.M. and Y. Masuchi, F.T. Robb, J.W. Ammerman, D.L. Maeder, M. Yanagibayashi, J. Tamaoka and C. Kato, 1998. *Pyrococcus horikoshii* sp. nov., a hyperthermophilic archaeon isolated from a hydrothermal vent at the Okinawa Trough. Extremophiles 2: 123–130.

Graham, D.E. and N. Kyrpides, I.J. Anderson, R. Overbeek and W.B. Whitman. 2001. Genome of *Methanocaldococcus (Methanococcus) jannaschii*. Methods Enzymol. 330: 40–123.

Grasset, O. 2001. Calibration of the R ruby fluorescence lines in the pressure range 0–1 GPa and the temperature range 250–300K. High Press. Res. 21: 139–157.

Gregg, C.J. and F.P. Stein, C.K. Morgan and M. Radosz. 1994. A variable volume optical pressure-volume-temperature cell for high-pressure cloud points, densities, and infrared spectra, applicable to supercritical-fluid solutions of polymers up to 2 Kbar. J. Chem. Engin. Data 39: 219–224.

Guillaume, D. and S. Teinturier, J. Dubessy and J. Pironon. 2003. Calibration of methane analysis by Raman spectroscopy in $H_2O$-NaCl-$CH_4$ fluid inclusions. Chem. Geol. 194: 41–49.

Harada, E. and G. Sarret, M.P. Isaure, N. Geoffroy, S. Fakra, M.A. Marcus, M. Birschwilks, S. Clemens, A. Manceau and Y.E. Choi. 2006. Detoxification of zinc in tobacco (*Nicotiana tabacum* L.) plants: Exudation of Zn as Ca-containing grains through the trichomes. Plant Cell Physiol. 47: S155–S155.

Harris, H.H. and I.J. Pickering and G.N. George. 2003. The chemical form of mercury in fish. Science. 301: 1203–1203.

Hartmann, M. and M. Breuss and K. Sommer. 2004. High pressure microscopy—A powerful tool for monitoring cells and macromolecules under high hydrostatic pressure. Cell Mol. Biol. 50: 479–484.

Hirsch, K.R. and W.B. Holzapfel. 1981. Diamond anvil high-pressure cell for Raman spectroscopy. Rev. Sci. Instru. 52: 52–55.

Hogan, P. and H. Änhagen and T. Doubt. 1981. Hyperbaric chamber for evaluating hydrostatic pressure effects on tissues and cells. Undersea Biomed. Res. 8: 51–58.

Ishii, A. and T. Sato, M. Wachi, K. Nagai and C. Kato . 2004. Effects of high hydrostatic pressure on bacterial cytoskeleton FtsZ polymers *in vivo* and *in vitro*. Microbiology 150: 1965–1972.

Jannasch, H.W. and C.D. Taylor. 1984. Deep-sea microbiology. Ann. Rev. Microbiol. 38: 487–514.

Jensen, M.J. and B.M. Tebo, P. Baumann, M. Mandel and K.H. Nealson. 1980. Characterization of *Alteromonas hanedai* (sp. nov.), a non-fermentative luminous species of marine origin. Curr. Microbiol. 3: 311–315.

Karlin, S. and J. Mrázek. 2000. Predicted highly expressed genes of diverse prokaryotic genomes. J. Bacteriol. 2182: 5238–5250.

Kato, C. Barophiles (Piezophiles). Pp. 91–111. *In*: K. Horikoshi and K. Tsujii (eds.). 1999. Extremophiles in Deep-Sea Environments. Springer-Verlag, Tokyo.

Kato, C. The handling of piezophilic microorganisms pp. 733–741. *In*: F. Rainey and A. Oren (eds.). 2006. Extremophiles, A Volume in the Methods in Microbiology Series. Vol. 35.

Kato, C. and Y. Nogi. 2001. Correlation between phylogenetic structure and function: examples from deep-sea *Shewanella*. FEMS Microbiol. Ecol. 35: 223–230.

Kato, C. and T. Sato and K. Horikoshi. 1995. Isolation and properties of barophilic and barotolerant bacteria from deep-sea mud samples. Biodiv. Conserv. 4: 1–9.

Kato, C. and L. Li, Y. Nakamura, Y. Nogi, J. Tamaoka and K. Horikoshi. 1998. Extremely barophilic bacteria isolated from the Mariana Trench, Challenger Deep, at a depth of 11,000 meters. Appl. Environ. Microbiol. 64: 1510–1513.

Kato, C. and K. Nakasone, M.H. Qureshi and K. Horikoshi. How do deep-sea microorganisms respond to changes in environmental pressure? Pp. 277–291. *In*: K. B. Storey and J.M. Storey (eds.). 2000. Cell and Molecular Response to Stress, Vol. 1. Environmental Stressors and Gene Responses Elsevier Science B.V., Amsterdam.

Kato, C. and T. Sato, Y. Nogi and K. Nakasone. 2004. Piezophiles: High pressure-adapted marine bacteria. Mar. Biotechnol. 6: S195–201.

Kelly, T.J. and P. Simancek and G.S. Brush .1998. Identification and characterization of a single-stranded DNA-binding protein from the archaeon *Methanococcus jannaschii*. Proc. Natl. Acad. Sci. USA 95: 14634–14639.

Koyama, S. and T. Miwa, T. Sato and M. Aizawa. 2001. Optical chamber system designed for microscopic observation of living cells under extremely high hydrostatic pressure. Extremophiles 5: 409–415.

Koyama, S. and M. Horii, T. Miwa and M. Aizawa. 2003. Tissue culture of the deep-sea eel *Simenchelys parasiticus* collected at 1,162 m. Extremophiles 7: 245–248.

Koyama, S. and H. Kobayashi, A. Inoue, T. Miwa and M. Aizawa. 2005. Effects of the piezo-tolerance of cultured deep-sea eel cells on survival rates, cell proliferation, and cytoskeletal structures. Extremophiles 9: 449–460.

Kuba, S. and H. Knözinger. 2002. Time-resolved *in situ* Raman spectroscopy of working catalysts: sulfated and tungstated zirconia. J. Raman Spectr. 33: 325–332.

Lange, R. and C. Balny. 2002. UV-visible derivative spectroscopy under high pressure. BBA Prot. Struct. Mol. Enzymol. 1595: 80–93.

Lemelle, L. and A. Simionovici, J. Susini, P. Oger, M. Chukalina, C. Rau, B. Golosio and P. Gillet. 2003. X-ray imaging techniques and exobiology. J. Phys. 104: 377–380.

Leonardo, M.R. and D.P. Moser, E. Barbieri, C.A. Brantner, B.J. MacGregor, B.J. Paster, E. Stackebrandt and K. H. Nealson. 1999. *Shewanella pealeana* sp. nov., a member of the microbial community associated with the accessory nidamental gland of the squid *Loligo pealei*. Int. J. Syst. Bacteriol. 49: 1341–1351.

MacDonell, M.T. and R.R. Colwell. 1985. Phylogeny of the Vibrionaceae, and recommendation for two new genera, *Listonella* and *Shewanella*. Syst. Appl. Microbiol. 6: 171–182.

Makemson, J.C. and N.R. Fulayfil, W. Landry, L.M. Van Ert, C.F. Wimpee, E.A. Widder and J.F. Case. 1997. *Shewanella woodyi* sp. nov., an exclusively respiratory luminous bacterium isolated from the Alboran Sea. Int. J. Syst. Bacteriol. 47: 1034–1039.

Margesin, R. and Y. Nogi. 2004. Psychropiezophilic microorganisms. Cell. Mol. Biol. 50: 429–436.

Marteinsson, V.T. and P. Moulin, J.L. Birrien, A. Gambacorta, M. Vernet and D. Prieur. 1997. Physiological responses to stress conditions and barophilic behavior of the hyperthermophilic vent archaeon *Pyrococcus abyssi*. Appl. Environ. Microbiol. 63 : 1230–1236.

Marteinsson, V.T. and J.L. Birrien, A.L. Reysenbach, M. Vernet, D. Marie, A. Gamabcorta, P. Messner, U.B. Slytr and D. Prieur. 1999. *Thermococcus barophilus* sp. nov., a new barophilic and hyperthermophilic archaeon isolated under high hydrostatic pressure from a deep-sea hydrothermal vent. Int. J. Syst. Bact. 49: 351–359.

Martinez, I. and C. Sanchez-Valle, I. Daniel and B. Reynard. 2004. High-pressure and high-temperature Raman spectroscopy of carbonate ions in aqueous solution. Chem. Geol. 207: 47–58.

Mattimore, V. and J.R. Battista. 1996. Radioresistance of *Deinococcus radiodurans*: Functions necessary to survive ionizing radiation are also necessary to survive prolonged desiccation. J. bacteriol. 178: 633–637.

Maruyama, A. and D. Honda, H. Yamamoto, K. Kitamura and T. Higashihara. 2000. Phylogenetic analysis of psychrophilic bacteria isolated from the Japan Trench, including a description of the deep-sea species *Psychrobacter pacificensis* sp. nov. Int. J. Syst. Evol. Microbiol. 50: 835–846.

Miller, J.F. and N.N. Shah, C.M. Nelson, J.M. Ludlow and D.S. Clark. 1988. Pressure and temperature effects on growth and methane production of the extreme thermophile *Methanococcus jannaschii*. Appl. Environ. Microbiol. 54: 3039–3042.

Mougel, C. and B. Cournoyer and X. Nesme. 2001. Novel tellurite-amended media and specific chromosomal and Ti plasmid probes for direct analysis of soil populations of *Agrobacterium* biovars 1 and 2. 67: 65–74.

Mountfort, D.O. and F.A. Rainey, J. Burghardt, F. Kasper and E. Stackebrant. 1998. *Psychromonas antarcticus* gen. nov., sp. nov., a new aerotolerant anaerobic, halophilic psychrophile isolated from pond sediment of the McMurdo ice shelf, Antarctica. Arch. Microbiol. 169: 231–238.

Mulvaney, S.P. and C.D. Keating. 2000. Raman spectroscopy. Anal. Chem. 72: 145R–157R.

Nakamoto, K. and J. Ferraro. 2002. Introductory Raman Spectroscopy. Academic Press. Amsterdam.

Nakasone, K. and A. Ikegami, C. Kato, R. Usami and K. Horikoshi. 1998. Mechanisms of gene expression controlled by pressure in deep-sea microorganisms. Extremophiles 2:149–154.

Nakasone, K. and A. Ikegami, H. Kawano, R. Usami, C. Kato and K. Horikoshi. 2002. Transcriptional regulation under pressure conditions by the RNA polymerase ?[54] factor with a two component regulatory system in *Shewanella violacea*. Extremophiles 6:89–95.

Nakasone, K. and H. Mori, T. Baba and C. Kato. 2003. Whole-genome analysis of piezophilic and psychrophilic microorganism. Kagaku to Seibutu 41: 32–39. (in Japanese)

Nogi, Y. and C. Kato. 1999. Taxonomic studies of extremely barophilic bacteria isolated from the Mariana Trench, and *Moritella yayanosii* sp. nov., a new barophilic bacterial species. Extremophiles 3: 71–77.

Nogi, Y. and C. Kato and K. Horikoshi. 1998a. *Moritella japonica* sp. nov., a novel barophilic bacterium isolated from a Japan Trench sediment. J. Gen. Appl. Microbiol. 44: 289–295.

Nogi, Y. and C. Kato and K. Horikoshi. 1998b. Taxonomic studies of deep-sea barophilic *Shewanella* species, and *Shewanella violacea* sp. nov., a new barophilic bacterial species. Arch. Microbiol. 170: 331–338.

Nogi, Y. and N. Masui and C. Kato. 1998c. *Photobacterium profundum* sp. nov., a new, moderately barophilic bacterial species isolated from a deep-sea sediment. Extremophiles 2: 1–7.

Nogi, Y. and C. Kato and K. Horikoshi. 2002. *Psychromonas kaikoae* sp. nov., a novel piezophilic bacterium from the deepest cold-seep sediments in the Japan Trench. Int. J. Syst. Evol. Microbiol. 52: 1527–1532.

Nogi, Y. and S. Hosoya, C. Kato and K. Horikoshi. 2004. *Colwellia piezophila* sp. nov., isolation of novel piezophilic bacteria from the deep-sea fissure sediments of the Japan Trench. Int. J. Syst. Evol. Microbiol. 54: 1627–1631.

Oger, P. and J. Gardès. unpublished results.

Oger, P.M. and I. Daniel, B. Cournoyer and A. Simionovici. 2004. *In situ* micro X-ray absorption near edge structure study of microbiologically reduced selenite($SeO_3^{2-}$). Spectrochimica Acta Part B-Atomic Spectroscopy. 59: 1681–1686.

Oger, P. and I. Daniel and A. Picard. 2006. Development of a low-pressure diamond anvil cell and analytical tools to monitor microbial activities *in situ* under controlled P and T. BBA Prot. Proteom. 1764: 434–442.

Oger, P. and I. Daniel, A. Simionovici and A. Picard. 2008. Micro-X-ray Absorption Near Edge Structure as a suitable probe to monitor live organisms. Spectrochimica Acta Part B-Atomic Spectroscopy 63: 512–517.

Owen, R. and R.M. Legros and S.P. Lapage. 1978. Base composition, size and sequence similarities of genome deoxyribonucleic acids from clinical isolates of *Pseudomonas putrefaciens*. J. Gen. Microbiol. 104: 127–138.

Pagliaro, L. and F. Reitz and J. Wang. 1995. An optical pressure chamber designed for high numerical aperture studies on adherent living cells. Undersea Hyperbar Med. 22: 171–181.

Pallikari, F. and G. Chondrokoukis, M. Rebelakis and Y. Kotsalas. 2001. Raman spectroscopy: A technique for estimating extent of polymerization in PMMA. Mat. Res. Innov. 4: 89–92.

Pelletier, M.J. 2003. Quantitative analysis using Raman spectrometry. Appl. Spectr. 57: 20A–42A.

Perrier-Cornet, J. and P. Marechal and P. Gervais. 1995. A new design intended to relate high-pressure treatment to yeast-cell mass-transfer. J. Biotechnol. 41: 49–58.

Perrier-Cornet, J.M. and M. Hayert and P. Gervais. 1999. Yeast cell mortality related to a high-pressure shift: occurrence of cell membrane permeabilization. J. Appl. Microbiol. 87: 1–7.

Picard, A. and P. Oger, I. Daniel, H. Cardon, J.-C. Chervin and B. Couzinet. 2006. A sensitive pressure sensor for diamond anvil cell experiments up to 2 GPa: FluoSpheres®. J. Appl. Phys. 100: 034915

Picard, A. and I. Daniel, G. Montagnac and P.M. Oger. 2007. *In situ* monitoring of alcohol fermentation by *Saccharomyces cerevisiae* under high pressure by quantitative Raman spectroscopy. Extremophiles. 11: 445–452.

Popescu, B.F.G. and I.J. Pickering, G.N. George and H. Nichol. 2007. The chemical form of mitochondrial iron in Friedreich's ataxia. J. Inorg. Biochem. 101: 957–966.

Prieur, D. and V.T. Marteinsson. 1998. Prokaryotes living under elevated hydrostatic pressure. Advances in biochemical Engineering/Biotechnology 61: 23–35.

Reck, T. and E. Sautter, W. Dollhopf and W. Pechhold. 1998. A high-pressure cell for optical microscopy and measurements on the phase diagram of poly (diethylsiloxane). Rev. Sci. Instru. 69: 1823–1827.

Rull Perez, F. and J. Martinez-Frias. 2006. Raman spectroscopy goes to Mars. Spectrosc. Eur. 18.

Salmon, E.D. and G.W. Ellis 1975. New miniature hydrostatic-pressure chamber for microscopy—Strain-free optical glass windows facilitate phase-contrast and polarized-light microscopy of living cells—Optional fixture permits simultaneous control of pressure and temperature. J. Cell Biol. 65: 587–602.

Sarret, G. and E. Harada, Y.E. Choi, M.P. Isaure, N. Geoffroy, S. Fakra, M.A. Marcus, M. Birschwilks, S. Clemens and A. Manceau. 2006. Trichomes of tobacco excrete zinc as zinc-substituted calcium carbonate and other zinc-containing compounds. Plant Physiol. 141: 1021–1034.

Sato, M. and H. Kobori. 1995. Ultrastructural effects of pressure stress to the nucleus in *Saccharomyces cerevisiae*: A study by immunoelectron microscopy using frozen thin sections. Microbiol. Lett. 132: 253–258.

Schmitz, W.J. Jr. 1995. On the interbasin-scale thermohaline circulation. Rev. Geophys. 33: 151–173.

Schweitzer-Stenner, R. 2001. Visible and UV-resonance Raman spectroscopy of model peptides. J. Raman Spectr. 32: 711–732.

Sebert, P. and M. Theron and A. Vettier. 2004. Pressure and temperature interactions on cellular respiration: A review. Cell Mol. Biol. 50: 491–500.

Seo, H.J. and S.S. Bae, J.-H. Lee and S.-J. Kim. 2005. *Photobacterium frigidiphilum* sp. nov., a psychrophilic, lipolytic bacterium isolated from deep-sea sediments of Edison Seamount. Int. J. Syst. Evol. Microbiol. 55: 1661–1666.

Sharma, A. and J.H. Scott, C.W. Cody, M.L. Fogel, R.M. Hazen, R.J. Hemley and W.T. Huntress. 2002. Microbial activity at gigapascal pressures. Science 295: 1514–1516.

322 *Comparative High Pressure Biology*

Shi, X. and S.J. Parus, K.D. Pennell and M.D. Morris. 1995. Detection of chlorinated hydrocarbons in aqueous surfactant solutions by near-IR Raman spectroscopy. Appl. Spectr. 49: 1146–1150.

Shimada, S. and M. Andou, N. Naito, N. Yamada, M. Osumi and R. Hayashi. 1993. Effects of hydrostatic pressure on the ultrastructure and leakage of internal substances in the yeast *Saccharomyces cerevisiae*. Appl. Microbiol. Biotechnol. 40: 123–131.

Skoulika, S.G. and C.A. Georgiou and M.G. Polissiou. 2000. FT-Raman spectroscopy— analytical tool for routine analysis of diazinon pesticide formulations. Talanta 51: 599–604.

Sors, T.G. and D.R. Ellis, G.N. Na, B. Lahner, S. Lee, T. Leustek, I.J. Pickering and D.E. Salt. 2005. Analysis of sulfur and selenium assimilation in Astragalus plants with varying capacities to accumulate selenium. Plant J. 42: 785–797.

Tatusov, R.L. and N.D. Fedorova, J.D. Jackson, A.R. Jacobs, B. Kiryutin, E.V. Koonin, D.M. Krylov, R. Mazumder, S.L. Mekhedov, A.N. Nikolskaya, B.S. Rao, S. Smirnov, A.V. Sverdlov, S. Vasudevan, Y.I. Wolf, J.J. Yin and D.A. Natale. 2003. The COG database: an update version includes eukaryotes. BMC Bioinformatics 4: 41.

Thavarajah, D. and A. Vandenberg, G.N. George and I.J. Pickering. 2007. Chemical form of selenium in naturally selenium-rich lentils (*Lens culinaris* L.) from Saskatchewan. J. Agri. Food Chem. 55: 7337–7341.

Urakawa, H. and K. Kita-Tsukamoto, S.E. Steven, K. Ohwada and R.R. Colwell. 1998. A proposal to transfer *Vibrio marinus* (Russell 1891) to a new genus *Moritella* gen. nov. as *Moritella marina* comb. nov. FEMS Microbiol. Lett. 165: 373–378.

Vezzi, A. and S. Campanaro, M. D'Angelo, F. Simonato, N. Vitulo, F.M. Laauro, A. Cestaro, G. Malacrida, B. Simionati, N. Cannata, C. Romualdi, D.H. Bartlett and G. Valle. 2005. Life at depth: *Photobacterium profundum* genome sequence and expression analysis. Science 307: 1459–1461.

Villalobos, M. and B. Lanson, A. Manceau, B. Toner and G. Sposito. 2006. Structural model for the biogenic Mn oxide produced by *Pseudomonas putida*. Am. Mineral. 91: 489–502.

Whitman, W.B. and D.C. Coleman and W.J.Wiebe. 1998. Prokaryotes: the unseen majority. Proc. NAtl. Acad. Sci. USA 95: 6578-6583.

Xu, K. and B.G. Ma. 2007. Comparative analysis of predicted gene expression among deep-sea genomes. Gene 397: 136–142.

Xu, Y. and Y. Nogi, C. Kato, Z. Liang, H-J. Rüger, D.D. Kegel and N. Glansdorff. 2003a. *Moritella profunda* sp. nov. and *Moritella abyssi* sp. nov., two psychropiezophilic organisms isolated from deep Atlantic sediments. Int. J. Syst. Evol. Microbiol. 53: 533–538.

Xu, Y. and Y. Nogi, C. Kato, Z. Liang, H-J. Rüger, D.D. Kegel and N. Glansdorff. 2003b. *Psychromonas profunda* sp. nov., a psychropiezophilic bacterium from deep Atlantic sediments. Int. J. Syst. Evol. Microbiol. 53: 527–532.

Yayanos, A.A. 1986. Evolutional and ecological implications of the properties of deep-sea barophilic bacteria. Proc. Natl. Acad. USA 83 : 9542–9546.

Yayanos, A.A. and A.S. Dietz and R. Van Boxtel. 1979. Isolation of a deep-sea barophilic bacterium and some of its growth characteristics. Science 205: 808–810.

Zeng, X. and J.L. Birrien, Y. Fouquet, G. Cherkasho, M. Jebbar, J. Querellou, P.M. Oger, M.A. Cambon-Bonavita, X. Xiao and D. Prieur. *Pyrococcus* CH1, an obligate piezophilic hyper thermophile: extending the upper pressure-temperature limits for life. ISME. J. 3: 873–876.

# 13

# Effects of the Deep-Sea Environment on Invertebrates

*Joseph F. Siebenaller*

## INTRODUCTION

### Historical Perspective

Since the 1800's biologists have been intrigued by the physical factors of the deep ocean. Edward Forbes (1844), based on the decrease in numbers of organisms sampled from the sea floor at increasing depths in the Aegean Sea, hypothesized that the ocean below depths of approximately 550 m would be azoic. This extrapolation seemed reasonable given the high pressures, low temperatures and darkness of the ocean depths. The hypothesis stimulated a debate on the role of environmental factors on the distribution of marine plants and animals (Anderson and Rice, 2006). Subsequent observations showed Forbes' hypothesis to be false; however, the role of environmental factors in setting distribution limits has remained a focus of interest (e.g., Siebenaller and Somero, 1978; Carney, 2005; Aquino-Souza et al., 2008). The physical factors of the deep ocean are one component of the parameters which interact to restrict species' depth distributions. Clearly, biological and trophic interactions and, for benthic species, substrate will play critical roles in establishing biogeographic distributions (Carney, 2005).

Organisms living in the deep ocean display strong vertical distribution patterns, with species restricted to specific depth bands (Haedrich, 1997; Carney, 2005). These patterns of zonation are reflected in abundance, diversity and community structure. These depth-related patterns may result

Department of Biological Sciences, Louisiana State University, Baton Rouge, Louisiana 70803, USA.
E-mail: zojose@lsu.edu

from a variety of factors, including biological and physical factors of the environment (Carney, 2005). Both hydrostatic pressure and temperature can directly affect the metabolism of ectotherms, influencing protein and membrane structure and function. Because hydrostatic pressure varies with depth, adaptation to this environmental factor may play an important role in establishing vertical distribution patterns. Species, depending on their depth distribution patterns, will experience different temperature and pressure regimes and individuals of species which migrate vertically will experience variable temperature and pressure. The pressure- and temperature-sensitivities of dispersal forms might restrict species distributions (e.g., Aquino-Souza et. al., 2008; Young et al., 1996).

This chapter explores the influences of factors of the physical environment, in particular hydrostatic pressure, on the invertebrate component of the deep-ocean fauna. For some pressure phenomena more extensive information is available from studies of vertebrates rather than invertebrates. However, where parallel studies have been performed, patterns similar to those observed for fishes are evident for invertebrate taxa as well. This examination focuses on invertebrates that are characteristic of the extensive pelagic and benthic components of the deep-sea and does not address the chemosynthetic deep-sea communities that were discovered in 1977 (Corliss et al., 1979; Koslow, 2007). These assemblages of animals harbor endosymbiotic chemosynthetic bacteria and are found at hydrothermal vents and hydrocarbon seeps (Van Dover et al., 2002). The animals at these sites occur at much higher biomass densities then the deep-sea fauna which depends ultimately on a photosynthetic food web (Gage and Tyler, 1991).

## ENVIRONMENTAL FACTORS

The deep sea is defined as the ocean depths below 200 m starting at the continental shelf-slope break which displays a gradient of species diversity (Hessler, 1974) or, alternatively, as the depths below 1,000 m where temperatures are relatively constant and sunlight does not penetrate (Angel, 1997), The deep sea is the largest habitat by volume on the planet. Ninety-two per cent of the volume of the ocean is below 200 m depth. The average depth of the ocean is 3,800 m; the deepest depth is approximately 11,000 m in the Challenger Deep of the Mariana Trench (Pickard and Emery, 1990). Pressure increases 0.101 MPa (= 1 atm) for every 10 m depth increase (Saunders and Fofonoff, 1976).

Sun light levels will affect primary productivity through photosynthesis and hence food availability at depth. The penetration of sun light is dependent on water clarity and latitude. Although sun light does not penetrate below approximately 1,000 m, there is biologically produced light in the deep sea; in the water column a number of invertebrates and fishes are bioluminescent (Angel, 1997). Bioluminescence is found at all depths in the sea (Herring, 2002). Below 1,000 m temperatures are approximately 2–4°C and relatively constant (Pickard and Emery, 1990). These environmental conditions and the nomenclature of pelagic realms are summarized in Fig. 1.

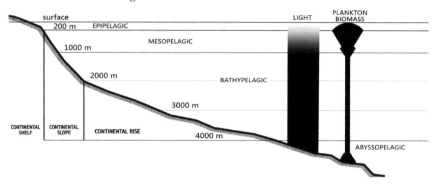

**Figure 1.** Section through the ocean indicating pelagic depth zones, the relative distribution of plankton biomass and the penetration of light. Virtually no sunlight penetrates below 1,000 m. After Marshall (1979).

## PRESSURE EFFECTS ON BIOLOGICAL MOLECULES

Pressure can potentially affect all of the biological molecules and metabolic processes of deep-living organisms. The effects of pressure result from volume changes. For biological systems, the relevant volume is that of the entire solute–solvent system. Pressure can differentially affect the non-covalent bonds which stabilize higher orders of macromolecular structure. Rates and equilibria can be affected. Variations in pressure may pose problems in regulating and coordinating metabolism if the components of a pathway have different signs and magnitudes of volume changes. According to Le Chatelier's principle, increased pressure will shift equilibria toward the lower volume state (Mozhaev et al., 1996). Hydrostatic pressure has been shown to affect ligand-protein interactions and catalytic rates (Siebenaller and Somero, 1989). There may be volume changes of the solute-solvent system associated with catalysis. For example, void volumes may change due to conformational changes of the enzyme, changes in the volumes of reactants and hydration changes of the components. The effects of pressure on the enzymes of marine teleost fishes have been explored more extensively than effects on the enzymes of invertebrates. However, the responses of invertebrate enzymes to increased pressure parallel the patterns observed for the teleost enzymes. Pressures typical of the bathyal habitat (2.03 to 20.26 MPa) perturb enzymes of shallow-living cold-adapted species.

NAD-dependent dehydrogenases have been studied in a number of deep- (common at depths greater than 550 m) and shallow-living species (Fig. 2). For the skeletal muscle-type lactate dehydrogenase ($A_4$-LDH) both the apparent Michaelis constant ($K_m$) values of substrate, pyruvate, and coenzyme, reduced nicotinamide adenine dinucleotide (NADH), are increased by moderate pressures, but to a much greater extent in the orthologs from shallow-occurring species (Siebenaller and Somero 1978, 1979). The $K_m$ of coenzyme is particularly sensitive to pressure perturbation in LDH homologs of shallow-living species. Adaptation to the moderate environmental pressures characteristic of the bathyal habitat appears to have been selected. Convergent evolution is seen in $A_4$–LDH orthologs from 4 different fish families. Each family has independently evolved resistance to perturbation by increased pressures. At pressures comparable to the *in situ* pressures experienced by the

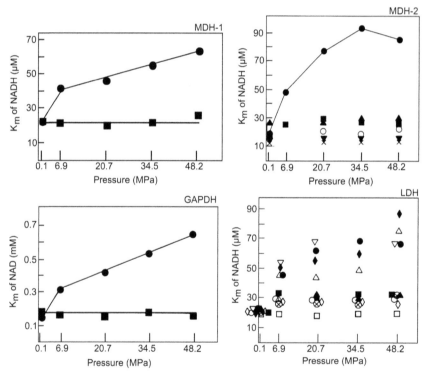

**Figure 2.** Effects of pressure on $K_m$ of coenzyme values for NAD-dependent dehydrogenases purified from white muscle tissue of deep- and shallow-living teleost fishes. Four families of deep-sea fish are represented. Determinations were made at 5°C. Deep-living species: *Coryphaenoides armatus* (□), *C. carapinus* (⊗), *C. acrolepis* (▲), *C. leptolepis* (▼), *Sebastolobus altivelis* (■), *Nezumia bairdii* (x), *Antimora microlepis* (○), *Halosauropsis macrochir* (◇). Shallow-living fishes: *Sebastolobus alascanus* (•), *Pagothenia borchgrevinki* (△), *Sebastes melanops* (▽), *Scorpaena guttata* (♦). Species common at depths greater than 550 m are considered deep-living. Redrawn from Siebenaller (1987).

species the $K_m$ values are conserved among the species. Additionally, it is apparent that the enzymes of cold-adapted shallow-living species are not necessarily pre-adapted for function at high pressure.

The pattern seen in the pressure adaptation of $A_4$-LDH orthologs from deep-sea species is also seen for three other dehydrogenases, cytoplasmic malate dehydrogenase-1 (MDH-1) isozymes, cytoplasmic malate dehydrogenase-2(MDH-2) and glyceraldehyde-3-phosphate dehydrogenase (GAPDH) (Fig. 2, Siebenaller, 1987). For each enzyme compared, the orthologs from the deeper-living species are resistant to pressure perturbation, but the ortholog of a cold-adapted shallow-living species, the sebastid *Sebastolobus alascanus* is not. The enzymes from its deeper living congener, *S. altivelis* are relatively unaffected by increased hydrostatic pressure (Fig. 2). In contrast, the $K_m$ of coenzyme values for the ortholog from *S. alascanus* double at pressures as low as 6.89 MPa. The $K_m$ values of the enzymes from *S. altivelis* were unperturbed by pressures up to 48.23MPa, the highest pressure tested. This range of pressure tolerance would insure that for a species with a broad depth distribution only a single form of an enzyme would be required.

Dahlhoff and Somero (1991) examined the effects of pressure and temperature on cytosolic malate dehydrogenase orthologs from deep- and shallow-living annelids, molluscs and arthropods (Fig. 3). These invertebrate

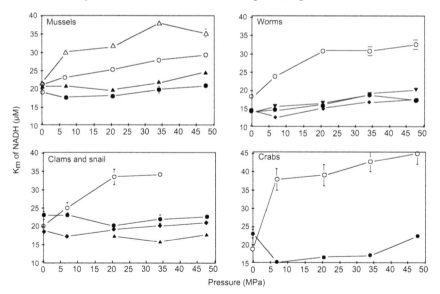

**Figure 3.** Effects of pressure on $K_m$ of NADH of partially purified cytosolic malate dehydrogenases from shallow- (open symbols) and deep-living (filled symbols) invertebrates. Determinations were made at 5°C. 95% confidence limits are shown. Redrawn from Dahlhoff and Somero (1991).

dehydrogenases displayed a pattern of conservation of $K_m$ of coenzyme values similar to that observed for teleost fishes (Fig. 2). Deep-living species had enzymes which were minimally affected by pressures up to 48.1 MPa (Fig. 3). Considering the studies of both vertebrate and invertebrate taxa, dehydrogenases of deep-living species of four phyla have evolved a resistance to increased pressure. This pressure resistance is not characteristic of the enzymes of shallow-living species. This is a remarkable example of convergent evolution at the molecular level.

Increased pressure affects the stability and susceptibility to proteolysis of teleost $A_4$-LDH (Hennessey and Siebenaller, 1985, 1987). Orthologs of deeper-living species are more resistant to denaturation by high pressure and are more resistant to inactivation by proteolysis than are the orthologs of shallower-living teleosts. Pressure may induce conformational shifts of proteins from shallow- living species resulting in increased susceptibility to proteolysis. More stable deep-sea enzymes may prevent too-rapid turnover of proteins which could be energetically wasteful in the food-poor deep-sea environment. LDH orthologs from deep-living species have lower $k_{cat}$ values than the enzymes of cold-water shallow species; this decreased catalytic efficiency may be the cost of a pressure-stable enzyme (Somero and Siebenaller, 1979).

Membrane fluidity will be altered by pressure increases and these fluidity changes may affect membrane functions such as permeability, ion transport and signal transduction (Siebenaller and Garrett, 2002). Increased pressure and low temperature will favor a more ordered acyl chain environment in membranes (Cossins and Macdonald, 1989). Evolutionary adaptations and short-term acclimatization of membranes to both temperature (e.g. Avrova, 1984; Logue et al., 2000) and hydrostatic pressure by changes in phospholipid and fatty acid composition have been documented (e.g., DeLong and Yayanos, 1985; Cossins and Macdonald, 1989; Yano et al., 1998). According to homeoviscous theory, adjustments of the composition of the phospholipid bilayer defend the functional characteristics of the membrane (e.g. Cossins 1994). Such adjustments of membrane microviscosity would also preserve microdomains in membranes and maintain the organization of signaling complexes (Ostrom et al., 2000; Muresan et al., 2001).

In teleost fish brain, transmembrane signaling by guanyl nucleotide binding protein (G protein) coupled receptors (GPCRs) is affected by pressure increases. A number of steps are affected: activation of the GPCR by agonists, the interaction of the receptor with the heterotrimeric G protein, the G protein GTPase cycle and the activation and function of the enzyme adenylyl cyclase (Siebenaller and Garrett, 2002). There are differences among species in the effects of increased hydrostatic pressure. Despite the fact that G protein-coupled signal transduction is sensitive to increased hydrostatic pressure in a number of species, some deep-living marine teleost fishes have evolved resistance to its disruptive effects.

Increased pressure may also affect function of the nervous system; mammals and other vertebrate taxa exhibit the high pressure nervous syndrome. Exposure to increasing pressure elicits a complex of physiological changes which include tremors and convulsion. This phenomenon has been observed in invertebrate taxa as well. Pressures of 10.13 MPA elicited locomotory behavioral changes interpreted as convulsions in representatives of six phyla (Brauer, 1984).

## PROTECTION BY ORGANIC OSMOLYTES

Although deep-sea species have evolved enzymes more resistant to high hydrostatic pressures than the enzymes of shallow-living, cold-adapted species, the resistance may not be complete and some proteins may be susceptible to perturbation by pressure increases. Yancey and colleagues have documented the accumulation of organic osmolytes in deep-sea teleost fishes and invertebrates that may serve to offset pressure perturbations (Yancey et al,. 2001; Yancey et al., 2004; Yancey, 2005; Samerotte et al., 2007). Osmoconforming marine organisms accumulate small organic compounds in response to water stress. Unexpectedly, some of these compounds, including methylamines (trimethylamine oxide (TMAO) and betaine) and polyols (myo- and scyllo-inositol), are accumulated in a depth-related pattern. For example, in both intra- and interspecific comparisons, sea anemones accumulate increasing concentrations of these organic osmolytes with increasing depth (Table 1).

**Table 1.** Major organic osmolytes (mmol/kg wet weight ± standard deviation) in species of sea anemone from four depths.

| Species Depth | Myo- Inositol | Scyllo- Inositol | Taurine | Glycine | Betaine | TMAO | Other Amino Acids |
|---|---|---|---|---|---|---|---|
| *Telia (Urticina) crassicornis* **Intertidal** | 1.6 ± 0.2 | nd | 24.6 ± 9.9 | 54.6 ± 7.5 | 3.9 ± 00.7 | nd | 21.7 ± 6.4 |
| *Stomphia didemon* **90 m** | 0.9 ± 0.2 | nd | 50.1 ± 4.4 | 50.4 ± 4.9 | 4.2 ± 0.5 | nd | 10.3 ± 1.5 |
| *Actinauge abyssorum* **1,900 m** | 2.0 ± 0.3 | 11.1 ± 1.6 | 31.7 ± 2.3 | 17.8 ± 2.0 | 12.0 ± 0.4 | 10.6 ± 1.7 | 32.1 ± 3.4 |
| *Actinauge abyssorum* **2,900 m** | 4.2 ± 0.5 | 16.9 ± 2.9 | 23.1 ± 2.1 | 12.7 ± 1.5 | 33.4 ± 1.7 | 31.6 ± 3.3 | 3.9 ± 1.4 |

*Data from Yancey et al., 2004 nd = not detected (< 0.1 mmol/kg wet weight).*

The methylamines belong to a class of osmolyte termed counteracting or compensatory (Yancey, 2005). These osmolytes counteract the perturbing effects of urea and inorganic salts. For example, TMAO, which is accumulated along with the chaotrope urea by marine elasmobranch fishes, counteracts many of the protein-perturbing effects of urea (Yancey, 2005). In deep-sea species these methylamines may serve to offset the effects of hydrostatic pressure. TMAO has been shown to counteract the perturbing effects of pressure on $K_m$ values (Yancey et al., 2004) and to stabilize lactate dehydrogenases of both fish and mammals at elevated pressures (Yancey and Siebenaller, 1999). Pressure increases may shift the equilibria between conformers of proteins and TMAO may alter those equilbria toward more stable forms (Yancey, 2005). Such a conformational shift by TMAO may explain the non-additive counteracting effects of TMAO on increased $K_m$ of pyruvate values induced by KCl for porcine $A_4$-LDH (Desmond and Siebenaller, 2006).

## METABOLIC RATES

There has been considerable interest in the metabolism of deep-living organisms. From an ecological perspective, much effort has been made to understand energy demands and fluxes in the deep-water habitat and to understand the adaptations that allow organisms to tolerate the extreme physical conditions in this food-poor environment. This section focuses on several studies of invertebrate taxa to illustrate the patterns observed and methodologies employed.

Amphipods are an abundant and widely-distributed component of the deep-sea fauna, and because of their scavenging life-style, they are readily attracted to baited traps (Blankenship et al., 2006). Yayanos (1978) successfully recovered amphipods in pressure-retaining traps from 5,700 m in the Central North Pacific Ocean. Pleopod beat rates at *in situ* temperature and pressure were similar to pleopod beat frequencies of shallow-living amphipods. Inadvertent decompression paralyzed the amphipods. Eight hours were required for individuals to recover from a 70% decompression.

In a subsequent study of *Paralicella caperesca* (Yayanos, 1981), amphipods recovered cold and undecompressed from a depth of 5,900 m (2°C, 601 bar = 60.1 MPa) were decompressed at a controlled rate (Fig. 4). Animals at 38.7 MPa displayed jerky pleopod movement, in contrast to the smooth and stable pleopod beats observed at 60.1 MPa. Amphipods lost locomotory activity at 21.5 MPa. After 4 min at atmospheric pressure the animals recovered upon recompression. However, long-term effects of decompression, e.g., on growth and reproduction, were not assessed. These amphipods tolerated a 30.0 MPa pressure decrease without a significant change in pleopod beat rate, at least suggesting that these animals might be capable of significant vertical migrations. In a study of the decompression tolerance of scavenging lysianassoid deep-sea amphipods, the majority of individuals recovered from depths between 3,950 and 4,420 m using a thermally protected amphipod trap were dead (Treude et al., 2002). Clearly there are limits to the decompression tolerance of deep-living amphipods. However, not only the magnitude of decompression but the rate of decompression may be important.

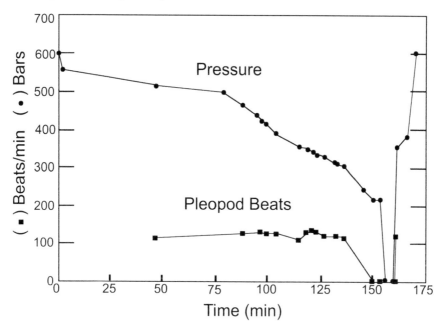

**Figure 4.** Activity, measured as pleopod beats min⁻¹, of the amphipod *Paralicella caperesca* undergoing controlled decompression from 60.1 MPa (= 601 bar) at 2°C. Cold and undecompressed (at 60.1 MPa) amphipods were collected from 5,900 m depth. Locomotor activity was arrested at pressures between 300 and 215 bars. After 4 min at atmospheric pressure, locomotor activity was almost immediately regained following recompression. Data were obtained for two individuals which behaved in a qualitatively similar manner; data are show for one individual. Redrawn from Yayanos (1981).

K.L. Smith, Jr. and collaborators have measured respiration rates of deep-sea animals *in situ* from submersibles and using free-vehicle instrumentation deployed from a surface vessel. Respiration rates (weight-specific oxygen consumption) of two amphipod species measured *in situ* had high variability, suggesting that the animals had both an active state and a resting or torpor-like state (Fig. 5, Smith and Baldwin, 1982). Exposure to the odor of bait appeared to elevate oxygen consumption rates for up to 8 h. The *in situ* resting-state oxygen consumption rates of the deep-sea species are similar to those of cold-adapted shallow-living amphipods (Table 2). The respiration

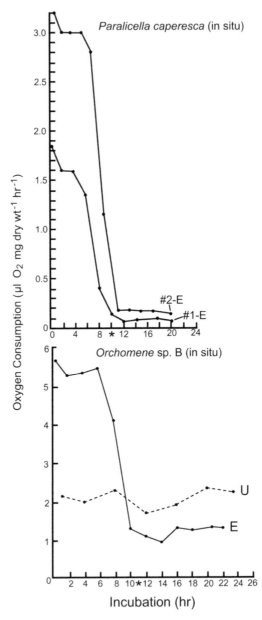

**Figure 5.** Weight-specific $O_2$ consumption of two amphipod species measured *in situ*. *Paralicella caperesca* was measured at a depth of 3,650 m in the North Atlantic Ocean; the two individuals were initially exposed to the odor of bait. *Orchomene* sp. B was measured at a depth of 1,300 m in North Pacific Ocean; there were two groups of individuals, one exposed to bait (E) and one unexposed (U). The asterisk on the abscissa indicates midnight. Redrawn from Smith and Baldwin (1982).

rates of deep-living amphipods, measured at atmospheric pressure, are in the range of values reported from *in situ* measurements (Table 2). This suggests that there may be no strong effect of pressure on the respiratory rates of these taxa.

**Table 2.** Weight-specific oxygen consumption rates of amphipods measured at atmospheric pressure (unless noted).

| Species | Respiration ($\mu$l O$_2$ mg$^{-1}$ wet wt hr$^{-1}$) | Temp (°C) | Depth (m) | Source |
|---|---|---|---|---|
| *Orchomene* sp. B | 0.21–0.98 | 3.0 | 1,300 (*in situ*) | Smith and Baldwin (1982) |
| *Paralicella caperesca* | 0.02–0.66 | 3.0 | 3,650 (*in situ*) | Smith and Baldwin (1982) |
| *P. caperesca* | 0.169–0.191 | 0–2 | 1,920 | Treude et al. (2002) |
| *Eurythenes gryllus* | 0.060–0.064 | 2 | 1,850 | George (1979) |
| *E. gryllus* | 0.051–0.118 | 0–2 | 3,950–4,050 | Treude et al. (2002) |
| *E. gryllus* | 0.0217 | −0.5– −1.0 | 3,168 | Takeuchi and Watanabe (1998) |
| *Abyssorchomene distincta* | 0.052–.074 | 0–2 | 4,420 | Treude et al. (2002) |
| *Paracallisoma coecus* | 0.04–0.05 | 5.5 | 500 + | Childress (1975) |
| *Bathyporcia pelagica* | 0.10–0.52 | 5.0 | 0 | Fish and Preece (1970) |
| *Gammarus oceanicus* | 0.04–0.12 | 5.0 | 0 | Halcrow and Boyd (1967) |
| *Parathemisto japonica* | 0.33–0.45 | 6.5–7.0 | 10 | Ikeda (1974) |
| *Boekosiumus affinis* | 0.58 | 0 | 6 | Busdosh and Atlas (1975) |
| *Paramoera walkeri* | 0.06–0.14 | 2.0 | 50 | Klekowski et al. (1973) |
| *Paramoera walkeri* | 0.03–0.06 | 1.9 | 5 | Opalinski (1974) |
| *Waldeckia obesa* | 0.0002–.006 | 0 | 60–100 | Chapelle et al. (1994) |

Echinoderms are a dominant component of the deep-sea epibenthic megafauna in soft-bottom habitats. These large animals generally occur in the sediment-water interface, but may burrow or swim short distances. Respiration rates of three ophiuroid species were determined *in situ* (Smith, 1983). *Ophiophthalmus normani* (1,300 m), *Ophiomusium lymani* (1,230 m) and *Ophiomusium armigerum* (3,650 m) displayed similar respiration rates. These brittle stars are of similar size and have similar feeding habits and locomotory abilities. Over the depth range of 1,230 to 3,650 m pressure does not seem to be a significant parameter in establishing oxygen consumption rates in ophiuroids. Smith (1978) observed a similar phenomenon with two benthopelagic teleost fish species of the genus *Coryphaenoides*. The respiration rates of *C. acrolepis* (1,230 m) were not significantly different from those of *C. armatus* (3,650 m), although the rates of both species were much lower than those of cold-adapted shallow-water fishes.

The holothurian *Scotoplanes globosa* is a conspicuous component of the bathyal benthic boundary layer community off the coast of California. *S. globosa*, measured *in situ* at 1,300 m, had a respiratory rate similar in magnitude to 14 species of shallow-water sea cucumbers with body temperatures of 6 to 25°C (Smith, 1983). The feeding habits and locomotor abilities of these species are similar. For echinoderms there is no apparent depth-related pattern in respiratory rate.

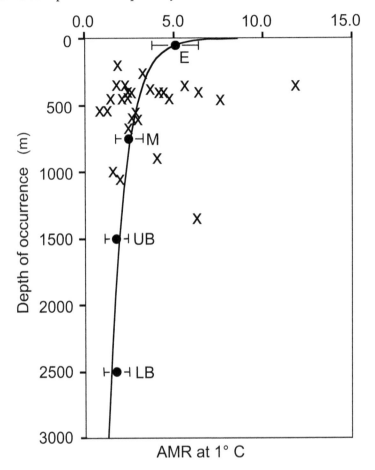

**Figure 6.** Relationship of adjusted metabolic rate (AMR) at 1°C and depth of occurrence in midwater copepods. Copepods were from the epipelagic (E, 0 to 200 m), mesopelagic (M, 500 to 1000 m), upper-bathypelagic (UB, 1000 to 2000 m) and lower-bathypelagic (LB, 2000 to 3000 m) depth zones. Respiration measurements were made at atmospheric pressure and *in situ* temperatures. Data were adjusted to a temperature of 1°C using a $Q_{10}$ value of 2. Experiments on respiration rates of copepods recompressed to 10.1 MPa found no statistical difference between respiration rates of groups at 0.1 and 10.1MPa. The figure includes data from Theusen et al. (1998). Redrawn from Ikeda et al. (2006).

In contrast, midwater Crustacea have lower metabolic rates with increasing depth (e.g., Cowles et al., 1991; Childress, 1995; Ikeda et al., 2006). In a comparison of over 50 species of pelagic copepods occurring from the surface to 2,500 m there is a decline in respiratory rates with depth (Fig. 6, Ikeda et al., 2006). Respiratory rates were measured at atmospheric pressure and *in situ* temperatures and standardized to a temperature of 1°C using a $Q_{10}$ of 2. The oxygen consumption rates decreased with increasing depth, but were also influenced by life history stage, sex and feeding type. To validate the oxygen consumption rates, electron transport system (ETS) enzyme activities were measured. These values correlated well with the expected 2:1 ratio of ETS enzyme activity to respiration, except for the copepods from the epipelagic zone (Fig. 7). The anomalous ratio (1.4) for the epipelagic species may result from the fact that the ETS measurement and the respiratory rates were from different sets of species. The copepod species in this study rely on chemo- and/or mechanoreceptors rather than visual perception for foraging and predator avoidance.

To test the effects of environmental parameters on respiration rates, Cowles et al. (1991) compared Pacific Ocean assemblages of midwater Crustacea (orders Decapoda, Mysidacea and Euphausiacea) from off Hawaii with Crustacea from off Southern California (Childress, 1975). Standardized oxygen consumption rates, measured at atmospheric pressure, decreased with increasing depth of occurrence in both assemblages (Fig. 8). Part of the decline in respiratory rate with depth could be accounted for by temperature, but a significant decrease remained after correction for this parameter. At depths shallower than 400 m these Crustacea probably rely on the visual environment in predator-prey interactions.

A study of 33 species of pelagic cephalopods from off California and off Hawaii found a strong decrease in respiratory rate with increasing depth of occurrence (Seibel et al., 1997). Over the 1,000 m depth range of the cephalopods studied, the rate of decrease with depth was steeper than that observed for crustaceans. The decline in metabolic rate with depth was interpreted as reflecting a decreased need for strong locomotory abilities in the deep sea because of a decreased reliance on visual interactions in this light-limited environment and because of reliance on more efficient locomotory styles, such as fin swimming versus jet propulsion by mantle contraction.

In contrast to crustaceans and cephalopods, other pelagic taxa, which are not reliant on visual orientation, showed no clear trend in respiratory rates with depth. Sixteen species of pelagic chaetognaths spanning a depth range of 2,400 m did not display a decline in respiratory rate with increased depth. (Fig. 9, Thuesen and Childress, 1993). Oxygen consumption rates of 14 species of hydromedusae and 5 species of coronate scyphomedusae with

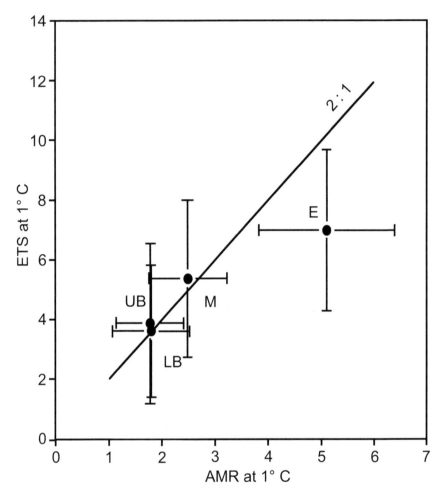

**Figure 7.** Relationship of adjusted metabolic rate (AMR) at 1°C and electron transport system enzyme activity (ETS) at 1°C in midwater copepods. Copepods were from the epipelagic (E, 0 to 200 m), mesopelagic (M, 500 to 1000 m), upper-bathypelagic (UB, 1000 to 2000 m) and lower-bathypelagic (LB, 2000 to 3000 m) depth zones. Respiration measurements were made at atmospheric pressure and *in situ* temperatures. ETS was measured at 10°C at atmospheric pressure. Data were adjusted to a temperature of 1°C using a $Q_{10}$ value of 2. Redrawn from Ikeda et al. (2006).

minimum depths of occurrence from 10 to 1,100 m also displayed no change in respiratory rate with depth (Fig. 10, Thuesen and Childress, 1994 ).

Childress (1995) and colleagues have advanced the visual interactions hypothesis to explain the decreased metabolic rates of deep-living fishes, crustaceans and cephalopods relative to their shallow-living counterparts. In the deep-sea environment where downwelling light is limiting and

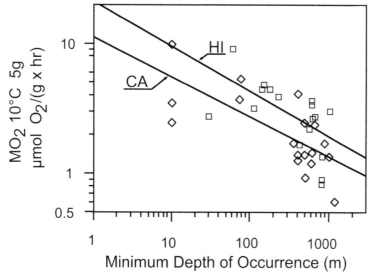

**Figure 8.** Regulated oxygen consumption normalized to 10°C and 5 g wet weight as a function of minimum depth of occurrence for midwater Crustacea from Hawaii (squares) and California (diamonds). The minimum depth of occurrence of a species is defined as the depth below which 90% of the population is found. Redrawn from Cowles et al. (1991).

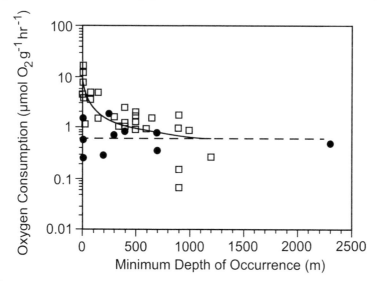

**Figure 9.** Oxygen consumption rates of pelagic crustaceans (squares) and chaetognaths (circles) versus minimum depth of occurrence. The minimum depth of occurrence of a species is defined as the depth below which 90% of the population is found. The specimens were collected off the coast of California. Measurements were made at atmospheric pressure; the temperatures employed reflect the temperatures the 16 chaetognath species and 28 crustaceans experience *in situ*. Redrawn from Thuesen and Childress (1993).

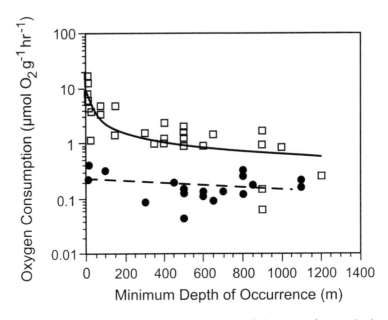

**Figure 10.** Oxygen consumption rates of medusae (circles) compared to respiration rates of 28 species of pelagic crustaceans (squares). The minimum depth of occurrence of a species is defined as the depth below which 90% of the population is found. The specimens were collected off the coast of California. The measurements of oxygen consumption rates of the medusae were made at atmospheric pressure, 5°C. Measurements of the crustaceans were as described in Fig. 9. Measurements were made on 14 species of hydromeduasae and 5 species of coronate scyphomedusae. Redrawn from Thuesen and Childress (1994).

bioluminescence restricts interactions to short distances, there is a reduced requirement for strong locomotor ability. The decrease in visually-cued predator— prey interactions with increasing depth is reflected in decreased locomotor abilities and decreased metabolic rates. The visual interaction hypothesis relates the decreased metabolic rates of visually-orienting deep-sea taxa to adaptations to predator—prey encounters rather than adaptations to temperature, pressure or food availability. Within these broad patterns, there may be other adaptations. Frank (2003) identified two groups of deep-sea crustaceans based on the flicker fusion frequency adaption of their photoreceptors to light. The differences among the two groups may reflect the bioluminescent mode and behavior of their preferred prey.

To test the effects of hydrostatic pressure on the respiratory rate of deep-living gelatinous zooplankton, Childress and Thuesen (1993) examined the effects of 0.101 and 10.1 MPa on six species from depths of 500 to 1,500 m. The species included three chaetognaths, two hydromedusans and a polychaete. Groups were tested at 0.101 MPa and 10.1 MPa, comparing a size range of individuals. With the exception of the data for the polychaete there was no significant effect of pressure on metabolic rate. For the polychaete species, the size range of individuals tested at elevated pressure was limited and there was not a consistent effect of pressure. However, the respiratory rates of the individuals tested at pressure bracketed the values obtained at atmospheric pressure for individuals of similar size. Thus, for these species there appears to be no significant effect of pressure on metabolic rate.

Other experiments with midwater species have shown that pressure changes can result in variations in activity and metabolic rate (e.g., Childress, 1977; Mickel and Childress, 1982). The copepod, *Gaussia princeps*, which migrates diurnally from below 400 m to 200 – 300 m has a daytime depth of 400–900 m. To examine the effects of vertical migration on metabolic rates, oxygen consumption rates were determined at 3 temperatures and 6 pressures (Fig. 11, Childress, 1977). Pressures as low as 2.837 MPa affected the oxygen consumption rate. Pressure and temperature effects interacted;

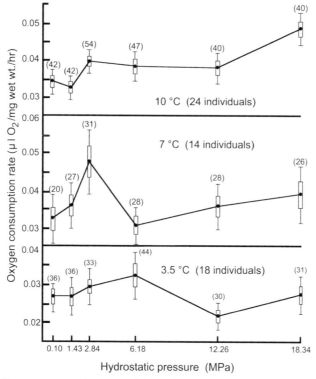

**Figure 11.** Oxygen consumption rates of the midwater copepod *Gaussia princeps* at different temperatures and pressures. Numbers in parentheses are the number of observations at each pressure. The blocks represent the standard error of the mean and the vertical lines represent the 95% confidence intervals. Redrawn from Childress (1977).

the $Q_{10}$ values are higher at elevated pressures. At all temperature and pressure combinations *G. princeps* has a low metabolic rate in comparison with shallow-living copepods.

The effects of pressure changes were studied on individuals of the bathypelagic mysid *Gnathophausia ingens* acclimated to either 0.101 or 7.599 MPa (Mickel and Childress, 1982). Activity was measured as pleopod beats per min. Pressure changes, both increases and decreases, caused transient increases in activity (Fig. 12). Pressure increases led to decreased activity rates. The activity of individuals acclimated to atmospheric pressure and 7.599 MPa were compared at 5 pressures (Fig. 13). The activity rates of individuals acclimated to 7.599 MPa was lower at pressures from 0.101 to 22.798 MPa. (Because of the low number of individuals, the results at 30.397 MPa were not analyzed.) Although there are differences in oxygen consumption rates for the groups acclimated to different pressures, this appears to result from differences in activity levels between the two groups rather than pressure effects per se.

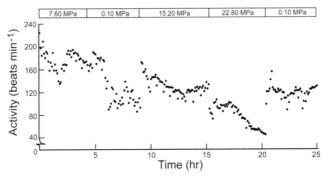

**Figure 12.** Activity, measured as pleopod beats min$^{-1}$, of an individual bathypelagic mysid, *Gnathophausia ingens* as a function of pressure at 5.5°C. The individual was acclimated to 7.60 MPa and exposed to 4 pressures over the time indicated. Redrawn from Mickel and Childress (1982).

## SUMMARY

The adaptations of membranes and proteins that are critical in allowing organisms to tolerate the high pressures and low temperatures of the deep ocean appear to play at most a minor role in establishing the metabolic rates of deep-sea species. Some characteristics of deep-sea enzymes, for example, the decreased catalytic capacities ($k_{cat}$ values) observed for LDH orthologs from deep-sea species may lead to some metabolic depression. However, if this decreased catalytic capacity results from a more pressure-stable enzyme, with an increased life time in the cell, this would be, overall, an energetic saving and advantage. Also, the fact that dehydrogenases of deep-sea fishes

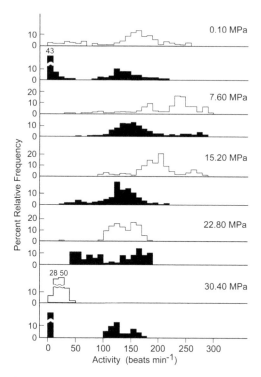

**Figure 13.** Activity of *Gnathophausia ingens* individuals at 5 pressures. Open bars represent individuals acclimated to atmospheric pressure (0.10 MPa). The number of observations at each pressure ranged from 60 to 356, with 1 to 5 individuals at each pressure. Dark bars represent individuals acclimated to 7.60 MPa. The number of observations at each pressure ranged from 54 to 401, with 2 to 5 individuals at each pressure. Redrawn from Mickel and Childress (1982).

and invertebrates have convergently evolved pressure resistance indicates that pressure changes need not alter metabolism. The accumulation of cytoprotectants which stabilize protein structure and function from pressure perturbation also would argue that deep-living organisms have evolved systems to protect against pressure inhibition of metabolism.

The light environment may play an important role for visually orienting taxa in establishing metabolic rates. Both downwelling sunlight and bioluminescence may influence the predator - prey interactions and selection for locomotory ability and the metabolism to power it.

For species which migrate vertically, either diurnally or over their life history, temperature changes may have significant effects on metabolism because $Q_{10}$ values are 2 or greater. Pressure changes may alter activity levels and through activity changes, oxygen consumption rates. However, these effects may be transient.

## ACKNOWLEDGMENTS

I thank David Siebenaller for his assistance in preparing the figures.

## REFERENCES

Anderson, T.R. and T. Rice. 2006. Deserts on the sea floor: Edward Forbes and his azoic hypothesis for a lifeless deep ocean. Endeavour 30: 131–137.

Angel, M.V. What is the deep sea? Pp. 1–41. *In*: D.J. Randall and A.P. Farrell (eds.). 1997. Deep-Sea Fishes. Academic Press, San Diego, USA.

Aquino-Souza, R. and S.J. Hawkins and P.A. Tyler. 2008. Early development and larval survival of *Psammechinus miliaris* under deep-sea temperature and pressure conditions. J. Mar. Biol. Assoc. UK 88: 453–161.

Avrova, N.F. 1984. The effect of natural adaptations of fishes to environmental temperature on brain ganglioside fatty acid and long chain base composition. Comp. Biochem. Physiol. 78B: 9003–909.

Blankenship, L.E. and A.A. Yayanos, D.B. Cadien and L.A. Levin. 2006. Vertical zonation patterns of scavenging amphipods from the hadal zone of the Tonga and Kermadec Trenches. Deep-Sea Res. I. 53: 48–61.

Brauer, R.W. 1984. Hydrostatic pressure effects on the central nervous system: perspectives and outlook. Phil. Trans. R. Soc. Lond. 304: 17–30.

Busdosh, M. and R.M. Atlas. 1975. Response of two Arctic amphipods, *Gammarus zaddachi* and *Boeckosimus (Onisimus) affinis*, to variations in temperature and salinity. J. Fish. Res. Bd. Canada 32: 2564–2568.

Carney, R.S. 2005. Zonation of deep biota on continental margins. Oceanogr. Mar. Biol. 43: 211–278.

Chapelle, G. and L.S. Peck and A. Clarke. 1994. Effects of feeding and starvation on the metabolic-rate of the necrophagous Antarctic amphipod *Waldeckia obesa* (Chevreux, 1905). J. Exp. Mar. Biol. Ecol. 183: 63–76.

Childress, J.J. 1975. The respiratory rates of midwater crustaceans as a function of depth of occurrence and relation to the oxygen minimum layer off southern California. Comp. Biochem. Physiol. 50A: 787–799.

Childress, J.J. 1977. Effects of pressure, temperature and oxygen on the oxygen-consumption rate of the midwater copepod *Gaussia princeps*. Mar. Biol. 39: 19–24.

Childress, J.J. 1995. Are there physiological and biochemical adaptations of metabolism in deep-sea animals? Trends Ecol. Evol. 10: 30–36.

Childress, J.J. and E.V. Thuesen. 1993. Effects of hydrostatic pressure on metabolic rates of six species of deep-sea gelatinous zooplankton. Limnol. Oceanogr. 38: 670–678.

Corliss, J.B. and J. Dymond, L.I. Gordon, J.M. Edmond, R.P. von Herzen, R.D. Ballard, K. Green, D. Williams, A. Bainbridge, K. Crane and T.H. van Andel. 1979. Submarine thermal springs on the Galápagos Rift. Science 203: 1073–1083.

Cossins, A.R. 1994. Temperature Adaptation of Biological Membranes, Portland Press, London.

Cossins, A.R. and A.G. Macdonald. 1989. The adaptations of biological membranes to temperature and pressure: fish form the deep and cold. J. Bioenerg. Biomembr. 21: 115–135.

Cowles, D.L. and J.J. Childress and M.E. Wells. 1991. Metabolic rates of midwater crustaceans as a function of depth of occurrence off the Hawaiian Islands: food availability as a selective factor? Mar. Biol. 110: 75–83.

Dahlhoff, E. and G.N. Somero. 1991. Pressure and temperature adaption of cytosolic malate dehydrogenases of shallow and deep-living marine invertebrates: evidence for high body temperatures in hydrothermal vent animals. J. Exp Biol. 159: 473–487.

DeLong, E.F. and A.A. Yayanos. 1985. Adaptation of membrane lipids of a deep-sea bacterium to changes in hydrostatic pressure. Science 228: 1101–1103.

Desmond, M.K. and J.F. Siebenaller. 2006. Non-additive counteraction of KCl-perturbation of lactate dehydrogenase by trimethylamine N-oxide. Protein Pept. Lett. 13: 555–557.

Fish, J.D. and G.S. Preece. 1970. The ecophysiological complex of *Bathyporeia pilosa* and *B. pelagic* (Crustacea: Amphipoda) 1. Respiration rates. Mar. Biol. 5: 22–28.

Forbes, E. 1844. On the light thrown on geology by submarine researches; being the substance of a communication made to the Royal Institution of Great Britain, Friday Evening, 23rd February 1844. Edinburgh New Philosophical Journal 36: 319–327.

Frank, T. 2003. Effects of light adaptation on the temporal resolution of deep-sea crustaceans. Integ. Comp. Biol. 43: 559–570.

Gage, J.D. and P.A. Tyler. 1991. Deep-sea Biology. Cambridge University Press, Cambridge, UK.

George, R.Y. 1979. Behavioral and metabolic adaptations of polar and deep-sea crustaceans: a hypothesis concerning physiological basis for evolution of cold adapted crustaceans. Bull. Biol. Soc. Wash. 3: 283–296.

Haedrich, R.L. Distribution and population ecology. Pp. 79–114. *In*: D.J. Randall and A.P. Farrell (eds.). 1997. Deep-Sea Fishes. Academic Press, San Diego USA.

Halcrow K. and C.M. Boyd. 1967. The $O_2$ consumption and swimming activity of the amphipod *Gammarus oceanicus* at different temperatures. Computers in Biochemistry and Physiology 23: 233–242.

Hennessey, J.P., Jr. and J.F. Siebenaller. 1985. Pressure inactivation of tetrameric lactate dehydrogenase homologues of confamilial deep-living fishes. J. Comp. Physiol. 155: 647–652.

Hennessey, J.P., Jr. and J.F. Siebenaller. 1987. Pressure-adaptive differences in proteolytic inactivation of M4-lactate dehydrogenase homologues from marine fishes. J. Exp. Zool. 241: 9–15.

Herring, P.J. 2002. The Biology of the Deep Ocean. Oxford University Press, New York, USA.

Hessler, R.R. The structure of deep benthic communities from central oceanic waters. Pp. 79–93. *In*: C.B. Miller (ed.). 1974. The Biology of the Oceanic Pacific. Oregon State University Press, Corvallis, OR, USA.

Ikeda, T. 1974. Nutritional ecology of marine zooplankton. Mem. Fac. Fish. Hokkaido Univ. 22: 1–97.

Ikeda, T. and F. Sano, A. Yamaguchi and T. Matsuishi. 2006. Metabolism of mesopelagic and bathypelagic copepods in the western North Pacific Ocean. Mar. Ecol. Prog. Ser. 322: 199–211.

Klekowski, R.Z. and K.W. Opalinski and S. Rakusa-Suszczewski. 1973. Respiration of Antarctic amphipoda *Paramoera walkeri* stebbing during the winter season. Polski Archiwum Hydrobiologii. 20: 301–308.

Koslow, T. 2007. The Silent Deep. University of Chicago Press, Chicago, USA.

Logue, J.A. and A.L. De Vries, E. Fodor and A.R. Cossins. 2000. Lipid compositional correlates of temperature-adaptive interspecific differences in membrane physical structure. J. Exp. Biol. 203: 2105–2115.

Marshall, N.B. 1979. Deep-Sea Biology. Garland STPM Press, New York, USA.

Mickel, T.J. and J.J. Childress. 1982. Effects of pressure and pressure acclimation on activity and oxygen consumption in the bathypelagic mysid *Gnathophausia ingens*. Deep-Sea Research. 29: 1293–1301.

Mozhaev, V.V. and K. Heremans, J. Frank, P. Masson and C. Balny. 1996. High pressure effects on protein structure and function. Proteins 24: 81–91.

Muresan, A. and H. Diamant and K.Y.C. Lee. 2001. Effect of temperature and composition on the formation of nanoscale compartments in phospholipid membranes. J. Am. Chem. Soc. 123: 6951–952.

Opalinski, K.W. 1974. Standard, routine and active metabolism of the Antarctic amphipod *Paramoera walkeri* Stebbing. Polskie Archwm Hydrobiol. 21: 423–429.

Ostrom, R.S. and S.R. Post and P.A. Insel. 2000. Stoichiometry and compartmentation in G protein-coupled receptor signaling: implications for therapeutic interventions involving $G_s$. J. Pharm. Exp. Ther. 294: 407–412.

Pickard, G.L. and W.J. Emery. 1990. Descriptive Physical Oceanography. 5th edit. Butterworth-Heinemann Oxford, UK.

Samerotte A.L. and J.C. Drazen, G.L. Brand, B.A. Seibel and P.H. Yancey. 2007. Correlation of trimethylamine oxide and habitat depth within and among species of teleost fish: an analysis of causation. Physiol. Biochem. Zool. 80: 197–208.

Saunders, P.M. and N.P. Fofonoff. 1976. Conversion of pressure to depth in the ocean. Deep-Sea Res. 23: 109–111.

Seibel, B.A. and E.V. Thuesen, J.J. Childress and L.A. Gorodezky. 1997. Decline in pelagic cephalopod metabolism with habitat depth reflects differences in locomotory efficiency. Biol. Bull. 192: 262–278.

Siebenaller, J.F. Pressure adaptation in deep-sea animals. Pp. 33–48. *In*: H.W. Jannasch, R.E. Marquis and A.M. Zimmerman. (eds.). 1987. Current Perspectives in High Pressure Biology. Academic Press, London, UK.

Siebenaller, J.F. and G.N. Somero. 1978. Pressure adaptive differences in lactate dehydrogenases of congeneric fishes living at different depths. Science 201: 255–257.

Siebenaller, J.F. and G.N. Somero. 1979. Pressure-adaptive differences in the binding and catalytic properties of muscle-type ($M_4$) lactate dehydrogenases of shallow- and deep-living marine fishes. J. Comp. Physiol. 129: 295–300.

Siebenaller, J.F. and G.N. Somero. 1989. Biochemical adaptation to the deep sea. Rev. Aquat. Sci. 1: 1–25.

Siebenaller, J.F. and D.J. Garrett. 2002. The effects of the deep-sea environment on transmembrane signaling. Comp. Biochem. Physiol. 131: 675–694.

Smith, K.L., Jr. 1978. Metabolism of the abyssopleagic *Coryphaenoides armatus* measured *in situ*. Nature 274: 362–362.

Smith, K.L., Jr. 1983. Metabolism of two dominant epibenthic echinoderms measured at bathyal depths in the Santa Catalina Basin. Mar. Biol. 72: 249–256.

Smith, K.L, Jr. and R.J. Baldwin. 1982. Scavenging deep-sea amphipods: effects of food odor on oxygen consumption and a proposed metabolic strategy. Mar. Biol. 68: 287–298.

Somero, G.N. and J.F. Siebenaller. 1979. Inefficient lactate dehydrogenases of deep-sea fishes. Nature 282: 100–102.

Takeuchi, I. and K. Watanabe. 1998. Respiration rate and swimming speed of the necrophagous amphipod *Eurythenes gryllus* from Antarctic deep waters. Mar. Ecol. Prog. Ser. 163: 285–288.

Thuesen, E.V. and J.J. Childress. 1993. Enzymatic activities and metabolic rates of pelagic chaetognaths: lack of depth-related declines. Limnol. Oceanogr. 38: 935–948.

Thuesen, E.V. and J.J. Childress. 1994. Oxygen consumption rates and metabolic enzyme activities of oceanic California medusa in relation to body size and habitat depth. Bio. Bull. 187: 84–98.

Theusen, E.V. and C.B. Miller and J.J. Childress. 1998. Ecophysiological interpretation of oxygen consumption rates and enzymatic activities of deep-sea copepods. Mar. Ecol. Prog. Ser. 168: 95–107.

Treude, T. and F. Janssen, W. Queisser and U. Witte. 2002. Metabolism and decompression tolerance of scavenging lysianassoid deep-sea amphipods. Deep-Sea Res. 49: 1281–1289.

Van Dover, C.L. and C.R. German, K.G. Speer, L.M. Parson and R.C. Vrijenhoek. 2002. Evolution and biogeography of deep-sea vent and seep invertebrates. Science 295: 253–257.

Yancey, P.H. 2005. Organic osmolytes as compatible, metabolic and counteracting cytoprotectants in high osmolarity and other stresses. J. Exp. Biol. 208: 2819–2830.

Yancey, P.H. and J.F. Siebenaller. 1999. Trimethylamine oxide stabilizes teleost and mammalian lactate dehydrogenises against inactivation by hydrostatic pressure and trypsinolysis. J. Ex. Biol. 76: 749–757.

Yancey, P.H. and AL. Fyfe-Johnson, R.H. Kelly, V.P. Walker and M.T. Auñón. 2001. Trimethylamine oxide counteracts effects of hydrostatic pressure on proteins of deep-sea teleosts. J. Exp. Zool. 289: 172–176.

Yancey, P.H. and M.D. Rhea, K.M. Kemp and D.M. Bailey. 2004. Trimethylamine oxide, betaine and other osmolytes in deep-sea animals: depth trends and effects on enzymes under hydrostatic pressure. Cell. Mol. Biol. 50: 371–376.

Yano, Y. and A. Nakayama, K. Ishihara and H. Saito. 1998. Adaptive changes in membrane lipids of barophilic bacteria in response to changes in growth pressure. Appl. Environ. Microbiol. 64: 479–485.

Yayanos, A.A. 1978. Recovery and maintenance of live amphipods at a pressure of 580 bars from an ocean depth of 5700 meters. Science 200: 1056–1059.

Yayanos, A.A. 1981. Reversible inactivation of deep-sea amphipods (*Paralicella capresca* [sic]) by a decompression from 601 bars to atmospheric pressure. Comp. Biochem. Physiol. 69: 563–565.

Young, C.M. and P.A. Tyler and J.D. Gage. 1996. Vertical distribution correlates with pressure tolerance of early embryos in the deep-sea asteroid *Plutonaster bifrons*. Mar. Biol. Assoc. UK 88: 453–161.

# 14

# Fish and Pressure

*Alfredo Damasceno-Oliveira*[1] and *Philippe Sébert*[2]

## INTRODUCTION

The diversity of fish is larger than for any other vertebrate group. There are more fish species (over 25,000) than of all the other vertebrates together. Fish also inhabit more diverse habitats than any other group of vertebrates, from Himalayan or Andean brooks at 4,000 meters to abyssal depths of at least 8,370 m (Nielsen, 1977), thus spanning an extremely wide range of hydrostatic pressures. This means that members of this taxonomic group experience a range of pressure from less than 0.1 MPa up to more than 82 MPa (1 atm = 1 bar = 14.7 psi = 0.10 MPa).

Since pressure increases by about 0.1 MPa for each 10 m of depth, it is by far the most variable factor in natural aquatic environments. It is important to highlight that even at low and moderate depths, fish are exposed to pressures several times higher than atmospheric pressure. In addition, many species experience daily or seasonal variations in pressure. A fish swimming from near surface to the sea bottom at 40 m will experience a rapid pressure increase from 0.1 to 0.5 MPa absolute pressure, i.e. a 5-fold increase. In general, other biologically important factors like temperature, salinity, oxygen levels and light are much more constant. In other words, high pressure environments (ocean depths greater than a few hundred meters) are generally characterized as having low temperatures (0–6°C), an absence of light, salinities (S) of 34.6–34.8 and relatively lower oxygen levels than those near the surface. There are a few exceptions where high pressure environments have particular characteristics, such as relatively high

---
[1] CIMAR-LA/CIIMAR, Universidade do Porto, Rua dos Bragas, 289, 4050-123 Porto, Portugal.
E-mail: aol@ciimar.up.pt
[2] Université Européenne de Bretagne, Université de Brest, Laboratoire ORPHY-EA 4324, 29238 Brest Cedex 3, France.
E-mail: Philippe.Sebert@univ-brest.fr

temperatures of ~13.5°C at thousands of meters in the Mediterranean (Yayanos, 1995), freshwater in the Baikal Lake or high temperatures and high mineral concentrations near hydrothermal vents.

Most of the biosphere is in fact composed of aquatic habitats characterized by pressures substantially different from that which is usually considered as normal. However, despite its importance as an ubiquitous environmental factor which may potentially affect living processes and organisms, the biological effects of hydrostatic pressure have been only a slowly growing research theme, mainly because of the difficulties in their study.

## A MODEL TO STUDY THE EFFECTS OF HYDROSTATIC PRESSURE: SHALLOW-WATER OR DEEP-SEA FISH?

In studying the effects of hydrostatic pressure on biological systems it is fundamental to experimentally dissociate those effects from those of other factors, though these interactions could be of obvious interest. Particularly difficult, at all levels of biological organization, is the distinction between hydrostatic pressure effects *per se* and increased gas partial pressures effects. When studying air breathing organisms under altered pressures, large biases are introduced as many physiological effects are related to the gas mixture present.

The only direct approach to study the effects of hydrostatic pressure *per se* on air-breathing animals is to use animals breathing liquids. From the first experiments by Kylstra (1962) to clinical trials with humans (Greenspan et al., 1999) lung ventilation with liquids like perfluorochemicals has been used in other research areas, but it remains a rather invasive technique with limited application in barobiology.

In fish and other aquatic, water-breathing organisms, however, those difficulties do not exist (Fenn, 1967). As pressurized water (in the absence of a gas phase) has a gas content which is independent of hydrostatic pressure in a wide range of values, fish are only submitted to hydrostatic pressure, and not to a gas mixture with a substantial increase in gas partial pressures as in air-breathing organisms. This allows the dissociation of the specific effects of hydrostatic pressure.

As outlined by Sébert (1997), other advantages of using fish as model organisms for the study of biological effects of pressure can be highlighted. One is the fact that fish breathe water, a medium with a density largely unaffected by changes in hydrostatic pressure thus not influencing the ventilatory mechanisms. Another advantage pointed out is their close physiological relationship with the environment: individual water parameters like temperature, salinity and pressure have a direct impact on susceptible organs, tissues and cells, but it also enables interactions between those factors and physiological changes to be studied.

Furthermore, most of the species in the subphylum Vertebrata are fish. Their diversity and wide range of habitats, along with their relative phylogenetic closeness and the similarity of most biochemical pathways, physiological mechanisms and organ systems with higher vertebrates (including humans) present incomparable advantages when fish are used as model systems. Most of the results and conclusions obtained on fish can be easily translated to other organisms. In addition, the use of fish as models in other research areas like genomics or proteomics has dramatically increased in recent years (Cossins and Crawford, 2005), potentially providing valuable tools to study the effects of hydrostatic pressure on living organisms. Even the existence of genome duplication events and the genomic complexity of fish, rather than complicate their use as models could help in understanding of the plasticity of many physiological processes (Meyer and van de Peer, 2005). Finally, some fish species are easy and inexpensive to obtain and maintain in the laboratory, which can be useful in some specific experimental works.

Adaptive differences between species are also useful to understand different biological processes of life under elevated pressure. Studies on deep-water fish reveal adaptations to elevated hydrostatic pressure (and also to other environmental factors in their habitat). On the other hand, studies in shallow-water fish, which normally live in relatively low hydrostatic pressure habitats, reveal the effects of elevated pressure *per se*. Some other species (like the European eel *Anguilla anguilla* Linnaeus) which during their life cycle occupy alternating habitats of low and high hydrostatic pressures could be extremely useful models in both of those two categories of studies.

It can still be argued that most fish species do not naturally face dramatic modifications of ambient pressure. However, due to the advantages stated earlier, both shallow-water fish and deep-sea fish could be considered good model species for studying the effects of altered hydrostatic pressure. Moreover, there are many fish species that face seasonal (or at least during a specific stage of their life cycle) changes in hydrostatic pressure. Of further relevance is the fact that even those shallow-water species that do not naturally enter great depths (like for example, the rainbow trout *Oncorhynchus mykiss* Walbaum or the goldfish *Carassius auratus auratus* Linnaeus) or those deep-sea species that are not observed at waters near to the surface cannot be easily classified as strictly stenobaric. In comparison, other species like the European eel can clearly be considered as eurybaric (Sébert and Theron, 2001). Although it is difficult to establish the limits for the distinction between stenobaric and eurybaric, there are fish species that are comparatively more tolerant to pressure changes. This could be the case of those species that face daily or seasonal, or at least during a specific stage of their life cycles, changes in hydrostatic pressure.

It can be concluded that, whenever using shallow-water fish to study the effects of elevated hydrostatic pressures or using deep-sea fish to study the effects of decreased hydrostatic pressures, fish diversity allows an extremely large range of choice of which species to use. Specific research objectives, financial costs and technical problems or solutions could overwhelm that choice.

## CHALLENGES TO STUDING FISH BAROBIOLOGY

As Sébert and Macdonald (1993) pointed out, it is a curious fact that few marine biologists in general, and fish biologists in particular, have responded to the extraordinary range of interesting hyperbaric phenomena which fish present. Even so, for more than 100 years biologists have studied pressure effects on fish using a number of different techniques (Sébert, 2002).

The work on the biological effects of pressure on fish was initiated more than 120 years ago, by Regnard and Bert (Regnard, 1884, 1885; Bert, 1878). Their work on aquatic animals like the goldfish and the European eel represents the starting point in this research field, providing the first results on the responses of shallow-water fish to increased pressures under experimental laboratory conditions and by using pressurized aquaria. Since then, two methodologies have been used in all the studies focusing on pressure effects on fish, involving either the use of field methodologies or laboratory high pressure equipment.

Field methodologies involve the study of fish living under hydrostatic pressure and ideally, under their overall native environmental conditions, i.e., at their native depths. This allows making observations of the behaviour and analyzing some physiological parameters in fish. On the other hand, costs and the logistics involved are considerable, mainly cost of ship-time and because undertaking such experiments requires research platforms like manned or remotely operated undersea vehicles and also autonomous free vehicles (also called lander vehicles).

These platforms have largely been used for the observation of the behaviour (feeding, swimming, reaction to disturbances like light, among others) of deep-sea fish (Barham et al., 1967; Cohen, 1977). They have also been used to deploy and operate respirometers as well as to capture fish (see below in this section). In the first successful deployment of a respirometer operated from a manned submersible, Smith (1978) measured the *in situ* oxygen consumption of the Abyssal grenadier (*Coryphaenoides armatus* Hector) at depths up to 3,650 m. Since then, respirometers have been sporadically used with both manned and remotely operated vehicles in order to pump or capture fish into the respirometry chamber (Smith and Baldwin, 1997). Respirometry *in situ* has also been used with autonomous vehicles (Bailey et al., 2002), deployed to depths up to 6,000 m, capturing fish attracted to bait and then measuring their oxygen consumption at ambient temperature and pressure. These autonomous free vehicles, equipped with cameras and instrumentation are also used to track and study the behaviour of deep-sea fish, with the aid of bait transponders (Collins et al., 1999; Vardaro et al., 2007). Nowadays, the technology incorporated into all of these platforms for field studies allows data obtained (like video recordings or oxygen sensor data) to be transferred and analyzed either after recovery of the platform or remotely aboard ship.

A different approach to field studies is comparing fish captured or maintained in cages at different depths, i.e. at different hydrostatic pressures. This technique has rarely been used, mainly due to practical reasons (like the inability to provide food to experimental fish) but even so, it can provide interesting results (Fontaine et al., 1985).

Field methods have yielded significant results, but are limited when compared with laboratory methodologies used in physiological studies of live animals (*in vivo*), at various levels, from gene expression to organismal physiology. In addition, the use of high pressure equipment to study *in vitro* effects of altered hydrostatic pressure provides data which allows a clearer interpretation than that obtained using samples from dead or moribund fish collected at depth.

Technical difficulties involved in the use of high pressure equipment in experiments with fish can be significant. To study the effects of hydrostatic pressure *per se* on whole organisms it is absolutely necessary to guarantee the complete absence of a gas phase in contact with the water where the aquatic organisms are maintained, otherwise the experimental animals will be simultaneously submitted to hydrostatic pressure and to considerably increased gas partial pressures. The gas concentration in an aqueous solution is unchanged when this solution is pressurized in the absence of a gas phase. On the contrary, a solution equilibrated with a high pressure of gases has a concentration of gas which is related to the pressure of the applied gas phase. In both situations the solutions have an elevated hydrostatic pressure.

In the presence of a gas phase it is impossible to discern the different effects and there could be toxic, neuronal or other effects on tissues, cells and their membranes directly provoked by the high concentrations of the dissolved gases. In order to avoid this, the water in the experimental chamber must be compressed from atmospheric pressure using hydraulic pumps or with compressed gases but isolated by a gas-proof membrane. Additionally, the accumulation of metabolites, hypercapnia and hypoxia limit the duration of *in vivo* experiments to some hours; to perform longer experiments (weeks or more) only a high pressure water circulation system can assure the maintenance of good water conditions, but without compromizing the fundamental requisite of not having significant pressure oscillations inside the chamber containing the experimental fish.

This methodology has been reviewed before (Sébert, 2002) but it must be emphasized that many of the results in the early periods of research were obtained by the use of systems that were not designed with these precautions. Nevertheless, high pressure equipment with volumes large enough to be used with adult fish were developed by several teams and described in previous works (Nishiyama, 1965; Brauer and Jordan, 1972; Avent, 1975; Quentin and Childress, 1980). Several solutions and designs were developed, but in general, these systems ranged from useful volumes of 350 litres and working pressures of 2 MPa (Blaxter and Tyler, 1972) to systems with useful volumes of 40 litres and working pressures of 50 MPa (Johnstone et al., 1989), all with only mechanical control over the pressure set-point values.

Currently, there are several designs/systems being used to study high pressure effects in the laboratory which fulfill the earlier mentioned requirements. They also have the characteristics necessary for the study of adult fish, namely a sufficiently large volume, flow-through water circulation and the possibility of fine-tuning control of the water parameters and to visualize the experimental animals inside (Sébert et al., 1990; Shillito et al., 2001; Damasceno-Oliveira et al., 2004). The pressurized aquariums in these systems have maximum working pressures of 8 to 30 MPa, a range of volumes from 15 to 150 litres and flow rates of the circulating water between 0.3 and 8 litres per minute.

The use of high pressure systems to study deep-sea fish under laboratory conditions has also been attempted and is still one of the more interesting methods to be explored. For fish living at depths higher than 10–20 m, decompression has profound and frequently lethal effects (particularly for fish with swimbladders), even if repressurized following the retrieval on-board or on shore. Successful *in vivo* laboratory studies imply specimens have to be captured at depth, recovered at their natural pressure and transferred under pressure to experimental vessels or alternatively, by connecting the capture/recovery device to an appropriate water recirculation system. Other environmental parameters (like temperature, oxygen and light) should also be maintained as close to those existing at capture locations.

Capture and recovery of deep-sea organisms under natural environmental conditions have been attempted in recent decades. This has been successfully achieved for small organisms (MacDonald and Gilchrist, 1972; Yayanos, 1978; Yoshiki et al., 2006) but the technical complexities increase with the size and activity of the target organisms, as in the case of fish. In addition, preventing major changes in temperature is relatively easy with proper insulation, whereas maintaining hydrostatic pressures during and after recovery within a small range could represent formidable difficulties. Nonetheless, some research teams were recently able to develop and to test this type of equipment. Koyama et al. (2002) designed a pressurized aquarium system/trap for capturing deep-sea organisms using suction with a submersible servomotor. This system allows a continuous flow of water (0.04 litre per minute) to be circulated through the pressurized chamber (20 litres of volume) and it was used to keep a deep-sea fish alive and fed under constant pressure (5 MPa) for 64 days. With improvements made to the design used earlier by Wilson and Smith (1985), Drazen et al. (2005) used a high-pressure fish trap-respirometer (volume 89.1 litres) with an autonomous free vehicle. Some fish (*Coryphaenoides acrolepis* Bean) were captured on a baited hook, mechanically pulled inside the trap and then brought to the surface alive. This system also allows for pumps and instrumentation to be connected in order to ensure a flow of water through the system (up to 0.247 litres per minute) and to maintain fish at desirable levels of pressure (up to 41.4 MPa) and temperature. Finally, Shillito et al. (2008) developed a prototype of a device operated by a manned submersible and successfully captured and retrieved a fish from 2,300 meters depth. Though limited by its relatively small size, the design of this device supposedly allows its evolution to an isobaric transfer device if connected to a peripheral pressurized circuit. In brief, the few individuals captured and maintained in the laboratory do not permit the conclusion that an efficient methodology enabling sustained research has yet been completely developed.

While the techniques have been steadily improved as we have tried to demonstrate, it is expected that the study on the effects of high pressure on fish will benefit from the increased use of fish as models in biological and life sciences, as well as from the advent of new technologies. One simple example would be the development and improvement of the procedures to collect and preserve biological samples from fish submitted to pressure, since the decompression of the sample material could be, in many cases, of fundamental importance.

## ADAPTATION AND ACCLIMATION TO ELEVATED HYDROSTATIC PRESSURE

### Acclimation Versus Adaptation

The two different approaches to the study of the biological effects of hydrostatic pressure (in laboratory or in the field) were originally related to the study of the two types of responses that organisms exhibit to adjust to a changed environment, namely acclimation and adaptation.

Acclimation is the compensatory response of physiological processes to altered hydrostatic pressure and allows organisms to function and survive under this challenging condition (Prosser, 1991). At this point, it is worthwhile highlighting that many fish species experience daily or seasonal changes in hydrostatic pressure, having developed versatile physiological processes in order to cope with these varying conditions.

In this field of hydrostatic pressure research, acclimation of fish can be clearly divided into two stages, short- and long-term responses. As mentioned before, most of the experimental work has been done on short-term responses, mainly because long-term response studies need more complex equipment and techniques. Short-term acclimation to altered hydrostatic pressure involves responses occurring within minutes to hours of compression or decompression and in general is characterized by the mobilization of pre-existing components of a physiological process. Long-term acclimation involves responses that may also begin within minutes of compression or decompression but which are much more evident within longer periods (days, weeks) following hydrostatic pressure changes. It could be characterized, for example, by changes in gene expression, increased or decreased efficiency of physiological processes or even morphological changes. Acclimation occurs if long-term responses improve performance and survival under the altered conditions of hydrostatic pressure.

While acclimation occurs at the organismal level, adaptation occurs at the population or at the species level and over larger time scales covering multiple generations. Adaptation involves evolutionary changes in the genotype of a population improving its fitness for the conditions of hydrostatic pressure in their environment, like for example pressure tolerance. Adaptive responses may involve growth, reproduction, development and other physiological processes as well as changes in morphology. Deep-sea fish are thus clearly adapted not only to hydrostatic pressure conditions in their environment, but also to other factors therein present.

## Acclimation and the Effects of Altered Hydrostatic Pressure in Shallow-Water and Deep-Sea Fish

In order to study acclimation to altered hydrostatic pressures it is essential to understand that an observed biological effect is always a result not only of a specific pressure level (magnitude) but also of the period that the organism was exposed to that level. Likewise, the effects elicited during compression or decompression will vary with the rate at which the pressure is altered and also with the time elapsed after this phase.

Tytler and Blaxter (1973) investigated the rate at which fish inflate the swimbladder and reabsorb gas from this organ. They found that the rate of pressure change experienced by physoclistous fish (Atlantic cod *Gadus morhua* Linnaeus and saithe *Pollachius virens* Linnaeus) needs to be maintained below well defined limits otherwise the fish were not able to adjust buoyancy and swimbladder rupture could occur. The influence of the rate of compression on fish responses was firstly shown by Brauer et al. (1979) in the swimbladderless pleuronectiform *Symphurus plagiusa* Linnaeus. At constant temperature, convulsion threshold pressures (Pc) in this species were 9.9 and 11.6 MPa when exposed to compression rates of 101.0 and 13.2 MPa per hour, respectively. Both the juveniles and adults of the physostomous European eel also show distinct responses with respect to the rate of compression. These fish have oxygen consumption ($MO_2$) rates of about 1 mmol per hour per kg, at 17°C and at 0.1 MPa. However, when exposed to 10.1 MPa hydrostatic pressure, $MO_2$ increases to 3.5 mmol per hour per kg if using a compression rate of 0.2 MPa per min, but increases to 6 mmol per hour per kg if using a compression rate of 1 MPa per min (Sébert and Barthélémy, 1985a; Simon et al., 1989; Sébert et al., 1997).

Results such as these and other observations justify that caution must be taken when comparing results from distinct experimental designs and methodologies or even results at different time points in one experiment. Generally, as pointed out by Brauer (1984) two categories of effects would coexist: rate independent effects related to the hydrostatic pressure change as such, and time dependant compensation for these pressure-induced effects.

## Transduction of Hydrostatic Pressure and Its Changes

There is an increased understanding on how pressure changes are sensed by aquatic animals and cells. The processes underlying the transduction by fish of the information inherent in this stimulus are far from being clear, though and at this point it is reasonable to consider two general mechanisms responsible for the hydrostatic pressure effects observed in fish.

The most common mechanism by which environmental stimuli affect physiological processes implies their conversion into electrochemical signals and the propagation of these signals through sensory neurons to specific areas within the brain, allowing their processing and eliciting stimulatory or inhibitory neurophysiological phenomena, usually involving changes in neurotransmitter synthesis, storage, release, action and/or deactivation. For such a mechanism a pressure sensor or a structure that could act as a pressure transducer must be present. The most obvious pressure transducer is a gas-filled organ and in 1932 Warner already hypothesized that the main organ involved was the swimbladder, although admitting that nothing was known about pressure sensing. Later studies minimize its role in monitoring absolute depth or pressure changes (Blaxter and Tytler, 1972), namely in physostomous fish (in which the swimbladder is connected to the intestinal tract by a special duct, the *ductus pneumaticus*). It was also shown that in the European eel many effects of hydrostatic pressure are still observed after removing the gas bladder or cutting the lateral line (Barthélémy et al., 1981). Nevertheless, when experiencing relatively rapid and substantial pressure changes, physoclistous fish (with no duct between the swimbladder and the intestinal tract) will unavoidably respond by expansion or contraction of the swimbladder and have their buoyancy affected (Tytler and Blaxter, 1973).

As there are many fish species without swimbladders or other obvious gas compartments but with high sensitivity to pressure, other pressure transducers should exist. In fact, it was found that some hair cell types also play a role in fish sensitivity to hydrostatic pressure, namely in the discrimination of pressure changes. Hair cells located in the labyrinth of the dogfish (*Scyliorhinus canicula* Linnaeus) sense small hydrostatic pressure changes of 0.03 MPa (Fraser and Shelmerdine, 2002, Fraser in this book). Rather interestingly, a model for the detection of pressure by a cellular pressure sensor, like hair cell, and which could be applied to fish, was recently suggested by Bell (2007, 2008).

The second mechanism for hydrostatic pressure transduction in fish involves a direct action of hydrostatic pressure at molecular or cellular levels (Siebenaller and Murray, 1995). Many of the effects of hydrostatic pressure, some of which are highlighted in the next sections, will inevitably influence physiological processes. Considering, for example, that neurons (at least some populations) are barosensitive even at relatively low pressures, modulation of neuronal function will allow the transduction of ambient pressures experienced by and within the organism (Dean et al., 2003). In general, components which are affected by pressure or its changes could be considered themselves as elements of a transduction system, allowing fish to respond to environmental conditions (Hazel and Williams, 1990). Thus, due to the numerous enzymes complexes and/or proteins located in cellular and sub-cellular membranes these last can be considered as pacemakers of metabolism (Hulbert and Else, 1999). Increase in hydrostatic pressure instantaneously modifies membrane fluidity and thus the functioning of the membrane itself and of its associated proteins, influencing cell function. In this context, membranes can be considered as pressure sensors.

It is reasonable to expect that referred mechanisms are employed simultaneously, although with dissimilar relative importance and inducing different responses in fish depending on the specific environmental conditions. Pressure sensitivity also depends on the interaction with other environmental factors. Pavlov et al. (2000), studying the effects of pressure, temperature and light regimes in juveniles of different physostomous and physoclistous fish species demonstrated that hydrostatic pressure as an individual factor is one of the strongest to influence fish behaviour and that a gradient of 0.06 MPa changes the fish's response to the other two factors. Moreover, pressure sensitivity also varies between species although it has been shown that fish can generally have remarkably low pressure sensitivity thresholds.

Some physoclistous species, like the Atlantic cod, haddock (*Melanogrammus aeglefinus* Linnaeus) and saithe have pressure discrimination thresholds as low as 0.5–2% of ambient pressure (Blaxter and Tytler, 1972). Adult blennies (*Lipophrys pholis* Linnaeus), although not having a swimbladder, modulate their activity in response to tidal cycles of hydrostatic pressure (0.03 MPa amplitude) (Northcott et al., 1991). Their larvae, which do possess a swimbladder have an even lower threshold of sensitivity of only approximately 0.0005 MPa. This influence of hydrostatic pressure cycles with low amplitudes on behaviour has also been confirmed in other teleost species, for example, the flounder (*Platichthys flesus* Linnaeus) and it was suggested that it could act as a zeitgeber for endogenous rhythms allowing fish to react to changes in their environment (Gibson, 1982, 1984). In addition, it has been shown that some fish, like the weakly electric fish (*Gnathonemus petersii* Günther), use this low pressure sensitivity threshold to sense small changes in the water column (some centimeters) and thus for orientation, relying on an internal representation oriented to hydrostatic pressure (Cain and Malwal, 2002). Similar low pressure sensitivities are also found in chondrichthyan species, as suggested by the behavioural changes in the blacktip juvenile shark (*Carcharhinus limbatus* Müller and Henle) in their shallow-water habitats, which were found to be triggered by small decreases in pressure (Heupel et al., 2003).

It is therefore clear that fish respond to pressure including considerably small hydrostatic pressure changes, although there is still much unknown about the considerable variability of the responses and about the mechanisms used in the transduction of absolute or relative pressure.

## Organismal Effects

Behaviour and survival are most probably the best studied phenomena since the first experiments on fish using laboratory high pressure equipment. Most of the work with this objective has been performed on shallow-water fish, either marine or freshwater, but there were also a few studies regarding deep-water species. The results obtained in all these earlier works show that when fish experience a change in ambient pressure, two major responses are generally observed (see Sébert and Macdonald, 1993): (1) hyperexcitability and increased activity during compression/ decompression, and (2) tremors, convulsions, paralysis and then death, when pressure is increased above a specific threshold (convulsion threshold pressure; Pc). These responses, once again, depend on the species and, of course, the experimental protocol followed.

In shallow-water fish a pressure increase invariably induces an increased motor activity, frequently associated with upwards swimming (Nishiyama, 1965). In some species, this behavioural pattern is induced even with small pressure increases equivalent to a few meters depth (blackspot seabream *Pagellus bogaraveo* Brünnich) and also when pressure decreases (flounder) (Damasceno-Oliveira, unpublished data).

The only data available concerning deep-water fish are from experiments with fish captured at depth, decompressed and then recompressed for a short period (hours). Understandably, their responses to increased pressures were distinct. When deep-water species (*Trachyscorpia cristulata echinata* Koehler and *Synuphobranchus kaupii* Johnson) were recompressed after being captured at 900 m depths (9 MPa) there was a noticeable improvement in their condition and some of the specimens recovered their normal swimming activity. It was also shown that in general, fish from deep water have a convulsion threshold pressure (Pc) higher than shallow-water species. They undergo convulsions above 15 MPa, while shallow-water fish (for example, Atlantic salmon *Salmo salar* Linnaeus, European plaice *Pleuronectes platessa* Linnaeus, European eel and Atlantic cod) convulse at 9.3–11.4 MPa and exhibit immobilization at 15 MPa (Macdonald et al., 1987). Pressure tolerances were also assessed in deep-water species from the freshwater Lake Baikal and with the same sequence of changes in behaviour: recovery of normal swimming activity with recompression, increased activity, hyperexcitability, convulsions and finally disruption of locomotion and immobilization at higher pressures (Brauer et al., 1984). In this work, comparisons of Pc between shallow-water species (cottids) and deep-water species (abyssocottids) also revealed that in these fishes convulsion threshold pressures were generally higher than for the shallow-water cottids.

Oxygen consumption rates ($MO_2$) in fish undergoing changes in hydrostatic pressure are directly related to the above mentioned changes in motor activity. This is clearly observed in freshwater fish exposed to short-term changes in hydrostatic pressure. The European eel, when pressurized to 10.1 MPa for 3 h showed a strong increase in $MO_2$, which was evident within minutes of the compression commencing (Sébert and Barthélémy, 1985a). In bloater (*Coregonus hoyi* Milner) compression and exposure to 0.4 MPa absolute pressure for 3 h increased $MO_2$ and motor activity whereas a subsequent pressure decrease of the same magnitude and duration caused a return of oxygen consumption and motor activity to baseline values observed at 0.1 MPa (Speers-Roesch et al., 2004).

On the other hand, oxygen consumption rate measurements of deep-water fish subjected to altered hydrostatic pressure is very scarce and it would be necessary to use efficient pressure-retaining traps to obtain reliable results for these species. In one of the few available works, fish from different species captured at depths up to 1,000 meters showed $MO_2$ values at atmospheric pressure which were similar to those considered normal for shallow-water species (Torres et al., 1979). In another work, Roer et al. (1984) observed that when fish (*Batrachocottus nikolskii* Berg, *Cottinella boulengeri*

Berg and *Abyssocottus korotneffi* Berg) from 660–1,120 meters depths were recompressed to their normal high pressure (after experiencing decompression during capture), a decrease in pressure by half significantly decreased $MO_2$. However, subsequent restoration of the normal high pressure did not restore the initial $MO_2$ values, revealing that recompression after capture and successive pressure changes could alter $MO_2$ of deep-water fish thereby generating unreliable results.

In most of the previously mentioned studies, experiments were performed using pressurized systems with no water circulation and were limited to short periods of a few hours. Hence only short-term responses could be observed, mainly resulting from the compression/decompression ($\Delta P$ over time) procedures and making the differentiation between the specific effects of each factor involved difficult. For example, changes in oxygen consumption during the first hours under altered hydrostatic pressure are not only a direct effect of hydrostatic pressure but are also related to variations in motor activity. Nevertheless, it should be stressed that presumably processes involved in long-term responses are being affected from the first minutes of compression or decompression but will be discernible only after longer periods (days, weeks).

When pressure is increased and maintained for more than several hours/days, it is then possible to observe pressure effects ($\Delta P$) and thus, long-term acclimation to altered hydrostatic pressure. There are no results on long-term acclimation to altered hydrostatic pressure in deep-water species. In shallow-water fish, longer periods at elevated pressure and after pressure stabilization induced a recuperation of normal activity and a $MO_2$ decrease from a maximum during the initial hours of changed pressure. These long-term responses were undoubtedly demonstrated in the European eel (Sébert and Macdonald, 1993; Sébert et al., 1991, 1995). Using a pressurized system (Sébert et al., 1990) with water circulation it was possible to observe the effects of elevated hydrostatic pressure for days/weeks, under normoxic conditions and at constant temperature, giving enough time for fish to overcome the effects of the initial compression period. During compression (0.2 MPa per minute) to 10 MPa and at 17°C, fish exhibited intense and disordered motor activity, this being more pronounced above 4 MPa. After some days at constant pressure (10.1 MPa) this motor activity returned to normal levels, as seen in fish at atmospheric pressure. During and after decompression to atmospheric pressure, and after 30 d at 10.1 MPa, fish showed a considerable decrease in motor activity. Actually, in decompressed fish normal motor activity was regained when fish were recompressed to 10.1 MPa pressure. Oxygen consumption followed a similar trend. There was a sharp increase (with an amplitude which is directly related to the compression rate) in $MO_2$ during the compression and after some hours $MO_2$ values returned to their initial levels. Finally, after 4–7 d of progressively decreasing $MO_2$, eels exposed to 10.1 MPa pressure showed a steady $MO_2$ value 35% lower than control fish at atmospheric pressure.

## Stress

The study of stress in fish has received extensive attention and their responses are well characterized (Wendelaar Bonga, 1997; Barton, 2002). Paradoxically, the definition of stress (and of stressor) is not consensual and varies between authors. A wide definition is that stress is a stimulus and the stress response is the reaction of the organism to that stimulus (Pickering, 1981). Based on this definition, we would need to consider every interaction between the fish and its environment a stress. A narrower and more evident concept of physiological stress is when stress hormones (catecholamines and adrenocortical steroids) are involved in regulatory processes (Hughes, 1981). At the cellular level, this could be accompanied by protein changes, namely in heat shock protein (HSP) levels (Tolgyesi and Böde, in this book). However, their regulation and the mechanisms underlying the sensing of a stressor are far from clearly understood (Iwama et al., 2004).

Hydrostatic pressure, being an environmental stimulus which can be perceived by fish and having a broad range of physiological effects on them, could also lead to stress, particularly during experimental compression and decompression phases or when hydrostatic pressure is rapidly altered.

Characteristically, physiological responses to stress occur immediately after the exposure to the stressor. The primary response is the quick release of catecholamines (mainly norepinephrine and epinephrine) from the chromaffin tissue and endings of adrenergic nerves and also of cortisol from interrenal tissue into the circulation (Randall and Perry, 1992). Following this, changes in blood and tissue chemistry and in hematology (e.g. glycemia) occur, which are considered as the secondary response. Generally, after an acute exposure, many of the physiological indicators of stress are cleared quickly from circulation, representing an additional difficulty in hydrostatic pressure experiments.

In cold-stressed fish (Chen et al., 2002), the elevated catecholamines in the plasma reflected the stress status and this increase in catecholamine levels was followed by the elevation of plasma cortisol levels. That acute stress response is similar to the response of fish during compression and short-term exposure to altered hydrostatic pressure. In European eels exposed to 10.1 MPa for 2 and 3 h (Sébert et al., 1986), circulating catecholamines (namely norepinephrine and epinephrine levels) were markedly increased though nothing is known about short-term effect on plasma levels of cortisol. The effects of altered hydrostatic pressure on brain catecholamines, as well as the importance of the changes in brain catecholamines in pressure induced stress are more difficult to interpret. Nevertheless, as in cold-stressed fish, changes in brain catecholamine content in pressure exposed fish were also observed. Short-term exposures to elevated hydrostatic pressure induced increases in epinephrine content in the brain of European eels (Sébert et al., 1986). In contrast, long-term exposure to cyclic pulses of hydrostatic pressure but not constant elevated hydrostatic pressure induce a general decrease in catecholamine content in the brain of flounders (Damasceno-Oliveira et al., 2006, 2007).

The effects of circulating catecholamine levels are associated with the effects on cardiac responses (Sébert and Barthélémy, 1985b). European eels submitted to compression (1 MPa per min) up to 10.1 MPa exhibited a rapid increase in cardiac frequency and reached a steady state at the end of the compression that was on average 80% higher than the reference value at 0.1 MPa. After 1 h at 10.1 MPa mean heart rate tended to slowly return to pre-compression reference levels. The most significant decrease in cardiac frequency was observed during decompression from 4.2 MPa to 0.1 MPa.

In long-term experiments of 14 d, flounders exposed to constant (0.8 MPa) or cyclic (0.7 MPa amplitude) hydrostatic pressures showed no differences in biliary levels of free and conjugated corticosteroids (Damasceno-Oliveira, 2007). This indirect method of measuring the accumulation of biliary metabolites is a more useful potential indicator of chronic stress than the cortisol plasma levels and illustrated an example of fish acclimation to altered hydrostatic pressures if sufficient time is given.

Other alternative approaches to monitoring physiological indicators of stress like plasma cortisol and catecholamines are the above mentioned disturbances in behaviour and $MO_2$ (Schjolden et al., 2005; Farrell et al., 1998). Actually, patterns of metabolic rate of European sea basses (*Dicentrarchus labrax* Linnaeus) exposed to acute salinity changes (Dalla Via et al., 1998) are remarkably similar to those observed during and after compression on European eels (Sébert, 1993). This similarity also highlights once again the importance of differentiating between the effects due to compression and related increased motor activity and those due to the prolonged stages under altered pressure conditions.

The study of the stress response in fish exposed to altered hydrostatic pressure is particularly challenging and the reason is the present technical difficulty in obtaining samples from fish under pressure (Sébert et al., 1987). On the other hand, the rapid onset of stress responses, as in all experimental work involving some degree of manipulation of organisms, makes it difficult to isolate stress effects. It is therefore clear that in short-term experiments (minutes, hours) it is impossible to isolate the effects induced by stress from the compression, decompression or manipulation from those of hydrostatic pressure.

## Cellular and Molecular Effects

Most of the available knowledge on the effects of altered hydrostatic pressure in fish is focused on particular cellular and molecular aspects mainly related to energy metabolism. When comparing the results and conclusions contained within this considerable body of published research, several issues should be taken into account, namely the fact that short-term experiments do not allow the assignment of one specific response in fish to stress or to hydrostatic pressure effects, due to the interaction of these factors and the different experimental designs and conditions. Some fish have been extensively used to study these cellular and molecular effects, like the eurybaric European eel, but others have also been used like the stenobaric rainbow trout and several deep-sea species.

In the European eel (yellow eel phase), exposure to 10.1 MPa induced a rapid decrease in aerobic metabolism. At this pressure the energetic requirements were necessarily ensured by anaerobic processes with a concomitant production of lactate. In fact, it was observed that plasma lactate levels were similar to those observed at 0.1 MPa when oxygen partial pressure is lower than the oxygen level considered as critical at atmospheric pressure (Sébert and Barthélémy, 1985a). After 3 h under those pressure conditions, at 17.6°C, and in freshwater, fish showed a decrease in white muscle and liver glycogen content (the anaerobic substrate) and an accumulation of total fatty acids (the exclusive aerobic substrate). The simultaneous decrease of cytochrome c oxidase and increase of lactate dehydrogenase activities were observed (Sébert et al., 1993), indicating an alteration of aerobic energy production and an activation of the anaerobic pathway and apparently inducing a state resembling histotoxic hypoxia. Long-term studies (weeks) also provided further support to this, and in European eels exposed to elevated hydrostatic pressure (10.1 MPa) for 1 mon, the metabolic design and namely the glycolytic pathway in the white muscle was altered. It appears that pressure may optimize energy production, allowing there fish to cope with the new environment and energy demand during migration in seawater and at depth (Sébert et al., 1998).

Improvement of the aerobic pathway in red muscle and of the anaerobic pathway in white muscle was indicated by variations in the activities of key enzymes like lactate dehydrogenase (LDH), phosphofructokinase (PFK), isocitrate dehydrogenase (IDH), and creatine phosphokinase (CPK), but no changes were observed in the liver. Further supporting those conclusions, muscle and gill $Na^+$ and $Cl^-$ contents increased with a simultaneous decrease in the activity of gill $Na^+/K^+$-ATPase and $Mg_2^+$-ATPase whereas in plasma, $Na^+$, $Cl^-$ but also $Mg_2^+$ levels were increased under pressure. Hydromineral balance disturbances, namely a $Na^+$ balance impairment occurring at the tissue level were also induced by a long-term exposure to hydrostatic pressure (Sébert et al., 1991).

Modifications in levels of energetic nucleotides were revealed in short-term exposure (3 h) to elevated hydrostatic pressure (Sébert et al., 1987), with AMP increases and ATP decreases in muscle, but they were not modified in brain. Since a failure in $O_2$ transport from ambient medium to the cell was not observed, it was concluded that the aerobic cellular metabolic pathways, the Krebs cycle and the respiratory chain coupled to oxidative phosphorylation were possibly perturbed.

When European eels were exposed for 30 d to increased hydrostatic pressure (10.1 MPa), enzyme activities, substrates and nucleotide tissue contents were similar or higher than in fish maintained at atmospheric pressure. These results clearly indicate that the eurybaric European eel is able to acclimate to elevated pressure and to improve oxidative phosphorylation (P/O coefficient). This pressure acclimation is not an adjustment to the energetic state observed in the first hours under altered pressure, but a return to that one before compression (Simon et al., 1992).

Further studies have contributed to a more complete understanding of the increased efficiency of oxidative phosphorylation in this species after acclimation to high hydrostatic pressure. Following 21 d exposure to 10.1 MPa, the activities of the enzyme complexes of the respiratory chain were affected. The activity of complex II was decreased to approximately 50% of the control value and the activity of complex IV (cytochrome c oxidase) increased to 149% of the control value. ADP/O ratios were also significantly increased (Theron et al., 2000).

Surprisingly, both male and female European eels (silver phase) exposed to 10.1 MPa hydrostatic pressure for 3 wk showed modifications in slow red muscle mechanical properties (Rossignol et al., 2006). These authors observed a three-fold decrease in maximum isometric stress of twitch and tetanic contractions, although they did not rule out the influence of potential decompression effects on these results. Nevertheless, the modifications do not impede the long spawning migration of the eel at depth and are partially compensated by improved efficiency of oxidative phosphorylation.

Acclimation to elevated pressure may also induce morphological changes in the tissues. A clear example was demonstrated by Dunel-Erb et al. (1996) using European eels subjected to 10.1 MPa for 1 mon. These fish showed a lower number of mucous cells within the gill epithelium and the density of chloride cells was significantly increased, similarly to what is observed in acclimation to low temperature.

It was suggested that recompression would allow, or at least partially restore functions in organs like in the isolated and perfused hearts from decompressed deep-sea fish (Pennec et al., 1988). However, subsequent work carried out at the cellular level has helped us to understand why recompression of fish to elevated pressure (for example, recompression of deep-sea fish after capture) does not allow correct conclusions from these *in vitro* preparations (Theron and Sébert, 2003). These authors recompressed mitochondrial extracts from pressure-acclimatized European eels and observed a decrease in mitochondrial performances under pressure when comparing with extracts from the same fish and measured at atmospheric pressure.

In *vitro* studies on the activities of lactate dehydrogenase (LDH) from three hagfish inhabiting different depths under increased pressure (Nishiguchi et al., 2008) revealed differences which are most probably related to small differences in the amino acid sequence between shallow- and deep-water species. LDH in *Eptatretus okinoseanus* Dean (captured at 1,000 m) showed high activity at 100 MPa pressure and maintained 70% of that activity at 0.1 MPa, whereas LDH activity in *Paramyxine atami* Dean (captured at 250–400 m) showed a 55% decrease at 15 MPa, and that in *Eptatretus burgeri* Girard (captured at 45–60 m) was undetected at 5 MPa.

Studies on the pressure tolerance of cells from deep-sea eel (*Simenchelys parasiticus* Gill) and conger eel (*Conger myriaster* Brevoort) showed that after 130 MPa hydrostatic pressurization for 20 min there were no living conger eel cells, whereas all deep-sea eel cells remained alive after being subjected to pressures up to 150 MPa for 20 min. It was further demonstrated that pressurization at 40 MPa for 20 min of the conger eel cells induced disruption of actin and tubulin filaments, resulting in profound cell-shape changes. On the other hand, in the deep-sea eel cells, disruption of actin and tubulin filaments only occurred after being subjected to hydrostatic pressures above 100 MPa for 20 min. Conger eel cells were remarkably sensitive to pressure under experimental conditions and did not grow at 10 MPa, but deep-sea eel cells were capable of growth in pressures up to 25 MPa, although they stopped growing at 30 MPa. However, deep-sea eel cells took four times less time to finish the M phase at atmospheric pressure when compared with cells maintained at 20 MPa (Koyama et al., 2005). Additionally, other *in vitro* studies showed that polymerization of the cytoskeletal proteins of shallow-water fish involves large increases in volume (approximately 100 mL per mol) making them particularly sensitive to elevated pressure. On the other hand, in deep-sea fish polymerization of actin entails a much smaller volume change making them relatively unaffected by pressure (Koyama and Aizawa, 2008).

Finally, it was shown that elevated hydrostatic pressure (13.6 MPa) also influences protein and enzyme conformation, either in shallow- or deep-water fish, although to a different degree in these two groups (Davis and Siebenaller, 1992; Hennessey and Siebenaller, 1987). Based on the low susceptibility of enzymes like malate dehydrogenase and glyceraldehyde-3-phosphate dehydrogenase and the opposite high susceptibility of others like lactate dehydrogenase, it was hypothesized that specific structural characteristics of some enzymes were responsible for the conformational perturbations observed.

Presently it is not yet clear what genes and to what extent they are involved in acclimation (and also in adaptation) of fish to hydrostatic pressure but gene responses are expected. Results on gene expression in fish in response to elevated pressure are scarce (J.M. Wilson, unpublished data). Studies on gene expression in fish in response to decreased pressure are non-existent.

## Differential Effects: Is There Just One Comprehensive Model?

In this section we will not address the entire spectrum of potential and more or less obvious interactions that can influence the responses of fish to altered hydrostatic pressure. It is reasonable to expect that interactions with temperature, pH, oxygen levels, or salinity make the interpretation of the experimental results extremely complex. In the same way, differential responses are expected when taking into account, for example, specific daily or seasonal rhythms of the experimental animals or the time taken to obtain tissue samples.

As can be noticed in this chapter, there is evidence that many of the observed effects of pressure on fish are not always expressed with the same pattern and intensity. They are clearly dependent, for example, on the native habitat of the model species (stenobaric or eurybaric, shallow-water or deep-water, and so on) and on experimental protocols used (rate of compression, duration and pressure levels). Several studies have specifically addressed these differences in fish responses to altered hydrostatic pressure, assessing the interactions between pressure and several variables and characteristics.

Differences between species are well illustrated in the work of Sébert and Theron (2001). Oxygen consumptions of red muscle permeabilized cells and mitochondria from rainbow trout and European eel were measured under 10.1 MPa hydrostatic pressure. The results showed a substantial alteration in oxidative phosphorylation in rainbow trout but not in European eel, with this later exhibiting mitochondrial pressure resistance. Such a mitochondrial functioning at pressure would then be essential for European eel to cope with the high pressure environment encountered during its spawning migration.

Comparisons between shallow- and deep-water fish point to different enzyme responses to compression. As an example, compression caused a continuous decrease in gill $Na^+/K^+$-ATPase activity in the shallow-water coho salmon (*Oncorhynchus kisutch* Walbaum). However, in the deep-water fish *Antimora rostrata* Günther compression from 0.1 MPa to 27.6 MPa increased the activity of that enzyme and only further compression induced a decrease in activity (Moon, 1975). These results provide evidence for different enzyme sensitivities to pressure between species which naturally experience different pressure conditions and reflect adaptation in each species. Most probably some specific properties of enzymes are maintained in order to ensure catalytic function, but other properties can be altered to benefit from the pressure conditions in the habitat, for example, by altering the enzyme-substrate interactions. Similar conclusions were also obtained from another deep-water fish (*Coryphaenoides* sp.) in which myosin $Na^+/K^+$-ATPase was nearly insensitive to pressure similar to pressure conditions found in its habitat at more than 2,000 meters depth (Dreizen and Kim, 1971).

At other organizational levels, Wardle et al. (1987) compared the effect of compression on the force and contraction time of the twitch of the lateral swimming muscle in the European eel and in the deep-sea eel, *Histiobranchus bathybius* Günther and found that for the same pressure the shallow-water fish have higher swimming speeds. Wilson and Knowles (1987) did not detect changes in the *in vitro* autoxidation rates of oxyhemoglobin ($HbO_2$) to methemoglobin from fish captured at depths between 1 to 3,800 m, concluding that in opposition to temperature, hydrostatic pressure does not affect autoxidation of $HbO_2$ in either shallow or deep-water fish. In another example (Phleger, 1971), optimal swimbladder lipid synthesis in the deep-sea fish *Coryphaenoides* sp. occurred at 34.5 MPa and 2°C, while in the shallow-water fish *Orthopristis forbesi* Jordan and Starks optimal conditions were 0.1 MPa and 15°C. Since these experiments were performed in the range between 0.1 and 69 MPa, that is very well correlated with the environmental conditions in the habitat of each of those fish.

Different responses to altered hydrostatic pressure exist even in the same species. For example, yellow and silver eels are actually consecutive phases of the life cycle in the European eel yet with some diverse responses to pressure. Indeed, the effects of the silvering process on pressure resistance are similar to those of pressure acclimatization in yellow eels (Vettier et al., 2005). Silvering appears to mimic pressure acclimatization and by increasing membrane fluidity at atmospheric pressure, this would allow silver eels to cope with pressure effects with minimal energy cost. Thus, silver eels appear to be pre-adapted to migrate under pressure and for long distances (see Sébert et al., 2008).

Surprisingly, different populations of the same species could also have different responses to pressure. Véttier and Sébert (2004), by studying red and white muscle aerobic metabolism, found that European eels from different locations have different energetic capacities to migrate. Responses to compression, mitochondrial respiration and cytochrome c oxidase activity measured under pressure (10.1 MPa) after 3 d were used as indicators of pressure resistance and revealed that only fish from some locations have the metabolic capacities to cope with long exposures to the high pressure encountered during migration.

Differential effects were also observed between males and females (Scaion et al., 2008), with males showing higher aerobic capacities than females at atmospheric pressure independent of temperature. However, an increase in temperature enhances pressure effects in males whilst diminishes those effects in females. This sex difference is of particular importance in understanding the migration strategies in males and females but also highlights how some variables could influence the results when carrying out experiments on the effects of hydrostatic pressure.

In conclusion, all models have limitations but intrinsic differences in their responses can be considered advantageous for furthering our knowledge of the effects of altered hydrostatic pressure not only on fish but also on general biological processes. Understanding and careful characterization of the differences and parallel studies using different species will overcome most of those limitations.

## Adaptation to Elevated Hydrostatic Pressure

Deep-water fish offer the opportunity to study the adaptations of organisms to elevated hydrostatic pressure, although in deep-water environments other factors like temperature, light, oxygen, and food availability could also play important roles.

The study of the mechanisms of adaptation to deep-water habitats is particularly complex due to the challenging physical characteristics present in those habitats, namely high pressures. In addition, deep-water organisms are relatively more difficult to obtain and also to maintain and these difficulties are exacerbated when studying fish or other higher organisms. Not surprisingly, research on the adaptation of fish to elevated hydrostatic pressure has been mainly dependant on biochemical analysis of some processes in individuals from a restricted number of species, captured at depth and which had suffered decompression during recovery.

Deep-water fish usually have a much wider depth distribution than shallow-water species. In other words, they are more likely to be adapted to a wider range of hydrostatic pressures. Several studies have shown that hydrostatic pressure has significant effects on proteins, enzymes included and although the effects depend on each particular protein, generally in deep-water fish there is no optimization for a habitat pressure but instead an insensitivity to pressure (Gibbs, 1997). Thus, in these deep-water species evolution would understandably favour a reduced pressure sensitivity of some biological molecules like enzymes. This insensitivity to pressure permits enzymes to maintain function under conditions which reduce activity or even inhibit the homologous enzyme of shallow-water fish.

The reduced pressure sensitivity was well illustrated in comparative studies on several muscle dehydrogenases from shallow- and deep-water fish. Enzymes involved in anaerobic and aerobic pathways like lactate dehydrogenase (LDH), isocitrate dehydrogenase (IDH), glyceraldehyde 3-phosphate dehydrogenase (GAPDH), and malate dehydrogenase (MDH) from deep-water species are comparatively more stable under elevated pressure (Hennessey and Siebenaller, 1985; Moon and Storey, 1975; Siebenaller, 1987). Enzymes involved in other pathways also exhibited reduced pressure sensitivity, as in the case of muscle and gill $Na^+/K^+$-ATPase (Kim and Dreizen, 1971; Gibbs and Somero, 1989). The higher structural stabilities and the maintenance of enzyme kinetics ($K_m$), and thus of pressure sensitivity, can be achieved through adaptational changes in the enzyme conformation, sometimes originating from minor differences in sequence of only one or a few aminoacids (Somero, 2003; Morita, 2008).

Many enzymes (for example the above mentioned gill $Na^+/K^+$-ATPase) are membrane-located and it is well established that membrane-based processes are particularly susceptible to perturbation by pressure. Perturbations by elevated hydrostatic pressure result mainly from the reduction on membrane fluidity, which decreases with increasing pressures and which is a function of the membrane composition. Adaptation to deep-water requires alteration in the phospholipid composition of the membranes in order to maintain its fluidity. In fact, evidence for this homeoviscous adaptation to pressure has been shown in deep-water fish (Cossins and Macdonald, 1984, 1986), which have greater proportions of unsaturated fatty acids than shallow-water fish. Unsaturated fatty acids in membrane phospholipids possess one or more double bounds and are less packed, guaranteeing the membrane fluidity and reducing rigidity in the plane of the membranes which affect proper enzyme function, thus counteracting the effects of pressure (Somero, 1991).

Likewise, transmembrane signal transduction systems are also susceptible to perturbations due to pressure. Among these, one major class that has been well studied in deep-water fish is the signal transduction system involving guanyl nucleotide binding protein-coupled receptors (GPCRs). Specific steps in this system have been identified as affected by those perturbations, namely the agonist activation of the receptor (integral membrane protein), its interaction with the heterotrimeric G protein (which is anchored to the membrane), the activation and function of the effector element, adenylyl cyclase (integral membrane protein), and finally, the G protein GTPase cycle (Siebenaller, 2000; Siebenaller and Muray, 1990; Murray and Siebenaller, 1993; reviewed by Siebenaller and Garrett, 2002). Comparative studies with species occurring at different depths, with some of them congeneric, and thus exposed to different ranges of hydrostatic pressures showed that deep-living species display more complete adaptation to high pressure and that only in *A. rostrata* (a species with a habitat extending to more than 2,500 meters depth) there is a clear adaptation to elevated pressure (Stevens and Siebenaller, 2000).

Another remarkable adaptation in deep-water fish is the accumulation of small organic solutes that stabilize protein traits (Yancey et al., 2002). Accumulation of solutes like trimethylamine-N-oxide (TMAO) helps counteract the deleterious effects of high urea concentrations on proteins in elasmobranchs (Yancey and Siebenaller, 1999), reducing osmoregulatory costs, increasing buoyancy (Kelly and Yancey, 1999). However, it is now well established that TMAO also conserves protein function and structure at elevated hydrostatic pressure counteracting its effects (Gillett et al., 1997, Yancey et al., 2001). In fact, it was observed that deep-water fish have greater TMAO content in most of their body tissues than shallow-water fish (Treberg and Driedzic, 2002). Tissue levels of TMAO in deeper-water teleosts (*A. rostrata* and *Synaphobranchus kaupii* Johnson) and elasmobranchs (*Centroscyllium fabricii* Reinhardt) are significantly greater than those in shallow-water teleosts (Atlantic cod, *Urophycis chuss* Walbaum, and *Merluccius bilinearis* Mitchill) and elasmobranchs (*Squalus acanthias* Linnaeus).

One aspect of the adaptation of fish to elevated pressure and other deep-sea conditions that has been particularly well studied and discussed during the last decades is the significance of the low oxygen consumption in deep-water species. The stability of enzymes and the decrease in their activities from deep-water fish are correlated with lower metabolic and respiratory rates which have been observed by many authors (reviewed by Seibel and Drazen, 2007) and include *in situ* measurements (Bailey et al., 2002). It is worth mentioning that the differences between oxygen consumption in shallow- and deep-water species may have been frequently overestimated as pointed by Sébert (1997). Keeping in mind that most of the environmental factors have a lower intensity at depth (for example temperature, light, and food availability) in opposition to pressure, the question of what factor or factors are responsible for the lower metabolism remains to be clearly answered. Seibel and Drazen (2007) concluded that low rates of metabolism in the deep-water fish do not result from food or oxygen limitation or from temperature or pressure constraint, but instead from a reduced motor activity established by the adaptation to the absence of visual predator-prey interactions. It was even suggested that the proximate composition of deep-sea fish (proteins, lipids, carbohydrates, and water) also reflects that adaptation (Drazen, 2007). However, these statements appear in contradiction with observations *in situ* of some deep-sea fish, as are the examples of *A. rostrata* (Bailey et al., 2003) and *S. kaupii* (Bailey et al., 2005). In the words of these authors "it is possible that the interactions between predators and prey may be every bit as furious in the deep-sea as in photic systems, with short bursts of activity necessary to attack or escape before disappearing into the darkness…"

## CONCLUSIONS

In spite of the advance in the experimental techniques and methods, the study of hydrostatic pressure effects on fish can still be considered a particularly difficult but also challenging task. More than a hundred years have passed since the first experimental work on aquatic animals and even though the present knowledge is fairly diversified, it remains relatively fragmented.

Shallow-water fish provide opportunities to observe effects of hydrostatic pressure and its rapid variations on behavioural and physiological functions. The particular case of pressure-acclimatized shallow-water fish, which can be considered as making the bridge between shallow- and deep-water fish, provide further opportunities to observe and understand phenotypic and genotypic responses to hydrostatic pressure. By studying deep-water fish it is possible to observe adaptation to high pressure, but also to other concurrent factors present in deep habitats. Finally, the attempted but not yet successfully achieved acclimation of deep-water fish to atmospheric pressure will surely address many other questions if overcome, namely by allowing the comparison of the physiological changes observed during that process with those occurring during the acclimation of shallow-water fish to high pressures similar to those found in the deep-sea.

Fish present many advantages over other organisms to study the effects of hydrostatic pressure and its changes. Further applied and fundamental studies will allow us to obtain a better understanding of many biological processes, from the adaptations that enable organisms to exploit deep-sea habitats to the variation in gene expression under elevated hydrostatic pressure. As vertebrates they provide excellent opportunities to extend the knowledge of how organisms, including humans, respond to this often forgotten environmental factor.

## ACKNOWLEDGMENTS

The authors wish to thank  Catarina Gomes, Jonathan Wilson, and Lisa Olohan for their help and advice in writing this manuscript. Damasceno-Oliveira is also grateful to Fundação para a Ciência e a Tecnologia (FCT) for providing opportunities to study the effects of hydrostatic pressure on fish through grants POCI/MAR/60365/2004 and SFRH/BPD/36213/2007.

# REFERENCES

Avent, R.M. 1975. Evidence for acclimation to hydrostatic pressure in *Uca Pugilator* (Crustacea, Decapoda, Ocypodidae). Mar. Biol. 31(2): 193–199.

Bailey, D.M. and A.J. Jamieson, P.M. Bagley, M.A. Coolins and I.G. Priede. 2002. Measurement of *in situ* oxygen consumption of deep-sea fish using an autonomous lander vehicle. Deep-Sea Res. Part I-Oceanogr. Res. Pap. 49(8): 1519–1529.

Bailey, D.M. and P.M. Bagley, A.J. Jamieson, M.A. Collins and I.G. Priede. 2003. *In situ* investigation of burst swimming and muscle performance in the deep-sea fish *Antimora rostrata* (Günther, 1878). J. Exp. Mar. Biol. Ecol. 285–286: 295–311.

Bailey, D.M. and B. Genard, M.A. Collins, J.-F. Rees, S.K. Unsworth, E.J.V. Battle, P.M. Bagley, A.J. Jamieson and I.G. Priede. 2005. High swimming and metabolic activity in the deep-sea eel *Synaphobranchus kaupii* revealed by integrated *in situ* and *in vitro* measurements. Physiol. Biochem. Zool. 78(3): 335–346.

Barham, E.G. and N.J. Ayer and R.E. Boyce. 1967. Macrobenthos of San Diego Trough: photographic census and observations from bathyscaphe, Trieste. Deep-Sea Res. 14(6): 773–784.

Barthélémy, L. and A. Belaud and A. Saliou. A study of the specific action of *per se* hydrostatic pressure on fish considered as a physiological model. Pp. 641–649. *In*: A.J. Bachrach and M.M. Matzen. (eds.). 1981. Proceedings of the Seventh Underwater Physiology Symposium. Undersea Medical Society, Bethesda, Maryland, USA.

Barton, B.A. 2002. Stress in fishes: a diversity of responses with particular reference to changes in circulating corticosteroids. Integ. Comp. Biol. 42: 517–525.

Bell, A. 2007. Detection without deflection? A hypothesis for direct sensing of sound pressure by hair cells. J. Biosci. 32: 385–404.

Bell, A. 2008. The pipe and the pinwheel: Is pressure an effective stimulus for the 9 + 0 primary cilium? Cell Biol. Int. 32: 462–468.

Bert, P. 1878. La pression barométrique. Masson, Paris, France. Reprinted in 1979 by Editions du Centre National de la Recherche Scientifique, Paris, France.

Blaxter, J.H.S. and P. Tyler. Pressure discrimination in teleost fish. Pp. 417–44. *In*: M.A. Sleigh and A.G. Macdonald. (eds.). 1972. The Effects of Pressure on Organisms. Symposia of the Society for Experimental Biology, XXVI. University Press, Cambridge, UK.

Brauer, R.W. 1984. Hydrostatic pressure effects on the central nervous system: perspectives and outlook. Phil. Trans. R. Soc. Lond. B304: 17–30.

Brauer, R.W. and M.R. Jordan. The double envelope concept in the design of high pressure aquaria. Pp. 394–412. *In*: R.W. Brauer. (ed.). 1972. Barobiology and the Experimental Biology of the Deep Sea. Chapel Hill, North Carolina, USA.

Brauer, R.W. and R.W. Beaver, S. Lahser, R.D. McCall and R. Venters. 1979. Comparative physiology of the high-pressure neurological syndrome: compression rate effects. J. Appl. Physiol. 46(1): 128–135.

Brauer, R.W. and V.G. Sidelyova, M.B. Dail, G.I. Galazii and R.D. Roer. 1984. Physiological adaptation of cottoid fishes of Lake Baikal to abyssal depths. Comp. Biochem. Physiol. Physiol. 77(4): 699–705.

Cain, P. and S. Malwal. 2002. Landmark use and development of navigation behaviour in the weakly electric fish *Gnathonemus petersii* (Mormyridae; Teleostei). J. Exp. Biol. 205: 3915–3923.

Chen, W.H. and L.T. Sun, C.L. Tsai, Y.L. Song and C.F. Chang. 2002. Cold-stress induced the modulation of catecholamines, cortisol, immunoglobulin M, and leukocyte phagocytosis in tilapia. Gen. Comp. Endocrinol. 126(1): 90–100.

Cohen, D.M. 1977. Swimming performance of gadoid fish *Antimora rostrata* at 2400 meters. Deep-Sea Res. 24(3): 275–277.

Collins, M.A. and I.G. Priede and P.N. Bagley. 1999. *In situ* comparison of activity in two deep-sea scavenging fishes occupying different depth zones. Proc. R. Soc. B-Biol. Sci. 266: 2011–2016.

Cossins, A.R. and A.G. Macdonald. 1984. Homeoviscous theory under pressure II. The molecular order of membranes from deep-sea fish. Biochim. Biophys. Acta 776: 144–150.

Cossins, A.R. and A.G. Macdonald. 1986. Homeoviscous adaptation under pressure. III. The fatty acid composition of liver mitochondrial phospholipids of deep-sea fish. Biochim. Biophys. Acta 860: 325–335.

Cossins, A.R. and D.L. Crawford. 2005. Fish as models for environmental genomics. Nat. Rev. Genet. 6: 324–333.

Dalla Via, J. and P. Villani, E. Gasteiger and H. Niederstatter. 1998. Oxygen consumption in sea bass fingerling *Dicentrarchus labrax* exposed to acute salinity and temperature changes: metabolic basis for maximum stocking density estimations. Aquaculture 169: 303–313.

Damasceno-Oliveira, A. 2007. Vertical migrations in flounder: effects of hydrostatic pressure on the control of the brain-pituitary-gonad axis. Ph.D. Thesis, Universidade do Porto, Portugal.

Damasceno-Oliveira, A. and J. Gonçalves, J. Silva, B. Fernández-Durán and J. Coimbra. 2004. A pressurising system for long-term study of marine or freshwater organisms enabling the simulation of cyclic vertical migrations. Sci. Mar. 68(4): 615–619.

Damasceno-Oliveira, A. and B. Fernández-Durán, J. Gonçalves, P. Serrão, P. Soares-da-Silva, M.A. Reis-Henriques and J. Coimbra. 2006. Effects of cyclic and constant hydrostatic pressure on norepinephrine and epinephrine levels in the brain of flounder. J. Fish Biol. 68: 1300–1307.

Damasceno-Oliveira, A. and B. Fernández-Durán, J. Gonçalves, P. Serrão, P. Soares-da-Silva, M.A. Reis-Henriques and J. Coimbra. 2007. Effects of cyclic hydrostatic pressure on the brain biogenic amines concentrations in the flounder, *Platichthys flesus*. Gen. Comp. Endocrinol. 153: 385–389.

Davis, B.J. and J.F. Siebenaller. 1992. Proteolysis at pressure and HPLC peptide mapping of M4-lactate dehydrogenase homologs from marine fishes living at different depths. Int. J. Biochem. 24(7): 1135–1139.

Dean, J.B. and D.K. Mulkey, A.J. Garcia, R.W. Putnam and R.A. Henderson. 2003. Neuronal sensitivity to hyperoxia, hypercapnia, and inert gases at hyperbaric pressures. J. Appl. Physiol. 95: 883–909.

Drazen, J.C. 2007. Depth related trends in proximate composition of demersal fishes in the eastern North Pacific. Deep Sea Res. Oceanogr. Res. Paper. 54: 203–219.

Drazen, J.C. and L.E. Bird and J.P. Barry. 2005. Development of a hyperbaric trap-respirometer for the capture and maintenance of live deep-sea organisms. Limnol. Oceanogr. Meth. 3: 488–498.

Dreizen, P. and H.D. Kim. 1971. Contractile proteins of a benthic fish. I. Effects of temperature and pressure on myosin ATPase. Am. Zool. 11(3): 513–522.

Dunel-Erb, S. and P. Sébert, C. Chevalier, B. Simon and L. Barthélémy. 1996. Morphological changes induced by acclimation to high pressure in the gill epithelium of the freshwater yellow eel. J. Fish Biol. 48(5): 1018–1022.

Farrell, A.P. and A.K. Gamperl and I.K. Birtwell. 1998. Prolonged swimming, recovery and repeat swimming performance of mature sockeye salmon *Oncorhynchus nerka* exposed to moderate hypoxia and pentachlorophenol. J. Exp. Biol. 201: 2183–2193.

Fenn, W.O. Possible role of hydrostatic pressure in diving. *In*: C.I. Lambertsen (ed.). 1967. Proceedings of the 3rd Symposium on Underwater Physiology. Williams and Wilkins. Washington D.C., USA.

Fontaine, Y.A. and S. Dufour, J. Alinat and M.A. Fontaine. 1985. A long immersion in deep-sea stimulates the pituitary gonadotropic function of the female European eel (*Anguilla anguilla* L.). Comptes Rendus Acad. Sci. Ser. III-Sci. Vie-Life Sci. 300(2): 83–87.

Fraser, P.J. and R.L. Shelmerdine. 2002. Dogfish hair cells sense hydrostatic pressure. Nature 415: 495–496.

Gibbs, A.G. Biochemistry at depth. pp. 239–277. *In*: D.J. Randall and A.P. Farrell. (eds.). 1997. Fish Physiology, Vol. 16, Deep-Sea Fishes. Academic Press, San Diego, USA.

Gibbs, A. and G.N. Somero, G.N. 1989. Pressure adaptation of Na+/K+-ATPase in gills of marine teleosts. J. Exp. Biol. 143(1): 475–492.

Gibson, R.N. 1982. The effect of hydrostatic pressure cycles on the activity of young plaice *Pleuronectes Platessa*. J. Mar. Biol. Assoc. UK 62: 621–635.

Gibson, R.N. 1984. Hydrostatic pressure and the rhythmic behavior of intertidal marine fishes. T. Am. Fish. Soc. 113: 479–483.

Gillett, M.B. and J.R. Suko, F.O. Santoso and P.H. Yancey. 1997. Elevated levels of trimethylamine oxide in muscles of deep sea gadiform teleosts: a high pressure adaptation? J. Exp. Zool. 279: 386–391.

Greenspan, J.S. and M.R. Wolfson and T.M. Shaffer. 1999. Liquid ventilation: Clinical experiences. Biomed. Instrum. Technol. 33(3): 253–259.

Hazel, J.R. and E.E. Williams. 1990. The role of alterations in membrane lipid composition enabling physiological adaptations of organism to their environment. Prog. Lipid Res. 29: 167–227.

Hennessey, J.P., Jr. and J.F. Siebenaller. 1985. Pressure inactivation of tetrameric lactate dehydrogenase homologues of confamilial deep-living fishes. J. Comp. Physiol. B Biochem. Syst. Environ. Physiol. 155: 647–652.

Hennessey, J.P., Jr. and J.F. Siebenaller. 1987. Inactivation of NAD-dependent dehydrogenases from shallow and deep-living fishes by hydrostatic pressure and proteolysis. Biochim. Biophys. Acta 913(3): 285–91.

Heupel, M.R. and C.A. Simpfendorfer and R.E. Hueter. 2003. Running before the storm: blacktip sharks respond to falling barometric pressure associated with Tropical Storm Gabrielle. J. Fish Biol. 2003, 63(5): 1357–1363.

Hughes, G.M. Effects of low oxygen and pollution on the respiratory systems of fish. pp. 121–146. *In*: A.D. Pickering. (ed.). 1981. Stress and Fish. Academic Press, London, UK.

Hulbert, A.J. and P.L. Else. 1999. Membranes as possible pacemakers of metabolism. J. Theor. Biol. 199: 257–274.

Iwama, G.K. and L.O.B. Afonso, A. Todgham, P. Ackerman and K. Nakano. 2004. Are hsps suitable for indicating stressed states in fish? J. Exp. Biol. 207: 15–19.

Johnstone, A.D.F. and A.G. Macdonald, W.R. Mojsiewicz and C.S. Wardle. 1989. Preliminary experiments in the adaptation of the European eel (*Anguilla anguilla*) to high hydrostatic pressure. J. Physiol.-London 417: P87.

Kelly, R.H. and P.H. Yancey. 1999. High contents of trimethylamine oxide correlating with depth in deep-sea teleost fishes, skates, and decapod crustaceans. Biol. Bull. 196: 18–25.

Kim, H.D. and P. Dreizen. 1971. Contractile proteins of a benthic fish. II. Composition and ATPase properties of actomyosin. Am. Zool. 11: 523–529.

Koyama, S. and M. Aizawa. 2008. Piezotolerance of the cytoskeletal structure in cultured deep-sea fish cells using DNA transfection and protein introduction techniques. Cytotechnology 56: 19–26.

Koyama, S. and T. Miwa, M. Horii, Y. Ishikawa, K. Horikoshi and M. Aizawa. 2002. Pressure-stat aquarium system designed for capturing and maintaining deep-sea organisms. Deep Sea Res. Oceanogr. Res. Paper. 49: 2095–2102.

Koyama, S. and H. Kobayashi, A. Inoue, T. Miwa and M. Aizawa. 2005. Effects of the piezo-tolerance of cultured deep-sea eel cells on survival rates, cell proliferation, and cytoskeletal structures. Extremophiles 9: 449–460.

Kysltra, J.A. 1962. Breathing fluid. Experientia 18: 68.

MacDonald, A.G. and I. Gilchrist. An apparatus for the recovery and study of deep sea plankton at constant temperature and pressure. Pp. 394–412. *In*: R.W. Brauer. (ed.). 1972. Barobiology and the Experimental Biology of the Deep Sea. Chapel Hill, North Carolina, USA.

Macdonald, A.G. and I. Gilchrist and C.S. Wardle. 1987. Effects of hydrostatic pressure on the motor activity of fish from shallow water and 900 m depths; some results of challenger cruise 6B/85. Comp. Biochem. Physiol. Physiol. 88(3): 543–547.

Meyer A. and Y. van de Peer. 2005. From 2R to 3R: evidence for a fish-specific genome duplication (FSGD). Bioessays 27: 937–945.

Moon, T.W. 1975. Effects of hydrostatic pressure on gill Na-K-ATPase in an abyssal and a surfacing-dwelling teleost. Comp. Biochem. Physiol. B Biochem. Mol. Biol. 52: 59–65.

Moon, T.W. and K.B. Storey. 1975. The effects of temperature and hydrostatic pressure on enzymes of an abyssal fish, *Antimora rostrata*: liver NADP-linked isocitrate dehydrogenase. Comp. Biochem. Physiol. B Biochem. Mol. Biol. 52: 51–57.

Morita, T. 2008. Comparative sequence analysis of myosin heavy chain proteins from congeneric shallow- and deep-living rattail fish (genus *Coryphaenoides*). J. Exp. Biol. 211: 1362–1367.

Murray, T.F. and J.F. Siebenaller. 1993. Differential susceptibility of guanine nucleotide-binding proteins to pertussis toxin-catalyzed ADP-ribosylation in brain membranes of two congeneric marine fishes. Biol. Bull. 185(3): 346–354.

Nielsen, J.G. 1977. The deepest living fish *Abyssobrotula galatheae*. A new genus and species of oviparous ophidioids (Pisces, Brotulidae). Galathea Rep. 14: 41–48.

Nishiguchi, Y. and T. Miwa and F. Abe. 2008. Pressure-adaptive differences in lactate dehydrogenases of three hagfishes: *Eptatretus burgeri*, *Paramyxine atami* and *Eptatretus okinoseanus*. Extremophiles 12: 477–480.

Nishiyama, T. 1965. A preliminary note on the effect of hydrostatic pressure on the behaviour of some fish. Bull. Fish. Sci. Hokkaido Univ. 15: 213–214.

Northcott, S.J and R.N. Gibson and E. Morgan. 1991. The effect of tidal cycles of hydrostatic pressure on the activity of *Lipophrys pholis* (L.) (Teleostei). J. Exp. Mar. Biol. Ecol., 148: 35–45.

Pavlov, D.S. and R.V. Sadkovskii, V.V. Kostin and A.I. Lupandin. 2000. Experimental study of young fish distribution and behaviour under combined influence of baro-, photo- and thermo-gradients. J. Fish Biol. 57: 69–81.

Pennec, J.-P. and C.S. Wardle, A.A. Harper and A.G. Macdonald. 1988. Effects of high hydrostatic pressure on the isolated hearts of shallow water and deep sea fish; results of Challenger cruise 6B/85. Comp. Biochem. Physiol. Physiol. 89(2): 215–218.

Phleger, C.F. 1971. Pressure effects on cholesterol and lipid synthesis by the swimbladder of an abyssal *Coryphaenoides* species. Am. Zool. 11: 559–570.

Pickering, A.D. Introduction: The concept of biological stress. Pp. 1–9. *In*: A.D. Pickering. (ed.). 1981. Stress and Fish. Academic Press, London, UK.

Prosser, C.L. Introduction: Definition of comparative physiology: Theory of adaptation. Pp. 1–11. *In*: C.L. Prosser. (ed.). 1991. Comparative Animal Physiology, 4th edit., Vol. 1, Environmental and Metabolic Animal Physiology. Wiley-Liss, New York, USA.

Quentin, L.B. and J.J. Childress. 1980. Observations on the swimming activity of two bathypelagic mysid species maintained at high hydrostatic pressure. Deep-Sea Res. Part A-Oceanogr. Res. Pap. 27: 383–391.

Randall, D.J. and S.F. Perry. 1992. Catecholamines. Pp. 255–300. *In*: D.J. Randall and A.P. Farrell. (eds.). 1992. Fish Physiology, Vol 12A. The Cardiovascular System. Academic Press, San Diego, USA.

Regnard, P. 1884. Effet des hautes pressions sur les animaux marins. C.R. Séances Soc. Biol. 36: 394–395.

Regnard, P. 1885. Phénomènes objectifs que l'on peut observer sur les animaux soumis aux hautes pressions. C.R. Séances Soc. Biol. 37: 510–515.

Roer, R.D. and V.G. Sidelyova, R.W. Brauer and G.I. Galazii. 1984. Effects of pressure on oxygen consumption in cottid fish from lake Baikal. Experientia 40: 771–773.

Rossignol, O. and P. Sébert and B. Simon. 2006. Effects of high-pressure acclimatization on silver eel (*Anguilla anguilla*, L.) slow muscle contraction. Comp. Biochem. Physiol. Mol. Integr. Physiol. 143: 234–238.

Scaion, D. and M. Belhomme and P. Sébert. 2008. Pressure and temperature interactions on aerobic metabolism of migrating European silver eel. Respir. Physiol. Neurobiol. 164: 319–322.

Schjolden, J. and T. Backström, K.G.T. Pulman, T.G. Pottinger and S. Winberg. 2005. Divergence in behavioural responses to stress in two strains of rainbow trout (*Oncorhynchus mykiss*) with contrasting stress responsiveness. Horm. Behav. 48: 537–544.

Sébert, P. 1993. Energy metabolism of fish under hydrostatic pressure: a review. Trends Comp. Biochem. Physiol. 1: 289–317.

Sébert, P. Pressure effects on shallow-water fishes. Pp. 279–323. *In*: D.J. Randall and A.P. Farrell. [eds.] 1997. Fish Physiology, Vol. 16, Deep-Sea Fishes. Academic Press, San Diego, USA.

Sébert, P. 2002. Fish at high pressure: a hundred year history. Comp. Biochem. Physiol. Mol. Integr. Physiol. 131(3): 575–85.

Sébert, P. and L. Barthélémy. 1985a. Effects of high hydrostatic pressure *per se*, 101 atm on eel metabolism. Resp. Physiol. 62: 349–357.

Sébert, P. and L. Barthélémy. 1985b. Hydrostatic pressure and adrenergic drugs (agonists and antagonists): effects and interactions in fish. Comp. Biochem. Physiol. C Comp. Pharmacol. 82: 207–212.

Sebert, P. and A.G. Macdonald. Fish. Pp. 147–196. *In*: A.G. Macdonald. (ed.). 1993. Advances in Comparative and Environmental Physiology, Effects of High Pressure on Biological Systems. Springer-Verlag, Berlin, Germany.

Sébert, P. and M. Theron. 2001. Why can the eel, unlike the trout, migrate under pressure. Mitochondrion. 1: 79–85.

Sebert, P. and L. Barthelemy and J. Caroff. 1986. Catecholamine content (as measured by the HPLC method) in brain and blood plasma of the eel: effects of 101 ata hydrostatic pressure. Comp. Biochem. Physiol. C. Comp. Pharmacol. 84: 155–157.

Sébert, P. and L. Barthélémy, J. Caroff and A. Hourmant. 1987. Effects of hydrostatic pressure *per se* (101 ATA) on energetic processes in fish. Comp. Biochem. Physiol. 86(3): 491–495.

Sébert, P. and L. Barthelemy and B. Simon. 1990. Laboratory system enabling long-term exposure (≥30 days) to hydrostatic pressure (≤100 atm) of fishes or other animals breathing water. Mar. Biol. 104: 165–168.

Sébert, P. and A. Pequeux, B. Simon and L. Barthélémy. 1991. Effects of long-term exposure to 101 ATA hydrostatic pressure on blood, gill and muscle composition and on some enzyme activities of the fw eel (*Anguilla anguilla* L.). Comp. Biochem. Physiol. B Biochem. Mol. Biol. 98(4): 573–577.

Sébert, P. and B. Simon and B. Barthélémy. 1993. Hydrostatic pressure induces a state resembling histotoxic hypoxia in *Anguilla anguilla*. Comp. Biochem. Physiol. Physiol. 105(2): 255–258.

Sébert, P. and A. Péqueux, B. Simon and L. Barthélémy, L. 1995. Effects of hydrostatic pressure and temperature on the energy metabolism of the chinese crab (*Eriocheir sinensis*) and the yellow eel (*Anguilla anguilla*). Comp. Biochem. Physiol. Physiol. 112(1): 131–136.

Sébert, P. and B. Simon and A. Péqueux. 1997. Effects of hydrostatic pressure on energy metabolism and osmoregulation in crab and fish. Comp. Biochem. Physiol. Physiol. 116: 281–290.

Sébert, P. and J. Peragon, J.B. Barroso, B. Simon and E. Melendez-Hevia. 1998. High hydrostatic pressure (101 ATA) changes the metabolic design of yellow freshwater eel muscle. Comp. Biochem. Physiol. B Biochem. Mol. Biol. 121: 195–200.

Sébert, P. and D. Scaion and M. Belhomme. 2008. High hydrostatic pressure improves the swimming efficiency of European migrating silver eel. Respir. Physiol. Neurobiol. DOI:10.1016/j.resp.2008.09.011.

Seibel, B.A. and J.C. Drazen. 2007. The rate of metabolism in marine animals: environmental constraints, ecological demands and energetic opportunities. Phil. Trans. Roy. Soc. Lond. B. 362(1487): 2061–2078.

Shillito, B. and D. Jollivet, P.-M Sarradin, P. Rodier, F. Lallier, D. Desbruyères and F. Gaill. 2001. Temperature resistance of *Hesiolyra bergi*, a polychaetous annelid living on deep-sea vent smoker walls. Mar. Ecol. Prog. Ser. 216: 141–149.

Shillito, B. and G. Hamel, C. Duchi, D. Cottin, J. Sarrazin, P.-M. Sarradin, J. Ravaux and F. Gaill. 2008. Live capture of megafauna from 2300m depth, using a newly designed pressurized recovery device. Deep-Sea Res. Part I-Oceanogr. Res. Pap. 55: 881– 889.

Siebenaller, J.F. Pressure adaptations in deep-sea animals. Pp. 33–48. *In*: H.W. Jannasch, A.M. Zimmerman and R.E. Marquis (eds.). 1987. Current Perspectives in High Pressure Biology. Academic Press, London, United Kingdom.

Siebenaller, J.F. 2000. The effects of hydrostatic pressure on signal transduction in brain membranes of deep-sea fishes of the genus *Coryphaenoides*. Fish Physiol. Biochem. 23: 99–106.

Siebenaller, J.F. and T.F. Murray. 1990. $A_1$ adenosine receptor modulation of adenylyl cyclase of a deep-living teleost fish, *Antimora rostrata*. Biol. Bull. 178(1): 65–73.

Siebenaller, J.F. and T.F. Murray. The effects of pressure on G protein-coupled signal transduction. Pp. 147–174. *In*: P.W. Hochachka and T.P. Mommsen. (eds.). 1995. Biochemistry and Molecular Biology of Fishes 5. Elsevier, Amsterdam, The Netherlands.

Siebenaller, J.F and D.J. Garrett. 2002. The effects of the deep-sea environment on transmembrane signaling. Comp. Biochem. Physiol. B Biochem. Mol. Biol. 131: 675–694.

Simon, B. and P. Sébert, and L. Barthélemy. 1989. Effects of long-term exposure to hydrostatic pressure *per se* (101 ATA) on eel metabolism. Can. J. Physiol. Pharmacol. 67: 1247–1251.

Simon, B. and P. Sébert, C. Cann-Moisan and L. Barthélémy. 1992. Muscle energetics in yellow fresh water eel (*Anguilla anguilla* L.) exposed to high hydrostatic pressure (101 ATA) for 30 days. Comp. Biochem. Physiol. B Biochem. Mol. Biol. 102(2): 205–208.

Smith, K.L. 1978. Metabolism of the abyssopelagic rattail *Coryphanoides armatus*, measured *in situ*. Nature 274: 362–364.

Smith, K.L. and R.J. Baldwin. Laboratory and *in situ* methods for studying deep-sea fishes. Pp. 351–378. *In*: D.J. Randall and A.P. Farrell. (eds.). 1997. Fish Physiology, Vol. 16, Deep-Sea Fishes. Academic Press, San Diego, USA.

Somero, G.N. Hydrostatic pressure and adaptations to the deep sea. Pp. 167–204. 1991. *In*: C.L. Prosser. (ed.). 1991. Comparative Animal Physiology. John Wiley and Sons, New York, USA.

Somero, G.N. 2003. Protein adaptations to temperature and pressure: complementary roles of adaptive changes in amino acid sequence and internal milieu. Comp. Biochem. Physiol. B. Biochem. Mol. Biol. 136: 577–591.

Speers-Roesch, B. and D. Lingwood and E.D. Stevens. 2004. Effects of temperature and hydrostatic pressure on routine oxygen uptake of the bloater (*Coregonus hoyi*). J. Great Lakes Res. 30(1): 70–81.

Stevens, C.K. and J.F. Siebenaller. 2000. The effects of hydrostatic pressure on pertussis toxin-catalyzed ribosylation of G proteins from deep-living macrourid fishes. Comp. Biochem. Physiol. B. Biochem. Mol. Biol. 125: 103–114.

Theron, M and P. Sébert. 2003. Hydrostatic pressure and cellular respiration: are the values observed post-decompression representative of the reality under pressure? Mitochondrion 3: 75–81.

Theron, M. and F. Guerrero and P. Sébert. 2000. Improvement in the efficiency of oxidative phosphorylation in the freshwater eel acclimated to 10.1 MPa hydrostatic pressure. J. Exp. Biol. 203: 3019–3023.

Torres, J.J. and B.W. Belman and J.J. Childress. 1979. Oxygen consumption rates of midwater fishes as a function of depth of occurrence. Deep Sea Res. Oceanogr. Res. Paper. 26: 185–197.

Treberg, J.R. and W.R. Driedzic. 2002. Elevated levels of trimethylamine oxide in deep-sea fish: evidence for synthesis and intertissue physiological importance. J. Exp. Zool. Comp. Exp. Biol. 293(1): 39–45.

Tytler, P. and J.H.S. Blaxter. 1973. Adaptation by cod and saithe to pressure changes. Netherlands J. Sea Res. 7: 31–45.

Vardaro, M.F. and D. Parmley and K.L. Smith Jr. 2007. A study of possible "reef effects" caused by a long-term time-lapse camera in the deep North Pacific. Deep-Sea Res. Part I-Oceanogr. Res. Pap. 54: 1231–1240.

Vettier, A. and P. Sébert. 2004. Pressure resistance of aerobic metabolism in eels from different water environments. Mitochondrion 3: 347–354.

Vettier, A. and A. Amérand, C. Cann-Moisan and P. Sébert. 2005. Is the silvering process similar to the effects of pressure acclimatization on yellow eels? Respir. Physiol. Neurobiol. 145: 243–250.

Wardle, C.S. and N. Tetteh-Lartey, A.G. Macdonald, A.A Harper and J.-P. Pennec. 1987. The effect of pressure on the lateral swimming muscle of the European eel *Anguilla anguilla* and the deep sea eel *Histiobranchus bathybius*; results of challenger cruise 6B/85. Comp. Biochem. Physiol. Physiol. 88(3): 595–598, 1987.

Warner, L.H. 1932. The sensitivity of fishes to sound and to other mechanical stimulation. Q. Rev. Biol. 7(3): 326–339.

Wendelaar Bonga, S.E. 1997. The stress response in fish. Physiol. Rev. 77: 591–625.

Wilson, R.R. and K.L. Smith. 1985. Live capture, maintenance and partial decompression of a deep-sea grenadier fish (*Coryphaenoides acrolepis*) in a hyperbaric trap-aquarium, Deep Sea Res. Oceanogr Res. Paper. 32: 1571–1582.

Wilson, R.R., Jr. and F.C. Knowles. 1987. Temperature adaptation of fish hemoglobins reflected in rates of autoxidation. Arch. Biochem. Biophys. 255(1): 210–213.

Yancey, P.H. and J.F. Siebenaller. 1999. Trimethylamine oxide stabilizes teleost and mammalian lactate dehydrogenases against inactivation by hydrostatic pressure and trypsinolysis. J. Exp. Biol. 202: 3597–3603.

Yancey, P.H. and A.L. Fyfe-Johnson, R.H. Kelly, V.P. Walker and M.T. Anuón. 2001. Trimethylamine oxide counteracts effects of hydrostatic pressure on proteins of deep-sea teleosts. J. Exp. Zool. 289: 172–176.

Yancey, P.H. and W.R. Blake and J. Conley. 2002. Unusual organic osmolytes in deep-sea animals: adaptations to hydrostatic pressure and other perturbants. Comp. Biochem. Physiol. Mol. Integr. Physiol. 133: 667–676.

Yayanos, A.A. 1978. Recovery and maintenance of live amphipods at a pressure of 580 bars from an ocean depth of 5700 meters. Science 200(4345): 1056–1059.

Yayanos, A.A. 1995. Microbiology to 10500 meters in the deep sea. Annu. Rev. Microbiol. 49: 777–805.

Yoshiki, T. and T. Toda, T. Yoshida and A. Shimizu. 2006. A new hydrostatic pressure apparatus for studies of marine zooplankton. J. Plankton Res. 28(6): 563–570.

# 15

# Pressure Tolerance in Diving Mammals and Birds

*Michael Castellini*

## INTRODUCTION

This section focuses on the concept that deep diving marine mammals and birds are adapted to withstand both the rapid and extreme pressure changes that are associated with diving. Even though they are air-breathing endotherms like humans, marine mammals and diving birds exploit the deep marine environment freely and with very few negative repercussions. Thus, while routine deep free diving is essentially impossible for humans, for these marine species, deep diving is an advantage that allows them to move, forage and interact in an environment that would be hostile to us. From the perspective of comparative biology and physiology, these animals have evolved and perfected the abilities that allow them to fit into a niche that for them, is normal, not dangerous and fully exploitable. Certainly, some species can dive deeper than others and for longer periods, but the essential adaptations that they employ are qualitatively similar and form the basis for this chapter.

### Which species are the divers?

There are three groups (Orders) of marine mammals that include the cetaceans (whales), pinnipeds (seals, sea lions, walrus) and sirenians (dugongs, manatees). In terms of deep diving species, we will exclude the sirenians and the walrus because they are routinely shallow divers. Of the

School of Fisheries and Ocean Sciences, University of Alaska Fairbanks, Fairbanks, Alaska, USA 99775.
E-mail: mikec@ims.uaf.edu

cetaceans, the toothed whales (odontocetes: sperm whales [*Physeter macrocephalus*], beaked whales, etc.) dive more deeply than the baleen whales (bowhead [*Balaena mysticetus*], humpback [*Megaptera novaeangliae*], etc.). In the pinnipeds, the deepest divers are the phocids (true seals: elephant seal, Weddell seal [*Leptonychotes weddellii*], etc.) while the sea lions (Steller sea lion [*Eumetopias jubatus*], fur seals, etc.) generally dive more shallow, though there are exceptions.

For marine birds, this review will confine itself to the truly extreme deep diving birds, the large king (*Aptenodytes patagonicus*) and emperor penguins (*Aptenodytes forsteri*). While there are many fresh water and marine diving avian species, most of these flying birds forage only near the surface or make dives to less than 100 m (Schreer and Kovacs, 1997; Halsey et al., 2006). The exception to this is the murre, which has been recorded diving to a depth of over 200 m (Croll et al., 1992a). Unfortunately, very little anatomical or physiological work has been conducted on murres. The king and emperor penguins dive as deeply as some pinnipeds and the comparative adaptations for deep diving make interesting contrasts between the birds and the marine mammals.

Table 1 lists the current diving depth records for a suite of marine mammals and birds. Most of these data have been obtained through attached dive recorders that can either be retrieved from the animals or that send diving data to satellites. The species list is not meant to be extensive, but it does set the depth parameters for the remaining discussions. Halsey et al. (2006) provide an appendix with the dive depths of over a 100 different species and Schreer and Kovacs (1997) provide depth and dive duration for almost 200 species along with correlations between mass, dive duration and depth. Berta et al. (2006) provide similar maximal dive data along with anatomical correlates for diving in marine mammals. For our purposes, it is evident from Table 1 that depths of 500 m are reachable by most of the species discussed in this chapter and that some can approach 2000 m dive depth.

### History of diving studies

Because all the marine mammals and birds dive far beyond what we can directly observe, studies of diving depth had to wait until we developed dive recorders that could be attached to the animals. Before that, there was only circumstantial evidence from whaling ships, diet studies or from animals that had been tangled in deep sea cables that suggested marine mammals might swim to great depths (Heezen, 1957; Santos et al., 1999). The modern field of the development of diving recorders started with Gerry Kooyman from Scripps in the mid 1960's when he built simple Bourdon tube type maximum depth recorders and placed them on Weddell seals in

**Table 1:** Maximum diving Depths.

| Common Name/<br>*Species Name* | Depth (m) | Reference |
|---|---|---|
| **Cetaceans** | | |
| Sperm whale | 1860 | Teloni et al., 2008 |
| *Physeter macrocephalus* | | |
| Blainville's beaked whale | 890 | Baird et al., 2004 |
| *Mesoplodon densirostris* | | |
| Bottlenose whale | 1453 | Hooker and Baird, 1999 |
| *Hyperoodon ampullatus* | | |
| Fin whale | 316 | Croll et al., 2001 |
| *Balaenoptera physalus* | | |
| **Pinnipeds** | | |
| Southern elephant seal | 1400 | Slip et al., 1994 |
| *Mirounga leonina* | | |
| Northern elephant seal | | |
| *Mirounga angustirostris* | 1530 | Stewart and De Long, 1991 |
| Weddell seal | >740 | Testa, 1994 |
| *Leptonychotes weddellii* | | |
| Harbor seal | 481 | Eguchi and Harvey, 2005 |
| *Phoca vitulina* | | |
| Steller sea lion | 424 | Swain, 1996 |
| *Eumetopias jubatus* | | |
| **Birds** | | |
| Emperor penguin | >500 | Kooyman and Kooyman, |
| *Aptenodytes forsteri* | | 1995; Wienecke et al., 2007 |
| King penguin | 318 | Sato et al., 2002 |
| *Aptenodytes patagonicus* | | |
| Blue-eyed shag | 74 | Wanless et al., 1999 |
| *Phalacrocorax atriceps (purpurascens)* | | |
| Thick billed murre | 210 | Croll et al., 1992a |
| *Uria lomvia* | | |

the Antarctic (Kooyman, 1966). He next developed simple mechanical timers that could also record dive time, then mechanical/electronic recorders and finally fully digital devices. Since that time, a host of others have built increasingly complex and ever smaller devices that communicate with satellites, have video capabilities and record dive behavior in 3D and inertial navigation modes (Stewart, 2002). Many such devices are now available essentially 'off the shelf' that are small and powerful enough to attach to smaller species. The older devices were large and therefore were applied to the larger diving species first (Kooyman et al., 1976). New devices have become even smaller and have allowed studies on ever smaller diving species (Naito, 2004).

Finally, attachment methods have also developed over the years from sutures and harnesses to glue. These direct attachment methods require close handling of animals and are usually confined to pinnipeds or birds that can be restrained and then set free. By contrast, attachment devices for whales have focused mainly on the ability to remotely apply a recorder that is shot from a bow or gun or attached with suction cups to a moving whale (Mate et al., 2007). Regardless of the device history or attachment method, all measure at least the depth and duration of dives and as such, have been referred to since their invention as Time-Depth-Recorders (TDR).

## What are the limitations to dive depth?

From first principles, one might expect that the ability to tolerate pressure collapse of internal air spaces (lungs), the ability to rapidly ascend and descend without suffering from the 'Bends' and the time necessary to swim to great depth would be important limiting factors for these mammals and birds. There is another aspect we will also consider, the ability to tolerate the direct impact of pressure on solid tissue, as manifested by apparent resistance to High Pressure Nervous Syndrome (HPNS) and by biochemical adaptations to pressure at the enzymatic and cellular level. From an ecological perspective, there is also the consideration of why animals would spend the energy to travel to depth in the first place. In almost all cases, extremely deep dives are not routine, but usually take place during foraging bouts of many shallower dives. For example, Weinecke et al. (2007) showed that less than 0.2% of free dives in emperor penguins exceeded 400 m depth and that most were less than 200 m. The concept of 'deep' is relative however, in that routine hunting dives of 200–250 m in a Weddell seal or emperor penguin are still extremely deep compared to many species, even though both species can reach 500 m during some dives. There are many excellent reviews and research articles that discuss the ecology of diving behavior in terms of the balance between the energy gained by diving deeply vs the cost of making the dive (Hiethaus and Dill, 2002; Costa and Sinervo, 2004; Halsey et al., 2006). For this chapter however, we will deal only with the impacts of deep diving on anatomy, biochemistry and physiology keeping in mind that dive depth may be determined not by these factors, but by foraging or social requirements.

### *Dive depth vs Dive Duration*

In some of the very first studies of diving depth and duration under natural conditions, Kooyman (1966) demonstrated that the longest dives of Weddell seals were usually 'shallow' (less than 200 m). That is, deep dives were not necessarily the longest dives. Kooyman suggested that long, shallow dives might be exploratory dives under the Antarctic sea-ice and that deeper, routine, foraging dives would be shorter. Weineke et al. (2007) found that no deep dives in emperor penguins were long and that no long dives were deep. This makes sense in that to maximize time underwater, a diving mammal/bird would seek to minimize its oxygen utilization rate. The easiest way to do that would be to swim to a very shallow depth and then just stay there, not working or swimming. This would maximize the dive time, but would not make much ecological sense. However, if one considers only the aspect that to dive deeper would require longer time periods, then within the deep dives, it could easily be concluded that a marine mammal/bird can only dive as deep as it has enough oxygen on board to make the return trip. However, this assumes that these species limit their extreme diving to the time utilization of oxygen. This limit, termed the Aerobic Diving Limit (ADL), reflects how long the animal can remain underwater using stored oxygen (Kooyman et al., 1980). It has been shown in several species however, that the ADL can be exceeded (Kooyman et al., 1980; Ponganis et al., 1997a, b, c) and this is marked by the appearance in the blood of the anaerobic metabolic end product of lactic acid, which will build up in the animal as it utilizes low oxygen (hypoxic/anaerobic) metabolic pathways to power the dive.

Consider this example: Assume that the underwater cruising speed of a pinniped is about 1.5 m/sec (Ponganis et al., 1993). Assume that the ADL for this species is about 15 min. Using only those values, we calculate that the animal could swim about 90m/min for a total of 1,350 m in 15 min. If the animal only swam straight down and straight back up, this would suggest the seal could dive to a depth of about 675 m based only on the amount of oxygen on board. Any swimming deviations from vertical would reduce that depth and any time spent at depth (for example, hunting for fish) would also reduce the maximum depth. In fact, for adult Weddell seals with an ADL of about 19 min, repetitive foraging diving to about 250–300 m is completely aerobic with no build-up of lactic acid in the blood.

For Emperor penguins, the ADL is about 8 min, only about 5% of dives exceed the ADL and routine diving is to less than 100 m depth (Kooyman and Kooyman, 1995). As suggested above for seals, the longest dives recorded in one study of emperor penguins were also very shallow (< 66 m) and the record dive time for an emperor penguin is just under 22 min (Weinecke et al., 2007). At about 3.0 m/sec swim speed (Kooyman et al., 1992), this would allow the bird to swim 180 m/min or about 3,960 m underwater. Again, if this was straight up and straight down, to a depth of almost 2,000 m, far beyond the depth records of penguins.

By contrast, there are developmental aspects of diving behavior to consider and it has been shown that as seal pups mature and develop both their oxygen stores and their swimming abilities, that the average depths of dives increase (Burns and Testa, 1997). This would imply that dive time (the development of oxygen stores), somehow plays a role in the development of dive depth in young animals. Perhaps this is the case for young seals, but by the time they become adults, their dive time abilities far exceed any restriction on dive depth.

If a marine mammal or a bird were to exceed the ADL, then the dives could be deeper as they could remain underwater for a longer time period. As noted above however, these animals are not diving simply to go as deep or as long as possible: They are hunting, navigating and interacting with one another. These activities utilize time and oxygen resources. For example, the maximum dive time of the southern elephant seal (*Mirounga leonina*) is almost 2 h (Slip et al., 1994). At 1.5m/ sec, the animal could cover about 11,000m of distance, or about 5,500m of vertical distance. The depth record for elephant seals is less than 2,000m (Table 1). Eguchi and Harvey (2005) concluded that the duration of harbor seal (*Phoca vitulina*) dives were impacted strongly by body oxygen stores, whereas only 40% of the variability of dive depth was attributed to body mass (and therefore oxygen stores). An important distinction is that routine diving depth appears to be set by ecological parameters, such as the depth of prey (Teloni et al., 2008). The anatomical, biochemical and physiological limits of the animal must allow it to move within this range of ecologically important depths with only minimal cost. In terms of how deep an animal could possibly dive based on oxygen stores and the ability to generate anaerobically fueled exercise, it appears that most of these diving species could dive much deeper. Clearly, there are other factors at work that set diving depth beyond just swimming speed and elapsed time that can be tolerated underwater.

## Mechanical Limitations to Diving

It was clear early in the studies of diving physiology that time underwater and swimming speed were not limiting dive depth and the research turned immediately to what was most obvious when comparing marine birds and mammals to humans: the impacts of pressure on air spaces, such as the lung (Kooyman, 1973, 1988). Recall that for every 10 m of depth, the pressure increases by 1 ATM and that the relationship between air volume and pressure is a constant (Boyles Law; $PV = K$). Thus, if pressure goes up by 20x (e.g. diving to a depth of 20 ATM or 200 m), then air space volume would only be 1/20 of the original surface volume. At the limits of current dive depth for marine mammals (about 2000 m or 200 ATM), then the airspaces in the body would be only 1/200 of air volume at the surface. In humans, lung volume decreases with diving depth and the vacuum created by the shrinking lung is taken up by intrathoracic blood pooling (Schaefer et al., 1968). This is because the chest wall of a human is not compressible (non-compliant) and forms a hard shell around the lung space. Thus, the shrinking lung volume creates a vacuum inside this hard shell and in order to equalize the pressure, blood vessels expand and pool into this space. For example, free dives to 70 m in humans induce about 980 ml of blood pooling into the thorax (Schaefer et al., 1968).

Marine mammals are different in their rib cage structure from humans in that their chest walls are extremely compliant (Leith, 1976; For reviews, see Kooyman and Ponganis, 1998 and Berta et al., 2006). Cetaceans have a small number of true ribs that are connected directly to the sternum and baleen whales are the extreme of this group with no connections to the sternum (Kooyman and Andersen, 1969).Thus, as the lung collapses from the increasing hydrostatic pressure, the chest wall follows along and collapses with the lung. There is no vacuum inside the thoracic space and the problem of blood pooling is avoided. Added to this is the behavioral observation that many diving pinnipeds species exhale upon diving (Kooyman, 1989) so that the lung volume at the time of dive initiation is already reduced. This has multiple implications to blood gas theory which we will discuss below, but it has been both calculated and now shown that the marine mammal lung alveoli demonstrate 'compression-absorption collapse' (Scholander, 1940; Kooyman et al., 1970, 1972; Falke et al., 1985). That is, between the combination of compression collapse due to the increased hydrostatic pressure and the utilization of the last remaining air volume in the alveoli, these terminal air structures are completely collapsed at depths below about 70–80 meters in marine phocids and small cetaceans. No such measurements have been made for large cetaceans, though their chest walls are also very compliant because of the minimal connections of ribs to the sternum (Kooyman and Andersen, 1969). The anatomy of the terminal airways in diving species was studied carefully by Kooyman over many years and he and his students/colleagues showed that the airways leading to the alveoli were strengthened with cartilage (Denison and Kooyman, 1973). They hypothesize that these reinforced, non-oxygen diffusing airways remain open even under high hydrostatic pressure which facilitates the re-opening of the lung tissue as the animal ascends. Recent work has demonstrated that the lung surfactants in diving species are adapted to facilitate the rapid and routine collapse and re-expansion of the alveolar tissue perhaps hundreds of times/day (Spragg et al., 2004).

What about the air spaces in the sinuses and the ear structures of marine mammals? Marine mammals have no facial sinuses, so the problems and pain faced by humans either diving or flying with blocked sinuses have been eliminated in marine mammals when they are diving (Kooyman, 2002). The space around the middle ear in pinnipeds is surrounded by vascularized soft tissue, so that as the air space collapses due to compression, the cavity fills with engorged blood vessels (Kooyman and Andersen, 1969). This not only relieves any ear pain associated with diving, but provides a liquid to liquid interface between the ear and the ocean water, which facilitates being able to hear sounds underwater (Repenning, 1972; Kastak and Schusterman, 1999). In cetaceans, the anatomy of air spaces associated with the ear is more complex. For example, the mandibular canal is filled with a fatty tissue ( 'fat pad' ) which achieves the same liquid-liquid interface as discussed in the phocids (Nummela et al., 2007). But, air spaces have not been eliminated in all species and toothed whales (odontocetes, for example sperm whales) have isolated the ears from the skull with air/lipid foam filled cavities (Fraser and Purves, 1954). Berta et al. (2006) provide an in-depth review of the anatomical structures associated with pinnipeds/cetacean hearing and possible impacts of diving.

Most of our discussion to this point has focused on the marine mammals because not as much is known about the same mechanical components of air space compression diving in penguins. Perhaps the most fundamental difference between the diving marine mammals and birds is that the lung is not a significant source of oxygen for the mammals during diving (due to lung collapse), but air spaces make up about 50% of the total available oxygen stored in penguins (Ponganis and Kooyman, 2000). By contrast with pinnipeds, the chest wall of a penguin is not compressible and lung collapse due to diving depth is not considered likely (Ponganis et al., 1999). Kooyman conducted laboratory compression tests on penguins in the 1970's (Kooyman et al., 1973) and both he and Ponganis have worked with deep diving king and emperor penguins examining the impacts of pressure (Kooyman et al. 1990, Ponganis and Kooyman, 1991). Working with Ponganis, Stockard et al. (2005) placed oxygen sensors in the air sacs of freely diving emperor penguins. They found that the partial pressure of oxygen increased during the initial phases of the dive as the hydrostatic pressure increased. As oxygen was utilized by the birds during the dive, the oxygen partial pressure declined. The decline in oxygen during diving implies that the air sacs were not collapsed, but because the birds in this experiment dived only to about 100m, the impacts of diving to extreme depth were not measured. Kooyman et al. (1973) showed that gas exchange continued in other species of diving penguins in simulated dives as deep as 136 m meaning that the air spaces remained open. Sato et al. (2002), discusses that the vigorous wing-beats of penguins as they descend implies that they are working against the buoyancy of air spaces in the body. In contrast, they appear to passively ascend, which also implies inflated air spaces. They also suggest that the birds alter the volume of air in the lungs before the dive in order to control buoyancy during the dive. How these birds avoid the absorption of nitrogen into the blood and avoid the bends if their air spaces remain open is not known (Kooyman and Ponganis, 1998). As we will discuss below, the problem of nitrogen absorption into the blood is minimized in diving seals due to lung collapse at depth. Finally, unlike flying birds, penguins do not have air spaces in their bones, so there would be no pressure effects on those spaces.

## Blood Gases and Diving

The discussion above on the mechanical aspects of increasing hydrostatic pressure on air spaces brings us to the next implication of collapsed or open respiratory surfaces: gas exchange between the blood and lung during diving.

As noted above, in diving mammals, the lung space collapses below about 80–100 m. This means that there is no gas exchange occurring during deep dives once the alveolar spaces are closed. Therefore, nitrogen does not equilibrate into the blood at depth and problems associated with nitrogen narcosis are effectively eliminated. It also means that deep diving marine mammals are unlikely to become 'bent' (Decompression sickness (DCS), the diver's disease, the bends, caisson disease) upon rapid ascent back to the surface. Nitrogen equilibrates into the blood during the initial phases of the descent and this has been measured in both laboratory and field diving conditions with seals and small cetaceans (Kooyman et al., 1972; Ridgway and Howard, 1979; Falke et al., 1985). However, the nitrogen levels are not high because the gas equilibrates throughout the entire blood and tissue volume during this time (Kooyman and Ponganis, 1998). Further, marine mammals also have a very high body fat content (up to 50% body fat) and nitrogen is 6–7 times more soluble in lipid than in water. Thus, the blubber and fat store may also make an excellent sink for the increased nitrogen (Kooyman and Andersen, 1969). Once the animal descends below the point of alveolar collapse, there is no further gas equilibration between the lung and the blood.

There are two exceptions to the bends problem that need to be considered. The first deals with marine mammals that dive routinely above the shallow depth of alveolar collapse. If these mammals (such as smaller fur seals) stay above the lung-collapse depth, then they will be equilibrating gases into the blood at higher pressure and then rapidly ascending. Even though they are not breathing compressed air at depth as is the case with human divers, there is the possibility that after many series of shallow dives, enough nitrogen would be equilibrated into their tissues to cause problems on ascent. Kooyman (1989) argues however that one solution to this problem is for each dive to be relatively short so that there is not enough time for full body equilibration of nitrogen. Further, short dives usually have full body blood circulation (long dives restrict blood flow to the central organs away from the periphery (For review see Castellini, 1991) which would increase the pool size for nitrogen distribution. From an empirical point of view, this problem has clearly been solved by shallow diving marine mammals because we have no evidence that they experience the bends.

The same theoretical problem exists for penguins because it appears that they don't collapse their lungs during diving. How they avoid the bends upon rapid ascent is not known. Sato et al. (2002) discuss this problem and notes that the birds change their ascent angle when nearing the surface to a more shallow approach. They propose that this may allow some time for nitrogen equilibration instead of coming to the surface vertically, but they do not have data to test this idea.

The second exception to the problem of the bends is the recent controversy about the beaching of beaked whales associated with Naval SONAR testing. Beaked whales are deep divers and one theory is that the SONAR exercises confuse or frighten the whales, they come to the surface too quickly, suffer acute decompression sickness and post-mortem evidence from beaked whales has shown gas emboli in tissues. (Fernandez et al., 2005; Cox et al., 2006). Kooyman (2006) however argues that because of the compression-absorption collapse of lung tissue, the resultant minimal rise in $PN_2$ would not necessarily cause the acute symptoms of decompression sickness upon ascent. There are also recent claims of erosion seen on the bone surfaces of sperm whales that is used to support the concept that whales suffer from decompression sickness (Moore and Early, 2004). However, this would be indicative of a chronic issue with nitrogen tension, not the acute conditions that would be seen because of SONAR exercises (Kooyman, 2006).

Perhaps there are chronic problems with nitrogen blood levels in diving species, but on a day to day basis, there is no evidence that any of these species suffer from decompression effects.

This same ultimate conclusion is relevant to the question of how diving species avoid High Pressure Nervous Syndrome (HPNS). This effect can be induced in other species during diving, is characterized by tremors and differs in its onset with types of animals (Brauer et al. 1979; Halsey 1982; Rostain et al, this book). Vertebrates such as newts and frogs are more resistant than small mammals (onset >150 ATM vs 80 ATM), but non-human primates are more susceptible (onset at about 50 ATM) (Brauer et al., 1979). Humans experience HPNS at depths as shallow as 100 m (10 ATM; Jain, 1994) but, as with decompression sickness, the ecological success of marine mammals and some penguins to exploit extreme depths is evidence that they avoid HPNS. Some have suggested that the equilibration of nitrogen into fatty tissue surrounding nerve cells (myelin sheath) protects against the onset of HPNS in marine mammals (LeBoeuf et al., 1989) on the basis that adding nitrogen to diving air mixtures can help avoid HPNS in humans (Bennett et al., 1981, 1982). However, there have been no experiments examining HPNS in marine mammals or their tissues and the question remains unanswered.

If marine mammals avoid mechanical problems with gas exchange at depth by collapsing the lung, this leads to the question of where they carry the oxygen that is necessary for them to dive within their Aerobic Dive

Limits? The answer is that they have enhanced their blood and tissue stores of oxygen far beyond terrestrial mammals. Generally, marine mammals have a larger blood volume per body mass, more red blood cells (RBC) per blood volume (elevated hematocrit; Hct) and more hemoglobin per RBC (For review, see Castellini, 1991) The net result is that phocid seals can carry three times more oxygen per body mass than terrestrial mammals, even without the use of the lung. However, they certainly can dive more than three times longer than humans and they do this through reduction in diving metabolic rate and increased tolerance to hypoxic or anaerobic metabolism for extremely long dives (Kooyman, 1989). This tissue level tolerance comes through a suite of biochemical and perhaps molecular adaptations for enzyme function, ion transport and pH buffering tolerance.

## Tissue Level Aspects Of Deep Diving

At the level of cellular adaptations for diving, most work has focused on the issues of oxygen storage, enzyme function under low oxygen conditions and tissue buffering capacity under anaerobic metabolism (Castellini and Castellini, 2004). However, these biochemical issues are not directly relevant to pressure tolerance.

The study of the direct impact of pressure on biochemical and cellular systems has only recently been considered in marine mammals and has not been examined in diving birds. The biochemical bases for how pressure alters enzyme or cellular function are covered extensively in other sections of this book. Essentially, pressure acts to increase the order of the phospholipid acyl chains of the bi-lipid membrane layer around cells and organelles (Cossins and Macdonald, 1989). This change in lipid fluidity in turn impacts cell membrane transport reactions and therefore acts to enhance or inhibit reactions that are integrally tied to cell lipid characteristics (Siebenaller and Garrett, 2002). Animals that live at depth have evolved mechanisms to counteract these pressure induced problems. These adaptations include modifications of proteins and membranes, including altering enzyme structure at the active sites so that pressure does not change the affinity of the enzyme for its substrate and in membranes substituting acyl chains that cause a more disordered phospholipid bi-layer (usually, more unsaturated) to counteract the ordering tendency of elevated pressure (Jaenicke, 1987; Siebenaller, 1987; Somero, 1992). Should marine mammals or penguins show any of these biochemical adaptations for pressure tolerance? Some species dive deeply enough, but both groups are warm bodied endotherms and could, at least theoretically, offset the structural ordering induced by high pressure because of their high body temperature. While there is a substantial amount of literature on the biochemical adaptations to pressure in fish and invertebrates, only a handful have been written on the same question in marine mammals.

The first such paper was by Croll et al. (1992b). They examined the affinity of the enzyme lactate dehydrogenase ($M_4$:LDH) for its co-factor Nicotinamide adenine dinucleotide (NADH) in several marine and terrestrial mammal species compared to fish as controls. The control of muscle LDH is critical in marine mammals since they produce large amounts of lactate during hypoxic diving conditions and then reverse the enzyme reaction direction and consume the lactate when oxygen is re-supplied to the tissues (Kooyman, 1989). Even though there is evidence that marine mammals may not have elevated LDH reaction rates for the production of lactate, they show an increased ability to process lactate and increased pH tolerance at this reaction site compared to terrestrial species (Castellini et al., 1981). Croll et al. (1992b) found that muscle LDH affinity for NADH was unchanged at 204 ATM and 37°C in both terrestrial and marine species. They concluded that this may be a result of the high body temperatures of marine mammals. They did not run the reaction in the lactate consumption direction (NAD and lactate converted to NADH and pyruvate) nor did they examine possible changes in affinity for pyruvate, therefore, we do not know if these aspects of the LDH reaction are pressure sensitive.

No work was subsequently done on biochemical aspects of diving in marine mammals until Williams et al. (2001) measured the order of red blood cell (RBC) membranes of marine and terrestrial mammals at both 37°C (core temperature) and 27°C (peripheral body temperature in diving species). They found that all the species showed high membrane order at lower temperature, which would be expected. But surprisingly, two marine species (harbor seal and northern elephant seal) had much higher membrane ordering than all other species at all pressures and both temperatures. Furthermore, the elephant and harbor seals were more sensitive to pressure induced alterations in membrane fluidity than the other mammalian species. These results are not what would be predicted if temperature provided some protective mechanism against increased membrane viscosity induced by elevated pressure. Though enzyme reaction chemistry and cell membrane fluidity cannot be directly compared, these results are at odds with the suggestion from Croll et al. (1992b) that enzymatic reactions in marine mammals are protected by high body temperature.

Castellini et al. (2001, 2002), examined the metabolic rates of marine and terrestrial mammal RBC and found differences in the glycolytic flux rate between the two groups at 37°C when they were pressurized to 136 ATM (2000 psi). Generally, marine mammal RBC either showed no reaction to pressure or increased their glycolytic rate while most terrestrial mammal RBC reduced glycolysis under pressure. Since they dealt with living cells, it was not possible to differentiate whether these changes were caused by membrane fluidity or enzymatic reaction rate responses to pressure. In this study, temperature did not provide protection against the changes in RBC metabolism induced by elevated pressure.

In a study on cellular function under elevated pressure conditions, Field et al. (2001) found that rapid pressure changes and temperatures lower than 20°C cause activation of human and terrestrial mammal platelets. Northern elephant seals routinely dive to pressures as high as 150 ATM, yet do not suffer from the thrombotic effects of platelet activation associated with rapid decompression. They concluded that decreased sensitivity of elephant seal platelets to agonists may be a protective mechanism developed in response to rapid pressure changes associated with deep diving.

Rivera (1997) is the only other study on marine mammal biochemical aspects of pressure tolerance. She worked on an oxidative tissue (heart) as opposed to the previous studies which had focused on peripheral muscle tissue. In the direction of lactate production by LDH, elevated pressure did not change the affinity of either terrestrial or marine mammal heart LDH for pyruvate, but decreased the affinity for NADH. This is different from the findings of Croll et al. (1992b) which suggested that the affinity for NADH by LDH was pressure insensitive. However, this may be due to the differences between LDH that is muscle type (M4) vs heart type (H4).

No further work has been conducted on marine mammal tissues examining the cellular or biochemical basis for pressure tolerance since these four studies. No tissue level biochemical work of any kind has been carried out on pressure tolerance in diving birds.

## CONCLUSIONS AND FUTURE STUDIES

Are there reasonable limits to the diving depths of marine mammals and birds? It is clear that the time necessary to dive to depth is not likely one of those limits. Marine mammals can stay underwater for much longer than it would take to extend their diving depths beyond current records. They have enough oxygen and can produce enough anaerobic energy to allow much deeper dives. The same scenario appears to apply to adult emperor penguins: They could dive much deeper based on their maximum dive times and average swim speeds. It is easy to suggest that the ecological and behavioral requirements on the animals set the actual depth patterns and that the anatomy, physiology and biochemical limits of the animals must allow the animals to easily move within that ecologically driven window. For example, if an animal has a maximum dive time of 20 min and it takes 5 min to dive to depth where it can forage for prey and 5 min to ascend, then it has about 10 min of bottom time to hunt. Deeper diving would reduce bottom time until the animal would get to the point where bottom time available to hunt would become zero. The research programs that have examined data from millions of recorded dives in birds and mammals have modeled how dive time, swim speed, bottom time and recovery time on the surface all relate to one another (Heithaus and Dill, 2002; Stewart, 2002; Costa and Sinervo, 2004). From these and measured or estimated forage success, they calculate dive efficiency in terms of energy gain vs energy cost.

These models are extremely useful for determining the ecological physiology of diving species, but are not sufficient to answer the physical or biochemical pressure limitation questions posed in this chapter: The animals may dive to a certain depth for ecological efficiency, but how deep are they capable of diving and what limits them? In theory, an emperor penguin could dive to almost 2,000 m based on maximum known dive time. Why have no dives to that depth been observed? Is it simply because there may be no reason to go that deep, or are there anatomical/physiological reasons?

For marine mammals, there may be no anatomical limitation on depths that are within range of their dive time limits. The problems with air spaces have been resolved, the lungs are collapsed and there are no problems with respiratory gases equilibrating into the blood. Maybe, seals and whales, which are anatomically capable of diving much deeper, have no ecological reason for doing so? This might be different for penguins because of their non-compliant chest wall and the air sacs that make up a large part of their diving air volume and the consequent problems associated with tissue compression and buoyancy control. Buoyancy through air space regulation is not an issue with seals because there are no significant air spaces remaining in them below 100 m.

If anatomy does not limit the diving marine mammals, then perhaps there are limits on physiology or biochemistry that keep them within the top 2000 m of the ocean. The physiology of diving, however does not have a component that would appear to be sensitive to elevated hydrostatic pressure. To dive as long as possible in order to get as deep as possible, they maximize the dive response, confine the blood to the central circulation and reduce circulation to the periphery. Control of that circulation is essential to the length of the dive. Perhaps, at extreme pressures, there may be neurological limitations that would compromise their ability to maintain the dive response? This would be suggestive of some type of neurological deficit similar to HPNS. We have no evidence of any manifestations of HPNS in marine mammals by the definition of how that syndrome is seen in other vertebrates. But, there are graded responses of HPNS in vertebrates and perhaps marine mammals are at the extreme end of that response. Or, some type of HPNS is characterized by different symptoms in marine mammals and we have not yet searched for those symptoms? If loss of fine circulatory control was one of those symptoms, this would limit the ability to restrict blood flow between the central and peripheral systems, which would limit dive time and therefore, dive depth.

What about the impact of extreme pressure on biochemical pathways? All data suggest that marine mammal biochemistry is either insensitive to pressure or that elevated pressure enhances some reactions. But, our data here are exceptionally sparse for marine mammals and non-existent for diving birds. There are thousands of metabolic reactions that could be critically compromised at high pressure and only 4–5 have been studied.

We are left then with the following: Behavioral data demonstrate that diving animals reach extreme depths with no negative consequences. All anatomical, physiological and biochemical data reinforce that these dives are not dangerous or at the limits of biological tolerance. But, the deepest known dives are about 2,000m and not beyond. Why? Is it truly a very simple case that there are no ecological/behavioral reasons to dive deeper and to do so would be a waste of energy? Most of the studies of Weddell seals have been done where the water is usually no more than 1,000 m deep. If we studied them in much deeper water, would we see dives that were deeper? These questions can be used to focus future studies on diving depth limitations in birds and mammals.

First, physiological measurements on diving birds have very little data for animals diving beyond 200 m. Because penguins have air spaces in them while diving, it is essential that we extend our physiological and anatomical observations to extreme depths. This is very hard to do in the field, but may be the only way to understand how deep these birds can dive and what happens to air spaces beyond 200 m.

Second, we need field and laboratory studies on the neural control of diving under elevated hydrostatic pressure. It has been many years since marine mammals were studied while diving in compression chambers, but we may need to return to that method. Additionally, perhaps marine mammal neural cells can be studied in isolation/culture under high pressure and their inherent tolerance to pressure changes can be examined (Grossman, this book).

Third, basic biochemical tolerance to elevated pressure at the level of ion transport (which also has neurological implications), cell membrane responses and metabolic regulation needs to be studied. These could begin with tissue samples or cells in culture and then extended to freely diving animals.

Finally, if routine diving depth is ecologically or behaviorally defined, then perhaps captive animals can be taught to dive even deeper and bypass these behavioral limits.

We began this discussion with the observation that deep diving is routine and adaptive in marine mammals and birds. Unlike humans, who have no adaptive pressure or selection to dive, the ability of marine mammals and diving birds to spend a great deal of time underwater and to tolerate extreme pressure is part of their ecological niche. Similar to the concept that some large marine fish have developed endothermy which helps them to exploit the 'coldness' of the deep ocean (Block, 1991), it is clear that marine mammals have adapted to the elevated pressures of the deep environment at anatomical, physiological, biochemical and behavioral levels. Their ability to function well in both worlds, at the surface and at extreme depth, defines their existence and as we watch them dive routinely at sea, we are left with our final question: How deep can they dive?

## REFERENCES

Baird, R.W. and D.J. McSweeney, A.D. Login and D.L. Webster. 2004. Tagging feasibility and diving of Cuvier's beaked whales (*Ziphius cabirostris*) and Blainville's beaked whales (*Mesoplodon densirostris*) in Hawaii. Report prepared under Order # AB133F-03-SE-0986) to the Hawai'I Wildlife Fund, Volcano, Hawaii, USA.

Bennett, P.B. and R. Coggin and J. Roby. 1981. Control of HPNS in humans during rapid compression with trimix to 650 m (2132 ft). Undersea Biomed. Res. 8(2): 85–100.

Bennett, P.B. and R. Coggin and M. McLeod. 1982. Effect of compression rate on use of trimix to ameliorate HPNS in man to 686 m (2250 ft). Undersea Biomed. Res. 9: 335–351.

Berta, A. and J.L. Sumich and K.M. Kovacs. 2006. Marine Mammals: Evolutionary Biology. 2nd edit. Elsevier, Amsterdam, The Netherlands.

Block, B.A. Endothermy in fish: Thermogenesis, ecology and evolution. Pp. 269–311. *In*: P.W. Hochachka and T.P. Mommsen (eds.). 1991. Biochemistry and Molecular Biology of Fishes. Vol. 1. Phylogenetic and biochemical perspectives. Elsevier Science Publishers, Amsterdam, The Netherlands.

Brauer, R.W. and R.W. Beaver, S. Lahser, R.D. McCall and R. Venters. 1979. Comparative physiology of the high-pressure neurological syndrome—compression rate effects. J. Appl. Physiol. 46: 128–135.

Burns, J.M. and J.W. Testa. Developmental changes and diurnal and seasonal influences on the diving behavior of Weddell seal (*Leptonychotes weddellii*) pups. Pp. 328–334. *In*: B. Battaglia, J. Valencia and D.W.H. Walton (eds.). 1997. Antarctic Communities: Species, Structure and Survival. Cambridge University Press, Cambridge, UK.

Castellini, M.A. The biology of diving mammals: Behavioral, physiological and biochemical limits. pp. 105–134 *In*: Advances in Comparative and Environmental Physiology 1991. Vol. 8. Springer-Verlag, Berlin, Germany.

Castellini, M.A. and J.M. Castellini. 2004. Defining the limits of diving biochemistry in marine mammals. Comp. Biochem. Physiol. 139(B): 509–518.

Castellini, M.A. and G.N. Somero and G.L. Kooyman. 1981. Glycolytic enzyme activities in tissues of marine and terrestrial mammals. Physiol. Zool. 54(2): 242–252

Castellini, M.A. and P.M. Rivera and J.M. Castellini. 2001. Adaptations to pressure in the RBC metabolism of diving mammals. Comp. Biochem. Physiol. 129(A): 751–757.

Castellini, M.A. and P.M. Rivera and J.M. Castellini. 2002. Biochemical aspects of pressure tolerance in marine mammals. Comp. Biochem. Physiol. 133(A): 893–899.

Cossins, A.R. and A.G. Macdonald. 1989. The adaptations of biological membranes to temperature and pressure: fish from the deep and cold. J. Bioenerg. Biomembr. 21: 115–135.

Costa, D.P. and B. Sinervo. 2004. Field physiology: Physiological insights from animals in nature. Ann. Rev. Physiol. 66: 209–238.

Cox, T.M. and T.J. Ragen, A.J. Read, E. Vos, R.W. Baird, K. Balcomb, J.Barlow, J. Caldwell, T. Cranford,L. Crum, A. D'amico, G. D'spain, A. Fernández, J. Finneran, R. Gentry,

W. Gerth , F. Gulland, J. Hildebrand, D. Houser, T. Hullar, P.D. Jepson, D. Ketten, C.D. Macleod, P. Miller, S. Moore, D.C. Mountain, D. Palka, P. Ponganis, S. Rommel, T. Rowles, B. Taylor, P. Tyack, D. Wartzok, R. Gisiner, J. Mead and L. Benner. 2006. Understanding the impacts of anthropogenic sound on beaked whales. J. Cetacean Res. Manage. 7(3): 177–187.

Croll, D.A. and A.J. Gaston, A.E. Burger and D. Konnoff. 1992a. Foraging behavior and physiological adaptation for diving in Thick-Billed Murres. Ecol. 73: 344–356.

Croll, D.A. and M.K. Nishiguchi and S. Kaupp. 1992b. Pressure and lactate dehydrogenase function in diving mammals and birds. Physiol. Zool. 65(5): 1022–1027.

Croll, D.A. and A. Acevedo-Gutierrez, B.R. Tershy and J. Urban-Ramirez. 2001. The diving behavior of blue and fin whales: is dive duration shorter than expected based on oxygen stores? Comp. Biochem. Physiol. 129(A): 797–809.

Denison, D.M and G.L. Kooyman. 1973. The structure and function of the small airways in pinniped and sea otter lungs. Resp. Physiol. 17: 1–10.

Eguchi, T. and J.T. Harvey. 2005. Diving behavior of the Pacific harbor seal (*Phoca vitulina richardii*) in Monterey Bay, California. Mar. Mamm. Sci. 21: 283–295.

Falke, K.J. and R.D. Hill, J. Qvist, R.C. Schneider, M. Guppy, G.C. Liggins, P.W. Hochachka, R.E. Elliott and W.M. Zapol. 1985. Seal lungs collapse during free diving: Evidence from arterial nitrogen tensions. Science 229: 556–558.

Fernandez, A. and J.F. Edwards, F. Rodriguez, A.E. de los Monteros, P. Herraez, P. Castro, J.R. Jaber, V. Martin and M. Arbelo. 2005. "Gas and fat embolic syndrome" involving a mass stranding of beaked whales (Family Ziphiidae) exposed to anthropogenic sonar signals. Vet. Path. 42: 446–457.

Field, C.L. and N.J. Walker and F. Tablin. 2001. Northern elephant seal platelets: Analysis of shape change and response to platelet agonists. Thromb. Res. 101: 267–277.

Fraser, F.C. and P.E. Purves. 1954. Hearing in cetaceans. Bull. Brit. Museum. Zool. 2(5): 103–113.

Halsey, M.J. 1982. Effects of high pressure on the central nervous system. Physiol. Rev. 62: 1341–1377.

Halsey, L.G. and T.M. Blackburn T.M. and P.J. Butler. 2006. A phylogenetic analysis of the allometry of diving. Am. Natural 67: 276–287.

Heezen, B.C. 1957. Whales entangled in deep sea cables. Deep-Sea Res. 4: 105–115.

Heithaus, M.R. and L.M. Dill. Feeding strategies and tactics. Pp. 412–422. In: W. Perrin, B. Wursig and J.G.M. Thewissen (eds.). 2002. Encyclopedia of Marine Mammals. Academic Press, San Diego, USA.

Hooker, S.K. and R.W. Baird. 1999. Deep-diving behaviour of the northern bottlenose whale, Hyperoodon ampullatus (Cetacea: Ziphiidae). Proc. Royal Soc. Lond. Series B-Biol. Sci. 66: 671–676.

Jaenicke, R. Cellular components under extremes of pressure and temperature: Structure-function relationship of enzymes under pressure. Pp. 257–272. In: H.W. Jannasch, R.E. Marquis and A.M. Zimmerman, A.M. (eds.). 1987. Academic Press, New York, USA.

Jain, K.K. 1994. High-pressure neurological syndrome (HPNS). Acta. Neurol. Scan. 90: 45–50.

Kastak, D. and R.J. Schusterman. 1999. In-air and underwater hearing sensitivity of a northern elephant seal (*Mirounga angustirostris*). Can. J. Zool. 77: 1751–1758.

Kooyman, G.L. 1966. Maximum diving capacities of the Weddell seal, (*Leptonychotes weddelli*). Science 151: 1553–1554.

Kooyman, G.L. 1973. Respiratory adaptations in marine mammals. Am. Zoolog. 13: 457–478.

Kooyman, G.L. 1988. Pressure and the diver. Can. J. Zool. 66: 84–88.

Kooyman, G.L. 1989. Diverse divers: Physiology and behavior Springer-Verlag, Berlin, Germany.

Kooyman, G.L. Diving Physiology. Pp. 339–344. *In*: W. Perrin, B. Wursig and J.G.M. Thewissen. (eds.). 2002. Encyclopedia of Marine Mammals. Academic Press, San Diego, USA.

Kooyman, G.L. 2006. Mysteries of adaptation to hypoxia and pressure in marine mammals—The Kenneth S. Norris lifetime achievement award lecture. Mar. Mamm. Sci. 22: 507–526.

Kooyman, G.L. and H.T. Andersen. Deep Diving. Pp. 65–94. *In*: H. T. Andersen (ed.). 1969 The Biology of Marine Mammals. Academic Press, New York, USA.

Kooyman, G.L. and T.G. Kooyman. 1995. Diving behavior of emperor penguins nurturing chicks at Coulman Island, Antarctica. Condor 97: 536–549.

Kooyman, G.L. and P.J. Ponganis. 1998. The physiological basis of diving to depth: Birds and mammals. Ann. Rev. Physiol. 60: 19–32

Kooyman, G.L. and D.D. Hammond and J.P. Schroeder. 1970. Bronchograms and tracheograms of seals under pressure. Science 169: 82–84.

Kooyman, G.L. and J.P. Schroeder, D.M. Denison, D.D. Hammond, J.J. Wright and W.P. Bergman. 1972. Blood nitrogen tensions of seals during simulated deep dives. Am. J. Physiol. 223: 1016–1020.

Kooyman, G.L. and J.P. Schroeder, D.G. Greene and V.A. Smith. 1973. Gas exchange in penguins during simulated dives to 30 and 68 m. Am. J. Physiol. 225: 1467–1471.

Kooyman, G.L. and R.L. Gentry and D.L. Urquhart. 1976. Northern fur seal diving behavior: A new approach to its study. Science 193: 411–412.

Kooyman, G.L. and E.A. Wahrenbrock, M.A. Castellini, R.W. Davis and E.E. Sinnett. 1980. Aerobic and anaerobic metabolism during voluntary diving in Weddell seals: Evidence of preferred pathways from blood chemistry and behavior. J. Comp. Physiol. 138: 335–346.

Kooyman, G.L. and P.J. Ponganis, L.S. Davis and J.T. Darby. Behavior and physiology of diving in emperor and king penguins. Pp. 229–242. *In*: L. S. Davis and J.T. Darby. (eds.). 1990. Penguin Biology. Academic Press, San Diego, USA.

Kooyman, G.L. and P.J. Ponganis, M.A. Castellini, E.P. Ponganis, K.V. Ponganis, P.H. Thorson, S.A. Eckert and Y. Le Maho. 1992. Heart rates and swim speeds of emperor penguins diving under sea ice. J. Exp. Biol. 165: 161–180.

Le Boeuf, B.J. and Y. Naito, A.C. Huntley and T. Asaga. 1989. Prolonged, continuous, deep diving by northern elephant seals. Can. J. Zool. 67: 2514–2519.

Leith, D. 1976. Comparative mammalian respiratory mechanics. Physiologist 19: 485–510.

Mate, B. and R. Mesecar and B. Lagerquist. 2007. The evolution of satellite-monitored radio tags for large whales: One laboratory's experience. Deep-Sea Research, Part II (Topical Studies in Oceanography) 54: 224–247.

Moore, M.J. and G.A. Early. 2004. Cumulative sperm whale bone damage and the bends. Science 306: 2215.

Naito, Y. 2004. Recent advances in studies of the diving behavior of marine birds and mammals with micro-data loggers. J. Yamashina Inst. Ornithol. 35: 88–104.

Nummela, S. and J.G.M. Thewissen, S. Bajpai, T. Hussain and K. Kumar. 2007. Sound transmission in archaic and modern whales: Anatomical adaptations for underwater hearing. Anatom. Rec.-Adv. Integ. Anat. Evol. Biol. 290: 716–733.

Ponganis, P.J. and G.L. Kooyman. 1991. Diving physiology of penguins. Pp. 1887–1892. Acta XX Congressus Internationalis Ornithologici. New Zealand Ornithological Congress Trust Board, Wellington, New Zealand.

Ponganis, P.J. and G.L. Kooyman. 2000. Diving physiology of birds: a history of studies on polar species. Comp. Biochem. Physiol. 126(A): 143–151.

Ponganis, P.J. and G.L. Kooyman, M.A. Castellini, E.P. Ponganis and K.V. Ponganis. 1993. Muscle temperature and swim velocity profiles during diving in a Weddell seal, *Leptonychotes weddellii*. J. Exp. Biol. 183: 341–346.

Ponganis, P.J. and G.L. Kooyman, L.N. Starke, C.A. Kooyman and T.G. Kooyman. 1997a. Post-dive blood lactate concentrations in emperor penguins, Aptenodytes forsteri. J. Exp. Bio. 200: 1623–1626.

Ponganis, P.J. and G.L. Kooyman, L.M. Winter and L.N. Starke. 1997b. Heart rate and plasma lactate responses during submerged swimming and trained diving in California sea lions, *Zalophus californianus*. J. Comp. Physio. 167(B): 9–16.

Ponganis, P.J. and G.L. Kooyman, E.A. Baranov, P.H. Thorson and B.S. Stewart. 1997c. The aerobic submersion limit of Baikal seals, *Phoca sibirica*. Can. J. Zool. 75: 1323–1327.

Ponganis, P.J. and G.L. Kooyman, R. Van Dam and Y. LeMaho. 1999. Physiological responses of king penguins during simulated diving to 136 m depth. J. Exp. Biol. 202: 2819–2822.

Repenning, C.A. 1972. Adaptive evolution of sea lions and walruses. Syst. Zool. 25: 375–390.

Ridgway, S.H and R. Howard. 1979. Dolphin lung collapse and intramuscular circulation during free diving: Evidence from nitrogen washout. Science 206: 1182–1183.

Rivera, P.M. 1997. Biochemical adaptations to asphyxia and increased hydrostatic pressure compared among seal and terrestrial species. M.S. Thesis, University of Alaska Fairbanks, Fairbanks, Alaska.

Santos, M.B. and G.J. Pierce, P.R. Boyle, R.J. Reid, H.M. Ross, I.A.P. Patterson, C.C. Kinze, S. Tougaard, R. Lick, U. Piatkowski and V. Hernandez-Garcia. 1999. Stomach contents of sperm whales *Physeter macrocephalus* stranded in the North Sea 1990–1996. Mar. Ecol. Prog. Ser. 183: 281–294.

Sato, K. and Y. Naito, A. Kato, Y. Niizuma, Y. Watanuki, J.B. Charrassin, C.A. Bost, Y. Handrich and Y. Le Maho. 2002. Buoyancy and maximal diving depth in penguins: do they control inhaling air volume? J. Exp. Biol. 205: 1189–1197.

Schaefer, K.E. and R.D. Allison, J.H. Dougherty, Jr., C.R. Carey, R. Walker, F. Yost and D. Parker. 1968. Pulmonary and circulatory adjustments determining the limits of depths in breathhold diving. Science 162: 1020–1023.

Scholander, P.F. 1940. Experimental investigations on the respiratory function in diving mammals and birds. Hvalradets Skrifter,Norske Videnskamp-Akad, Oslo 22: 1–131.

Schreer, J.F. and K.M. Kovacs. 1997. Allometry of diving capacity in air-breathing vertebrates. Can. J. Zool. 75: 339–358.

Siebenaller, J.F. Pressure adaptation in deep-sea animals. Pp. 33–48. *In*: H.W. Jannasch, R.E. Marquis and A.M. Zimmerman, A.M. (eds.). 1987. Academic Press, New York, USA.

Siebenaller, J.F. and D.J. Garrett. 2002. The effects of the deep-sea environment on transmembrane signalling. Comp. Biochem. Physiol. B. 131: 675–694.

Slip, D.J. and M.A. Hindell and H.R. Burton. Diving behavior of Southern Elephant seals from Macquarie Island: an overview. Pp. 253–270. *In*: B.J. Le Boeuf and Bbehavior, and Pysiology. University of California Press, Berkeley, USA.

Somero, G.N. 1992. Adaptations to high hydrostatic pressure. Ann. Rev. Physiol. 54: 557–577

Spragg, R.G. and P.J. Ponganis, J.J. Marsh, G.A. Rau and W. Bernhard. 2004. Surfactant from diving aquatic mammals. J. App. Physiol. 96: 1626–1632.

Stewart, B.S. Diving Behavior. Pp. 333–339. *In*: W. Perrin, B. Wursig and J.G.M. Thewissen. (eds.). 2002. Encyclopedia of Marine Mammals. Academic Press, San Diego, USA.

Stewart, B.S.and R.L. DeLong. 1991. Diving patterns of northern elephant seal bulls. Mar. Mamm. Sci. 7(4): 369–384.

Stockard, T.K. and J. Heil, J.U. Meir, K. Sato, K.V. Ponganis and P.J. Ponganis. 2005. Air sac P-O2 and oxygen depletion during dives of emperor penguins. J. Exp. Biol. 208: 2973–2980.

Swain, U.G. 1996. Foraging behaviour of female Steller sea lions in Southeastern Alaska and the Eastern Gulf of Alaska. Alaska Department of Fish and Game, Anchorage, Alaska. Wildlife Technical Bulletin 13: 135–166.

Teloni, V. and J.P. Mark, M.J.O. Patrick and M.T. Peter. 2008. Shallow food for deep divers: Dynamic foraging behavior of male sperm whales in a high latitude habitat. J. Exp. Mar. Biol. Ecol. 354: 119–131.

Testa, J.W. 1994. Over-Winter Movements and Diving Behavior of Female Weddell Seals (*Leptonychotes-Weddellii*) in the Southwestern Ross Sea, Antarctica. Can. J. Zool. 72: 1700–1710.

Wanless, S. and S.K. Finney, M.P. Harris and D.J. McCafferty. 1999. Effect of the diel light cycle on the diving behaviour of two bottom feeding marine birds: the blue-eyed shag (*Phalacrocorax atriceps*) and the European shag (*P. aristotelis*). Mar. Ecol. Prog. Ser. 188: 219–224.

Wienecke, B. and G. Robertson, R. Kirkwood and K. Lawton. 2007. Extreme dives by free-ranging emperor penguins. Polar Biol. 30: 133–142.

Williams, E.E. and B.S. Stewart, C.A. Beuchat, G.N. Somero and J.R. Hazel. 2001. Hydrostatic-pressure and temperature effects on the molecular order of erythrocyte membranes from deep-, shallow-, and non-diving mammals. Can. J. Zool. 79: 888–894.

# PRESSURE AND MAN

# 16

# Introduction to the Effects of Pressure on Human Organisms and Other Mammals

*J.C. Rostain*

Increasing hydrostatic pressure or gaseous pressure on organisms induces disruption of their metabolism. All mammals including man exposed to increasing pressure of breathing gas mixture show disturbances at the level of the nervous system, which differ according to which gas is used. Compressed air or nitrogen-oxygen mixture produces from 3 bars, nitrogen narcosis (Bennett and Rostain, 2003a). When nitrogen is replaced by a gas less narcotic such as helium, the breathing mixture induces from 10 bars the High Pressure Nervous Syndrome (HPNS) (Bennett and Rostain, 2003b).

For a better understanding of the effects of the hyperbaric environment on human physiology, the basic physical principles will be briefly reviewed. At sea level, atmospheric pressure is 0.1 MPa (1 megapascal ~1Kg/cm² ~1 atmosphere or 1 bar). Underwater, the greater the depth, the more the weight of water increases. At 10 metres of sea water (msw) each cm² of the body is under an increase of pressure that corresponds to a weight of a column of water of 1000 cm of high on 1 cm² of area i.e. 1000 cm³ of water weighting 1 kg. Thus, the pressure due to water at 10 m is 1 kg/cm² or 1 bar. When we go progressively from 10 m to 20 m, the pressure gradually increases from 1 bar to 2 bars. As a result, each time a man goes 10 metres under water he is exposed to an increase of 1 bar of pressure. Thus at a depth of 10 metres the absolute pressure equals 2 atmospheres (1 of atmospheric pressure and 1 for the 10 metres of water). At 20 metres, the absolute pressure increases to 3

Université de la Méditerranée, UMR-MD2—Physiologie et physiopathologie en condition d'oxygénation extrême, Institut de Neurosciences Jean Roche, Faculté de Médecine Nord, Bd P. Dramard, 13015 Marseille, France.
E-mail: jean-claude.rostain@univmed.fr

bars (1 of atmospheric pressure and 2 for the 20 metres of water), and at 450 m.s.w., it will be 46 bars (1 of atmospheric pressure and 45 bars for the 450 metres of water). The relationship between pressure and volume (Boyle-Mariotte's law) requires that the man must breathe a gaseous mixture the pressure of which is equal to the ambient pressure.

The quantity of gas dissolved within the body tissues is a function of the partial pressure of each gas making up the mixture (Henry's law), the partial pressure of each gas being equal to the product of the ambient pressure and the percentage of this gas in the respiratory mixture.

For instance at atmospheric pressure, air is composed of 21% of oxygen and 79% of nitrogen. Consequently the partial pressure of oxygen is 0.210 bar and that of nitrogen is 0.790 bar. At 30 m.s.w., the absolute pressure is 4 bars and the partial pressure of oxygen is $4 \times 0.21 = 0.840$ bar and that of nitrogen is $4 \times 0.79 = 3.16$ bars.

Both the pressure itself (pressure per se) and the partial pressures of the gases in the respiratory mixture, will affect the organism. Indeed, numerous gases have effects which change when their partial pressure is increased. Thus, increasing the pressure of air produces disturbances in the central nervous system characterized by behavioural troubles and changes in mental processes from an absolute pressure of 4 bars (Bennett and Rostain, 2003a). This is due to nitrogen which becomes narcotic when its partial pressure is increased and is called nitrogen narcosis.

Oxygen at pressure is itself toxic: pressure of 2–3 bars affect the nervous system, and over a long period of time even pressure higher than 0.5 bars will affect the respiratory system. To avoid these physiological problems, it is necessary to use synthetic gas mixtures which include oxygen and diluent gas other than nitrogen. Such mixtures are made up in part of oxygen, the partial pressure of which must remain within acceptable limits, thus requiring a decrease in its percentage with the increase in pressure. In addition, the diluent gas should be as inert as possible for the organism and should have a narcotic potency lower than nitrogen. Three inert gases (i.e. gases which are not metabolized) have a narcotic potency lower than that of nitrogen: hydrogen, neon and helium. The studies on the narcotic potency of these three gases agree that helium has almost no narcotic activity (Behnke et al., 1935; Case and Haldane, 1941; Marshall, 1951; Bennett, 1966; Brauer et al., 1968). The use of synthetic helium-oxygen mixture in which the oxygen partial pressure is maintained at 0.04 MPa, has allowed man to reach pressure higher than 1.1 MPa of absolute pressure. Nevertheless, helium-oxygen which one breathes at high pressure causes a group of disturbances different from those of nitrogen narcosis, which are nervous in origin. They are attributed to a pressure effect and are grouped together under the heading High Pressure Nervous Syndrome (H.P.N.S.) (Brauer et al., 1969; Fructus et al., 1969; Rostain and Naquet, 1974).

The effect of the gas pressure on the nervous system, although important, is not the only one to consider in the choice of a respiratory mixture. In order to be suitable it should not be too 'heavy' or it will reduce gaseous exchange in the lungs. One therefore chooses a gas that is as light as possible (helium; hydrogen) since density is directly a function of the pressure (at 61 MPa, the mass of a helium-oxygen mixture is approximately 10 grams per litre).

Finally, each exposition to gas pressure includes a decompression phase whose duration is a function of the value of the pressure reached, the degree of saturation of gas dissolved in the organism, and of the physical characteristics of the gases used.

## REFERENCES

Behnke, A.R. and R.M. Thomas and E.P. Motley. 1935. The psychologic effects from breathing air at 4 atmospheres pressure. Am. J. Physiol. 112: 554–558.

Bennett, P.B. 1966. The Aetiology of Compressed Air Intoxication and Inert Gas Narcosis. International Series of Monographs in Pure and Applied Biology. Zoology Division, Vol. 31, Pergamon Press, Oxford, UK.

Bennett, P.B. and J.C. Rostain. Inert gas narcosis. Pp. 300–322. *In*: A.O. Brubakk and T.S. Neuman (eds.) 2003a. Bennett and Elliott's Physiology and Medicine of Diving. 5th edit. Saunders, London, UK.

Bennett, P.B. and J.C. Rostain. The High Pressure Nervous Syndrome. Pp. 323–357. *In*: A.O. Brubakk and T.S. Neuman (eds.). 2003b. Bennett and Elliott's Physiology and Medicine of Diving. 5th edit. London: Saunders.

Brauer R.W. and R.O. Way and T.A. Perry. Narcotic effects of helium and hydrogen and hyperexcitability phenomenon at simulated depths of 1500 to 4000 ft of sea water. Pp. 241–255. *In*: B.R. Fink (ed.). 1968. Toxicity of Anesthetics. Williams and Wilkins, Baltimore, USA.

Brauer, R.W. and S. Dimov, X. Fructus, A. Gosset and R. Naquet. 1969. Syndrome neurologique et électrographique des hautes pressions. Rev. Neurol. 121: 264–265.

Case, E.M. and J.B.S. Haldane. 1941. Human physiology under high pressure. J. Hyg. Camb. 41: 225–249.

Fructus, X. and R. Naquet, A. Gosset, P. Fructus and R.W. Brauer. 1969. Syndrome Nerveux des Hautes Pressions. Marseille Med. 6: 509–512.

Marshall, J.M. 1951. Nitrogen narcosis in frogs and mice. Am. J. Physiol. 166: 699–711.

Rostain, J.C. and R. Naquet. 1974. Le Syndrome Nerveux des Hautes Pressions: caractéristiques et évolution en fonction de divers modes de compression. Rev. EEG. Neurophysiol. 4: 107–124.

# 17

# Ventilatory and Circulatory Responses to Diving

*Yves Jammes*

## INTRODUCTION

This short chapter focuses on the consequences of diving on the thorax including the changes in hemodynamics and pulmonary mechanics, and the increased respiratory heat loss, both factors playing key roles in the limitation of cardiopulmonary function at depth. The physiological aspects of breath-hold diving will not be commented. The reader will find more information in previous general reviews (Jammes and Roussos, 1995; Jammes et al., 2006). It must be pointed out that all data in humans and experimental animal models were obtained in pressure chambers where individuals were surrounded by gas and not water, the ambient temperature being increased to maintain a relative thermoneutrality. This is never the case in deep sea diving where the water temperature is always below that of the thermoneutral one.

To breathe during diving there must be a measure of agreement between the pressure of the breathing source and the hydrostatic pressure on the diver's chest. Thus, the hydrostatic pressure increases in the respiratory tract in parallel to the elevated ambient pressure. The lungs, heart, and great vessels being included in same cavity, heart-lung interactions occur during diving.

## BIOPHYSICAL CONSEQUENCES OF DENSE GAS BREATHING

Due to the constant value of the pressure by volume product, measured at constant temperature in perfect gases, the density of a gas mixture increases linearly with the absolute pressure, expressed in ATA (Air Liquide, 1976). Then, the density of compressed air measured at 11 ATA (100 msw) is

UMR MD2 P2COE, Faculty of Medicine, University of Mediterranee.
E-mail: jammes.y@jean-roche.univ-mrs.fr

multiplied by around 10. The use of dilutant gases of lower density than nitrogen ($N_2$), as Helium (He) or Hydrogen ($H_2$), allows to notably reduce the slope of the density versus ambient pressure relationship. Nevertheless, at 46 ATA (450 msw) (experimental human dive ENTEX XI, Marine Nationale Française), the density of the He-$O_2$ and He-$N_2$-$O_2$ gas mixtures which human subjects breathe was respectively 8.7 and 11 g.$L^{-1}$, i.e., very high. The gas density is the major factor in the Reynolds number which determines the magnitude of turbulent air flow stream and constitutes the main reason for the limitation of the pulmonary mechanics at depth. Other physical parameters characterize a breathing mixture and depend on the choice of the dilutant gases. Thus, the specific heat (Cp) and thermal conductivity (K), which determine the thermal exchanges through the respiratory apparatus, are elevated in He and $H_2$ compared to $N_2$. Baseline Cp values for He and $H_2$ are 5 and 14 times higher than that of $N_2$, and their K coefficient is 6 and 7 times higher than for $N_2$, respectively. Because the convective respiratory heat loss depends on the gas density and Cp, and the K coefficient intervenes in the conductive heat loss, the combined effects of elevated gas density and relatively high Cp and K values of He and $H_2$, severely increase the respiratory heat losses. The convective respiratory heat loss has been measured until 25 ATA in He-$O_2$ (Jammes et al., 1988), 45 ATA in He-$N_2$-$O_2$ (Burnet et al., 1990), and 30 ATA with gas mixtures containing hydrogen (Burnet et al., 1992b). Due to physical factors, the heat loss required to increase the temperature of one degree in He-$O_2$ gas mixture is multiplied by 20 at 11 ATA and by 59 at 31 ATA compared to 1 ATA. This effect is accentuated when $H_2$ is used as a dilutant gas (Burnet et al., 1992b).

## LIMITATION OF LUNG MECHANICS

### Air flow limitation

Numerous human studies, conducted in subjects inhaling different gas mixtures containing $N_2$, He, Neon (Ne) or sulfur hexafluoride ($SF_6$), at normal or elevated ambient pressure, have clearly demonstrated that the maximal mechanical performances of the ventilatory system (maximal voluntary ventilation, forced expiratory volume measured at 1 s and peak expiratory flow rate) decrease as the square root of the increased gas density (Spaur et al., 1977). This is mostly due to enhanced turbulent air flow stream in the large airways (Broussolle et al., 1976). Thus, the diver's capacity for performing work is limited by the respiratory system. The more significant observations are that airway resistances increase with elevated pressure in the inspired gas even during eupneic ventilation (Varène et al., 1967; Clarke et al., 1982; Jammes et al., 1988; Burnet et al., 1990). The total lung resistance (RL) increases linearly with the density of the breathing mixture (Jammes et al., 1988; Burnet et al., 1990). In consequence, the work of breathing increases in proportion to the elevated pressure of the inspired gas mixture (Lenoir et al., 1990). When subjects hyperventilate, the work of breathing (W) increased in proportion to the value of minute ventilation (VE) in hyperbolic manner, limiting the maximal ventilation and exercise performances.

Two other factors may accentuate the alterations in lung mechanics under hyperbaric condition: 1) the occurence of hyperbaric tremor recorded in skeletal muscles and the diaphragm (Lenoir et al., 1990), and 2) mostly, a reflex bronchospasm, often associated with mucous hypersecretion, produced by an insufficient warming of the inspired gas mixture (Jammes et al., 1988; Burnet et al., 1990).

## Cold-induced bronchospasm

As commented above, the respiratory heat loss severely increases during deep diving. This enables in warming the inspired gas and also the ambient gas mixture in the diving chamber. The use of hydrogen as a dilutant gas requires enhanced warming of the inspired gas. Thus, thermal comfort is obtained for an ambient temperature of 30°C to 31°C in He-$O_2$ mixture (Webb, 1970) and 34°C in $H_2$-He-$O_2$ mixture (Hydra Comex Dives). Dyspnea and excessive upper respiratory tract secretions have been reported in divers breathing cold (+1° C) He-$O_2$ mixture under 27 ATA (Hoke et al. 1976). The inhalation of relatively fresh He-$O_2$ or He-$N_2$-$O_2$ gas mixtures(+7°C to +8°C) at pressure up to 8 ATA enhanced RL value and this effect is proportional to the elevated pressure (Jammes et al., 1988; Burnet et al., 1990). The same subjects did not present any RL change when they were breathing room air at the same temperature at sea level. On the basis of animal studies (Jammes et al., 1983; 1986) and some experiments in healthy humans (Heaton et al., 1983), the airway response to the inhalation of cold gas is attributed to a vagal reflex due to the stimulation of thermosensitive units in the larynx and the upper trachea. The airway response to cold was found to correlate directly to the magnitude of the respiratory heat loss and not to the absolute value of inspired temperature (Burnet et al., 1990).

## ADAPTIVE AND NON ADAPTIVE VENTILATORY RESPONSES TO ALTERED LUNG MECHANICS

Breathing high density gas mixtures constitutes a condition of resistive internal ventilatory loading. In resting humans, VE increases in proportion to the density of the breathing mixture, whatever the nature of the dilutant gas, including $N_2$, He, $H_2$, Ne or $SF_6$ (Dwyer et al., 1977; Lenoir et al., 1990; Saltzman et al., 1971; Wood et al., 1976), In parallel, the respiratory frequency slows down and the tidal volume increases. These observations are confirmed in conscious cats compressed at 101 ATA in He-$O_2$ or He-$N_2$-$O_2$ (Burnet and Naraki, 1988; Burnet et al., 1992a). On the other hand  in anesthetized animals breathing dense gas mixture ($SF_6$ plus $O_2$) at sea level, we measured a hypoventilation, combining a bradypnea and no change in tidal volume (Barrière et al., 1993): this response disappeared after bivagotomy. Thus, the hyperventilation reported in  individuals who were awake (man and animals) during dives, accentuated during periods of compression, is attributed to the general excitatory influence exerted by hydrostatic pressure changes on the respiratory centres and the cortical modulation of their spontaneous activity (Gelfand et al., 1980 and 1983; Linnarson and Hesser, 1978). The biochemical properties of the dilutant gases may play also a role. Indeed, in cats the hyperventilation measured at 101 ATA is higher with He-$O_2$ than He-$N_2$-$O_2$ gas mixture, i.e., in condition of severe High Pressure Nervous Syndrome (Burnet and Naraki, 1988). The diving bradypnea may correspond to a reflex adaptive response, which opposes the elevated impedance of the respiratory system due to the increased gas density, but the accompanying enhanced  amplitude of breath accentuates the respiratory muscle work, already increased by dense gas breathing. Thus, these changes in the breathing pattern might shorten the occurrence of diaphragmatic fatigue due to contractions at a high strength, a phenomenon suspected during deep diving.

Electromyographic (EMG) recordings of spontaneous diaphragmatic activity in human volunteers compressed at pressure up to 46 ATA (gas density 8 to 11 g.$L^{-1}$) (Lenoir et al., 1990) and in cats studied until 101 ATA (gas density equal 25 g.$L^{-1}$) (Burnet et al., 1992a) have shown a marked recruitment of diaphragmatic motor units in parallel to the increased amplitude of tidal breaths. This is associated with the changes in EMG power spectrum, indicating an enhanced recruitment of low frequency motor units which are more resistant to fatigue. These EMG changes persisted, with no accentuation, during a 12-d stay at 46 ATA in man (Lenoir et al., 1990) or a 3-d stay at 101 ATA in cats (Burnet et al., 1992a) and are not proportional to the increased gas density. They are followed by a reduction, then a suppression, of the diaphragmatic response to direct electrical muscle stimulation which attests the contraction failure, i.e., muscle fatigue (Burnet et al., 1992a). These EMG changes are not found in skeletal muscles nor in the respiratory muscles which do not participate to the resting ventilation at depth, whereas they are present in intercostal muscles which rhythmically contracted under high pressure (Burnet et al., 1992a; Lenoir et al., 1990). Normobaric studies have clearly demonstrated that the mechanical failure of an exercising skeletal or respiratory muscle is preceded by an increase in the low frequency EMG component of that muscle, attributed to a reflex modulation of the neuromuscular drive (Woods et al., 1987). Thus, EMG changes observed at depth are attributed to an alert signal indicating the risk of respiratory muscle fatigue. In the record human dive by Comex (–710 msw) (Fontanari et al., 2000), the maximal pressure generated by inspiratory muscle contraction (Pimax) decreased with increasing the depth and altered respiratory muscle contractions persisted during decompression, recovering control pre-dive value 1 mon after the dive. Thus, long-term ventilatory loading due to dense gas breathing seems to alter the contractile properties of inspiratory muscles. This may have functional consequences if deep dives are too frequently repeated.

Figure 1 schematizes the direct consequences of elevated ambient pressure with depth on the pulmonary mechanics and the accentuation of altered pulmonary function by bronchomotor and secretory reflexes, activated by an insufficient warming of the inspired gas.

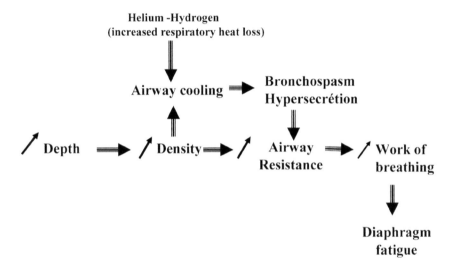

**Figure 1.** Consequences of elevated ambient pressure on the pulmonary mechanics: accentuation by airway cooling.

## CARDIOVASCULAR FUNCTION UNDER HIGH PRESSURE

The changes in heart rate and systemic blood pressure were often studied at depth. Under hyperbaric environment, several studies have reported bradycardia in man (Kerem and Salzano, 1974; Wilson et al., 1977) and animals (Shida and Lin., 1981; Stuhr et al., 1990), and also in isolated heart exposed to high pressure (Doubt and Hogan, 1978; Gennser and Ornhagen, 1989). Bradycardia was attributed to the direct effect of elevated hydrostatic pressure on the nodal tissue. On the other hand, inconsistent cardiac responses have been reported in man (Bühlman et al., 1970; Joulia et al.,

1992). Some studies have shown that the changes in heart rate are independent on the pressure level and elevated gas density (Flynn et al., 1972; Joulia et al., 1992). Moreover, bradycardia is not detected when *in vivo* or *in vitro* hyperbaric experiments are performed under normoxic conditions (Hordness and Tyssebotn, 1985). Thus, bradycardia seems to result from hyperoxia, the arterial oxygen partial pressure reaching 400 to 500 mmHg under routine hyperbaric condition.

Abberrent cardiac rhythms, ranging from coupled extrasystoles to conduction blocks, have been sometimes observed during human saturation dives in air (Wilson et al., 1977) and He-O$_2$ (Argeles et al., 1977) and in isolated myocardial strips exposed to high hydrostatic pressure (Doubt and Evans, 1982). Tall, peak T waves are reported by Wilson et al. (1977) during a 7-h human dive at 7 ATA of compressed air with a normal oxygen pressure (175 mmHg). In divers participating in a 73-d saturation dive at 25.5 ATA (Hydra IX Comex) and in conscious dogs compressed until 91 ATA in He-N$_2$-O$_2$ mixture (Joulia et al., 1992), the T wave amplitude increases in proportion to the gas density but not to that of hyperoxia. No satisfactory explanation of these T wave changes is proposed.

Measurements of cardiac output and regional blood flow distribution were performed in rats using radioactive labelled microspheres (Aanderud et al., 1985) and in rats and dogs with chronically implanted Doppler flow meters (Barthelemy, 1990) at pressure up to 71 ATA (rats) or 91 ATA (dogs). There are no significant changes in aortic blood flow, but reduced perfusion to the kidneys, surrenals, skeletal muscles, and brain. The decreased blood flow in limb muscles, splanchnic organs, and brain is markedly accentuated during dives with the He-O$_2$ mixture than with He-N$_2$-O$_2$ mixture (Barthelemy, 1990). Thus, the excitatory effects of high helium pressure on the nervous system and perhaps also the sympathetic motor control to blood vessels may be responsible for the reduced organ perfusion under high ambient pressure.

The baroreceptor reflex has been studied in man at 46 ATA using a non-invasive technique (Bowser-Riley, 1990). The reflex heart rate response to applied positive pressure to the carotid sinus is accentuated at depth. Thus, the elevated hydrostatic pressure, exerting on the surrounding tissues, may modify the gain of the sinusal baroreflex arch. However, in none of the cases does the systemic blood pressure significantly vary during human deep diving.

Thus, apart from minor changes in cardiac excitability which seem to have no functional consequences, experimental human deep diving has no significant influence on the cardiac blood flow and systemic blood pressure.

# REFERENCES

Aanderud, L. and J. Onarheim and I. Tyssebotn. 1985. Effects of 71 ATA He-O2 on organ blood flow in the rat. J. Appl. Physiol. 59: 1369–1375.

Air Liquide. 1976. Encyclopédie des Gaz. Elsevier, Amsterdam.

Argeles, H.J. and M.D. Freezor and H.A. Saltzman. 1977. Human vector electrocardiographic responses at increased pressure to 1000 msw. Undersea Biomed. Res. 4: A 30.

Barrière, J.R. and S. Delpierre, M.J. DelVolgo and Y. Jammes. 1993. Comparisons among external resistive loading, drug-induced bronchospasm and dense-gas breathing in cats: roles of vagal and spinal afferents. Lung 171: 125–136.

Barthélemy, P. 1990. La distribution des débits circulatoires régionaux sous haute pression. Arch. intern. Physiol. Bioch. 98: A 389–A 392.

Bowser-Riley, F. 1990. Cardiovascular function and vasomotor reflexes during deep diving. Arch. intern. Physiol. Bioch. 98: A 385–A 388.

Broussolle, B. and J. Chouteau and R. Hyacinthe R. Respiratory function during a simulated saturation dive to 51 ATA with a helium-oxygen mixture. Pp. 79–89. *In*: C.J. Lambertsen (ed.). 1976. Underwater Physiology V. FASEB, Bethesda MD.

Bühlman, A.A. and H. Matthys, G. Overrath, P.B. Bennett, D.H. Elliott and S.P. Gray. 1970. Saturation exposures at 31 ATA in an oxygen-helium atmosphere with excursions to 36 ATA. Aerospace Med. 41: 394–403.

Burnet, H. and N. Naraki. 1988. Ventilatory failure in cats during prolonged exposure to very high pressure. Undersea Biomed. Res. 15: 19–30.

Burnet, H. and M. Lucciano and Y. Jammes Y. 1990. Respiratory effects of cold gas breathing in humans under hyperbaric environment. Respir. Physiol. 81: 413–424.

Burnet, H. and P. Lenoir and Y. Jammes Y. 1992a. Changes in respiratory muscle activity in conscious cats during experimental dives at 101 ATA. J. Appl. Physiol. 73: 465–472.

Burnet, H. and M. Raynaud-Gobert, M. Lucciano and Y. Jammes. 1992b. Relationships between inspired and expired gas temperature under hyperbaric environment. Respir. Physiol. 90: 377–386.

Clarke, J.R. and M.J. Jaeger, J.L. Zumrick, R. O'Bryan and W.H. Spaur. 1982. Respiratory resistance from 1 to 46 ATA measured with the interrupter technique. J. Appl. Physiol. 52: 549–555.

Doubt, T.J. and P.M. Hogan. 1978. Effects of hydrostatic pressure on conduction and excitability in rabbit atria. J. Appl. Physiol. 45: 24–32.

Doubt, T.J. and D.E. Evans. 1982. Hyperbaric exposures alter cardiac excitation-contraction coupling. Undersa Biomed. Res. 9: 131–145.

Dwyer, J. and H.A. Saltzman and R. O'Bryan. 1977. Maximal physical work capacity of man at 43.4 ATA. Undersea Biomed. Res. 4: 359–372.

Flynn, E.T. and T.E. Berghage and E.F. Coil. 1972. Influence of increased ambient pressure and gas density on cardiac rate in man. NEDU.

Fontanari, P. and M. Badier, C. Guillot, C. Tomei, H. Burnet, B. Gardette and Y. Jammes. 2000. Changes in maximal performances of inspiratory and skeletal muscles during and after the 7.1 Mpa Hydra 10 record human dive. Eur. J. Appl. Physiol. 81: 325–328.

Gelfand, R. and C.J. Lambertsen and R.E. Peterson. 1980. Human respiratory control at high ambient pressures and inspired gas densities. J. Appl. Physiol. 48: 528–539.

Gelfand, R. and C.J. Lambertsen, R. Strauss, J.M. Clark and C.D. Puglia. 1983. Human respiration at rest in rapid compression and high pressures and gas densities. J. Appl. Physiol. 54: 290–303.

Gennser, M. and H.C. Ornhagen. 1989. Effects of hydrostatic pressure, $H_2$, He and $N_2$ on beating frequency of rat atria. Undersea Biomed. Res. 16: 153–164.

Heaton, R.W. and   A.F. Henderson, B.J. Gray and J.F. Costello. 1983. The bronchial response to cold air challenge: evidence for different mechanisms in normal and asthmatic subjects. Thorax 38: 506–511.

Hoke, B. and D.L. Jackson, J.M. Alexander and E.T. Flynn. Respiratory heat loss and pulmonary function during cold gas breathing at high pressures. Pp. 725–740. *In*: C.J. Lambertsen (ed.). 1976. Underwater Physiology V. FASEB, Bethesda MD.

Hordness, C. and I. Tyssebotn. 1985. Effect of high ambient pressure and oxygen tension on organ blood flow in conscious trained rats. Undersea Biomed. Res. 12: 115–128.

Jammes, Y. and C. Roussos. Respiration during diving. Pp. 1557–1584. *In*: C. Roussos (ed.). 1995. The Thorax 2nd edit. Marcel Dekker Inc., New-York.

Jammes, Y. and P. Barthélemy and S. Delpierre S. 1983. Respiratory effects of cold air breathing in anesthetized cats. Respir. Physiol. 54: 41–54.

Jammes, Y. and P. Barthélemy, M. Fornaris and C. Grimaud. 1986. Cold-induced bronchospasm in normal and sensitized rabbits. Respir. Physiol. 63: 347–360.

Jammes, Y. and H. Burnet, P. Cosson and M. Lucciano. 1988. Bronchomotor response to cold air or helium-oxygen at normal and high ambient pressures. Undersea Biomed. Res. 15: 179–192.

Jammes, Y. and P. Giry and P. Hyacinthe. Physiologie respiratoire et plongée. Pp. 86–114. *In*: B. Broussolle, J.L. Méliet and M. Coulange (eds.). 2006. Physiologie et Médecine de la Plongée. Ellipses, Paris.

Joulia, F. and P. Barthélemy, F. Guerrero and Y. Jammes. 1992. T wave changes in humans and dogs during experimental dives. J. Appl. Physiol. 73: 1708–1712.

Kerem, D. and J. Salzano. 1974. Effect of high ambient pressure on human apneic bradycardia. J. Appl. Physiol. 37: 108–111.

Lenoir, P. and Y. Jammes, P. Giry, J.C. Rostain, H. Burnet, C. Tomei and C. Roussos. 1990. Electromyographic study of respiratory muscles during human diving at 46 ATA. Undersea Biomed. Res. 17: 121–137.

Linnarson, D. and C.M. Hesser. 1978. Dissociated ventilatory and central respiratory responses to CO2 at raised N2 pressure. J. Appl. Physiol. 45: 758–761.

Saltzman, H.A. and J.V. Salzano, G.O. Blenkarn and J.A. Kylstra. 1971. Effects of pressure on ventilation and gas exchange in man. J. Appl. Physiol. 30: 443–449.

Shida, K.K. and Y.C. Lin. 1981. Contribution of environmental factors in development of hyperbaric bradycardia. J. Appl. Physiol. 50: 731–735.

Spaur, W.H. and L.W. Raymond, M.M. Knott, J.C. Crothers, W.R. Braithwaite, E.D. Thalmann and D.F. Uddin. 1977. Dyspnea in divers at 49.5 ATA: mechanical, not chemical in origin. Undersea Biomed. Res. 4: 183–198.

Stuhr, L.E.B. and J.A. Ask and I. Tyssebotn. 1990. Cardiovascular changes in anesthetized rats during exposure to 30 bar. Undersea Biomed. Res. 17: 383–393.

Varène, P. and J. Timbal and C. Jacquemin. 1967. Effect of different ambient pressures on airway resistance. J. Appl. Physiol. 22: 699–706.

Webb, P. 1970. Body heat loss in undersea gaseous environments. Aerosp. Med. 41: 1282–1288.

Wilson, J.M. and P.D. Klighfield, G.M. Adams, C. Harvey and K.E. Schaeffer. 1977. Human ECG changes during prolonged hyperbaric exposures breathing $N_2-O_2$ mixtures. J. Appl. Physiol. 42: 614–623.

Wood, L.D.H. and A.C. Bryan, S.K. Bau, T.R. Weng and H. Levison. 1976. Effects of increased gas density on pulmonary gas exchange in man. J. Appl. Physiol. 41: 206 –210.

Woods, J.J. and F. Furbush and B. Bigland-Ritchie. 1987. Evidence for a fatigue induced reflex inhibition of motoneuron firing rates. J. Neurophysiol. 58: 125–137.

# 18

# Inert Gas Narcosis

*J.C. Rostain*[1] and *C. Lavoute*[2]

## INTRODUCTION

Exposure to pressures of air higher than 4 ATA (4.05 bar) induced in man dangerous signs and symptoms of euphoria, intoxication and narcosis. Compressed air or nitrogen-oxygen mixture produces from 4 ATA, disturbances at the level of the nervous system called nitrogen narcosis (Bennett and Rostain, 2003a).

The first reports on the occurrence of the narcosis were written in the 19th century. Junod in 1835 noted that when breathing compressed air "the functions of the brain are activated, imagination is lively, thoughts have a peculiar charm and, in some persons, symptoms of intoxication are present". Green in 1861 described a feeling of sleepiness, accompanied by hallucinations and impaired judgement. Paul Bert in 1878, also reported some narcotic properties of air when breathing at increased pressures. With the development of work in compressed air , similar signs and symptoms were further observed by Hill and Greenwood (1906) and Hill and McLeod (1903) in tunnel workers or caisson workers breathing air at 46 m. At 91m, Damant in 1930 noted that divers had cognitive and psychological impairments. At 100m, Hill et al. (1933) described a phenomenon entitled 'semi-loss of consciousness'. However, it was only with the work of Behnke

[1] Université de la Méditerranée et Institut de Médecine Navale du Service de Santé des Armées., UMR-MD2—Physiologie et physiopathologie en condition d'oxygénation extrême, Institut de Neurosciences Jean Roche, Faculté de Médecine Nord, Bd P. Dramard 13015 Marseille, France.
E-mail: jean-claude.rostain@univmed.fr

[2] Université de la Méditerranée et Institut de Médecine Navale du Service de Santé des Armées., UMR-MD2—Physiologie et physiopathologie en condition d'oxygénation extrême, Institut de Neurosciences Jean Roche, Faculté de Médecine Nord, Bd P. Dramard 13015 Marseille, France.
E-mail: cécile.lavoute@univmed.fr

et al. (1935) that the phenomenon were related to the narcotic potency of nitrogen that composes air at 79%. This work established that the phenomenon observed with compressed air was a particular manifestation of a general phenomena induced by all the inert gases. In fact, the inert gases induce narcotic disturbances from a partial pressure that is specific of each gas, depending of its narcotic potency.

Many attempts have been made to correlate the narcotic potency of helium, neon, argon, krypton and xenon to properties such as lipid solubility, partition coefficients and molecular weight (Behnke and Yarbrough, 1939), adsorption coefficients (Case and Haldane, 1941), thermodynamic activity (Ferguson, 1939; Brink and Posternak, 1948) and the formation of clathrates (Miller, 1961; Pauling, 1961). From these numerous attempts, it seems that the most satisfactory correlation is produced by lipid solubility (Table 1).

**Table 1.** Molecular weight and lipid solubility of inert gases and their rank from the least narcotic to the most narcotic according to their lipid solubility (data extracted from Bennett and Rostain, 2003a).

| Gas | Molecular weight | Lipid Solubility | Rank *(narcotic potency)* |
|---|---|---|---|
| | | | Least narcotic |
| He | 4 | 0.015 | 1 (0.2) |
| Ne | 20 | 0.019 | 2 (0.3) |
| $H_2$ | 2 | 0.036 | 3 (0.6) |
| $N_2$ | 28 | 0.067 | 4 (1.0) |
| A | 40 | 0.14 | 5 (2.3) |
| Kr | 83.7 | 0.43 | 6 (2.5) |
| Xe | 131.3 | 1.7 | 7 (25.6) |
| | | | Most narcotic |

According to their lipid solubility, three gases are more narcotic than nitrogen: xenon is anaesthetic at atmospheric pressure (Lawrence et al., 1946; Lazarev et al., 1948; Cullen and Gross, 1951; Featherstone, 1960; Pittinger, 1962), krypton causes dizziness (Lawrence et al., 1946; Cullen and Gross, 1951) and argon will be narcotic and about twice as potent as nitrogen (Behnke and Yarbrough, 1939). Three other gases are less narcotic than nitrogen: hydrogen which would be between two to three times less narcotic than nitrogen (Brauer et al., 1968), neon which would be at least three times less narcotic than nitrogen (Marshall, 1951) and at last, helium which is the least narcotic.

## INERT GAS NARCOSIS

### Nitrogen narcosis

When men are exposed to pressure of air higher than 3 bars, they present signs and symptoms recapitulated in Table 2.

**Table 2.** Psychophysiological troubles of nitrogen narcosis in man from 4 bars of absolute pressure.

| NITROGEN NARCOSIS IN MAN |
| :---: |
| Temporo–spatial disorientation |
| Memory troubles |
| Euphoria |
| Hallucinations |
| Mood changes |
| Impaired neuromuscular coordination |
| Psychomotor and intellectual decrements |

Edmonds et al. (1981) have proposed the following development concerning the signs and symptoms of nitrogen narcosis:

2-4 ATA : decrease of task ability and light euphoria.

4 ATA : alteration of mental efficiency, of short term memory and reaction time;

4–6 ATA : euphoria, tendency to fixate on an idea, overfilling, difficulty to perform mathematical tests.

6 ATA : sleepiness, hallucinations, impaired judgment.

6–8 ATA : mood changes.

8 ATA : deterioration of intellectual function and manual dexterity.

8–9 ATA : mental confusion.

10 ATA : memory troubles, hyperexcitability, euphoria, hallucinations, loss of consciousness.

These troubles are reversible when pressure is decreased.

When experimental animals are exposed to compressed air or to pressure of nitrogen-oxygen mixture, they present signs and symptoms of narcotic type for pressure higher than 8 to 10 bars. Anaesthesia occurs in rat for pressure around 40 bars. (Abraini et al., 1998)

The inert gases cause similar signs and symptoms but vary in their potency (for review see Bennett and Rostain, 2003a).

**Helium Narcosis**

On the base of lipid solubility, the narcotic effect of helium would occur around 40 bars (Bennett and Rostain, 2003a). However the pressure reversal effect (Miller et al., 1973) counteract this weak narcotic potency and the symptoms that occur are different from those observed in narcosis; they are called the High Pressure Nervous Symptoms (HPNS) (Brauer et al., 1969; Fructus et al., 1969; Rostain and Naquet, 1974). There is no helium narcosis…

However, from recent results obtained during experimental dives with narcotic gases added to helium at great pressure (Stoudemire et al., 1984; Fructus et al., 1984), mood changes or sensorial hallucinations reported in some cases in helium-oxygen dives for pressure greater than 40 bars, could be a narcotic effect of helium rather than a pressure effect (Rostain, 1994, 1999). Moreover, hallucinatory behavior has been also reported in monkeys in helium-oxygen mixture for pressure of 80 bars and above (Rostain, 1987; Gardette and Gortan, 1994; Rostain et al., 1994), which could be due to a narcotic effect of helium at high pressure (Rostain et al., 1994).

### Neon Narcosis

The high density of neon relative to helium will cause respiratory embarrassment at much lower pressure. On the basis of animal experiments, Marshall (1951) predicted that this gas would be three times less narcotic than nitrogen which is in good correlation with the lipid solubility theory.

Few experiments have been made with neon in man and the experimental dives performed in man at depths that did not exceed 20 bars have not shown signs or symptoms of narcosis (Bennett, 1966; Shreiner et al., 1972; Towsend et al., 1971). Other studies performed at 37 bars by breathing neon by mask did not report signs or symptoms of narcosis (Lambertsen, 1976). According to Miller et al. (1967) the loss of righting reflex in mice will occur for partial pressure of neon above 110 bars.

### Hydrogen Narcosis

Hydrogen is another inert gas which has been considered and used for deep diving (Case and Haldane, 1941; Zetterstrom, 1948; Bjurstedt and Severin, 1948; Zal'tsman, 1961, 1968; Fructus, 1987; Gardette et al., 1990; Rostain et al., 1988). Hydrogen has a lower density than helium and thus would be better for breathing mechanisms. It also has a greater narcotic potency than helium, which may, in accordance with the critical volume hypothesis, reduce some of the symptons of HPNS. It is, however, explosive in mixtures of more than 4% oxygen, and work carried out by Brauer and Way (1970) have established that its narcotic potency would be in agreement with its lipid solubility.

In the past, several groups have studied the effects of hydrogen at pressure in man and in animals (Zetterstrom, 1948; Brauer and Way, 1970; Michaud et al., 1969; Rostain and Naquet, 1972; Halsey et al., 1975). The results have been contradictory. However, Edel et al. (1972) and Fife (1979) suggested that the use of hydrogen in diving could be beneficial (see Bennett and Rostain, 2003b for review). In the last 25 years, COMEX has carried out several experiments with hydrogen in mice, rats, monkeys and man (Gardette, 1987, 1989). In man, significant narcotic sensations (different from those reported with nitrogen) were reported by divers from 240 metres, when breathing hydrogen oxygen mixture during HYDRA IV dive. According to Fructus et al. (1984) and Fructus (1987), the self observation of the divers indicated that at 240 metres and for a $PH_2$ of 24.5 bars, the hydrogen narcosis was characterized chiefly by sensory and somesthetic hallucinations which are accentuated by relative sensory rest and affect mood more than intellect. Further experiments performed in hydrogen-oxygen mixture (HYDRA VII, IX) or in hydrogen-helium-oxygen mixture (HYDRA V, VI, X) have confirmed these results: narcotic effects of psychotropic type occurred when partial pressure of hydrogen is higher than 25 bars (Table 3). Indeed in HYDRA VI, 520 metres with a $PH_2$ of 24 bars or HYDRA IX 300 metres with a $PH_2$ of 30.6 bars, psychotic  type of disorders have been observed in some subjects, which consisted of hallucinations, mood disturbances, agitation, delirium and paranoid thoughts (Raoul et al., 1988; Rostain et al., 1990). These results which indicate that pressures of hydrogen higher than 24–25 bars induce narcosis, are consistent with the work of Brauer et al. (1968) and Brauer and Way (1970) which predicted hydrogen narcosis between 25 and 30 bars. However, the use of helium-hydrogen-oxygen mixture with a partial pressure of hydrogen which did not exceed 25 bars reduces the clinical symptoms of HPNS and the absolute pressure of 71 bars has been reached with this mixture (Rostain et al., 1994, 1999).

**Table 3.** Signs and symptoms observed from 26 ATA in hydrogen-oxygen mixture.

| **HYDROGEN NARCOSIS** |
| :---: |
| **From 25 bars** |
| Sensory and somesthetic hallucinations |
| Mood changes |
| Agitation |
| Delirium |
| Paranoid thoughts |

## ORIGINS AND MECHANISMS OF INERT GAS NARCOSIS

Little is known about the origins and mechanisms of the troubles produce by inert gas breathing at pressure. The cause is complex and cannot be related to any one factor. Numerous variables have led to many suggestions or theories for the cause of  narcosis, most of which are incorrect. Two or three major theories could be evoked as to the cause of the narcosis.

### The Theories

#### *The inert gas theory*

From the work of Benhke et al. (1935), based on the Meyer-Overton hypothesis (Meyer, 1899; Overton, 1901), the nitrogen and inert gas theory suggest that there is a parallel between the affinity of a narcotic or anaesthetic gas for lipid and its narcotic potency. Consequently the traditional view was that anaesthetics dissolve in the lipid bilayer of the cellular membrane, expand its volume. Anaesthesia occurs when the volume of a hydrophobic site expands beyond a critical amount by the absorption of molecules of a narcotic gas; if the volume of this site is restored by increasing pressure, then the anaesthesia will be removed. The observation of the pressure reversal effect on general anaesthesia (Miller et al., 1973) that has been reported for different anaesthetics including inert gases, has for a long time supported the lipid theory.

However, Franks and Leib (1978) and Simon et al. (1979) report that at anaesthetic concentrations there is no significant increase in membrane thickness. Moreover, the critical volume hypothesis suggests that the anesthetics act on the same molecular site, but the work of Halsey et al. (1978) have suggested a multi-site expansion theory

### The aqueous phase theories

This theory is based on the formation of hydrates. Pauling (1961) suggests that clathrates are formed in which the inert gas atoms called 'guests' are held by van der Waals forces in crystalline cages formed by the 'hosts' or molecules of a second agent. However, owing to their high dissociation pressures, such hydrates would be unstable under body conditions.

Miller (1961) postulated an alternative theory which implied that the inert gases or anaesthetics may increase the area of highly ordered water or 'icebergs' surrounding a dissolved gas molecule. In this manner the conductance of the brain tissue would be lowered, the lipid membranes stiffened and the membranes occluded.

However the experiments of Dawe et al. (1964), Miller et al. (1965, 1967) and Eger et al. (1969) did not provide support for the aqueous medium as the critical phase for anaesthetic action.

### The protein theory

Some experiments have suggested a 'bonding mechanism' in which inert gases are bound to specific sites within protein molecules (Featherstone et al., 1961; Featherstone and Meuhlbaecher, 1963; Schoenborn et al., 1965; Schoenborn, 1965; Eyring et al., 1973; Katz  and Simon, 1977). Recently, protein theories have  been considered increasingly since results obtained from experiments with inhalational anaesthetics have been interpreted as evidence for a direct anaesthetic-protein interaction (Franks and Lieb, 1982, 1984, 1991, 1994). The question is to know whether inert gases exert binding processes on proteins at raised pressure. Data obtained by Abraini et al. (1998) with two inert gases and one anaesthetic gas seemed to indicate that inert gases bind directly to a modulatory site of protein receptor and act as allosteric modulators. This theory seems to be supported by the results obtained by Colloc'h et al. (2007) with protein crystallography study under xenon and nitrous oxide pressure indicated that nitrous oxide molecule or xenon atom binds to the same main site in two soluble proteins as structural models: urate oxidase, which is a prototype of a variety of intracellular globular proteins, and annexinV, which has structural and functional characteristics that allow it to be considered as a prototype for the N-Methyl-D-Aspartate receptor.

## Cellular and neurochemical mechanisms

### Ion transports

Bennett et al. (1967) have demonstrated an electrolytic imbalance with several narcotic gases (nitrogen, argon, neon) that were attributed to changes in membrane permeability to cation. Other works have indicated that narcotic gas induced an activation of potassium and sodium ion transports (Johnson and Miller, 1970; Galey and Van Nice, 1980).

### Neurotransmitters

McLeod et al. (1988) studied the pattern of norepinephrine (NE) and dopamine (DA) neurotransmitter release in the caudate nucleus and hypothalamus of rats exposed to 20 ATA of nitrogen oxygen mixture. High pressures of nitrogen resulted in significant decreases in both NE and DA concentrations in the hypothalamus whereas in the caudate there were moderate increases.

More recently, neurochemical studies have been carried out with voltametry or microdialysis technology on the effect of inert gas narcosis at the level of basal ganglia, and particularly at the level of the nigro-striatal pathway. These structures are implicated in the regulation of motor, locomotor and cognitive processes that are disrupted by pressure of inert gases. The results (Fig. 1) indicated that dopamine at the striatal level decreased when rats are exposed to increased pressure of nitrogen or nitrous oxide, an anaesthetic gas at atmospheric pressure, (Turle et al., 1998, 1999; Balon et al., 2002; Lavoute et al., 2005, 2006) and also argon (Balon et al., 2002). Risso et al. (1999) using microdialysis technology noted a 70% decrease in dopamine at the synaptic cleft under 0.9 Mpa (8.9 ATA) of nitrogen-oxygen mixture. Behavioural studies showed an associated 50% decrease in locomotor and vigilance tasks using a special skinner box. Balon et al. (2003b) reported a biphasic pattern in motor and locomotor behaviour. In the rat striatum, Vallée et al. (2009) recorded beside the decrease in dopamine, a decrease in glutamate.

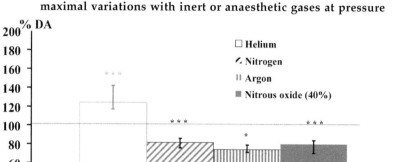

**Figure 1.** Dopamine levels recorded by voltametry in the striatum of rat with 3 inert gases and one anesthetic gas. The change in dopamine levels are expressed in percentage from basal values (100%).

In contrast, with helium, striatal dopamine level is increased (Abraini and Rostain, 1991; Rostain and Forni, 1995; Darbin et al., 1997a) with serotonine and glutamate (Darbin et al., 1997b, 2001). These studies demonstrated at least at the level of dopamine release in the striatum, the reverse effect of pressure and narcotic gases. Thus, the nitrogen-induced decrease of dopamine level is attributed to a narcotic effect rather than to a pressure per se effect.

The GABA (gamma amino butyric acid) neurotransmission could be implicated in these changes. We know that general anaesthetics including inhalational agents, have been shown to enhance $GABA_A$ receptor activities (Franks and Lieb, 1991; Macdonald and Olsen, 1994; Sincoff et al., 1996). Moreover the administration of $GABA_A$ antagonist resulted in an increase of the onset pressure of argon required to produce 100% loss of righting reflex in rats (Abraini et al., 2003). Consequently the narcotic gas could have an action on GABA and reciprocally on glutamate neurotransmission.

The nigrostriatal pathway is under the control of excitatory glutamatergic afferents (Gerfen and Wilson, 1996), from prefrontal cortex and subthalamic nucleus, acting on NMDA receptors located on dopaminergic neurons (Albers et al., 1999; Wedzony et al., 2001), as demonstrated by the increased activity of dopaminergic neurons (Christoffersen and Meltzer, 1995) and by the enhanced striatal DA release (Balon et al., 2003a; Westerink et al., 1992) when NMDA is administered in the substantia nigra pars compacta (SNc). In addition, most afferents in SNc are GABAergic (Bolam and Smith, 1990) and mediate an inhibitory control on the nigrostriatal pathway preferentially through $GABA_A$ receptors (Paladini et al., 1999) localized on DA cells (Sugita et al., 1992).

Studies performed on $GABA_A$ and NMDA receptors at the level of the nigro-striatal pathway of rats (Lavoute et al. 2007) have indicated that nitrogen at pressure mainly exerts its narcotic potency by enhancing $GABA_A$ receptor-mediated inhibition. In addition, NMDA receptors located on DA cells remained functional while the direct excitatory glutamatergic input mediated by these receptors is reduced suggesting a decrease in glutamate release confirm by the microdialysis studies at the level of striatum (Vallée et al., 2009).

Vjotosh et al. (1999) have suggested that nitric oxide (NO) may be involved in nitrogen narcosis. In their experiments, control rats showed motor activity at 5–12 ATA (4.9–11.8 bar), ataxia at 10–34 ATA (9.86–33.6 bar) and side body position at 26–41 ATA (25.7–40.5 bar). In animals treated with L-NAME inhibitor of NO synthesis, the nitrogen induced signs of narcosis occurred at higher pressure.

*Pharmacological studies*

In 1963, Bennett showed that Carbachol, Frenquel, Doriden, Phenacitine, Aspirine protected against nitrogen narcosis. In contrast, Scopolamine, Methedrine, Megimide and Leptazol enhanced nitrogen narcosis.

Cationic and anionic detergents (Stearylamine and Ceyltrimethyl ammonium bromide) were also given to rats exposed to raised pressures of nitrogen (Bennett and Dossett, 1970). The cationic detergents significantly prevented the depression of evoked potentials seen in the animals without the drugs. Anionic compounds had no effect. Intraperitoneal lithium carbonate (Radomski and Wood, 1970), lithium chloride or sodium chloride (Bennett et al., 1980) administered in rats prevented some nitrogen narcosis induced symptoms.

Lithium ion, like the cationic detergents, is believed to produce its effects either by stabilizing nerve membranes or alternatively by active displacement of intracellular sodium (Meltzky and Blachly, 1971).

From the beginning of this century, new pharmacological studies were carried out on the effects of the inert gas at pressure on some membrane proteins or on neurotransmission with agonists or antagonists of GABAergic, glutamatergic and monoaminergic systems (Abraini et al., 2003; Balon et al., 2003a; David et al., 2001; Lavoute et al., 2007) and on NO implication.

## Effect of Recurrent Exposures to Nitrogen Narcosis Pressure

Men exposed to nitrogen pressure develop motor and cognitive disturbances related to inert gas narcosis. After repetitive exposures, adaptation to narcosis was subjectively reported. However, some studies failed to detect an improvement of motor disturbances, following 5 successive exposures to relative pressure of 0.55 MPa in man (Hamilton et al., 1995).

Recently, Lavoute et al. (2005) have shown that repetitive exposures to pressure that induce nitrogen narcosis, change the release of striatal dopamine in rats (Fig. 2). The first nitrogen exposure at 3 MPa led to a significant decrease of striatal extracellular level of dopamine up to –20% as previously observed in voltammetric or microdialysis studies (Balon et al., 2002; Dedieu et al., 2004). Successive exposures at 1 MPa of nitrogen modified the dopaminergic neurotransmission in the striatum. DA level did not return to basal value but was increased by +15%. This fact and the lack of improvement of motor disturbances do not support the hypothesis of a physiological adaptation and suggest an eventual neurotoxicity. This possible neurotoxicity have been studied through the activities of excitatory and inhibitory pathways of the central nervous system particularly via NMDA and $GABA_A$ receptors (Lavoute et al., 2008 a and b). The results obtained with agonist or antagonists of NMDA and $GABA_A$ receptors indicate that, repetitive nitrogen exposure induces a decrease in the activity of the dopaminergic pathway, reflecting increased GABAergic neurotransmission linked to alteration of $GABA_A$ receptors. These findings demonstrate a parallel long-lasting decrease of glutamate release coupled to an increase in NMDA receptor sensitivity (Lavoute et al., 2008a). However, the increase of dopamine that occurred during recurrent exposure to nitrogen results in

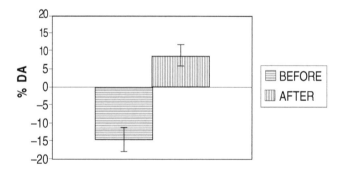

**Figure 2.** Changes in stiatal dopamine level in rats exposed a first time to 31 ATA of nitrogen-oxygen mixture (before) and a second time after several exposition (5) to nitrogen narcosis (1MPa). (Lavoute et al., 2005). Changes are expressed in percentage difference from basal value and calculated on the median values obtained during a 2 h stay at 31 ATA.

disinhibition of the nigrostriatal pathway, while excitatory control mediated by NMDA receptors is disrupted (Lavoute et al., unpublished data). Investigations are in progress to determine the existence of nitrogen neurotoxicity or nitrogen addiction induced by recurrent exposures.

## REFERENCES

Abraini, J.H. and J.C. Rostain and B. Kriem. 1998. Sigmoidal compression rate-dependence of the narcotic potency of inert gases in rats: implication for lipid vs protein theories of inert gas action in the central nervous system. Brain Res. 808: 300–304.

Abraini, J.H. and B. Kriem, N. Balon, J.C. Rostain and J.J. Risso. 2003. Gamma-aminobutyric acid neuropharmacological investigations on narcosis produced by nitrogen, argon, or nitrous oxide. Anesthesia and Analgesia 96: 746–749.

Albers, D.S. and S.W. Weiss, M.J. Iadarola and D.G. Standaert. 1999. Immunohistochemical localization of N-methyl-D-aspartate and alpha-amino-3-hydroxy-5-methyl-4-isoxazolepropionate receptor subunits in the substantia nigra pars compacta of the rat. Neuroscience 89: 209–220.

Balon, N. and B. Kriem, E. Dousset, M. Weiss and J.C. Rostain. 2002. Opposing effects of narcotic gases and pressure on the striatal dopamine release in rats. Brain Res. 947: 373–379.

Balon, N. and L. Dupenloup, F. Blanc, M. Weiss and J.C. Rostain. 2003a. Nitrous oxide reverses the increase in striatal dopamine release produced by N-methyl-D-aspartate infusion in the substantia nigra pars compacta in rats. Neurosci. Lett. 343: 147–149.

Balon, N. and J.J. Risso, F. Blanc, J.C. Rostain and M. Weiss. 2003b. Striatal dopamine release and biphasic pattern of locomotor and motor activity under gas narcosis. Life Science 72: 2731–2740.

Behnke, A.R. and O.D. Yarbrough. 1939. Respiratory resistance, oil-water solubility and mental effects of argon compared with helium and nitrogen. Am. J. Physiol. 126: 409–415.

Behnke, A.R. and R.M. Thomas and E.P. Motley E.P. 1935. The psychologic effects from breathing air at 4 atmospheres pressure. Am. J. Physiol. 112: 554–558.

Bennett, P.B. 1963. Prevention in rats of the narcosis produced by inert gases at high pressures. Am. J. Physiol. 205: 1013–1018.

Bennett, P.B. 1966. The Aetiology of Compressed Air Intoxication and Inert Gas Narcosis. International Series of Monographs in Pure and Applied Biology. Zoology Division, Vol. 31, Pergamon Press, Oxford, UK.

Bennett, P.B. and A.N. Dossett. 1971. Studies of cationic and anionic compounds on the effects of high pressures of nitrogen and oxygen. Pp. 69–71. In: Proc 42nd Annual Scientific Meeting. Aerospace and Undersea Medical Society. Washington DC.

Bennett, P.B. and J.C. Rostain. Inert gas narcosis. Pp. 300–322. In: A.O. Brubakk and T.S. Neuman (eds.). 2003a. Bennett and Elliott's Physiology and Medicine of Diving. 5th edit. Saunders, London, UK.

Bennett, P.B. and J.C. Rostain. The High Pressure Nervous Syndrome. Pp. 323–357. In: A.O. Brubakk and T.S. Neuman (eds.). 2003b. Bennett and Elliott's Physiology and Medicine of Diving 5th edit. Saunders. London.

Bennett, P.B. and D. Papahadjopoulos and A.D. Bangham. 1967. The effect of raised pressures of inert gases on phospholipid model membranes. Life Sci. 6: 2527–2533.

Bennett, P.B. and B.L. Leventhal, R. Coggin and L. Racanska. 1980. Lithium effects: protection against nitrogen narcosis, potentiation of HPNS. Undersea Biomed. Res. 7: 11–16.

Bert, P. 1878. La Pression Barométrique. Masson, Paris, France.

Bolam, J.P. and Y. Smith Y. 1990. The GABA and substance P input to dopaminergic neurons in the substantia nigra of the rat. Brain Res. 529: 57–78.

Brauer, R.W. and R.O. Way. 1970. Relative narcotic potencies of hydrogen, helium, nitrogen and their mixtures. J. Appl. Physiol. 29: 23–31.

Brauer R.W. and R.O. Way and T.A. Perry. Narcotic effects of helium and hydrogen and hyperexcitability phenomenon at simulated depths of 1500 to 4000 ft of sea water. Pp. 241–255. *In*: B.R. Fink (ed.). 1968. Toxicity of Anesthetics. Williams and Wilkins, Baltimore, USA.

Brauer, R.W. and S. Dimov, X. Fructus, A. Gosset and R. Naquet. 1969. Syndrome neurologique et électrographique des hautes pressions. Rev. Neurol. 121: 264–265.

Brink, F. and J. Posternak. 1948. Thermodynamic analysis of relative effectiveness of narcotics. J. Cell Comp. Physiol. 32: 211–233.

Bjurstedt, T. and G. Severin. 1948 The prevention of decompression sickness and nitrogen narcosis by the use of hydrogen as a substitute for nitrogen. Milit. Surg. 103: 107–116.

Case, E.M. and J.B.S. Haldane. 1941. Human physiology under high pressure. J. Hyg. Camb. 41: 225–249.

Christoffersen, C.L. and L.T. Meltzer. 1995. Evidence for N-methyl-D-aspartate and AMPA subtypes of the glutamate receptor on substantia nigra dopamine neurons: possible preferential role for N-methyl-D-aspartate receptors. Neuroscience 67: 373–381.

Colloc'h, N. and J. Sopkova-de Oliveira Santos, P. Retailleau, D. Vivarès, F. Bonneté, B. Langlois d'Estainto, B. Gallois, A. Brisson, J.J. Risso, M. Lemaire, T. Prangé and J.H. Abraini. 2007. Protein crystallography under xenon and nitrous oxide pressure: comparison with in vivo pharmacology studies and implications for the mechanism of inhaled anesthetic action. Biophys. J. 92: 217–224.

Cullen, S.C. and E.G. Gross. 1951. The anesthetic properties of xenon in animals and human beings with additional observations on krypton. Science, NY 113: 580–582.

Damant, G.C.C. 1930. Physiological effects of work in compressed air. Nature, Lond. 126: 606–608.

Darbin, O. and J.J. Risso and J.C. Rostain. 1997a. A new system analysis of motor and locomotor activities associated with a midrodialysis study of pressure-induced dopamine increase in rats. Physiol. Behav. 62: 367–371.

Darbin, O. and J.J. Risso and J.C. Rostain. 1997b. Pressure induces striatal serotonin and dopamine increases: a simultaneous analysis in free moving microdialysed rats. Neuroscience Lett. 238: 69–72.

Darbin, O. and J.J. Risso and J.C. Rostain. 2001. Helium-oxygen pressure induces striatal glutamate increase: a microdialysis study in freely-moving rats. Neuroscience Lett. 297: 37–40.

David, H.N. and N. Balon, J.C. Rostain and J.H. Abraini. 2001. Nitrogen at raised pressure interacts with the GABAA receptor to produce its narcotic pharmacological effects in the rat. Anesthesiology 95: 921–927.

Dawe, R.A. and K.W. Miller and E.B. Smith. 1964. Solubility relations of fluorine compounds and inert gas narcosis. Nature, Lond. 204: 798.

Dedieu, D. and N. Balon, M. Weiss, J.J. Risso, R. Kinkead and J.C. Rostain. 2004. Microdialysis study of striatal dopaminergic dysfunctions induced by 3 MPa of nitrogen- and helium-oxygen breathing mixtures in freely moving rats. Brain Res. 998: 202–207.

Edel, P.O. and J.M. Holland, C.L. Fischer and W.P. Fife. 1972. Preliminary studies of hydrogen-oxygen breathing mixtures for deep sea diving. Pp. 257–270. *In*: Proc Symp The Working Diver, Marine Technology Society. Washington, DC, USA.

Edmonds, C. and C. Lawry and J. Pennefather. Nitrogen narcosis. Pp. 197–208. *In*: C.Edmonds (ed.). 1981. Diving and Subaquatic Medicine. 2nd edit. Diving Medical Center Publication, Australia.

Eger, E.I and C. Lundgren, S.L. Miller and W.C. Stevens. 1969. Anesthetic potencies of sulfur hexafluoride, carbon tetrafluoride, chloroform and ethane in dogs: Correlation with the hydrate and lipid theories of anesthetic action. Anesthesiology 30: 129–135.

Eyring, H. and J.W. Woodbury and J. D'Arrigo. 1973. A molecular mechanism of general anesthesia. Anesthesiology 38: 415-424.

Featherstone, R.M. Anesthesia: xenon. p. 22. *In*: O. Glasser (ed.). 1960. Medical Physics, Vol 3, Yearbook, Chicago, USA.

Featherstone, R.M. and C.A. Muehlbaecher. 1963. The current role of inert gases in the search for anesthesia mechanisms. Pharm. Rev. 15: 97–121.

Featherstone, R.M. and C.A. Muehlbaecher, F.L. Debon and J.A. Forsaith. 1961. Interactions of inert anesthetic gases with proteins. Anesthesiology 22: 977–981.

Ferguson, J. 1939. The use of chemical potentials as indices of toxicity. Proc. R. Soc. B197: 387–404.

Fife, W.P. 1979. The Use of Non Explosive Mixtures of Hydrogen and Oxygen for Diving. Texas A & M University, Hyperbaric Lab. TAMU-SG-70-201. College Station, USA.

Franks, N.P. and W.R. Lieb. 1978. Where do anesthetics act. Nature, Lond 274: 339–342.

Franks NP, Lieb WR. 1982. Molecular mechanisms of general anaesthesia. Nature 300: 487–493.

Franks, N.P. and W.R. Lieb. 1984. Do general anaesthetics act by competitive binding to specific receptors? Nature 310: 599–601.

Franks, N.P. and W.R. Lieb. 1991. Stereospecific effects of inhalational general anaesthetic optical isomers on ion nerve channels. Science 254: 427–430.

Franks, N.P. and W.R. Lieb. 1994. Molecular and cellular mechanisms of general anaesthesia. Nature 367: 607–614.

Fructus, X.R. Hydrogen, pressure and HPNS. Pp. 125–138. *In*: R.W. Brauer (ed.). 1987. Hydrogen as a Diving Gas. Undersea and Hyperbaric Medical Society, Bethesda, MD, USA.

Fructus, X. and R. Naquet, A. Gosset, P. Fructus and R.W. Brauer. 1969. Syndrome Nerveux des Hautes Pressions. Marseille Med. 6, 509–512.

Fructus, X. and B. Gardette, M. Carlioz and Y. Giran. 1984. Hydrogen Narcosis. Pp. 87–96. *In*: T. Nome, G. Susbielle, M. Comet, M. Jacquin and R. Sciarli. (eds.). 1984. Proceeding of Xe Congress of European Undersea Biomedical Society. Marseille, France.

Galey, W.R. and P.S. Van Nice. The effects of hyperbaric and elemental narcotic gases on cellular membrane ion transport. Pp. 410–415. *In*: B.R. Fink (ed.). 1980. Progress in Anesthesiology: Molecular Mechanisms of Anesthesia, Vol II, Raven Press, New-York, USA.

Gardette, B. Hydra IV and Hydra V: human deep hydrogen dives 1983–1985. Pp. 109–117. *In*: R.W. Brauer (ed.). 1987. Hydrogen as a diving gas. UHMS publication 69 (WS-HYD) Bethesda, Md, USA.

Gardette, B. Compression procedures for mice and human hydrogen deep diving: COMEX HYDRA program. Pp. 217–231. *In*: J.C. Rostain, E. Martinez and C. Lemaire (eds.). 1989. High Pressure Syndrome 20 Years Later. ARAS-SNHP Publications. Marseille, France.

Gardette, B. and C. Lemaire, J.C. Rostain and X. Fructus. The french deep diving scientific program on oxygen-helium, trimix, and oxygen-hydrogen gas mixtures. Pp. 69–100. *In*: Y.C. Lin and K.K. Shida (eds.). 1990. Man in the Sea. Best Publishing Compagny, San Pedro Ca, USA.

Gardette, B. and C. Gortan. Mice and monkeys deep dives in heliox, hydrox and hydreliox gas mixtures—synthesis of COMEX "Hydra" programme. Pp. 173–184. *In*: P.B. Bennett and R.E. Marquis (eds.). 1994. Basic and Applied High Pressure Biology. University Press of Rochester, Rochester, USA.

Gerfen, C.R. and C.J. Wilson. The Basal Ganglia. Pp. 371–468. *In*: L.W. Swanson, A. Björklund, and T. Hökfeft (ed.). 1996. Integrated Systems of the CNS, Part III: Cerebellum, Basal Ganglia, Olfactory System. Handbook of Chemical Neuroanatomy, Vol. 12. Elsevier, Amsterdam, The Netherland.

Green, J.B. 1861. Diving With and Without Armour. Leavitt, Buffalo, USA.

Halsey, M.J. and E.L. Eger, D.W. Kent and P.J. Warne. High pressure studies of anaesthesia. Pp. 353–361. *In*: B.R. Fink (ed.). 1975. Molecular Mechanisms of Anesthesia (Progress in Anesthesiology), Raven press New-York, USA.

Halsey, M.J. and B. Wardley-Smith and C.J. Green. 1978. Pressure reversal of general anesthesia—a multi-site expansion hypothesis. Br J. Anaesth 50: 1091–1097.

Hamilton, K. and M.F. Laliberte and B. Fowler. 1995. Dissociation of the behavioural and subjective components of nitrogen narcosis and diver adaptation, Undersea Hyperbaric Med. 22: 41–49.

Hill, L. and J.J. McLeod. 1903. The influence of compressed air on respiratory exchange. J. Physiol. 29: 492–510.

Hill, L. and M. Greenwood. 1906. The influence of increased barometric pressure on man. Proc. R. Soc. B77: 442–453.

Hill. L. and R.H. David, R.P. Selby, A. Pridham and A.E. Malone. 1933. Deep Diving and Ordinary Diving. Report of a Committee Appointed by the British Admiralty.

Johnson, S.M. and K.W. Miller. 1970. Antagonism of pressure and anaesthesia. Nature, Lond 228: 75–76.

Junod, T. 1835. Recherches sur les effets physiologiques et thérapeutiques de la compression et de raréfaction de l'air, tant sur le corps que les membres isolés. Ann. Gen. Med. 9: 157.

Katz, Y. and S.A. Simon. 1977. Physical parameters of the anesthetic site. Biophys. Acta 471: 1–15.

Lambertsen, C.J. Collaborative investigation of limits of human tolerance to pressurization with helium, neon and nitrogen.Simulation of density equivalent to helium-oxygen respiration at depth to 2000, 3000, 4000 and 5000ft of sea water. Pp. 35–52. *In*: C.J.Lambertsen (ed.). 1976. Underwater Physiology V. FASEB, Bethesda, USA.

Lavoute, C. and M. Weiss and J.C. Rostain. 2005. Effects of repeated hyperbaric nitrogen-oxygen exposures on the striatal dopamine release and on motor disturbances in rats. Brain Res. 1056: 36–42.

Lavoute, C. and M. Weiss and J.C. Rostain. 2006. Effects of NMDA administration in the subtantia nigra pars compacta on the striatal dopamine release before and after repetitive exposures to nitrogen narcosis in rats. Undersea Hyperb. Med. 33: 175–179.

Lavoute, C. and M. Weiss and J.C. Rostain. 2007. The role of NMDA and GABA(A) receptors in the inhibiting effect of 3 MPa nitrogen on striatal dopamine level. Brain Res. 1176: 37–44.

Lavoute, C. and M. Weiss and J.C. Rostain. 2008a. Alterations in nigral NMDA and GABAA receptor control of the striatal dopamine level after repetitive exposures to nitrogen narcosis. Exp. Neurol. 212: 63–70.

Lavoute, C. and M. Weiss, J.M. Sainty, J.J. Risso and J.C. Rostain. 2008b. Post effect of repetitive exposures to pressure nitrogen-induced narcosis on the dopaminergic activity at atmospheric pressure. Undersea Hyperb. Med. 35: 21–25.

Lawrence, J.H. and W.F. Loomis, C.A. Tobias and F.H. Turpin. 1946. Preliminary observations on the narcotic effect of xenon with a review of values for solubilities of gases in water and oils. J. Physiol. 105: 197–204.

Lazarev, N.V. and Y.I. Lyublina and R.Y Madorskaya. 1948. Narcotic action of xenon [in Russian]. Fiziol. Zh. SSSR 34: 131–134.

Macdonald, R.L. and R.W. Olsen. 1994. GABAA receptor channels. Annu. Rev. Neurosci. 17: 569–602.

McLeod, M. and P.B. Bennett and R.L. Cooper. 1988. Rat brain catecholamine release at 1, 10, 20 and 100 ATA heliox, nitrox and trimix. Undersea and Hyperbaric Med. 15: 211–221.

Maletzky, B. and P.H. Blachly. 1971. The Use of Lithium in Psychiatry. Butterworths. London, UK.

Marshall, J.M. 1951. Nitrogen narcosis in frogs and mice. Am. J. Physiol. 166: 699–711.

Meyer, H.H. 1899. Theoris der alkoholnarkose. Arch. Exp. Path. Pharmak. 42: 109.

Michaud, A. and J. Parc, L. Barthelemy, J. Le Chuiton, J. Corriol, J. Chouteau and F. Le Boucher. 1969. Premières données sur une limitation de l'utilisation du mélange oxygène-hydrogène pour la plongée profonde à saturation. C.R. Acad. Se. (Paris) 269: 497–499.

Miller, S.L. 1961. A theory of gaseous anesthetics. Proc. Natl. Acad. Sci. USA 47: 1515–1524.

Miller, K.W. and W.D.M. Paton and E.B. Smith. 1965. Site of action of general anesthetics. Nature, Lond 206: 575–577.

Miller, K.W. and W.D. Paton, W.B. Streett and E.B. Smith. 1967. Animals at very high pressures of helium and neon. Science 157: 97–98.

Miller, K.W. and W.D.M. Paton, R.A. Smith and E.B. Smith. 1973. The pressure reversal of general anesthesia and the critical volume hypothesis. Mol. Pharmacol. 9: 131–143.

Overton, E. 1901. Studien über die Narkose. Jena: Fischer.

Paladini, G. and P. Celada and J.M. Tepper. 1999. Striatal, pallidal and pars reticulata evoked inhibition of nigrostriatal dopaminergic neurons is mediated by GABA(A) receptors in vivo. Neuroscience 89: 799–812.

Pauling, L. 1961. A molecular theory of anaesthesia. Science, NY 134: 15–21.

Pittinger, C.B. 1962. Mechanisms of anesthesia. Xenon as an anesthetic. In: Proc 22nd Int Congr. Physiol. Sci. 1. Excerpta Medical Foundation. London.

Radomski, M.W. and J.D. Wood. 1970. Effect of metal ions on oxygen toxicity. Aerospace Med. 41: 1382–1387.

Raoul, Y. and J.L. Meliet and B. Broussolle. 1988. Troubles psychiatriques et plongée profonde. Médecine et armées 16: 269–270.

Risso, J.J. and A. Saget, N. Turle, B. Zouani and O. Darbin. 1999. Neurochemical and behavioural studies on narcosis. Undersea and hyperbaric Med. 26: 81.

Rostain, J.C. The high pressure nervous syndrome at the central nervous system level. Pp. 137–148. In: H.W. Jannasch, R.E. Marquis and A.M. Zimmerman (eds.). 1987. Current Perspectives in High Pressure Biology. Academic Press, London, UK.

Rostain, J.C. Nervous system at pressure. Pp. 157–172. In: P.B. Bennett and R.E. Marquis (eds.). 1994. Basic and Applied High Pressure Biology. University Press of Rochester, Rochester, USA.

Rostain, J.C. The high pressure nervous syndrome and behavioral adaptations. Pp. 286-293. In: K.Pandloff, N. Takeda and P.K. Singal (eds.). 1999. Adaptation Biology and Medicine Volume 2: Molecular Basis. Narosa Publishing House, New Delhi, India.

Rostain, J.C. and R. Naquet. Résultats préliminaires d'une étude comparative de l'effet des mélanges oxygène-hélium et oxygène-hydrogène et des hautes pressions sur le babouin Papio papio. Pp. 44–49. In: X. Fructus (ed.). 1972. Proc. des troisièmes journées internationales d'hyperbarie et de physiologie subaquatique. Doin. Marseille, France.

Rostain, J.C. and R. Naquet. 1974. Le Syndrome Nerveux des Hautes Pressions: caractéristiques et évolution en fonction de divers modes de compression. Rev. EEG. Neurophysiol. 4, 107–124.

Rostain, J.C. and C. Forni. 1995. The effects of high pressures of various gas mixtures on rat striatal dopamine detected in vivo by voltammetry. J. Appl. Physiol. 78: 1179–1187.

Rostain, J.C. and M.C. Gardette-Chauffour, C. Lemaire and R. Naquet. 1988. Effects of hydrogen-helium-oxygen mixture on high pressure nervous syndrome up to 450 msw. Undersea Biomed. Res. 15: 257–270.

Rostain J.C. and M.C. Gardette-Chauffour and R. Naquet. 1990. Studies of neurophysiological effects of hydrogen-oxygen mixture in man up to 30 bars. Undersea Biomed. Res. 17 (Suppl): 159.

Rostain, J.C. and M.C. Gardette-Chauffour and B. Gardette. Neurophysiological studies in macaca fascicularis during exposures with breathing mixtures containing hydrogen up to 1200 msw. Pp. 243–252. *In*: P.B. Bennett and R.E. Marquis (eds.). 1994a. Basic and Applied High Pressure Biology. University Press of Rochester, New-York, USA.

Rostain, J.C. and M.C. Gardette-Chauffour and B. Gardette. 1994b. HPNS during a deep hydrogen-helium-oxygen dive up to 701 meters. Undersea and Hyperbaric Med 21 (suppl.): 40.

Rostain, J.C. and M.C. Gardette-Chauffour and B. Gardette. 1999. Hydrogen, a gas for diving: a mini review. Undersea and Hyperbaric Med. 26 (suppl.): 62.

Schoenborn, B.P. 1965. Binding of xenon to horse haemoglobin. Nature, Lond. 208: 760–762.

Schoenborn, B.P. and H.C. Watson and J.C. Kendrew. 1965. Binding of xenon to sperm whale myoglobin. Nature, Lond 207: 28–30.

Schreiner, H.R. and R.W. Hamilton and T.D. Langley. 1972 Neon: An attractive new commercial diving gas. *In*: Proc Offshore Technology Conference. May 1–3. Houston, TX.

Simon, S.A. and T.J. McIntosh, P.B. Bennett and B.B. Shrivastav. 1979. Interaction of halothane with lipid bilayers. Mol. Pharmacol. 16: 163–170.

Sincoff, R. and J. Tanguy, B. Hamilton, D. Carter, E.A. Brunner and J.Z. Yeh. 1996. Halothane acts as a partial agonist of the alpha6 beta2 gamma2S GABA(A) receptor. FASEB J. 10: 1539–1545.

Stoudemire, A. and J. Miller, B.S.F. Schmitt, P. Logue, D. Shelton, P.A.G. Latson and P.B. Bennett. 1984 Development of an organic affective syndrome during a hyperbaric diving experiment. Am. J. Psychiatry 141: 1251–1254.

Sugita, S. and S.W. Johnson and R.A. North. 1992. Synaptic inputs to GABAA and GABAB receptors originate from discrete afferent neurons. Neurosci. Lett. 134: 207–211.

Townsend, R.E. and L.W. Thompson and I. Sulg. 1971. Effect of increased pressures of normoxic helium, nitrogen and neon on EEG and reaction time in man. Aerospace Med. 42: 843–847.

Turle, N. and A. Saget, B. Zouani and J.J. Risso. 1998. Neurochemical studies of narcosis: a comparison between the effects of nitrous oxide and hyperbaric nitrogen on the dopaminergic nigro-striatal pathway. Neurochemical Res. 14: 999–1005.

Turle-Lorenzo, N. and B. Zouani and J.J. Risso. 1999. Narcotic effect produced by nitous oxide and hyperbaric nitrogen narcosis in rats performing a fixed ratio test. Physiol. Behav. 67: 321–325.

Vallée, N. and J.C. Rostain, A. Boussuges and J.J. Risso. 2009. Comparison of Nitrogen Narcosis and Helium Pressure Effects on Striatal Amino Acids: A Microdialysis Study in Rats. Neurochem. Res. 34: 835–844.

Vjotosh, A. and A. Popov, O.S. Alekseeva, A. Boso and I.T. Demchenko. 1999. Role of nitric oxide in the mechanism of nitrogen narcosis. Undersea and Hyperbaric Med. 26: 81.

Wedzony, K. and K. Czepiel and K. Figal. 2001. Immunohistochemical evidence for localization of NMDAR1 receptor subunit on dopaminergic neurons of rat substantia nigra pars compacta. Pol. J. Pharmacol. 345: 523–529.

Westerink, B.H. and M. Santiago and J.B. De Vries. 1992. The release of dopamine from nerve terminals and dendrites of nigrostriatal neurons induced by excitatory amino acids in conscious rat. Naunyn-Schmiedberg's Arch. Pharmacol. 345: 523–529.

Zal'tsman, G.L. 1961. Physiological Principles of a Sojourn of a Human in Conditions of Raised Pressure of the Gaseous Medium. English translation, Foreign Technology Division, Wright-Patterson Air Force Base 1967. Ohio, AD655 360.

Zal'tsman, G.L. 1968. Hyperbaric Epilepsy and Narcosis (Neurophysiological Studies). Sechenov Institute of Evolutionary Physiology and Biochemistry, Academy of Sciences, Leningrad, USSR.

Zetterstrom, A. 1948. Deep sea diving with synthetic gas mixtures. Mil. Surg. 103: 104–106.

# 19

# The High Pressure Nervous Syndrome

*J.C. Rostain,[1] J.J. Risso[2] and J.H. Abraini[3]*

## INTRODUCTION

Commercial diving in compressed air is not feasible to depths greater than 50 m.s.w. (i.e. 6 bars of absolute pressure) since the nitrogen in the air becomes narcotic at partial pressure of about 3 to 4 bars and induces nitrogen narcosis constituted by behavioural problems and changes in mental processes (Table 2 of previous chapter). In order to enable man to dive deeper than 50 to 70 m.s.w., researchers have tried to replace the nitrogen by another inert gas which has fewer narcotic properties. The inert gases are not metabolized by the organism and thus will be inactive.

Correlation trials between certain physical properties of such inert gases and their narcotic potencies have been carried out by numerous authors, among them Behnke et al. (1935), Case and Haldane (1941), Marshall (1951), Bennett (1966) and Brauer et al. (1968), and they all agree that helium is preferable to the other gases. In fact, if one links the narcotic potency of the so called inert gases to their physical properties and notably to their solubility

[1] Université de la Méditerranée et Institut de Médecine Navale du Service de Santé des Armées, UMR-MD2—Physiologie et physiopathologie en condition d'oxygénation extrême, Institut de Neurosciences Jean Roche, Faculté de médecine Nord Bd P. Dramard, 13015 Marseille, France.
E-mail: jean-claude.rostain@univmed.fr

[2] Université de la Méditerranée et Institut de Médecine Navale du Service de Santé des Armées., UMR-MD2—Physiologie et physiopathologie en condition d'oxygénation extrême, IMNSSA, BP 84, 83800 Toulon Armées, France.
E-mail: j.j.risso@imnssa.net

[3] Université de Caen, UMR 6232, CI-NAPS, Centre CYCERON, BP 5229, Boulevard Becquerel, 14074 Caen cedex, France.
E-mail: abraini@cyceron.fr

in lipids, three gases show, in decreasing order, a narcotic potency less than that of nitrogen: hydrogen, neon and helium.

Hydrogen, with a molecular weight equal to half of that of helium ($N_2$ = 28; Ne = 20; He = 4; $H_2$ = 2) could advantageously replace the latter when considering density of a mixture: for a given pressure a hydrogen-oxygen mixture would be half as dense as one of helium-oxygen. However, being more soluble in lipids than helium it has a greater narcotic potency. According to Brauer et al. (1968), Brauer and Way, (1970) hydrogen narcosis appears at about 250 to 300 msw.

While neon, though ten times heavier than hydrogen, has physical characteristics which are often inferior to those of hydrogen, thus its narcotic potency could be weaker. This hypothesis has, however, not been demonstrated because tests of this gas remain limited (Thorne et al., 1974; Lambertsen, 1976). In addition the cost of its production is high.

The choice has thus turned to helium which seems to have almost no narcotic potency. The use of a synthetic helium-oxygen respiratory mixture where the oxygen partial pressure is maintained at values close to those of the atmospheric pressure has allowed divers to reach great depths. The first deep dives (equal or greater than 100 m.s.w.) were carried out in the sixties. After the Precontinent operations (Fructus and Chouteau, 1963) numerous dives in hyperbaric chambers and at sea were carried out between 100 and 300 m.s.w. each one going a bit deeper (Cabarrou et al., 1966; Aquadro and Chouteau, 1967; Bennett, 1967; Fructus and Delauze, 1968; Waldvogel and Bulhmann, 1968). Dives reached and went beyond 300 m.s.w. between 1968 and 1970 (Fructus et al., 1969; Overfield et al., 1969; Overrath et al., 1969; Salzano et al., 1970).

Nonetheless during these dives in a helium-oxygen atmosphere a certain number of new facts appeared. During a rapid compression to 180 m.s.w., Bennett (1966) described the existence of nausea, vertigo and most markedly tremors which the author attributed to helium (helium tremor). Later, during dives to between 300 and 365 m.s.w. Brauer et al. (1969) and Fructus et al. (1969) showed a whole series of disturbances, different from those of nitrogen narcosis, which those authors attributed to pressure and named the High Pressure Nervous Syndrome (HPNS), events later confirmed by Bennett and Towse (1971) during a dive to 457 m.s.w.

## CHARACTERISTICS OF HPNS IN HELIUM-OXYGEN MIXTURE

HPNS is composed of neurological disturbances and of electrophysiological changes, the most important of which are shown in Table 1. The first signs generally appear between 10 and 20 bars.

Table 1. Signs and symptoms of the high pressure nervous syndrome in helium-oxygen mixture.red about 11 ATA.

| Pressure effects in Man Helium-Oxygen Mixture Up to 62 ATA | |
|---|---|
| **Behavioural Symptoms** | **Electrophysiological Changes** |
| Termor | EEG: Increase in slow waves, decrease in fast activities |
| Fasciculations, Myoclonia | |
| Dysmetria | Change in evoked potentials, in cortical excitability cycle |
| Somnolence | |
| Cognitive Impairments up to –20% | SLEEP: Increase in stage 1 and 2 Decrease in stage 3 and 4, REM |
| | Hyper-Reflexia |

## Neurological Disturbances

A tremor is first observed at the extremities of the limbs during periods of activity. It is analyzed during the 'serment' test with an accelerometer placed on the middle finger of the dominant hand. In man it begins between 15 and 20 bars. It increases with the depth and progressively affects all the limbs and the trunk. The power spectra carried out on the signals recorded from the accelerometer give a frequency of 8 to 12 Hz (Rostain and Lemaire, 1973; Rostain and Naquet, 1974, 1978; Fructus and Rostain, 1978) (Fig. 1).

Fasciculation and myoclonus (muscular jerks) first appear at the extremities of the limbs (at about 400 to 500 meters). They then reach the limbs and finally the muscles of the neck, face and body trunk (Fructus and Rostain, 1978).

Dysmetria is seen during classical neurological tests ('fleuret' and finger to nose tests) and appears at depths greater than 400–500 m.s.w. (Fructus and Rostain, 1978).

Under some conditions drowsiness occurs when the subject is left at rest, but it is easily overcome by activity.

Beside these neurological symptoms, decreases in performance have been recorded. A decrement up to 20% was recorded in psychomotor and intellectual tests.

## Electrophysiological Disturbances

Electroencephalographic (EEG) changes have been recorded from 10 bars. These changes (Fig. 2.) consist at first by a decrease of alpha waves (8–14 Hz) which occurred in the posterior region of the scalp from 10 bars (Rostain 1993, Bennett and Rostain 2003) and of an increase of slow wave activities particularly that of the theta frequency (4–7 Hz) sometimes accompanied by that of the delta frequency (1–4 Hz) (Fig. 2). These changes predominate in the anterior region of the scalp (Fig. 3). Often there is also a decrease of the

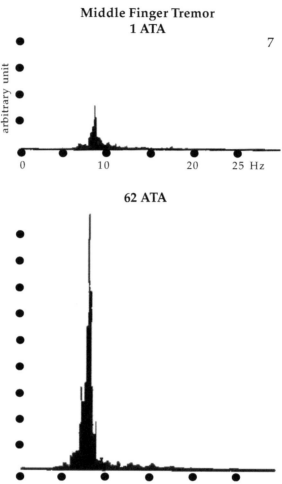

**Figure 1.** Power spectra of the middle finger tremor at 1 ATA and at 62 ATA in helium-oxygen mixture.

faster activities (beta 14–40 Hz) (Rostain and Naquet, 1974, 1978; Rostain, 1993; Bennett and Rostain, 2003).

There are also changes of the cortical evoked potentials brought about by different sensory modalities (visual, somatic, auditory) (Rostain and Dimov, 1976; Bennett and Rostain, 2003b). The cortical excitability cycles are also modified and indicated a cortical hyperexcitability (Rostain and Dimov ,1976). Some reflexes are modified and show generally hyper-reflexia (Roll et al., 1978; Harris, 1979a, b; Harris and Bennett, 1983; Parmentier et al.,1985; Bennett and Rostain, 2003).

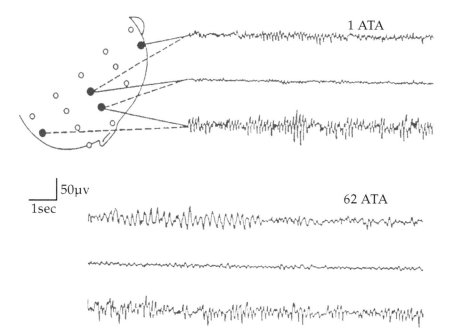

**Figure 2.** EEG recording of one subject during a control period at 1 ATA and at 62 ATA. In each condition, three leads are represented: fronto palar-central, central-mid temporal, mid temeporal-occipital. One can see the increase of slow waves on the fronto central lead at 62 ATA and the decrease of the alapha activity in the posterior lead.

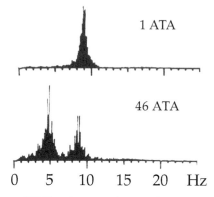

**Figure 3.** Power spectra of EEG activities in fronto-polar-central lead at 1 ATA (top) and 46 ATA (Bottom). At 1 ATA, the EEG is characterized by an activity predominant in the the alpha frequency (8–10 Hz). At 46 ATA this activity is decreasing and we recorded an increase in theta EEG frequency (4 to 6 Hz).

Studies of sleep (electroencephalographic) have also shown changes: awake periods, stages 1 and II of sleep (light sleep) increase; stages III and IV (deep sleep) lessens ; periods of paradoxical sleep are unstable and decrease after 40 bars. (Rostain and Lyagoubi, 1971; Rostain et al., 1973, 1988, 1991, 1997).

Other symptoms such as nausea and vertigo have also been described under certain conditions (Bennett and Rostain, 2003).

## Characteristics of HPNS in Experimental Animal Models

Most of these symptoms have been described in animals exposed to high pressures of helium (Brauer, 1975; Halsey, 1982) (Table 2). In the rat and monkey neurological disturbances have also been noted. Tremors at 10 to 14 Hz and myoclonus were reported. In free moving animals, motor and hyperlocomotor activities have been recorded. For pressures greater than those reached by man (higher than 70 bars), hypertonia and muscular spasms have been recorded together with convulsions (Rostain et al., 1984a, 1986a).

**Table 2.** Signs and symptoms of the high pressure nervous syndrome in rats and monkeys in helium-oxygen mixture.

| HPNS in animal models (Rats and Monkeys) in helium-oxygen mixture Up to 101 ATA | |
|---|---|
| **Behavioral Changes** | **Electrophysiological Disruptions** |
| Tremor (10–14 Hz) | |
| Muscular hypertonus | EEG changes: Increase of slow waves, |
| Myoclonus | paroxystic activity, epileptic seizure. |
| Muscular spasms | Changes in evoked potentials, cortical |
| Convulsions | excitability and reflexes (monkeys) |

Electrophysiological changes have also been recorded. An increase of the slow wave EEG activities (theta and delta) has been found together with a decrease of the activities of more rapid activity frequencies (alpha and beta). EEG activities of a paroxysmal type with a pointed character or a spike and wave have also been recorded for pressure never reached by man (greater than 70 bars.). Such paroxysmal activities are usually followed by a generalized epileptic seizure of the tonic-clonic type (Rostain et al., 1984a, 1986a).

Somatic evoked potentials, cortical excitability cycles and reflexes are also modified in monkeys (Hugon et al., 1981, 1983). The electrical activity of diverse brain structures has been studied *in vivo* in the rat. These include the hippocampus, thalamus and caudate nucleus and show an increase in slow waves (Fagni et al., 1985; Rostain et al., 1986a). However, epileptic seizures do not appear first in any of these three structures (Rostain et al., 1986a); when they occur they are secondary to the electrographic seizure recorded at the cortical level.

## THE ROLE OF HYPERBARIC VARIABLES IN THE APPEARANCE AND DEVELOPMENT OF HPNS

### Influence of Pressure

The effects of pressure itself have been described during hydrostatic compression of aquatic animals. The works of Regnard (1891), Ebbecke (1936), Cattell (1936), for example, have shown that the principal effect of pressure is a stimulation of the central nervous system from 50 bars. With pressure of 200 to 300 bars spontaneous muscular contractions appear, followed by paralysis of the animal. Pressure of 400 to 500 bars is lethal. Miller (1972) compressed tritons with increasing hydrostatic pressure or with pressures of mixed gases. He observed in the first case behavioural changes at about 100 bars and paralysis of the animal at 200 bars. In the second case, with helium, the effects were similar but appeared at a slightly greater depth; protection by the gas being linked to its narcotic potency. Fish under compression (Barthelemy and Belaud, 1972; Barthelemy et al., 1981) show pressure induced changes from 20 bars.

In mammals, the effects of pressure per se are difficult to define because the effects of compression are so closely associated with the pressurizing respiratory mixture. Nevertheless, the phenomena which one observes during a prolonged stay at constant pressure might be in part linked to the pressure effect itself.

### Influence of Variations of Pressure

The compression and the type of compression (speed as a function of depth, stops at specific levels) play a large part in the appearance of HPNS both in the intensity of its diverse symptoms and in the magnitude of the changes of performance.

The first very deep dives by man with a helium-oxygen mixture were performed with very fast rates of compression. The COMEX Physalie 1 to IV dive series in 1968, were performed with a compression rate of 18 bars per 1 hour. They induced neurological disturbances beyond 20 bars, and the pressure of 36.5 bars was not exceeded because these symptoms were too severe (Brauer et al., 1969; Fructus et al., 1969). Subsequent experiments were carried out with slower speeds of compression, and pressures up to 61 bars were reached (Bennett and Towse, 1971; Rostain and Naquet, 1974; Fructus et al., 1976; Bennett and Rostain, 2003) (Fig 4). These experiments have shown that the pressure at which the symptoms of HPNS occur tends to increase while their intensity tends to decrease if compression follows a slow exponential time course interrupted by holds at intermediate pressure. HPNS is enhanced by continuous and/or rapid compression.

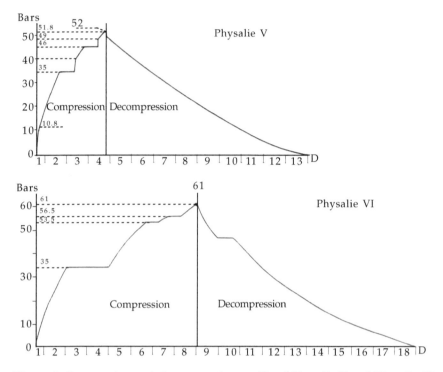

**Figure 4.** Compression and decompression profile of Physalie V and Physalie VI experiments up to 53 and 62 ATA performed in helium-oxygen mixture. In Y axis pressure expressed in bar; X axis, the days (D).

However the slowing of the rates and the prolonging of the compression periods beyond certain values does not reduce HPNS. This demonstrates that in addition to compression there are other variables of the hyperbaric environment which could bring about certain of the observed changes. The composition of the gas mixture could well be an important element.

## Influence of the Breathing Mixture

### *HPNS in a helium-nitrogen-oxygen mixture*

Numerous experiments have been performed throughout the world in order to study the effects of various gaseous mixtures. Studies of anaesthetics in the 1940s demonstrated that anaesthesia was reversed by an increase in pressure (Johnson et al., 1942a, 1942b). In contrast, as a function of this anaesthesia-pressure antagonism it was proposed that the presence of narcotic or anaesthetic substances would allow the reduction of the effects of an increase of pressure. Various authors have shown that the depth at which HPNS symptoms appear becomes greater with narcotic substances

(Lever et al., 1971; Miller, 1972; Miller et al., 1973; Brauer et al., 1974; Belaud et al., 1977). Thus Miller et al. (1973) proposed the critical volume hypothesis according to which the membrane expansion brought about by the narcotics compensated for the reduction produced by pressure.

According to the antagonism between anaesthetics or narcotic agents and pressure as was suggested by the critical volume hypothesis (Miller et al., 1973), it might be supposed that the use of nitrogen, which is narcotic under pressure, might in a helium-oxygen mixture reduce or suppress HPNS. A reduction or even a suppression of HPNS has been reported in animals (Miller, 1972; Brauer, 1975) and in man (Bennett et al., 1974) with a helium-nitrogen-oxygen mixture (trimix).

However, the studies carried out with the helium-nitrogen-oxygen mixture in *Papio papio* monkeys and in man have shown that the effects of nitrogen are more complex than simple antagonism (Naquet et al., 1975; Rostain et al., 1980). Some neurological symptoms such as tremors were reduced or suppressed but others such as EEG changes were increased.

Experiments performed on the baboon (*Papio papio*) have indicated that the introduction of nitrogen at the beginning of the compression did not reduce the HPNS and even enhanced some symptoms such as EEG changes. In contrast, its introduction at the end of the compression did reduce the HPNS which had developed during the compression in helium-oxygen. The beneficial antagonism between nitrogen and pressure was enhanced when nitrogen was added progressively to helium-oxygen mixtures during slow exponential compression (Rostain, 1980; Rostain et al., 1978, 1984a).

From these results Rostain et al. (1984b) have proposed a compression schedule in the monkey which consists of :

1) an exponential compression profile with compression speeds that become increasingly slow as the final depth is reached;
2) stages every 10 bars lasting a relatively short time in relation to the total duration of the compression;
3) nitrogen added during compression before every stage to readjust the concentration to 5.5%.

With this method monkeys have been compressed up to 110 bars. Hyperbaric tremor occurred around 6 bars as opposed to 20–30 bars when the same compression curve was used without nitrogen addition. In contrast, the EEG changes were similar with or without nitrogen injection and the synergistic effect of nitrogen and pressure on these symptoms was eliminated by this method of compression. However, beyond 80 bars this method with nitrogen injection induced HPNS disturbances some of which differed from those classically described during rapid compression, and which had never been seen, since epileptic seizures provoked by the rapid compressions masked them.

Such disturbances consisted of :

a)  an overall depression of EEG activity beyond 80–90 bars ;
b)  periods of muscular disturbances: hypertonus, muscular spasms, motor agitation and myoclonus with no cortical component (EEG paroxystic activities) beyond 90–100 bars;
c)  palpebral clonus and ocular movements associated with spike discharges in the posterior area of the skull and which could develop into an epileptic seizure localized in this area around 110 bars;
d)  hallucinatory behaviour beyond 100 bars;
e)  sleep deprivation beyond 90–100 bars.

As a result this method of compression was not able to reach or exceed 100 bars without serious nervous disorders. Nonetheless it did allow baboons to reach a depth of 80 bars without any epileptic seizures and with reduced HPNS.

The extrapolation of this method of compression from monkey to man has been carried out at a depth of 45 bars and includes (Gardette and Rostain, 1981):

a)  compression to 45 bars in 38 h where speeds followed an exponential function of the depth;
b)  stages of 150 min every 10 bars.

With this compression protocol, the neurological symptoms were reduced and even suppressed: little or no tremor was recorded; no dysmetria or fasciculation were observed. The EEG changes were of low intensity, and the synergistic effect of nitrogen and pressure on EEG changes was not recorded. The same compression curve without nitrogen injection induced neurological symptoms (Rostain et al., 1984c, 1984d, 1987). Consequently, the method of compression in a helium-nitrogen-oxygen mixture studied in the monkey and extrapolated to man has improved the neurological symptoms of HPNS (Table 3). Nevertheless, no visible improvement was recorded for sleep disturbances in a helium-nitrogen-oxygen mixture versus one of helium-oxygen. The decrement in performance of psychomotor and intellectual tests was around 10%.

This compression protocol with nitrogen addition reduces the duration of the compression used generally in helium-oxygen mixture and improves the life and the works under pressure by reducing some HPNS symptoms. With similar methods men were compressed up to 68.5 bars (Bennett et al., 1982) and several dives have been performed between 300 and 600 m (31 and 61 bars of absolute pressure) (Bennett and Schafstall, 1990; Bennett and Rostain, 2003).

Table 3. Signs and symptoms of the high pressure nervous syndrome in helium-nitrogen-oxygen mixture (N2 = 5%).

| High preassure Nervous Syndrome Helium-Nitrogen-Oxygen Mixture up to 46 ATA | |
|---|---|
| Behavioural Symptoms | Electrophysiological Changes |
| | EEG: Increase in slow waves, decrease in fast activities |
| Somnolence | Change in evoked potentials, |
| Cognitive impairments | in cortical excitability cycle |
| around –10% | SLEEP: Increase in stages 1 and 2 Decrease in stages 3 and 4, REM |
| | Hyper-Reflexia |

## HPNS in a hydrogen-helium-oxygen mixture

Nitrogen is relatively heavy and causes problems in pulmonary ventilation at great pressure by increasing the density of the respiratory mixtures, thereby reducing ventilatory flow. Thus, it would be preferable to use a gas with a lower density.

Another 'inert gas' that is possible to use is hydrogen. Indeed, hydrogen is a physiologically inert gas which has been considered and used for deep diving (Case and Haldane, 1941; Zetterstrom, 1948; Bjurstedt and Severin, 1948; Zal'tsman, 1961, 1968). It is, however, explosive in mixtures of more than 4% oxygen, and works by Brauer et al. (1968, 1971) have established that its narcotic potency is in agreement with its lipid solubility. With a relative narcotic potency of 0.6, it is more narcotic than either helium or neon but fewer narcotics than nitrogen.

Hydrogen has a lower density than helium and thus would be better for breathing with a molecular weight equal to half that of helium; a hydrogen-oxygen mixture is half as dense as helium-oxygen. Being more soluble in lipids than helium, it also has a greater narcotic potency at pressure which may, in accordance with the critical volume hypothesis and the antagonism of pressure on general anesthesia, reduce some of the symptoms of HPNS. According to Brauer et al. (1968) and Brauer and Way (1970), hydrogen narcosis will occur at about 24–30 bars. In the past, different groups have studied the effects of hydrogen under pressure in man (Case and Haldane, 1941; Zetterstrom, 1948; Edel et al., 1972; Fife, 1979, 1987) and in animals (Michaud et al., 1969; Brauer & Way, 1970; Rostain and Naquet, 1972; Halsey et al., 1975; Fife, 1979, 1987). The results were contradictory. Some reported that hydrogen improved HPNS in mice, monkeys and dogs (Brauer & Way, 1970; Halsey et al., 1975; Fife, 1979, 1987); others reported no difference between the two gases in the baboon (Rostain and Naquet, 1972) and others reported that rabbits died (Michaud et al., 1969). A possible reason for the discrepancy is lack of chamber heating, which may have caused the deaths and the speed of compression used at this time which were very fast (20 bars/hour). However, Edel et al. (1972) who performed experiments on humans to 6.1 bars and 10.8 bars and Edel (1971) and Fife (1979) with experiments in dogs up to 30.5 bars, suggested that the use of hydrogen in diving could be beneficial.

The technical progress in these past few years has enabled the manipulation of hydrogen to be more safe. A more systematic study of the effects of this gas has recently been undertaken in Sweden (Adolfson et al., 1984) and in France by COMEX.

From 1983, a French company of diving, COMEX, has carried out several experiments with hydrogen in mice, rats, monkeys and man (Gardette, 1987, 1989). The first experiment in man (HYDRA IV) was performed with 6 divers compressed in a helium-oxygen mixture but who were breathing hydrogen-oxygen by mask at 18, 24 and 30 bars. Significant narcotic sensations (different from those reported with nitrogen) were related by some divers at 24 bars and by all the divers at 30 bars. According to Fructus et al. (1984) and Fructus (1987), the self observation of the divers indicated that at 240 msw and for a $PH_2$ of 2.45 MPa (24.5 bar), the hydrogen narcosis was characterized chiefly by sensory and somesthetic hallucinations which are accentuated by relative sensory rest and affect mood more than intellect. These results, which indicate that pressures of hydrogen higher than 2.4–2.5 MPa (24–25 bars) induce narcosis are consistent with the work of Brauer et al. (1968) and Brauer and Way (1970) which predicted hydrogen narcosis around 2.5 and 3 MPa (25–30 bars).

Hydrogen-oxygen mixture breathing during 30 min at 24 bars and 30 bars induced EEG changes similar to those recorded with other mixtures. However, the narcotic potency of hydrogen was too high to permit its use alone in the mixture (Carlioz et al., 1984; Fructus et al., 1984).

Consequently hydrogen-helium-oxygen mixture was used with partial pressure of hydrogen which did not exceed 25 bars. Experiments were performed at 45 and 50 bars (HYDRA V and VI) with the same compression curve to 45 bars described earlier in this chapter for the helium-nitrogen-oxygen mixtures. The compression was performed with helium till 20 bars, then with hydrogen till 45 bars (54 to 56% of H2). During the compression and stay at 45 bars, no significant tremor was recorded. The EEG changes were similar to those recorded with other mixtures during dives performed with the same compression curve to this depth (between 100 and 300% increase of theta waves, around 80% decrement of alpha waves). The sleep disruptions recorded in six divers were analogous to those recorded in helium-oxygen or helium-nitrogen-oxygen mixtures at pressure of 45 bars. The decrement of the performance in psychomotor and intellectual tests did not exceed 5%. (Table 4).

Table 4. Signs and symptoms of the high pressure nervous syndrome in hydrogen-helium-oxygen mixture ($H_2$ = 50%).

| High preassure Nervous Syndrome Helium-Nitrogen-Oxygen Mixture up to 46 ATA | |
| --- | --- |
| Behavioural Symptoms | Electrophysiological Changes |
| Cognitive impairments less than 5% | EEG: Increase in slow waves, decrease in fast activities SLEEP: Increase in stage 1 and 2 Decrease in stage 3 and 4, REM |

The experiment carried out to 50 bars where hydrogen was introduced from 23 bars to give 50% $H_2$ at 45 bars and 49% at 50 bars, confirmed these results (Rostain et al., 1989). Nevertheless, at 50 bars, tremor suddenly occurred in one subject who developed psychotic disorders a few hours later (Raoul et al., 1988). The appearance of tremors followed by psychotic disorders was found again in two divers during an experiment with hydrogen-oxygen at 30 bars (HYDRA IX) (Rostain et al., 1990). Psychotic disorders have also been reported by Stoudemire et al., (1984) during an experiment using a helium-nitrogen-oxygen mixture and changes in mood have often been observed with helium-oxygen mixtures. These results seem to indicate that high pressure may induce a behavioural syndrome which can be potentiated by the presence of a 'narcotic gas'.

At last during experiment with a hydrogen-oxygen mixture the principal EEG changes consisted of a decrease in all the EEG frequencies (Rostain et al., 1990). This depression in all the frequency bands would lead to 'monorhythmicity' or EEG activities described previously in the *Papio papio* baboons with this mixture (Rostain and Naquet, 1972).

These studies in man indicate that the use of hydrogen at partial pressures higher than 25 bars (2.5 Mpa) may cause problems in some susceptible subjects. This is in agreement with the work of Brauer et al. (1968) and Brauer and Way (1970), which predicted hydrogen narcosis around 25–30 bars.

Nevertheless, the hydrogen-helium-oxygen mixture with a partial pressure of 50% hydrogen for a total pressure of 46 bars (450 m) appears to be the best mixture to use, since most of the neurological symptoms of HPNS, including tremors, are reduced or suppressed in man.

To study the narcotic or toxic effects of hydrogen, experiments were performed in monkeys *(Macaca fascicularis)* with a hydrogen-oxygen mixture, hydrogen-helium-oxygen mixture and also a helium-oxygen mixture. With hydrogen-oxygen mixture experiment were performed up to 100 bars and with hydrogen-helium-oxygen mixture up to 120 bars. Narcotic symptoms occurred in these animals around 100 bars (Gardette and Gortan, 1994). With hydrogen-helium-oxygen mixtures, the occurrence of significant cognitive impairment was recorded between 20 bars to 40 bars deeper than in the helium-oxygen mixture. With a hydrogenated mixture, the behavioural symptoms (motor and locomotor hyperactivity, tremor, myoclonia) were not significantly observed before 80 bars compared with 50 bars in helium-oxygen (Rostain et al., 1994a, 1995).

From these animal studies, a final exposure with hydrogen-helium-oxygen mixture was performed to 70.1 bars (Gardette et al., 1993; Rostain et al., 1994b) by three professional COMEX divers. Compression took 13 d and they stayed 3 d between 65 bars and 67.5 bars with an excursion of a few hours to 70.1 bars. Compression was made with helium between 10 and 20 bars, then with hydrogen between 20 bars and 40 bars and helium again between 40 bars and 71 bars. Tremors were observed from 65 bars and EEG studies indicated a decrease in alpha EEG frequency from 20 bars and an increase in theta activity which reached 200–300% between 65 bars and 70.1 bars which was lower than the values obtained during heliox dives at 61 bars. It was concluded that $H_2$-He-$O_2$ with 2 MPa $H_2$, reduced the clinical symptoms of HPNS. EEG theta activity increased to 65 bars but alpha was still decreased and sleep disturbances persisted. Gardette et al. (1993) concluded that "Hydreliox, not only enhances the diver's effectiveness and working capabilities, but also provides improved safety. By optimization of the following parameters : compression rate, hydrogen partial pressure, selection and training of divers and chamber comfort, pressure of 30 bars to 65 bars can be considered accessible to a large number of professional divers".

COMEX has performed an offshore dive, HYDRA VIII, with a helium-hydrogen-oxygen mixture at 500 m.s.w. (49% $H_2$) with 6 excursion dives in the open sea to 520 m (53 ATA) and one to 530 m (54 ATA); 26 h of underwater work were performed by 6 divers in greater comfort than a helium-oxygen mixture (Gardette, 1989). According to Gardette (1989), man breathing hydrogen will be more efficient, less tired and more comfortable, and thus much safer while working in the water.

The way in which nitrogen and hydrogen work is not yet clearly understood. It is probably related to their narcotic potency, but their action is certainly more complex than a simple antagonism between the narcotic potency and the pressure. Research carried out on anaesthetic and non-anaesthetic substances which interact with the hyperbaric environment show the complexity of anti-HPNS effects (Rowland James et al., 1981; Angel et al., 1984; Smith et al., 1984; Little and Thomas, 1986; Rostain et al., 1986a, b; Little, 1996; Abraini et al., 1998; David et al., 2001; Balon et al., 2002).

## ORIGINS AND MECHANISMS OF HPNS

The neural origins and the mechanisms of HPNS were begun to be understood from the neurochemical studies performed during the last 20 years.

In order to try to explain the events which occur at pressure, several years ago various authors proposed the hypothesis of hypoxia (Bennett, 1966; Chouteau and Imbert, 1971; Hyacinthe et al., 1973) or have postulated that there are disturbances of the membrane structure (Bennett et al., 1975; Miller, 1975) or of the energetic process (Rostain and Charpy, 1976).

Electrophysiological studies have shown that the cerebellum is probably not the origin of hyperbaric tremor (Kaufmann et al., 1978; Fagni, 1987) but that the spinal cord is affected by the hyperbaric environment and could produce, in the spinal animal, rhythmic muscular activities, but which differ from those recorded in the intact animal (Kaufmann et al., 1979; Fagni 1987; Fagni et al., 1982). This suggested a role for the higher nervous centres in the aetiology of hyperbaric tremor.

The muscular spasms and myoclonus without electroencephalographic changes of a paroxystic type (spikes, spike and wave) recorded in the rat and in the baboon *Papio papio* (Rostain et al., 1984a, 1986a) suggest that they have a non-cortical origin. The muscular disturbances could originate in the peripheral nervous system or in the lower part of the brain (spinal cord, brain stem) and could be similar to the non epileptic myoclonus described by Hallett et al. (1977). The results of Kaufmann et al. (1979) in the spinal rat and those of Fagni et al. (1982) on the spinal cat show that the spinal cord plays a role in HPNS. In addition, Brauer et al. (1981) have shown with 5-deoxyglucose that Type I seizures in the mouse (which are similar to muscular seizures in the baboon *Papio papio*, Rostain, 1980) would have a sub-cortical origin found in the ventral region of the thalamus, the posterior hypothalamus and the pontine and medullar reticular formation.

All these studies have demonstrated the complexity of HPNS. The variables of the hyperbaric environment have differential actions on the different nervous structures which react according to their own sensitivity (Rostain, 1980; Bennett and Rostain, 2003). In general, the behavioural and neurophysiological results obtained during experiments carried out both in animals and humans suggest dysfunctions by pressure of numerous neurochemical mechanisms.

Thus in the baboon *Papio papio* the epileptic seizure localized in the posterior region of the cortex in a helium-nitrogen-oxygen mixture, while generalized and of tonic-clonic type in the absence of nitrogen, could be linked to hyperexcitability of this region in relation to a change in the level of gamma amino butyric acid (GABA), an inhibitory neurotransmitter. Similar phenomena have been observed in photosensitive baboons when the level of this neurotransmitter is modified by pharmacological manipulations (Meldrum and Horton, 1974; Meldrum et al., 1979).

Again, in the baboon *Papio papio* changes in cortical excitability recorded during the study of somatic evoked potentials could also be related to disturbances of the GABAergic system, as could cycles of excitability. The decrease in vigilance and the disturbances of sleep suggest the possibility of disruption of the noradrenaline-serotonin system, insofar as it can be implicated in the regulation of vigilance.

Likewise, the increase of tremor could be related to an increase in catecholamines (increase in physiological tremor, Marsden, 1978). The disturbances of the monoamines could be related to behavioural changes and hallucinatory type episodes.

In other respects, the study of reflexes carried out on the baboon *Papio papio* and on man (Fagni et al., 1980; Hugon et al., 1980, 1981; Harris, 1979a and b; Harris and Bennett, 1983) demonstrate a certain number of variations which suggests that there are disturbances of the cholinergic system at least in the peripheral nervous system.

At last in the mouse and rat the use of various anesthetics under pressure has shown contradictory effects. Althrough the doses of all these drugs must be increased under pressure in order to maintain anesthesia, their effect on the HPNS differ (Bailey et al., 1977; Beaver et al., 1977; Green et al., 1977; Halsey, 1982). Methohexitone and Propanidid enhance certain symptoms of HPNS, Thiopentone has no effect, Althesin, Ketamine, Phenobarbitone and Pentobarbitone delay the appearance of seizures and tremors. Another drug, barbituric acid, which has no anesthetic potency, did have an anti-HPNS effect (Rostain et al., 1986b). These results show that, in addition to the physical effect on the membrane, there is a specific effect of anaesthetics or narcotics. When these drugs bind to specific protein sites, it may bring about changes in certain types of neurotransmission. These sites may show differential sensitivity to pressure depending on their physico-chemical characteristics. It should be noted that the anesthetic substances having a marked anti-HPNS action on any of the symptoms such as Ketamine are known to enhance GABA-mediated transmission or antagonize NMDA neurotransmission (Kemp et al., 1987). One can thus deduce that narcotic or anesthetic substances, and by extrapolation nitrogen and hydrogen, reduce certain symptoms of HPNS by increasing the activity of the GABAergic neurotransmission and by lessening or blocking the aspartergic type activity.

Therefore, the study of HPNS mechanisms and neurophysiological effects of compression proceeds via a neuropharmacological and neurochemical approach associated with an electrophysiological study of the nervous structures likely to be implicated in the appearance of HPNS.

## Amino-Acid Neurotransmitters

The results obtained from several experiments have suggested that there are disturbances of GABA mediated neurotransmission.

Benzodiazepine that enhanced GABA activity and is known to be anticonvulsive has been studied at pressure. Flurazepam and diazepam seem to have an anti-SNHP effect (Bichard et al., 1981; Bichard and Little, 1982; Gran et al., 1980; Halsey and Wardley-Smith, 1981); ethosuximide and carbamazepine have no effect.

The increase in GABA level in the rat was studied with drugs which blocked the GABA-transaminase and glial and/or neuronal re-uptake. These drugs were injected intraperitoneally or intracerebroventricularly into rats prior to compression (Rostain et al., 1986a). These drugs reduced the increase of slow waves recorded in the EEG of control rats from 20 bars. However, only one, the sodium valproate also increased the threshold pressure for tremor onset (from 40 to 80 bars) and myoclonus (from 70 to 90 bars). Sodium valproate is known to have complex pharmacological actions, and the effect of this drug on several HPNS symptoms might be the result of these various actions. Indeed, sodium valproate produces increased cortical content of GABA (Chapman et al., 1982). The mechanism includes some inhibition of GABA transaminase (Anlezark et al., 1976; Simler et al., 1973) and some inhibition of succinic-semialdehyde dehydrogenase (Van der Laan et al., 1979) in addition to a possible small stimulation of glutamic acid decarboxylase, an enzyme involved in the synthesis of GABA (Loscher, 1981). More recently, it has been shown that sodium valproate leads to a decrease in the concentration of aspartate in the cortex and hippocampus of the rat brain (Chapman et al., 1982).

Aspartic acid is another putative amino acid neurotransmitter which has an excitatory action. The action of sodium valproate on tremor and myoclonus could be related to its action on aspartate (Rostain et al., 1986). To test this hypothesis, a compound which is an antagonist of excitation produced by dicarboxylic amino acids with preferential activity at N-methyl-D-aspartate preferring receptors was used: 2-amino-7-phosphonoheptanoic acid (2APH), (Perkins et al., 1981). The injection of 2APH increased the pressure threshold for tremor in rats from 40 bars to 80 bars (Meldrum et al., 1983; Wardley-Smith et al., 1986).

These results indicate that GABA does not play an exclusive role in the appearance of HPNS. The injection of the drugs which increase the GABA level reduces the general excitability of the nervous system and could have a beneficial effect on certain symptoms. The enhancement of GABA by a pharmacological agent during *in vitro* studies on the isolated spinal cord (Bichard and Little, 1985) or on the superior cervical ganglion (Little and Thomas, 1986) or on hippocampal slices (Zinebi et al., 1988a, 1988b) support this conclusion by showing that hyperexcitability brought on by pressure can be reduced, but is insufficient to suppress HPNS symptoms.

Changes in neurotransmission of aspartate type brought about by the increase in pressure of the helium-oxygen mixture could be implicated in the occurrence of hyperbaric tremor. Indeed, focal injection of NMDA into the ventrolateral thalamic nucleus and more particularly into the substantia nigra significantly lowered the threshold pressure for tremor by about 30–40% and increased its intensity. In contrast, the injection of 2 APH significantly increased the threshold pressure by about 40–50% and decreased its intensity (Millan et al., 1989, 1990).

Such a change produced by a hyperbaric environment has been confirmed by Fagni et al. (1987) and Zinebi et al. (1990) who have recorded an increase in the sensitivity of the receptors to N-methyl-D-aspartate in hippocampal slices exposed to high pressures of helium. Indeed, this *in vitro* preparation includes excitatory neurotransmission (Schaffer-commissural afferents from hippocampic pyramidal cells CA3), which activates two types of amino acid receptors of pyramidal cells CA1: receptors sensitive both to quisqualate and kainate, and receptors sensitive to NMDA. It also includes inhibitory neurotransmission with receptors sensitive to GABA (pyramidal cells CA1).

These studies have shown that high pressure produced a decrease of excitatory and inhibitory amino acid neurotransmissions which reached 50% at 80–100 bars (8–10 Mpa) (Fagni et al., 1987a; Zinebi et al., 1988a). Similar observations have been reported by Kendig and Grossman (1987) in neuromuscular junction in crustacea. However, an intrinsic hyperexcitability was recorded in the target cell (pyramidal cells CA1) which may induce epileptiform activities (Fagni et al., 1987a).

Application of quisqualate or kainate performed *in vitro* in hippocampal slices show that excitatory responses were not modified under pressure. In the same way, pressure did not modify the inhibitory response obtained by application of GABA or muscinol ($GABA_A$ receptor agonist) (Fagni et al., 1987b; Zinebi et al., 1988b). Similar results have been obtained from the rat superior cervical ganglion *in vitro* (Little and Thomas, 1986). These results suggest that the decreases in excitatory and inhibitory neurotransmission would be related to a decrease in the release of glutamate and GABA and not to a deficit of receptors.

Moreover, the excitatory effects of NMDA on pyramidal cells CA1 were enhanced under pressure (Fagni, 1987). In contrast, the use of specific antagonists of NMDA receptors (5-aminophosphonovalerate, 5APV or 2APH) and the use of substances which enhanced the GABA inhibition (nipecotic acid, muscinol or diazepam) when administered together suppressed the excitatory effects of pressure in hippocampal slices *in vitro* (Fagni et al., 1989; Zinebi et al., 1990).

Consequently the hypersensitivity of NMDA receptor and the decrease of GABA activity recorded under pressure may explain the intrinsic hyperexcitability of CA1 cells (Zinebi et al., 1988b) and the epileptiform activities recorded in rat hippocampal slices *in vitro* (Fagni et al., 1985). Pressure produced an increase in the sensitivity of NMDA receptors and a decrease of GABAergic neurotransmission which led to hyperexcitability.

Similar results have been reported by Grossman and Kendig (1988), in excitatory and inhibitory synapses of isolated nerve-muscle preparation of 2nd and 4th segments of a lobster. These results and the explanation of the effects of pressure on hippocampal cortex could be extended to other structures (neocortex, spinal cord), which have the same excitatory and inhibitory amino acid controls. These results suggest that drugs which enhance GABA neurotransmission and antagonize NMDA activities could reduce HPNS in whole animals.

Recent experiments performed on GABA neurotransmission with the agonist of $GABA_A$ and $GABA_B$ receptors (muscimol and baclofen) or GABA reuptake inhibitor (nipecotic acid), or GABA transaminase inhibitor, (Gamma Vinyl GABA GVG) injected in the substantia nigra, suggested that changes in GABA neurotransmission in the substantia nigra reticulata (SNr) but not in the substantia nigra compacta (SNc), play a crucial role in the control of motor and locomotor hyperactivity (Kriem et al., 1998a; Balon et al., 1999, 2002). Moreover, the administration of NMDA receptor antagonists (2-APH) into the SNr reduced the LMA (Kriem et al., 1999) and the myoclonia while the NMDA receptor agonist (PDA) increase the myoclonia. The administration of the glutamate uptake inhibitor L-trans-PDC in the substantia nigra attenuated the locomotor hyperactivity. These results, obtained at the level of SNr, suggest that high helium pressure induced a decrease in GABAergic inputs to the SNr (Kriem et al., 1998a) with a concomitant more severe reduction in activity of the glutamatergic inputs in the SNr, so that the main effect would be a reduction in activity of the GABAergic output from the SNr to the thalamus. It is likely that local infusion into the SNr of L-trans-PDC, which selectively increases glutamate level by preventing its uptake could have restored the neurochemical balance between GABA and glutamate neurotransmissions and thus attenuated the development of LMA (Abraini et al., 1999). This coincides with data obtained by Millan et al. (1989) that has shown that injection into the SNr of NMDA, a glutamate receptor agonist, had a sedative effect. Furthermore, the works of Abraini et al. (1999) confirm that LMA and myoclonia have different neural origins and further suggest that glutamate in the SNr would be involved in the development of myoclonia. Several authors have shown the implication of other neural structures such as pedunculopontine nucleus, ventrolateral thalamic nucleus, red nucleus, entopeduncular nucleus (Milan et al., 1989, 1990, 1991) and the globus pallidus (Kriem et al., 1999; Abraini et al., 1999; Darbin et al., 2000) in the development of the different symptoms of HPNS. Furthermore, experiments using non-NMDA receptor antagonists indicate that they are sometimes effective against tremors but generally ineffective against seizures (Wardley -Smith and Meldrum, 1984; Wardley-Smith et al., 1987; Pearce et al., 1994) and that non-NMDA receptor agonists administered in substantia nigra or the globus pallidus reduce LMA (Kriem et al., 1999).

However, the microdyalisis studies in the striatum of free moving rats have shown an increase of glutamate concentrations under helium pressure. Darbin et al. (2001) and Vallée et al. (2009) indicated that at least glutamate is not reduced in this structure

All these studies demonstrated the complexity of the effects of pressure with helium on the central nervous system and did not exclude that disruption of other neurotransmissions by pressure could be implicated in the occurrence of some symptoms of HPNS.

### Monoamine Neurotransmitters

Cerebral monoamines have been implicated in the occurrence of some HPNS symptoms by several authors. Thus Brauer et al. (1978) and Koblin et al. (1980) decreased the pressure threshold for hyperbaric epileptic seizure by the depletion of monoamines. Bowser-Riley et al. (1982) affected HPNS by administering L-DOPA (precursor of dopamine) systemically or by intracerebroventricularly injecting neurotroxins of catecholaminergic neurons such as 6-hydroxydopamine (6-0HDA) or of indolaminergic neurons such as 5–6 and 5–7 dihydroxytryptamine (DHT). However, increasing dopamine level by various pharmacological agents did not increase the pressure threshold of the different symptoms of HPNS (Brauer, 1975; Halsey and Wardley Smith, 1981).

Measurements of monoamines have been carried out on mice brain homogenates at the end of a pressure experiment (Koblin et al., 1980; Daniels et al., 1981; Aanderud and Broch, 1987). However, it is difficult to draw conclusions from these measurements as :

- the measurements include several variables resulting from the effects of compression and decompression;
- the measurements were carried out on mouse brain homogenates, but neurophysiological experiments have shown that HPNS is the result of diverse actions on different structures of the nervous system (Rostain, 1980; Brauer et al., 1981; Rowland-James et al., 1981); data reported by McLeod et al. (1988) demonstrated the opposite effects of pressure in two different structures of the brain. Consequently, analysis performed on the whole brain may show no changes in overall levels if some struc-

tures have increased monoamines, for example, and others a comparable decrease.

In addition, these methods  do not take into account disorders of neurotransmission either pre or post -synaptically.

The use of electrochemical techniques developed *in vivo* for animals (Adams, 1978; Gonon et al., 1978; Marsden, 1979) enables continuous measurement of monoamines at a predetermined point in the nervous system. A multifiber carbon electrode developed by Forni (1982) which is sensitive to dopamine and allows *in vivo* measurement for several weeks as described by Forni and Nieoullon (1984) and El Ganouni et al. (1987) has been implanted in the striatum of hamsters and rats. This structure is rich in dopaminergic endings and is implicated in the control of motor, locomotor and some cognitive processes that are disrupted by pressure. The differential pulse voltammetry (DPV) studies demonstrated an increase in extracellular dopamine level which occurred around 30 bars (Forni and Rostain, 1989; Abraini and Rostain, 1991; Rostain and Forni, 1995). The increase of dopamine at 80 or 90 ATA (8 or 9 MPa) was around 25–30% with a return to basal value during the decompression. This increase is independent of changes of temperature or partial pressure of oxygen, which occur in the hyperbaric experiments (Forni and Rostain 1989, Rostain and Forni 1995). Consequently, the increase of extracellular DA in the caudate nucleus is pressure-dependent; it was also found in the nucleus accumbens (Abraini and Rostain, 1991d). This increase is correlated with the locomotor and motor hyperactivity (LMA) described in free-moving animals (Tomei et al., 1991; Abraini and Rostain, 1991a). Similar changes have been reported in monkeys exposed to high pressure of a helium-oxygen mixture (Forni and Rostain, 1984; Cosson et al., 1989).

Neuropharmacological studies have indicated that the pressure increased dopamine level is related to an increase of dopamine synthesis and release (Abraini et Rostain, 1991b and c; Fechtali et al., 1994). Moreover, studies with agonists and antagonists of D1 and D2 receptors (Abraini and Rostain, 1991c; Abraini et al., 1992) have indicated a different action of pressure on D1 and D2 receptors. Abraini and Fechtali (1992) suggested that high pressure would produce interactive disorders in the link mechanisms of regulation between the dopaminergic and amino-acidergic systems.

Studies using microdialysis in free-moving rats have confirmed the pressure induced DA striatal increase (Requin and Risso, 1992; Darbin et al., 1997a, b) and the implication of  D1, D2 and NMDA receptor in this change (Darbin et al., 1999). The development of LMA requires striatal D1, D2 and NMDA receptor activities (Darbin et al., 1999, 2000) and pressure enhanced NMDA activity in the striatum and also in the globus pallidus which appeared to be implicated in the occurrence of these motor disturbances and to be involved in myoclonia (Kriem et al., 1999; Darbin et al., 2000).

The GABA gamma amino butyric acid (GABA) neurotransmission is one of the neurotransmission implicated in these changes as demonstrated by the use of agonist drugs of GABAa or GABAb receptors when injected in the substantia nigra reticulata or compacta (Kriem et al., 1998; David et al., 2001; Balon et al., 2002 b, c). These studies indicated that the activation of GABAb receptors in the SNr decreased both DA and LMA by the inhibition of the nigro-striatal pathway (NSP) and the thalamo-cortical pathway (TCP) and suggested an implication of these receptors in the regulation of the NSP and in the development of LMA. In other words, the activation of GABAa receptors directly inhibited the NSP but also induced a desinhibition of the nigro-thalamic pathway (NTP) which produced an activation of the TCP and consequently a decrease of striatal DA release without LMA disappearance. Consequently, at pressure, the results obtained with the agonists of GABAa and GABAb receptors, suggest a change in the sensibility of the GABAa and GABAb receptors in the SNr, and in the SNc, with a greater response of the GABAa postsynaptic receptors on GABAergic nigrothalamic pathway in the SNr and in the GABAergic interneurones of the SNc. Helium pressure would act by a stimulation of GABAa receptors on GABA neurones of the SNr which induced a desinhibition of DA neurons of the NSP and glutamatergic neurons of the TCP (Balon et al., 2002c) resulting in the increase of DA release.

Other monoamines such as serotonin (5-HT) could be implicated in HPNS and some behavioural symptoms in rats seem to be related to activation of the 5-HT1a receptor subtype (Wardley-Smith et al., 1990). David et al. (1988) suggest the involvement of serotonin in the enhancement of rat spinal excitability under hyperbaric pressure. The hyperbaric alteration in spinal excitability in rats pretreated with the classical 5-HT receptor antagonist drug metergoline was suppressed.

Microdialysis studies have indicated that helium pressure induces an increase of serotonin in the striatum (Darbin et al., 1997b). Studies performed on the interaction serotonin-dopamine at the level of the nigro striatal or meso-limbic pathway have demonstrated that the 5-HT1B and 5-HT3 receptors are activated by pressure and are implicated in part in the DA increase and LMA development (Kriem et al., 1995, 1996a). Moreover, the antagonists of these receptors blocked the occurrence of myoclonia. At the level of nucleus accumbens, the use of 5-HT2-C receptors agonist has indicated that this receptor plays a crucial role in the development of LMA and in the pressure-induced increase in DA accumbens release. In contrast, the administration of 5-HT2A receptors antagonist had no effect on LMA and DA increase but reduce myoclonia (Kriem et al., 1996b; 1998b). These results suggest that helium pressure may simultaneously induce an increase in 5-HT transmission at the level of 5-HT2A receptors which would be located at the post synaptic level on GABA neurons and a decrease in 5-HT transmission at the level of 5-HT2C receptors that would be an autoreceptor (Kriem et al., 1998b).

This research have increased the data on the origin and mechanisms of HPNS. Helium pressure disrupts the neurochemical interactions between the systems and the structures of the central nervous system, changes specifically some neurotransmission by acting on synthesis or receptor activities. Daniels et al. (1998) have shown in Xenopus oocytes, that NMDA receptors are sensitive to pressure and Mor and Grossman (2006; 2007) have reported that isolated NMDAR response was augmented at high pressure. Helium pressure could have an action on ion channels (Etzion and Grossman, 1999).

It is evident that to increase our understanding, more basic research must be done *in vivo* or *in vitro* on the effect on pressure on the central nervous system at neurotransmitter and receptor levels and at the cellular and molecular levels.

## REFERENCES

Aanderud, L. and O.J. Broch. 1987. The effect of high pressure on the kinetics of monoamine transmitters in the rat brain. Undersea Biomed. Res. 14: 85–91.

Abraini, J.H. and J.C. Rostain. 1991a. Pressure-induced striatal dopamine release correlates hyperlocomotor activity in rats exposed to high pressure. J. Appl. Physiol. 71: 638–643.

Abraini, J.H. and J.C. Rostain. 1991b. Effects of the administration of a-methyl-p-tyrosine on the striatal dopamine increase and the behavioral motor disturbances in rats exposed to high pressure. Pharmacol. Biochem. Behav. 40: 305–310.

Abraini, J.H. and J.C. Rostain. 1991c. Effects of helium-oxygen pressure on dopaminergic receptors. Undersea Biomed. Res. 18 (Suppl.) 55.

Abraini, J.H. and J.C. Rostain. 1991d. Dopamine increase in the nucleus accumbens of rats exposed to high pressure. Neuroreport 2: 233–235.

Abraini, J.H. and T. Fechtali. 1992. A hypothesis regarding possible interactions between the pressure-induced disorders in dopaminergic and amino-acidergic transmission. Neurosci. and Biobehav. Rev. 16: 597–602.

Abraini, J.H. and T. Fechtali and J.C. Rostain. 1992. Pressure reversed extracellular striatal dopamine decrease produced by D1 receptor agonist SKF 38393, and D2 receptor agonist LY 171555, but failed to change the effect of the activation of both Dl and D2 receptors. Neuroscience 50: 395–402.

Abraini, J.H. and J.C. Rostain and B. Kriem. 1998. Sigmoidal compression rate-dependence of the narcotic potency of inert gases in rats: implication for lipid vs protein theories of inert gas action in the central nervous system. Brain Res. 808: 300–304.

Abraini, J.H. and B. Kriem and J.C. Rostain. 1999. Administration of the glutamate uptake inhibitor L-trans-PDC in the globus pallidus and the substantia nigra reticulata, but not in the striatum attenuates the psychostimulant effect of high helium pressure on locomotor activity in the rat. Neurosci. Res. 35: 273–279.

Adams, R.N. 1978. *In vivo* electrochemical recording: A new neurophysiological approach. Trends Neurosci. 1: 160–163.

Adolfson, J. and H. Ornhagen and D.H. Ingvar. Psychophysiological performance during breathing of 1.3 MPA (13 ATA) Hydrox. Pp. 61–73. *In*: T. Nome, G. Susbielle, M. Comet, M.Jacquin and R. Sciarli (eds.). 1984. Proceeding of Xth Congress EUBS. EUBS, Marseille. France.

Aquadro, C.F. and J. Chouteau. Problems of extreme duration in open sea saturation exposure. Pp. 98–108. *In*: C.J. Lambertsen (ed.). 1967. Underwater Physiology III. Williams and Wilkins, Baltimore, USA.

Angel, A. and M.J. Halsey, H.J. Little., B.S. Meldrum, J.A.S. Ross, J.C. Rostain and B. Wardley-Smith. 1984. Specific effects of drugs at pressure: animal investigations. Phil. Trans. R. Soc. Lond. B304: 85–94.

Anlezark, G. and R.W. Horton, B.S. Meldrum and M.C.B. Samaya. 1976. Anticonvulsivant action of ethanolamine-a-sulphate and di-n-propylacetate and the metabolism of y-aminobutyric acid (GABA) in mice with audiogenic seizures. Biochem. Pharmacol. 25: 413–417.

Balon, N. and B. Kriem and J.C. Rostain. 1999. Implication of GABA receptors in regulation of the nigro striatal pathway of rats exposed to helium-oxygen pressure. Undersea and Hyperbaric Med. 26 (suppl): 82.

Balon, N.and B. Kriem, E. Dousset, M. Weiss and J.C. Rostain. 2002a. Opposing effects of narcotic gases and pressure on the striatal dopamine release in rats. Brain Res. 947: 373–379.

Balon, N. and B. Kriem, M. Weiss and J.C. Rostain. 2002b. GABAergic modulation in the substantia nigra of the striatal dopamine release and of the locomotor activity in rats exposed to helium pressure. Brain Res. 948: 82–92.

Balon, N. and B. Kriem, M. Weiss and J.C. Rostain. 2002c. Indirect presynaptic modulation of striatal dopamine release by GABA B receptors in the rat substantia nigra. Neuroscience Lett. 325: 33–36.

Bailey, C.P. and C.J. Green, M.J. Halsey and B. Wardley-Smith. 1977. High pressure and 'intravenous steroid anaesthesia in rats. J. Appl. Physiol. 43: 183–188.

Barthélémy, L. and A. Belaud. 1972. Constatations physiologiques et physiopathologiaques faites sur un poisson (*Anguilla anguilla* L.) en conditions hyperbares. Bull. Med. Sub. Hyp. 8: 33.

Barthelemy, L. and A. Belaud and A. Saliou. A study of the specific action of "per se" hydrostatic pressure on fish considered as a physiological model. Pp. 641–649. *In*: A.J. Bachrach and M.M. Matzen (eds.). 1981. Underwater Physiology VII. Undersea Medical Society, Bethesda MD, USA.

Beaver, R.W. and R.W. Brauer and S. Lasher. 1977. Interaction of central nervous system effects of high pressure with barbiturates. J. Appl. Physiol. 43: 221–229.

Behnke, A.R. and R.M. Thomas and E.P. Motley. 1935. The psychologic effects from breathing air at 4 atmospheres pressure. Am. J. Physiol. 112: 554–558.

Belaud, A. and L. Barthelemy and C. Peyraud. 1977. Temperature and per se hydrostatic pressure reversal of pentobarbital anaesthesia in fish. J. Appl. Physiol. Respirat. Environ. Exercise Physiol. 42 : 329–334.

Bennett, P.B. 1966. The Aetiology of Compressed Air Intoxication and Inert Gas Narcosis. International Series of Monographs in Pure and Applied Biology; Zoology Division, Vol. 31, Pergamon Press, Oxford, U.K.

Bennett, P.B. Performance impairment in deep diving due to nitrogen, helium, neon and oxygen. Pp. 327–340. *In*: C.J. Lambertsen (ed.). 1967. Underwater Physiology III. Williams and Wilkins, Baltimore, USA.

Bennett, P.B. and E.J. Towse. 1971. The High Pressure Nervous Syndrome during a Simulated Oxygen-helium Dive to 1500 fl. Electroenceph. clin. Neurophysiol. 31: 383–393.

Bennett P.B. and H. Schafstall. The value of TRIMIX 5 to control HPNS. Pp. 101–115. *In*: L. Y. Chong and K. Shida (eds.). 1990. Man in the Sea, Vol 1 Best Publications. San Pedro, CA, USA.

Bennett, P.B. and J.C. Rostain. The High Pressure Nervous Syndrome. Pp. 323–357. *In*: A.O. Brubakk and T.S. Neuman (eds.). 2003. Bennett and Elliott's Physiology and Medicine of Diving 5th edit. Saunders, London.

Bennett, P.B. and G.D. Blenkarn, J. Roby and D. Youngblood. 1974. Suppression of the High Pressure Nervous Syndrome in Human Deep Dives by He-N$_2$-O$_2$. Undersea Biomed. Res. 1: 221–237.

Bennett, P.B. and S. Simon and Y Katz. High pressure of inert gases and anesthesia mechanisms. Pp. 367–403. *In*: B.R. Fink (ed.). 1975. Molecular Mechanism of Anesthesia. Progress in Anesthesiology. Raven Press, New-York, USA.

Bennett, P.B. and R. Coggin and M. Mcleod. 1982. Effect of compression rate on use of trimix to ameliorate HPNS in Man to 686 m (2250 ft). Undersea Biomed. Res. 9: 335–351.

Bichard, A.R. and H.J. Little and W.D.M. Paton. 1981. The involvement of GABA in the high pressure neurological syndrome (HPNS). Br. J. Pharmacol. 74: 221.

Bichard, A.R. and H.J. Little. 1982. Drugs that increase y-aminobutyric acid transmission protect against the High Pressure Neurological Syndrome. Br. J. Pharmacol. 76: 447–452.

Bichard, A.R. and H.J. Little. 1985. Effects of high pressure helium on gamma amino butyric acid release from isolated spinal cord. J. Neurochemistry 44: 999–1005.

Bjurstedt, T. and G. Severin. 1948. The prevention of decompression sickness and nitrogen narcosis by the use of hydrogen as a substitute for nitrogen. Milit. Surg. 103: 107–116.

Bowser-Riley, F. and J.A. Dobbie, W.D.M. Paton and E.B. Smith. 1982. A possible role for monoaminergic inhibition in the high pressure nervous syndrome. Undersea Biomed. Res. 9 (Suppl): 32–33.

Brauer, R.W. The high pressure nervous syndrome: animals. Pp. 231–247. *In*: P.B. Bennett and D.H. Elliott (eds.). 1975. Physiology and medicine of diving and compressed air work. Bailliere Tindall, London, UK.

Brauer, R.W. and R.O. Way. 1970. Relative narcotic potencies of hydrogen, helium, nitrogen and their mixtures. J. Appl. Physiol. 29: 23–31.

Brauer R.W. and R.O. Way and T.A. Perry. Narcotic effects of helium and hydrogen and hyperexcitability phenomenon at simulated depths of 1500 to 4000 ft of sea water. Pp. 241–255. *In*: B.R. Fink (ed.). 1968.Toxicity of Anesthetics. Williams and Wilkins, Baltimore, USA.

Brauer, R.W. and S. Dimov, X. Fructus, A. Gosset and R. Naquet. 1969. Syndrome neurologique et électrographique des hautes pressions. Rev. Neurol. 121, 264–265.

Brauer, R.W. and R.O. Way, M.R. Jordan and D.E. Parrish. Experimental studies on the high pressure nervous syndrome in various mammalian species. Pp. 487–500. *In*: C.J. Lambertsen CJ (ed.). 1971. Underwater Physiology IV. Academic Press. New York, USA.

Brauer, R.W. and S.M. Goldman, R.W. Beaver and M.B. Sheehan. 1974. N2, H2 and N2-O antagonism of high pressure neurological syndrome in mice. Undersea Biomed. Res. 1: 59–72

Brauer, R.W. and R.W. Beaver and M.E. Sheehan. The role of monoamine neurotransmitters in the compression rate dependence of HPNS convulsions. Pp. 49–59. *In*: C.W. Shilling and M.W. Beckett (eds.). 1978. Underwater Physiology VI. Fed. Am. Socs. Exp. Biol. Bethesda, MD, USA

Brauer, R.W. and W.M. Mansfield, R.W. Beaver and H.W. Gillen. The HPNS as a composite entity-consequences of an analysis of the convulsion stage. Pp. 391–399. *In*: A.J. Bachrach and M.M. Matzen (eds.). 1981. Underwater Physiology VII. Undersea Med. Soc., Bethesda, MD, USA.

Cabarrou, P. and H. Harmann, K.H. Weiner, P. Alinat and H.D. Fust. 1966. Introduction à la physiologie de "Homo aquaticus". Presse Med. 74: 2771–2773.

Carlioz, M. and M.C. Gardette-Chauffour, J.C. Rostain and B. Gardette. Hydrogen narcosis: psychometric and neurophysiological study. Pp. 97–109. *In*: T. Nome, G, Susbielle, M. Comet, M. Jacquin and R. Sciarli (eds.). 1984. Proceeding of Xth Congress of European Undersea Bio-medical Society, EUBS. Marseille.

Case, E.M. and J.B.S. Haldane. 1941. Human physiology under high pressure. J. Hyg. Camb. 41: 225–249.

Cattel, M.C.K. 1936. The physiological effects of pressure. Biol. Rev. 11: 441–475.

Chapman, A. and P.E. Keane, B.S Meldrum, J. Simiand and J.C. Vernieres. 1982. Mechanisms of anticonvulsant action of valproate. Prog. Neurobiol. 19: 315–359.

Chouteau, J. and G. Imbert. 1971. La limitation hypoxique de la plongée profonde de longue durée. Maroc Med. 52: 229–236.

Cosson, P. and J. Etienne and J. Gillard. Voltammetry in the striatum of chronic implanted monkey during exposure to hydrogen-oxygen mixtures at 51 ATA: Detection of dopamine. Pp. 239–249. *In*: J.C. Rostain, E. Martinez and C. Lemaire (eds.). 1989. High Pressure Nervous Syndrome, 20 Years Later. ARAS-SNHP Publications, Marseille, France.

Daniels, S. and A.R. Green, D.D. Koblin, R.G. Lister, H.J. Little, W.D.M. Paton, F. Bowser Riley, W.S.G. Sha and E.B. Smith. Pharmacological investigation of the High Pressure Neurological Syndrome: Brain Monoamine concentrations. Pp. 329–336. *In*: A.J. Bachrach and M.M. Matzen (eds.). 1981. Underwater Physiology VII. UMS, Bethesda, MD, USA.

Daniels, S. and R. Roberts and N. William. Effects of high pressure on post synaptic ionotrophic receptors. Pp. 22–31. *In*: P.B. Bennett, I. Demchenko and R.E. Marquis (eds.). 1998. High Pressure Biology and Medicine: University of Rochester Press, New York, USA.

Darbin, O. and J.J. Risso and J.C. Rostain. 1997a. A new system analysis of motor and locomotor activities associated with a midrodialysis study of pressure-induced dopamine increase in rats. Physiol. Behav. 62: 367–371.

Darbin, O. and J.J. Risso and J.C. Rostain. 1997b. Pressure induces striatal serotonin and dopamine increases: a simultaneous analysis in free moving microdialysed rats. Neuroscience Lett 238: 69–72.

Darbin, O. and J.J. Risso and J.C. Rostain. 1999. The full expression of locomotor and motor hyperactivities induced by pressure requires both striatal dopaminergic and NMDA receptors activities in rat. Neuroscience Lett. 267: 149–152.

Darbin, O. and J.J. Risso and J.C. Rostain. 2000 High pressure enhanced NMDA activity in the striatum and the globus pallidus: relationships with myoclonia and locomotor hyperactivity in rat. Brain Res. 852: 62–67.

Darbin, O. and J.J. Risso and J.C. Rostain. 2001. Helium-oxygen pressure induces striatal glutamate increase: a microdialysis study in freely-moving rats. Neuroscience Lett. 297: 37–40.

David, H.N. and N. Balon, J.C. Rostain and J.H. Abraini. 2001. Nitrogen at raised pressure interacts with the GABAA receptor to produce its narcotic pharmacological effects in the rat. Anesthesiology 95: 921–7.

David, J.M. and J.J. Risso and J. Pellet. 1988. Involvement of serotonin in the enhancement of the ra t spinal excitability by high pressures. Undersea Biomed. Res. 15: 1–11.

Ebbecke, U. 1936. Uber das verhalten des Zentralnervensystems (Ruckenmarksfrosh) unter Einwirkung hoher Drucke. Pflug. Arch. 237: 785–789.

Edel, P.O. 1971. Dog breathes $H_2$-$O_2$ in 1000 ft dive. Ocean Ind., May 21–22.

Edel, P.O.and J.M. Holland, C.L. Fischer and W.P. Fife. 1972. Preliminary studies of hydrogen-oxygen breathing mixtures for deep sea diving. Pp. 257–270. *In*: Proc Symp The Working Diver,. Marine Technology Society. Washington, DC, USA.

El Ganouni, S. and C. Forni and A. Nieoullon. 1987. In vitro and in vivo characterization of the properties of a multifiber carbon electrode allowing long-term electrochemical detection of dopamine in freely moving animals. Brain Res. 404: 239–256.

Etzion, Y. and Y. Grossman. 1999. Spontaneous NA+ and CA2+ spike firing of cerebellar Purkinje neurons at high pressure. Pflugers Arch. 437: 276–284.

Fagni, L. 1987. Effets des hautes pressions d'hélium sur l'excitabilité cellulaire et la transmission synaptique dans le système nerveux central. DSc Thesis, University of Mediterranée, Marseille, France.

Fagni, L. and M. Hugon and J.C. Rostain. 1980. Facilitation des potentiels évoqués somesthésiques (PES) en plongée pfofonde a saturation. J. Physiol. (Paris) 1980, 76: 17A.

Fagni, L. and M. Weiss, J. Pellet and M. Hugon. 1982. The possible mechanism of the pressure-induced motor disturbances in the cat. Electroenceph. clin. Neurophysiol. 53: 590–601.

Fagni, L. and B. Soumireu-Mourat, E. Carlier and M. Hugon. 1985. A study of spontaneous and evoked activity in the rat hippocampus under helium-oxygen high pressure. Electroenceph. clin. Neurophysiol. 60: 267–275.

Fagni, L. and F. Zinebi and M. Hugon. 1987a. Evoked potential changes in rat hippocampal slices under helium pressure Exp. Brain Res. 65: 513–519.

Fagni, L. and F. Zinebi and M. Hugon. 1987b. Helium pressure potentiates the N-methyl-D-aspartate and D, L-homocysteate induced decreases of field potentials in the rat hippocampal slice preparation. Neurosciences letters. 81: 285–290.

Fagni, L. and F. Zinebi, M. Hugon and J.C. Rostain. 1989. Hypothèses sur les mécanismes du syndrome nerveux des hautes pressions. Rev. Scient. Tech. Defence 2: 115–128.

Fechtali, T. and J.H. Abraini, B. Kriem and J.C. Rostain. 1994. Pressure increases de novo synthesized striatal dopamine release in free moving rats. Neuroreport 5: 725–728.

Fife, W.P. 1979. The Use of Non Explosive Mixtures of Hydrogen and Oxygen for Diving. Texas A & M University, Hyperbaric Lab. TAMU-SG-70-201. College Station, USA.

Fife, W.P. 1987. The toxic effects of hydrogen-oxygen mixtures. Pp.13–23. *In*: R.Brauer (ed.). 1987. Hydrogen as a Diving Gas: 33rd Undersea and Hyperbaric Medicine Workshop. Undersea and Hyperbaric Medical Society. Bethesda, MD, USA.

Forni, C. 1982. Realization of a new multifiber electrochemical device allowing continuous *in vivo* measurements of neuromediators. J. Neurosci. Meth. 5: 167–171.

Forni, C. and A. Nieoullon. 1984. Electrochemical detection of dopamine release in the striatum of freely moving hamsters. Brain Res. 297: 11–20.

Forni, C. and J.C. Rostain. Influence des hautes pressions de mélange hélium-oxygène sur la libération de dopamine. Pp. 31–36. *In*: T. Nome, G. Susbielle, J. Comet, M. Jacquin and R. Sciarli (eds.). 1984. Proc Xth Congress of European Undersea Biomedical Society EUBS. Marseille, France.

Forni, C. and J.C. Rostain. 1989. Effect of helium-oxygen pressure on dopamine detected *in vivo* in the striatum of hamsters. J. Appl. Physiol. 67: 1617–1622.

Fructus, X.R. Hydrogen, pressure and HPNS. Pp. 125–138. *In*: R.W.Brauer (ed.). 1987. Hydrogen as a Diving Gas. Undersea and Hyperbaric Medical Society, Bethesda, MD, USA.

Fructus, X. and J. Chouteau. 1963. Aspects physiologiques de la vie sous pression. Médecine de la plongée. 1–17

Fructus, X. and H.G. Delauze. 1968. Aspects opérationnels de la plongée profonde. Rev. Pysiol. Sub. Med. Hyp. 1: 50–58.

Fructus, X. and J.C. Rostain. HPNS: A clinical study of 30 cases. Pp. 3–8. *In*: C.W. Shilling and M.W. Beckett (eds.). 1978. Underwater Physiology VI. FASEB, Bethesda, MD, USA.

Fructus, X. and R. Naquet, A. Gosset, P. Fructus and R.W. Brauer. 1969. Syndrome Nerveux des Hautes Pressions. Marseille Med. 6, 509–512.

Fructus, X. and C. Agarate, R., Naquet and J.C. Rostain. Postponing the "High Pressure Nervous Syndrome" (HPNS) to 1640 feet and beyond. Pp. 21–33. *In*: C.J. Lambertsen (ed.). 1976. Underwater Physiology V. FASEB, Bethesda, Maryland, USA.

Fructus, X. and B. Gardette, M. Carlloz and Y. Giran. 1984. Hydrogen narcosis pp. 87-96. *In*: T. Nome, G. Susbielle, M. Comet, M. Jacquin and R. Sciarli (eds.). 1984. Proceeding of Xth Congress EUBS. EUBS. Marseille, France.

Gardette, B. Hydra IV and Hydra V: human deep hydrogen dives 1983–1985. Pp. 109–117. *In*: R.W. Brauer (ed.). 1987. Hydrogen as a diving gas. UHMS publication 69 (WS-HYD) Bethesda, MD, USA.

Gardette, B. Compression procedures for mice and human hydrogen deep diving: COMEX HYDRA program. Pp. 217–231. *In*: J.C. Rostain, E. Martinez and C. Lemaire (eds.). 1989. High Pressure Syndrome 20 Years Later. ARAS-SNHP Publications. Marseille, France.

Gardette, B. and J.C. Rostain. 1981. Compression technique in deep dives for man and animals in He-N2-O2 mixture. Undersea Biomed. Res. 8 (suppl.): 15–16.

Gardette, B. and C. Gortan. Mice and monkeys deep dives in heliox, hydrox and hydreliox gas mixtures—synthesis of COMEX "Hydra" programme. Pp. 173–184. *In*: P.B. Bennett and R.E. Marquis (eds.). 1994. Basic and applied high pressure biology. University Press of Rochester, New-York, USA.

Gardette, B. and J.Y. Massimelli, M. Comet, C. Gortan and H.G. Delauze. 1993. Hydra 10: a 701 msw on shore dive on "hydreliox". Pp. 32–37. *In*: R.E.Reinertsen, A.O. Brubakk and G.Bulstad (eds.). Proceeding of XIXth EUBS Congress EUBS. Trondheim, Norway.

Gonon, F. and R. Cespuglio, J.L. Ponchon, J. Buda, M. Jouvet, R.N. Adams and J.F. Pujol. 1978. Mesure électrochimique continue de la libération de dopamine réalisée *in vivo* dans le neostratum du rat. CR Acad. Sci. (Paris) 286: 1203–1206.

Gran, L. and R. Coggin and P.B. Bennett. 1980. Diazepam under hyperbaric conditions in rats. Acta Anaesth. Scand. 24: 407–411.

Green, C.J. and M.J. Halsey and B. Wardley-Smith. 1977. Possible protection against some of the physiological effects of high pressure. J. Physiol. London 267: 46–47.

Grossman, Y. and J.J. Kendig. 1988. Synaptic integrative properties at hyperbaric pressure. J. Neurol. 60: 1497–1511.

Hallet, M. and D. Chadwick, J. Adam and C.D. Marsden. 1977. Reticular reflexe myoclonus. A physiological type of human post hypoxic myoclonus. J. Neurol. Neurosurg. Psychiat. 40: 253–264.

Halsey, M.J. 1982. Effects of High Pressure on the Central Nervous System. Physiological Rev. 62: 1341–1377.

Halsey, M.J. and B. Wardley-Smith. 1981. High pressure neurological syndrome: Do anti-convulsants prevent it? Br. J. Pharmacol. 72: 502–503.

Halsey, M.J. and E.L. Eger, D.W. Kent and P.J. Warne. High pressure studies of anaesthesia. Pp. 353–361. *In*: B.R. Fink (ed.). 1975. Molecular Mechanisms of Anesthesia (Progress in Anesthesiology), Raven Press New-York, USA.

Harris, D.J. 1979a. Observations on the knee-jerk in Oxy-Helium at 31 and 43 bar. Undersea Biomed Res. 6: 55–74.

Harris, D.J. 1979b. Hyperbaric hyperreflexia : tendon-jerk and Hoffmann reflexes in man at 43 bars. Electroenceph. Clin. Neurophysiol. 47: 680–692.

Harris, D.J. and P.B. Bennett. 1983. Force and duration of muscle twitch contractions in humans at pressures up to 70 bar. J. Appl. Physiol.: Respirat. Environ. Exercise Physiol. 54: 1209–1215.

Hugon, M. and L. Fagni and J.C. Rostain. 1980. Cycle d'excitabilité des reflexes monosynaptiques et mécanismes cholinergiques. Observations de physiologie hyperbare chez les primates. J. Physiol. (Paris) 76: 19a.

Hugon, M. and L. Fagni, J.C. Rostain and K. Seki. Somatic evoked potentials and reflexes in monkey during saturation dives in dry chamber. Pp. 381–390. *In*: A.J. Bachrach and M.M. Matzen (eds.). 1981. Underwater Physiology VII. Undersea Medical Society, Bethesda, MD. USA.

Hugon, M. and L. Fagni and K. Seki. Deep sea diving: human performance and motor control under hyperbaric conditions with inert gas. Pp. 829–849. *In*: J.L. Desmedt (ed.). 1983. Motor control Mechanisms in Health and Disease. Raven Press. New-York, USA.

Hyacinthe, R. and B. Broussolle and D. Desbruyeres. 1973. Affinité de l'hémoglobine pour l'oxygène chez l'homm en milieu hyperbare. Bull. Med. Sub. Hyp. 10: 44–46.

Johnson, F.H., D.E.S. Brown and D.A. Marsland. 1942a. Pressure reversal of the action of certain narcotics. J. Cell. Physiol. 20: 269–276.

Johnson, F.H. and D.E.S. Brown and D.A. Marsland. 1942b. A basic mechanism in the biologic effects of temperature, pressure and narcotics. Science 95: 200–203.

Kaufmann, P.G. and P.B. Bennett, and J.C. Farmer. 1978. Effect of cerebellar ablation on the high pressure nervous syndrome in rats. Undersea Biomed. Res. 5: 63–70.

Kaufmann, P.G. and C.C. Finley, P.B. Bennett and J.C. Farmer. 1979. Spinal cord seizures elicited by high pressure of helium. Electroenceph. clin. Neurophysiol. 47: 31–40.

Kemp, J.A. and A.C. Foster and E.H.F. Wong. 1987. Non-competitive antagonists of excitatory amino acid receptors. TINS 10: 294–298.

Kendig J.J. and Y. Grossman. How can hyperbaric pressure increase central nervous system excitability? Pp. 159–169. *In*: H.W. Jannash, R.E. Marquis and A.M. Zimmerman [eds.] 1987. Current Perspectives in High Pressure Biology. Academic Press. New York, USA.

Koblin, D.D. and H.J. Little, A.R. Green, S. Daniels, E.G. Smith and W.B.M. Paton. 1980. Brain monoamines and the high pressure neurological syndrome. Neuropharmaco. 19: 1031–1038.

Kriem, B. and J.C. Rostain and J.H. Abraini. 1995. Involvement of 5HT3 receptor in the pressure-induced increase in striatal and accumbens dopamine release and the occurrence of behavioral disorders in free-moving rats. Neurosciences Lett. 197: 57–60.

Kriem, B. and J.H. Abraini and J.C. Rostain. 1996a. Role of 5-HT1B receptor in the pressure-induced behavioral and neurochemical disorders in rats. Pharmacol. Biochem. Behav. 53: 257–264.

Kriem, B. and J.C. Rostain and J.H. Abraini. 1996b. Contribution of central 5-HT2 receptors in the occurrence of locomotor activity and myoclonia in freely-moving rats exposed to high pressure. Neuroreport 7: 2687–2690.

Kriem, B. and B. Cagniard, J.C. Rostain and J.H. Abraini. 1998a. Modulation by GABA transmission in the substantia nigra compacta and reticulata of locomotor activity in rat exposed to high pressure. Neuroreport 9: 1343–1347.

Kriem, B. and J.C. Rostain and J.H. Abraini. 1998b. Crucial role of the 5-HT2C receptor, but not of the 5-HT2A receptor, in the down regulation of simulated dopamine release produced by pressure exposure in freely moving rats. Brain Res. 796: 143–149.

Kriem, B. and J.C. Rostain and J.H. Abraini. 1999. Administration of either non-NMDA receptor agonists or NMDA receptor antagonists into the substantia nigra reticulata or the globus pallidus reduces the psychostimulant effect of high helium pressure on locomotor activity in rats. Neuroreport 10: 3777–3783.

Lambertsen, C.J. Collaborative investigation of limits of human tolerance to pressurization with helium, neon and nitrogen: Simulation of density equivalent to helium-oxygen respiration at depths to 2000, 3000, 4000 and 5000 ft of seawater. Pp. 35–52. *In*: C.J. Lambertsen (ed.). 1976. Underwater Physiology V. Fed Am. Soc. Exp. Biol. Bethesda, MD, USA.

Lever, M.J. and K.W. Miller, W.D.M Paton. and E.B. Smith. 1971. Pressure reversal of anaesthesia. Nature 231: 368–371.

Little, H.J. 1996. How has molecular pharmacology contributed to our understanding of the mechanisms of general anesthesia? Pharmacol. Ther. 69: 37–58.

Little, H.J. and D.L. Thomas. 1986. The effects of anaesthetics and high pressure on the responses of the rat superior cervical ganglion in vitro. J. Physiol. (London) 374: 387–399.

Loscher, W. 1981. Valproate induced changes in GABA metabolism at the subcellular level. Biochem. Pharmacol. 30: 1364–1366.

McLeod, M. and P.B. Bennett and R.L. Cooper. 1988. Rat brain catecholamine release at 1, 10, 20 and 100 ATA heliox, nitrox and trimix. Undersea Biomed. Res. 15: 211–221.

Marsden, C.D. The mechanisms of physiological tremor and their signifiance for pathological tremors. Pp. 1–16. *In*: J.E. Desmedt (ed). 1978. Physiological tremors, pathological tremors and clonus. Prog. clin. Neurophysiol. 5.

Marsden, C.D. 1979. Functional aspects of 5-hydroxytryptamine neurones: Application of electrochemical monitoring *in vivo*. Trends Neurosci. 5: 1–16.

Marshall, J.M. 1951. Nitrogen narcosis in frogs and mice. Am. J. Physiol. 166: 699–711.

Meldrum, B.S. and R.W. Horton. Neuronal inhibition mediated by GABA and patterns of convulsions in baboons with photosensitive epilepsy (Papio papio). Pp. 55–64. *In*: P. Harris and C. Mandley (eds.). 1974. The Natural History and Management of Epilepsy. Livingstone, Edinburgh, UK.

Meldrum, B.S. and C. Menini, R. Naquet, H. Laurent and J.M. Stutzmann. 1979. Proconvulsivant, convulsivant and other actions of the baboon Papio papio. Electroenceph. clin. Neurophysiol. 47: 383–395.

Meldrum, B.S. and B. Wardley-Smith, M.J. Halsey and J.C. Rostain. 1983. 2–amino-phosphonoheptanoic acid protects against the high pressure neurological syndrome. European J. Pharmacol. 1983, 87: 501–502.

Michaud, A. and J. Parc, L. Barthelemy, J. Le Chuiton, J. Corriol, J. Chouteau and F. Le Boucher. 1969. Premières données sur une limitation de l'utilisation du mélange oxygène-hydrogène pour la plongée profonde à saturation. C.R. Acad. Se. (Paris) 269: 497–499.

Millan, M.H. and B. Wardley-Smith, M.J. Halsey and B.S. Meldrum. 1989. Studies on the role of the NMDA receptor in the substantia nigra pars reticulata and entopedeuncular nucleus in the development on the high pressure neurological syndrome in rats. Exp. Brain Res. 78: 174–178.

Millan, M.H. and B. Wardley-Smith, M.J. Halsey and B.S. Meldrum. 1990. Effect of NMDA and 2-amino-7-phosphonoheptanoate focal injection into the ventrolateral thalamic nucleus on the high pressure neurological syndrome in the rat. Brain Res. 507: 354–356.

Millan, M.H. and B. Wardley-Smith, M.J. Halsey and B.S. Meldrum. 1991. Brain nuclei and neurotransmitters involved in the regulation of the high pressure neurological syndrome in the rat. Neuropharmacology 30: 1351–1355.

Miller, K.W. Inert gas narcosis and animals under high pressure. Pp. 363–378. *In*: M.A. Sleigh and A.G. Macdonald (eds.). 1972. The Effects of Pressure on Organisms. Symposia of the Society for Experimental Biology. Cambridge University Press, UK.

Miller, K.W. The pressure reversal of anesthesia and the critical volume hypothesis. Pp. 341–351. *In*: B.R. Fink (ed.). 1975. Molecular Mechanisms of Anesthesia. Raven Press, New York, USA.

Miller, K.W. and W.D. Paton, R.A. Smith and E.B. Smith. 1973. The pressure reversal of general anesthesia and the critical volume hypothesis. Molecular Pharmacol. 9: 131–143.

Mor, A. and Y. Grossman. 2006. Modulation of isolated N-methyl-d-aspartate receptor response under hyperbaric conditions. Eur. J. Neurosci. 24: 3453–3462.

Mor, A. and Y. Grossman. 2007. High pressure modulation of NMDA receptor dependent excitability. Eur. J. Neurosci. 25: 2045–2052.

Naquet, R. and J.C. Rostain and X. Fructus. High pressure nervous syndrome: clinical and electrophysiological studies in man. Pp. 62–25. *In*: M.J. Halsey, W. Settle and E.B. Smith (eds.). 1975. The Strategy for Future Diving to Depths greater than 1000 feet. UMS report number WS : 6-15-75. Bethesda, MD, USA.

Overfield, E.M. and H.A. Saltzman, J.A. Kylstra and J.V. Salzano. 1969. Respiratory gas exchange in normal men breathing 0.9 0J0 oxygen in helium at 31.3 ATA. J. Appl. Physiol. 27: 471–475.

Overrath, G. and H. Matthys and A.A. Buhlmann. 1969. Saturation experiment at 31 ATA in an oxygen-helium atmosphere. Helv. Med. Acta 35: 180–200.

Parmentier, J.L. and D.J. Harris and P.B. Bennett. 1985. Central and peripheral causes of hyperreflexia in humans breathing 5% trirnix at 650 m. J. Appl. Physiol. 58: 1239–1245.

Pearce, P.C. and C.J. Maclean, H.K. Shergil, E.M. Ward, M.J. Halsey, G. Tindley, J. Pearson and B.S. Meldrum. 1994. Protection from high pressure induced hyperexcitability by the AMPA/Kainate receptor antagonists GYKI 52466 and LY 293558. Neuropharmacology 33: 605–612.

Perkins, M.N. and T.W. Stone, J.F. Collins and K. Curry. 1981. Phosphonate analogues of carboxylic acids as amino acid antagonists on rat cortical neurones. Neurosci. letters 23: 333–336.

Raoul, Y. and J.L. Meliet and B. Broussolle. 1988. Troubles psychiatriques et plongée profonde. Médecine et armées 16: 269–270.

Regnard, P. 1891. Recherches experimentales sur les conditions physiques de la vie sous les eaux. Masson, Paris, France.

Requin, M. and J.J. Risso. 1992. Effects of high pressure on dopamine release in freely moving rats: a microdialysis study. Neuroscience Lett. 146: 211–214.

Roll, J.P., M. Lacour, M. Hugon and M. Bonnet. Spinal reflex activity in man under hyperbaric heliox conditions (31 and 62 ATA). Pp. 21–28. *In*: C.W. Shilling and M. W. Beckett (eds.). 1978. Underwater Physiology VI. FASEB, Bethesda MD, USA.

Rostain, J.C. 1980. Le syndrome nerveux des hautes pressions chez l'homme et le singe Papio papio. D.Sc. Thesis. Université d' Aix-Marseille I, France.

Rostain, J.C. The nervous system: man and laboratory mammals. Pp. 198–238. *In*: A.G. Macdonald (ed.). 1993. Advances in comparative and environmental physiology: effects of high pressure on biological systems: Springer Verlag, Berlin, Germany.

Rostain, J.C. and S. Lyagoubi. 1971. Etude électroencephalographique de la veille et du sommeil sous hyperbarie prolongée en atmosphère hélium-oxygène (opération Janus 11). Rev. EEG. Neurophysiol. 1: 239–241.

Rostain, J.C. and R. Naquet. Résultats préliminaires d'une étude comparative de l'effet des mélanges oxygène-hélium et oxygène-hydrogène et des hautes pressions sur le babouin Papio papio. Pp. 44–49. *In*: X. Fructus (ed.). 1972. Proc. des troisièmes journées internationales d'hyperbarie et de physiologie subaquatique. Doin. Marseille, France.

Rostain, J.C. and C. Lemaire. 1973. Evolution du tremblement au repos et pendant l'effort au cours de plongées profondes en atmosphère hélium-oxygène. Revue EEG. Neurophysiol. Paris. 3: 203–206.

Rostain, J.C. and R. Naquet. 1974. Le Syndrome Nerveux des Hautes Pressions : caractéristiques et évolution en fonction de divers modes de compression. Rev. EEG. Neurophysiol. 4, 107–124.

Rostain, J.C. and J.P. Charpy. 1976. Effects upon the EEG of psychometric performance during deep dives in helium-oxygen atmosphere. Electroenceph. clin. Neurophysiol. 40: 571–584.

Rostain, J.C. and S. Dimov. 1976. Potentiels évoqués visuels et cycle d'excitabilité au cours d'une plongée simulée à –610 m en atmosphère hélium-oxygène (physalie VI). Electroenceph. clin. Neurophysiol. 41: 287–300.

Rostain, J.C. and R. Naquet. Human neurophysiological data obtained from two simulated dives to a depth of 610 meters. Pp. 9–19. *In*: C.W. Shilling and M.W. Beckett (eds.). 1978. Underwater Physiology VI. FASEB, Bethesda MD, USA.

Rostain, J.C. and C. Forni. 1995. The effects of high pressures of various gas mixtures on rat striatal dopamine detected in vivo by voltammetry. J. Appl. Physiol. 78: 1179–1187.

Rostain, J.C. and G. Regesta and R.Naquet. 1973. Study of the organisation of sleep of divers staying at 300 meter depths in an oxy-helium atmosphere. Electroenceph. clin. Neurophysiol. 34: 788.

Rostain, J.C. and J.P. Imbert, B. Gardette, C. Lemaire, J.C. Dumas and R. Naquet. 1978, Compression methods: 11. Study of the effects of profiles and N2 injections on HPNS of the baboon Papio papio. Undersea Biomed. Res. 5: 46.

Rostain, J.C. and M.C. Gardette-Chauffour and R. Naquet. 1980. HPNS during rapid compression of men breathing He-O$_2$ and He- N$_2$-O$_2$ at 300 m and 180 m. Undersea Biomed. Res. 7: 77–94.

Rostain, J.C. and J.C. Dumas, B. Gardette, J.P. Imbert and C. Lemaire. 1984a. Effects of addition of nitrogen during rapid compression of baboon. J. Appl. Physiol. Respirat. Environ. Exercise. Physiol. 57: 332–340.

Rostain, J.C. and B. Gardette, M.C. Gardette-Chauffour and C. Forni. 1984b. HPNS of the baboon during helium-nitrogen-oxygen slow exponential compressions. J. Appl. Physiol.: Respirat. Environ. Exercise Physiol. 57: 341–350.

Rostain, J.C. and C. Lemaire, M.C. Gardette-Chauffour and R. Naquet. Evolution of HPNS in 16 divers breathing He-N$_2$-O$_2$ during long stays at 45 bars. Pp. 665–672. *In*: A.J. Bachrach and M.M. Matzen (eds.). 1984c. Underwater Physiology VIII. Undersea Medical Society, Bethesda MD, USA.

Rostain, J.C. and C. Lemaire, M.C. Gardette-Chauffour and R. Naquet. Le SNHP au cours d'une plongée au mélange hélium-oxygène jusqu'à 610 m (ENTEX 9) pp. 19–30. *In*: T. Nome, G. Susbielle, M. Comet, M. Jacquin and R. Sciarli (eds.). 1984d. Proceeding of the Xth congress of European Undersea Biomedical Society. EUBS. Marseille, France.

Rostain, J.C. and B. Wardley-Smith, C. Forni and M.J. Halsey. 1986a. Gamma amino butyric acid and the high pressure neurological syndrome. Neuropharmacol. 25: 545–554.

Rostain, J.C. and B. Wardley-Smith, M.J. Halsey and J.F. Nunn. 1986b. Effect of barbituric acid on the high pressure nervous syndrome in the rat. Undersea Biomed. Res. 13: 407–411.

Rostain, J.C. and B. Gardette and R. Naquet. 1987. Effects of exponential compression curves with nitrogen injections in humans. J. Appl. Physiol. 63: 421–425.

Rostain, J.C. and M.C. Gardette-Chauffour, J.P. Gourret and R. Naquet. 1988. Sleep disturbances in man during different compression profiles up to 62 bars in helium-oxygen mixture. Electroenceph. clin. Neurophysiol. 69: 127–135.

Rostain J.C. and M.C. Gardette-Chauffour and R. Naquet. Comparative studies of the effects of hydrogenated breathing mixtures and helium-oxygen or helium-nitrogen-oxygen mixtures up to 450 msw in man. Pp. 195–202. *In*: J.C. Rostain, E. Martinez and C. Lemaire (eds.). 1989. High Pressure Nervous Syndrome, 20 Years Later. ARAS-SNHP Publication, Marseille, France.

Rostain, J.C. and M.C. Gardette-Chauffour and R. Naquet. 1990. Studies of neurophysiological effects of hydrogen-oxygen mixture in man up to 30 bars. Undersea Biomed. Res. 17 (Suppl): 159.

Rostain, J.C. and G. Regesta, M.C. Gardette-Chauffour and R. Naquet. 1991. Sleep organization in man during long stays in helium-oxygen mixture at 30 and 40 bars. Undersea Biomed. Res. 18: 21–36.

Rostain, J.C. and M.C. Gardette-Chauffour and B. Gardette. Neurophysiological studies in macaca fascicularis during exposures with breathing mixtures containing hydrogen up to 1200 msw. Pp. 243–252. *In*: P.B. Bennett and R.E. Marquis (eds.). 1994a. Basic and Applied High Pressure Biology, University of Rochester Press, New York, USA.

Rostain, J.C. and M.C. Gardette-Chauffour, and B. Gardette. 1994b. HPNS during a deep hydrogen-helium-oxygen dive up to 701 meters. Undersea and Hyperbaric Med. 21 (suppl) 40.

Rostain J.C. and T. Alexinsky, B. Gardette, M.C. Gardette-Chauffour, J. Gillard and J. Hardouin. 1995. Cognitives studies in macaca fascicularis exposed to high pressure of helium-oxygen mixtures. Medsubhyp Int. 5: 25–30.

Rostain, J.C. and M.C. Gardette-Chauffour and R. Naquet. 1997. EEG and sleep disturbances during dives at 450 msw in helium-nitrogen-oxygen mixture. J. Appl. Physiol. 83: 575–582.

Rowland-James, P. and M.W. Wilson and K.W. Miller. 1981. Pharmacological evidence for multiple sites of action of pressure in mice. Undersea Biomed. Res. 8: 1–11.

Salzano, J.V. and D.C. Rausch and H.A. Saltzman. 1970. Cardiorespiratory responses to exercise at a simulated seawater depth of 1000 feet. J. Appl. Physiol. 28: 34–41.

Simler, S. and L. Ciesielski, M. Maitre, H. Randrianarisoa and P. Mandel. 1973. Effect of sodium-n-dipropylacetate on audiogenic seizures and brain y-aminobutyric acid level. Biochem. Pharmacol. 22: 1701–1708.

Smith, R.A. and B.A. Dodson and K.W. Miller. 1984. The interactions between pressure and anaesthetics. Phil. Trans. R. Soc. Lond. B304 : 69–84.

Stoudemire, A. and J. Miller, B.S.F. Schmitt, P. Logue, D. Shelton, P.A.G. Latson and P.B. Bennett. 1984. Development of an organic affective syndrome during a hyperbaric diving experiment. Am. J. Psychiatry. 141: 1251–1254.

Thorne, D.R. and A. Findling and A.J. Bachrach. 1974. Muscle tremors under helium, neon, nitrogen and nitrous oxide at 1 to 37 atm. J. Appl. Physiol. 37: 875–879.

Tomei, C. and J.H. Abraini and J.C. Rostain. 1991. A new device for behavioral analysis on rats exposed to high pressure. Physiol. Behav. 49: 393–396.

Vallée, N. and J.C. Rostain, A. Boussuges and J.J. Risso. 2009. Comparison of Nitrogen Narcosis and Helium Pressure Effects on Striatal Amino Acids: A Microdialysis Study in Rats. Neurochem. Res. 34: 835–844.

Van Der Laan, J.W. and Th. De Boer and J. Brulnvels. 1979. Di n-propylacetate and GABA degradation. Preferential inhibition of succinic semi aldehyde dehydrogenase and indirect inhibition of GABA-T. J. Neurochem. 32: 1769–1780.

Waldvogel, W. and A.A. Buhlmann. 1968. Man's reaction to long lasting overpressure exposure. Examination of the saturated organism at a helium-pressure of 21–22ATA. Helv. Med. Acta 34: 130–150.

Wardley-Smith, B. and B.S. Meldrum. 1984. Effect of excitatory amino-acid antagonists on the high pressure neurological syndrome in rats. Eur. J. Pharmacol. 105: 351–354.

Wardley-Smith, B. and J.C. Rostain, B.S. Meldrum and M.J. Halsey. 1986. Effect of 2-aminophosphonoheptanoic acid on the EEG of rats exposed to high pressure. Undersea. Biomed. Res. 13: 155–163.

Wardley-Smith, B. and B.S. Meldrum and M.J. Halsey. 1987. Effect of two novel dipeptide antagonists of excitatory amino acid neurotransmission on the high pressure neurological syndrome in the rat. Eur. J. Pharmacol. 138: 417–420.

Wardley-Smith, B. and S. Hudson, C.J. Dore, A. Charlett, A. Fletcher, N.T. Brammer, M.C.W. Minchin and K.T. Wann. 1990. Exposure to high pressure may produce the 5-HT behavioral syndrome in rats. Undersea Biomed. Res. 17: 275–286.

Zal'tsman, G.L. 1961 Physiological Principles of a Sojourn of a Human in Conditions of Raised Pressure of the Gaseous Medium. English translation, Foreign Technology Division, Wright-Patterson Air Force Base, Ohio, AD655 360, 1967.

Zal'tsman, G.L. (ed.). 1968. Hyperbaric Epilepsy and Narcosis (Neurophysiological Studies). Sechenov Institute of Evolutionary Physiology and Biochemistry, Academy of Sciences, Leningrad, USSR.

Zetterstrom, A. 1948. Deep sea diving with synthetic gas mixtures. Mil. Surg. 103: 104–106.

Zinebi, F. and L. Fagni and M. Hugon. 1988a. The influence of helium pressure on the reduction induced in field potentials by various amino acids and on the GABA—mediated inhibition of the CA1 region of hippocampal slices in the rat. Neuropharmacol. 27: 57–65.

Zinebi, F. and L. Fagni and M. Hugon. 1988b. Decrease in recurrent and feed-forward inhibitions under high pressure of helium in rat hippocampal slices. Eur. J. Pharmaco. 153: 191–199.

Zinebi, F. and L. Fagni and M. Hugon. 1990. Excitatory and inhibitory amino-acidergic determinants of the pressure induced neuronal hyperexcitability in rat hippocampal slices. Undersea Biomed. Res. 17: 487–493.

# Hyperbaric Oxygen Therapy

*Stephen R. Thom*

## INTRODUCTION

Hyperbaric oxygen ($HBO_2$) therapy is a treatment modality in which a person breathes 100% $O_2$ while exposed to increased atmospheric pressure. Treatments are performed in either a monoplace (single patient) or a multiplace (typically 2–14 patients) chamber (Fig. 1). Pressures applied while patients are in the chamber usually are 2 to 3 atmospheres absolute (ATA), but may reach as high as 6 ATA and rarely even higher when treating extreme cases of bubble-related illnesses. At pressures above 3 ATA gas mixes are utilized so that a patient never breathes more than 3 ATA $O_2$. This is done to limit the development of $O_2$ toxicity. Treatments last for 1.5–8 (or more) h, depending on the indication, and may be performed 1–3 times daily. Monoplace chambers usually are compressed with pure $O_2$, and at pressures no greater than 3 ATA. Multiplace chambers are pressurized with air, and patients breathe pure $O_2$ through a tight-fitting face mask, a head tent, or endotracheal tube.

## MECHANISMS OF ACTION

### Primary effects

Therapeutic mechanisms of action for $HBO_2$ are based on elevation of both hydrostatic pressure and the partial pressure of $O_2$. Elevation of the hydrostatic pressure raises gas partial pressures in the body and causes a reduction in the volume of gas-filled spaces according to Boyle's law. This action has direct relevance to pathologic conditions in which gas bubbles

Professor of Emergency Medicine, Chief, Hyperbaric Medicine, University of Pennsylvania, 3620 Hamilton Walk, Philadelphia, PA 19104 USA.
E-mail: sthom@mail.med.upenn.edu

**Figure 1a.** A typical multiplace hyperbaric chamber with patients breathing $O_2$ via tight-fitting masks. Also note that treatments are typically conducted with an inside chamber attendant to assist with medical care (attendant does not wear the $O_2$ mask).

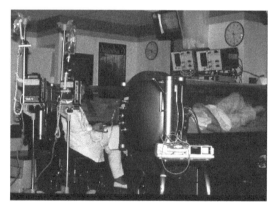

**Figure 1b.** Typical monoplace hyperbaric chambers are shown with patients lying recumbent inside the acrylic tubes. Intravenous pumps (left foreground) are used to maintain parenteral administration of fluids and medications during the hyperbaric treatment.

are present in the body, such as arterial gas embolism and decompression sickness (DCS). During treatment, the arterial $O_2$ tension typically exceeds 1500 mm Hg, and levels of 200–400 mm Hg occur in tissues (Thom, 1989). All perfused tissues are subjected to elevated partial pressures of $O_2$ in association with $HBO_2$ exposure. Under normal environmental conditions, hemoglobin is virtually saturated with $O_2$ on passage through the pulmonary microvasculature, so the primary effect of $HBO_2$ is to increase dissolved $O_2$ content of plasma. Application of each additional atmosphere of $O_2$ increases the dissolved oxygen concentration in the plasma by 2.2 mL $O_2$/dL (vol%).

Over the past decade, experimental evidence has demonstrated that therapeutic mechanisms related to the hyperoxia achieved during $HBO_2$ exposure are linked to production of reactive species (Fig. 2). It is well accepted that reactive $O_2$ species (ROS) mediate $O_2$ toxicity, and that adverse effects of $HBO_2$ are related to excessive ROS production (Clark and Thom, 2003). ROS and reactive nitrogen species (RNS) also serve as signaling molecules in transduction cascades, or pathways, for a variety of growth factors, cytokines and hormones (Allen and Balin, 1989; Maulik, 2002; Ushio-Fukai and Alexander, 2004; Rojas et al., 2006; Calabrese et al., 2007). As such, reactive species can generate either 'positive' or 'negative' effects depending on their concentration and intracellular localization.

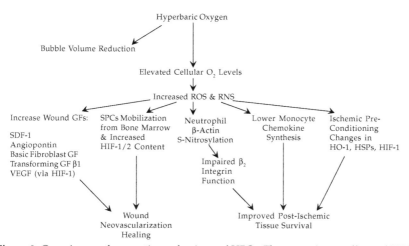

**Figure 2.** Overview on therapeutic mechanisms of $HBO_2$. The two primary effects of $HBO_2$ are to reduce the volume of bubbles in the body and elevate tissue oxygen tensions. The figure outlines effects that occur due to increased production of reactive oxygen species (ROS) and reactive nitrogen species (RNS) because of hyperoxia. Other abbreviations: GFs=growth factors, VEGF=vascular endothelail growth factor, HIF-1= hypoxia inducible factor-1, SPCs=stem/progenitor cells, HO-=heme oxygenase-1, HSPs=heat shock proteins.

ROS are generated as natural by-products of metabolism and they include superoxide ($O_2 \cdot$), hydrogen peroxide ($H_2O_2$), hypochlorous acid (HClO) and hydroxyl ($\cdot OH$). ROS are increased in many organs by hyperoxia (Jamieson et al., 1986). Scavenging antioxidants combat the overproduction of reactive species. Enzymatic antioxidants include superoxide dismutase, catalase, thioredoxin- and glutathione-dependent peroxidase(s) and reductase(s). Acting in conjunction with these enzymes are the non-enzymatic antioxidants vitamin C, vitamin E, thioredoxin, glutathione, uric acid, β-carotene and carotene (Valko et al., 2007). Because typical exposures to hyperoxia in clinical $HBO_2$ protocols is rather brief, studies show that antioxidant defenses are adequate so that biochemical stresses are reversible (Narkowicz et al., 1993; Dennog et al., 1996; Dennog et al., 1999a; Rothfuss et al., 2001).

RNS include nitric oxide (·NO) and agents generated by reactions between ·NO, or its oxidation products, and ROS. There are three nitric oxide synthase enzymes responsible for synthesizing ·NO while converting L-arginine to L-citrulline; NOS-1 (neuronal ·NO synthase, nNOS), NOS-2 (inducible/inflammatory ·NO synthase, iNOS) and NOS-3 (endothelial ·NO synthase, eNOS). Peroxynitrite (ONOO-) is the product of a reaction between $O_2·$ and ·NO (Beckman and Koppenol, 1996). Additionally, peroxide enzymes, and especially myeloperoxidase, can catalyze reactions between nitrite ($NO_2$), a major oxidation product of ·NO, and hydrogen peroxide or hypochlorous acid to generate oxidants such as nitryl chloride and nitrogen dioxide that are capable of nitration and S-nitrosylation reactions (Sampson et al., 1998; Brennan et al., 2002; Lakshmi et al., 2005).

There are a variety of conditions or disorders where $HBO_2$ is used worldwide, and indications vary in different countries. Presumably, this variance is related to cultural differences in the requisite weight of scientific evidence as well as empiric observation required for a therapeutic to achieve acceptance. In order to simplify this discussion and focus on mechanisms of $HBO_2$, we believe that treatment protocols can be divided into two major groups. Wound healing disorders is one group where $HBO_2$ is used in current practice; the principal ones being refractory diabetic wounds and delayed radiation injuries. A typical wound healing treatment protocol is daily exposures to 2.0 to 2.4 atmospheres absolute (ATA) for 90 to 120 min for 20 to 40 d. Treatments often include a so-called air break, where a patient breathes just air for 5 minutes once or twice through the course of a treatment. This intervention has been demonstrated to enhance pulmonary $O_2$ tolerance (Hendricks et al., 1977). The second major grouping for $HBO_2$ treatment includes a variety of inflammatory or post-ischemic disorders. Treatment protocols for these conditions are typically much shorter than for wound healing. The course of treatments is just a few days rather than weeks, they are performed at higher $O_2$ partial pressures (~2.5 to 3.0 ATA), and may occur multiple times in the same day.

### Wound Healing

Discussion of the pathophysiology of diabetic wounds and delayed radiation injuries is beyond the scope of this chapter, and the reader is referred to several recent publications (Denham and Hauer-Jensen, 2002; Falanga, 2005). Common elements shared by both disorders include depletion of epithelial and stromal cells, chronic inflammation, fibrosis, an imbalance or abnormalities in extracellular matrix components and remodeling processes, and impaired keratinocyte functions (Martin et al., 2000; Denham and Hauer-Jensen, 2002; Falanga, 2005; Stojadinovic et al., 2005; Brem and Tomic-Canic, 2007; Ueno et al., 2007). Diabetic wound healing is also impaired by deceased growth factor production, while in post-radiation tissues there appears to be an imbalance between factors mediating fibrosis versus normal tissue healing (Denham and Hauer-Jensen, 2002; Brem and Tomic-Canic, 2007; Ueno et al., 2007).

The effectiveness of $HBO_2$ as an adjuvant therapy for the treatment of diabetic lower extremity ulcerations is supported by six randomized trials and evaluations from a number of independent evidence-based reviews (1999a; 1999b; Wood, 2002; Hess et al., 2003; Kranke et al., 2004; Hailey et al., 2007). The pathophysiology of radiation injury is obviously different than diabetic wounds, but the varied tissue abnormalities have been described as a chronic wound (Denham and Hauer-Jensen, 2002). The benefit of $HBO_2$ for radiation injury also has been shown in randomized trials and its utilization supported by independent evidence-based reviews (Marx et al., 1985; Bennett et al., 2008; Clarke et al., 2008). It is important to state that for both diabetic wounds and radiation injuries, $HBO_2$ is used in conjunction with standard surgical management techniques. That was the format followed in clinical trials demonstrating its efficacy. By itself, or if used only in a post-operative period, $HBO_2$ is frequently inadequate treatment (Maier et al., 2000; Annane et al., 2004). Animal trials have also documented benefits of $HBO_2$ (Marx et al., 1990; Goldstein et al., 2006; Gallagher et al., 2007; Zhang et al., 2008). The basis for its efficacy is only partially understood, but appears to be a combination of systemic events as well as local alterations within the wound margin (see Fig. 2).

Neovascularization occurs by two processes. Regional angiogenic stimuli influence the efficiency of new blood vessel growth by local endothelial cells (termed angiogenesis) and they stimulate the recruitment and differentiation of circulating stem/progenitor cells (SPCs) to form vessels *de novo* in a process termed vasculogenesis (Carmeliet, 2000; Hattori, 2001; Tepper et al., 2005). Clinical $HBO_2$ has effects on both these processes.

$HBO_2$ does not alter circulating levels of insulin, insulin-like growth factors, or pro-inflammatory cytokines [e.g. tumor necrosis factor (TNF)-$\alpha$, interleukin (IL)-6 and IL-8] in normal healthy humans (Fildissis et al., 2004; Chen et al., 2006). $HBO_2$ does reduce pro-inflammatory cytokine production under stress [e.g. endotoxin challenge (Fildissis et al., 2004)] and in wounded tissues or isolated cells $HBO_2$ increases synthesis of many growth factors. Vascular endothelial growth factor (VEGF) and angiopoietin, as well as stromal derived factor-1 (SDF-1) influence SPCs homing to wounds and SPCs differentiation to endothelial cells (Peichev et al., 2000; Hildbrand et al., 2004). Synthesis of VEGF has been shown to be increased in wounds by $HBO_2$, and it is the most specific growth factor for neovascularization (Sheikh et al., 2000). $HBO_2$ also stimulates synthesis of basic fibroblast growth factor and transforming growth factor β1 by human dermal fibroblasts (Kang et al., 2004), angiopoietin-2 by human umbilical vein endothelial cells (Lin et al., 2002), and it up-regulates platelet derived growth factor (PDGF) receptor in wounds (Bonomo et al., 1998).

Oxidative stress at sites of neovascularization will stimulate growth factor synthesis by augmenting synthesis of hypoxia inducible factor (HIF)-1 (Hunt et al., 2007; Milovanova et al., 2008, 2009). Hypoxia inducible factors are heterodimers of HIF-α and a constitutively expressed HIF-β (also called the aryl hydrocarbon receptor nuclear translocator [ARNT] subunit). Enhanced growth factor synthesis by $HBO_2$ is due at least in part to augmented synthesis and stabilization of HIFs (Sheikh et al., 2000; Thom and Milavonava, 2008a,b; Milovanova et al., 2008, 2009). While this clearly sounds paradoxical, even under normoxic conditions HIF activity is regulated by a variety of cellular micro-environmental modifications. It is well recognized that expression and activation of HIF-α subunits are tightly regulated, and their degradation by the ubiquitin-proteasome pathway typically occurs when cells are replete with $O_2$ (Salceda and Caro, 1997; Semenza, 2001). However, whether hypoxic or normoxic conditions prevail, free radicals are required for HIF expression (Brauchle et al., 1996; Nemoto et al. 2000; Semenza, 2001; Schroedl et al., 2002; Dulak and Jozkowicz, 2003; Semenza, 2003). In addition to ROS, synthesis of ·NO is required for VEGF-mediated angiogenesis (Fukumura et al ., 2001), and many down-stream effects of VEGF are stimulated via ·NO (Parenti et al., 1998; Aramoto et al., 2004).

There are three distinct HIF-α proteins: HIF-1α, 2α and 3α. HIF-1 and 2 coordinate many cell responses involved with neovascularization by regulating gene transcription and, while there is substantial overlap in their activity, there are also a number of genes preferentially regulated by either HIF-1 or -2 (Wang et al., 2005). The biological function of HIF-3 is unclear, and at least one splice variant negatively modulates HIF-1α and –2α, although its expression is restricted to specific tissues and subject to hypoxic conditions (Makino et al., 2002; Maynard et al., 2007).

The influence $HBO_2$ has on HIF isoform expression appears to be conflicting and further work is needed to elucidate what are likely to be variations based on tissue-specific responses. Additionally, higher or lower levels of HIF isoforms may vary based on chronology [e.g. looking early or late after wounding or an ischemic insult]. One recent model showing accelerated wound healing by $HBO_2$ reported lower HIF-1 levels at wound margins, along with reduced inflammation and fewer apoptotic cells (Zhang et al., 2008). In contrast, higher levels of HIF-1 have been linked to elevated VEGF in wounds in response to hyperoxia (Sheikh et al., 2000; Hunt et al., 2007). Recently, exposure to $HBO_2$ was shown to elevate HIF-1 and -2 levels in vasculogenic SPCs. The basis for this effect is related to augmented production of the antioxidant, thioredoxin and one of its regulatory enzymes, thioredoxin reductase, in response to oxidative stress (Milavonava et al., 2008, 2009). Among other actions, thioredoxin has been shown to promote the expression and activity of HIFs (Ema et al., 1999; Welsh et al., 2002; Kaga et al., 2005). HIF-1 and -2 then secondarily can stimulate transcription of many genes involved with neovascularization including SDF-1 and its counterpart ligand, CXCR4, as well as VEGF. A physiological oxidative stress that triggers the same pathway is lactate metabolism (Milovanova et al., 2008).

Bone marrow NOS-3 activity is required for SPCs mobilization (Aicher et al., 2003). SPCs mobilization is compromised by diabetes, apparently because NOS activity can be impaired due to responses related to hyperglycemia and a reduced presence of insulin (Du et al., 2001; Bivalacqua et al., 2004; Bucci et al., 2004; Du et al., 2006). In addition, radiation and chemotherapy, along with other factors such as age, female gender and coronary artery disease, are known to diminish SPCs mobilization (Vasa et al., 2001; Seggewiss et al., 2003; Ikeda 2004; Platzbecker et al., 2005). By stimulating NO synthesis in bone marrow, $HBO_2$ mobilizes SPCs in normal humans and patients previously exposed to radiation (Thom et al., 2006a) and preliminary observations suggest the same is true for diabetic patients (Yang et al., 2007; Thom and Milovanova, 2008a). In animal models SPCs mobilized by $HBO_2$ home to wounds and accelerate healing (Goldstein et al., 2006; Gallagher, 2007; Thom and Milovanava, 2008b; Milovanova et al., 2008, 2009). $HBO_2$ also improves clonal cell growth of SPCs from humans and animals (Thom et al., 2006b). Functional enhancements of SPCs by $HBO_2$ appear to be related to augmentation of HIF-1 and -2 levels (Milavonava et al., 2008, 2009).

Therefore, to summarize, $HBO_2$ can stimulate healing in refractory wounds and irradiated tissues. One oxidative stress response that triggers improved function, at least for SPCs, involves elevations of thioredoxin and thioredoxin reductase which secondarily increase HIF-1 and HIF-2. The influence of $HBO_2$ on HIFs in other cell types or tissues is variable, but increased growth factor synthesis has been demonstrated. A separate free radical-based mechanism for augmentation of neovascularization by $HBO_2$ is bone marrow SPCs mobilization, which increases the number of circulating SPCs that may home to injured tissues.

## Reperfusion/inflammatory injuries and HBO$_2$

For this introductory section, we will group a variety of disorders together to facilitate the discussion on mechanisms of HBO$_2$, although we admit this approach grossly simplifies complex pathophysiological processes. Skin graft and flap failures may be due to ischemia-reperfusion injuries. A prospective, blinded clinical trial found that administration of HBO$_2$ prior to and for 3 d following the procedure led to a significant 29% improvement in graft survival (Perrins and Cantab, 1967). This is the only randomized clinical trial on skin grafts, but numerous animal studies support its conclusions [see citations in (Kindwall et al., 1991)]. Clinical studies have also documented significant survival enhancement with HBO$_2$ for extremity re-implantation and free tissue transfer, and following crush injury (Waterhouse et al., 1993; Bouachour et al., 1996). Other clinical trials have shown reductions in coronary artery re-stenosis after balloon angioplasty / stenting (Sharifi et al., 2002; Sharifi et al., 2004), decreased muscle loss after thrombolytic treatment for myocardial infarction (Shandling et al., 1997; Stavitsky et al., 1998; Dekleva et al., 2004), improved hepatic survival after transplantation and more rapid return of donor liver function (Mazariegos et al., 1999; Suehiro et al., 2008) and reduced incidence of encephalopathy seen after cardiopulmonary bypass and following carbon monoxide poisoning (Weaver et al., 2002; Alex et al., 2005).

As was the case with our discussion of wound healing, there appear to be complex and perhaps overlapping mechanisms for therapeutic effects of HBO$_2$ (see Fig.1). An early event associated with tissue reperfusion is adherence of circulating neutrophils to vascular endothelium by $\beta_2$ integrins. When animals or humans are exposed to HBO$_2$ at 2.8 to 3.0 ATA (but not to just 2.0 ATA), the ability of circulating neutrophils to adhere to target tissues is temporarily inhibited (Thom, 1993; Zamboni et al., 1993; Thom et al., 1997; Labrouche et al., 1999; Kalns et al., 2002). In animal models, HBO$_2$-mediated inhibition of neutrophil $\beta_2$ integrin adhesion has been shown to ameliorate reperfusion injuries of brain, liver, heart, lung, skeletal muscle and intestine, as well as smoke-induced lung injury and encephalopathy due to carbon monoxide (CO) poisoning (Thom, 1993; Zamboni et al., 1993; Wong et al., 1996; Ueno et al., 1999; Atochin et al., 2000; Tahepold et al., 2001; Yang et al., 2001; Thom et al., 2002; Kihara et al., 2005). It also appears that benefits of HBO$_2$ in decompression sickness are related to temporary inhibition of neutrophil $b_2$ integrins, in addition to the Boyle's Law-mediated reduction in bubble volume as discussed in the introduction (Martin and Thom, 2002).

Exposure to HBO$_2$ inhibits neutrophil $\beta_2$ integrin function because hyperoxia increases synthesis of reactive species derived from NOS-2 and myeloperoxidase, leading to excessive S-nitrosylation of $\beta$ actin (Thom et al., 2008). This is a highly localized process occurring within neutrophils and not observed in other leukocytes, probably because of a paucity of myeloperoxidase. This modification increases the concentration of short, non-cross-linked F-actin, alters F-actin distribution within the cell, and it inhibits $\beta_2$ integrin clustering on the membrane surface. $HBO_2$ does not reduce neutrophil viability, and functions such as degranulation, phagocytosis and oxidative burst in response to chemoattractants remain intact (Thom, 1993; Thom et al., 1997; Juttner et al., 2003). Inhibiting $\beta_2$ integrins with monoclonal antibodies will also ameliorate ischemia-reperfusion injuries but in contrast to $HBO_2$, antibody therapy causes profound immunocompromise (Mileski et al., 1990; Mileski et al., 1993). Probably the most compelling evidence that $HBO_2$ does not cause immunocompromise comes from studies in sepsis models, where $HBO_2$ has a beneficial effect (Ross and McAllister, 1965; Thom et al., 1986; Buras et al., 2006). $HBO_2$ does not compromise neutrophil antibacterial functions because the G-protein coupled 'inside-out' pathway for activation remains intact, and actin nitrosylation is reversed as a component of this activation process (Thom et al., 2008). The 'de-nitrosylation' mechanism in neutrophils is an area of current investigation.

Monocyte-macrophages exhibit lower stimulus-induced pro-inflammatory cytokine production after exposure to $HBO_2$. This is seen with cells removed from humans and animals exposed to $HBO_2$ and also when cells are exposed to $HBO_2$ *ex vivo* (Lahat, 1995; Weisz et al., 1997; Benson et al., 2003). The $HBO_2$ effect on monocyte/macrophages may be the basis for reduced circulating cytokine levels after endotoxin stress, as was mentioned above (Fildissis et al., 2004). The mechanism is unknown, but could be related to $HBO_2$-mediated enhancement of heme oxygenase-1 and heat shock proteins (HSP) [e.g. HSP 70] (Dennog et al., 1999b; Rothfuss et al., 2001). Hence, once again, an oxidative 'stress' process seems to occur. There are additional mechanisms involved with beneficial $HBO_2$ effects in reperfusion models. $HBO_2$ augments ischemic tolerance of brain, spinal cord, liver, heart and skeletal muscle by mechanisms involving induction of antioxidant enzymes and anti-inflammatory proteins (Gregorevic et al. 2001; Kim et al., 2001; Yu et al., 2005; Cabigas et al., 2006; Nie et al., 2006; Hirata et al., 2007).

HIF-1 is responsible for induction of genes that facilitate adaptation and survival from hypoxic stresses (Semenza, 2001), and so has been a focus of interest when examining $HBO_2$ therapeutic mechanisms in ischemia-reperfusion models. HIF-1 is involved with pro- as well an anti-apoptotic pathways and in the brain, promotes astrocyte mediated-chemokine synthesis (Acker and Acker, 2004; Mojsilovic-Petrovic et al., 2007). In several models, exposure to $HBO_2$ appears to ameliorate post-ischemic brain injury by decreasing HIF-1 expression (Li et al., 2005; Calvert et al., 2006). When $HBO_2$ is used in a prophylactic manner to induce ischemic tolerance, however, its mechanism appears related to up-regulation of HIF-1 and at least one of its target genes, erythropoietin (Gu et al., 2008). Thus, as was the case in wound healing models, timing of $HBO_2$ application appears to influence cellular responses.

There has been a long tradition of considering $HBO_2$ therapy for a variety of highly virulent infectious diseases, such as necrotizing fasciitis and clostridial myonecrosis, with a view that the microorganisms involved were particularly sensitive to elevated partial pressures of $O_2$. Several retrospective cohort trials indicate there is a benefit to including $HBO_2$ with antibiotics and surgery for necrotizing fasciitis (Riseman et al., 1990; Wilkinson and Doolette, 2004; Escobar et al., 2005). There is only one multi-center review where a mixed recommendation on its use resulted, however, largely because of the marked heterogeneity in clinical status among the patients (Brown et al., 1994). Retrospective comparisons examining efficacy of $HBO_2$ in clostridial myonecrosis support its use, but again there is on-going debate (Hart et al., 1974b).

With regard to mechanisms, most clinically significant anaerobic organisms are actually rather aerotolerant and thus tissue $O_2$ tensions, even those achievable with $HBO_2$, are likely to be only bacteriostatic to these organisms (Korhonen, 2000). Exotoxin production, which is $O_2$-sensitive, can be inhibited at tissue partial pressures achievable with $HBO_2$ (Hart et al., 1974b) and leukocyte killing will improve at progressively higher $O_2$ tensions (Mader et al., 1980). We suggest that a broader focus may be required to elucidate the as yet unclear pathophysiology of these serious infections and the role of $HBO_2$. A recent study of streptococcal myonecrosis showed that host responses to sometimes minor traumatic injuries increase expression of vimentin in muscle tissue which mediates adhesion/ sequestration of microorganisms (Bryant et al., 2006). There is also a role for intravascular platelet-neutrophil aggregation with vascular occlusion in these infectious processes (Bryant et al., 2005; Hickey et al., 2008). These issues are much closer to the pathophysiological events seen with disorders such as ischemia/reperfusion injuries than traditional ideas in infectious diseases. There is ample room for further investigation.

In summary, oxidative stress responses triggered by $HBO_2$ improve outcome from a wide variety of post-ischemic/inflammatory insults and $HBO_2$ improves ischemic tolerance when used in a prophylactic manner. The basis for these effects is only partially understood. Augmented synthesis of reactive species temporarily inhibits endothelial sequestration of neutrophils by inhibiting $\beta_2$ integrin function, and in many tissues $HBO_2$ will induce antioxidant enzymes and anti-inflammatory proteins. Figure 1 provides a summary of mechanisms, all of which appear to stem from elevations in reactive species. The rest of this chapter will present short discussions on some of the clinical indications for $HBO_2$ although this is not an exhaustive list of world-wide usage (Table 1).

Table 1. Common conditions treated with hyperbaric oxygen therapy.

| |
|---|
| ARTERIAL GAS EMBOLISM & DECOMPRESSION SICKNESS |
| CARBON MONOXIDE |
| CARBON MONOXIDE COMBINED WITH CYANIDE |
| CARBON TETRACHLORIDE |
| CLOSTRIDIAL MYONECROSIS |
| CRUSH INJURY/LIMB REPERFUSION |
| OSTEOMYELITIS REFRACTORY TO STANDARD MANAGEMENT |
| PROGRESSIVE NECROTIZING INFECTIONS |
| THERMAL BURNS |
| THREATENED SKIN FLAPS AND GRAFTS |
| WOUND HEALING |
|    - REFRACTORY DIABETIC LOWER EXTREMITY WOUNDS |
|    - RADIATION INJURIES (SOFT TISSUE & OSTEORADIONECROSIS) |

## INDICATIONS FOR HBO$_2$ THERAPY

### Arterial Gas Embolism (AGE) & Decompression Sickness (DCS)

Among the earliest applications of hyperbaric therapy was to treat disorders related to gas bubbles in the body. In the nineteenth century, laborers working in pressurized compartments (caissons) were noted to frequently experience joint pains, limb paralysis, or pulmonary compromise when they returned to ambient pressure. This condition—DCS, caisson disease, or bends—was later attributed to nitrogen bubbles in the body, and recompression was found to relieve symptoms (Pol and Wattelle, 1854; Snell, 1896; Keays, 1909). The mechanism was thought to be based purely on Boyle's law, gas bubble volume was reduced and inert gas partial pressure increased due to pressure. Later on, treatment was improved by adding supplemental O$_2$ with the idea that this would hasten inert gas diffusion out of the body (Zuntz, 1897; Yarbrough and Behnke, 1939; Bert, 1978).

AGE can arise whenever disruption of a vascular wall occurs, as in major trauma. In scuba divers, AGE is due to pulmonary overexpansion injuries on decompression. Iatrogenic AGE has been reported in association with cardiovascular, obstetrical/gynecological, neurosurgical, and orthopedic procedures. Non-surgical processes reported to cause AGE include over expansion during mechanical ventilation, hemodialysis, and with accidental opening of central venous catheters.

Emergency treatment of gas bubble disorders includes standard support of airway, breathing and circulation plus application of HBO$_2$. Recommendation is for referral as soon as possible. Because gas bubbles can persist in tissues for several days, many reports note success when HBO$_2$ is begun even after long delays (Takita et al., 1968; Conahan, 1979; Bray et al., 1983). Therefore, even when treatments cannot be applied rapidly, a trial of therapy is often recommended. Controlled animal trials support efficacy of $HBO_2$ but randomized clinical trials have not been done (Levin et al., 1981). Effects of pressure and oxygen on bubble volume and rate of diameter reduction represent cornerstones to the use of $HBO_2$ in bubble-mediated disorders. Bubbles also damage the vascular endothelium and leukocyte activation with endothelial adherence is causally linked with DCS in animal trials (Nossum et al., 1999; Martin and Thom, 2002; Nossum et al., 2002). Hence, $HBO_2$-mediated inhibition of neutrophil-endothelial interactions via inhibition of $\beta_2$ integrin adhesion molecules on neutrophils appears to play a role in bubble-related $HBO_2$ therapy (Thom et al., 1997; Martin and Thom, 2002).

## Carbon Monoxide

Animal studies indicate that CO poisoning initiates a progressive neuropathological process with overlapping elements of hypoxic/ischemic, excitotoxic and immune-mediated brain injury (Piantadosi et al., 1997; Thom et al., 2004a; Thom et al., 2004b; Thom et al., 2006). Hence, CO-mediated brain injury does not fit into a single pathophysiological category is has often been considered in traditional toxicology. Administration of supplemental oxygen is the cornerstone for treatment of CO poisoning. Historically, its use was based on the affinity of CO for heme proteins and formation of carboxyhemoglobin (COHb). Elevated COHb can precipitate tissue hypoxia, and $O_2$ inhalation hastens dissociation of CO from hemoglobin and provides enhanced tissue oxygenation. $HBO_2$ causes COHb dissociation to occur at a rate greater than that achievable by breathing pure $O_2$ at sea-level pressure (Pace et al., 1950). Additionally, $HBO_2$ accelerates restoration of mitochondrial oxidative processes (Brown and Piantadosi, 1991). CO poisoning causes perivascular inflammatory injuries, especially in the brain, and animal studies have shown that $HBO_2$ ameliorates this process by impeding neutrophil adherence to the vasculature (Thom, 1993; Chen et al., 1996; Thom et al., 1997).

Survivors of CO poisoning are faced with potential impairments to cardiac and neurological function. CO poisoning can cause an acute cardiac compromise and survivors exhibit an increased risk for cardiovascular-related death in the following 10 yr (Satran et al., 2005; Henry et al., 2006). As yet, no investigations have been reported examining whether $HBO_2$ alters long-term cardiac risk. With regard to neurological impairments, there is an historical precedence for dividing disorders into 'acute/persistent' and 'delayed' forms. Some patients exhibit acute abnormalities wherein they have an abnormal level of consciousness and/or focal neurological findings from the time of initial presentation and never recover (Anderson et al., 1973; Ginsberg and Myers, 1974; Cramlet et al., 1975; Coburn and Forman,

1987). Other patients seemly recover from acute poisoning but then manifest neurological or neuropsychiatric abnormalities from 2 d to about 5 wk after poisoning (Meyer, 1928; Choi, 1983; Myers et al., 1985; Raphael et al., 1989; Maeda et al., 1991; Gorman et al., 1992; Shimosegawa et al., 1992; Silverman et al., 1993; Ducasse et al., 1995; Thom et al., 1995; Mathieu et al., 1996; Weaver et al., 2002; Hsiao, 2004). Events occurring after a clear or 'lucid' interval have been termed 'delayed' neurological sequelae. The multiple pathological processes triggered by CO occur in close proximity and in some cases concurrently. Therefore, one can expect that pathological insults for 'acute/persistent' and 'delayed' sequelae overlap. Thus, there may be more of a continuum of clinical disorders versus distinctly different syndromes.

Animals poisoned with CO and treated with $HBO_2$ have more rapid improvement in cardiovascular status (End and Long, 1942), lower mortality (Peirce et al., 1972), and lower incidence of neurological sequelae (Thom, 1997). Benefits likely are based on both improved oxygenation and secondary effects pertaining to inhibition of neutrophil adhesion. Since 1960, clinical use of $HBO_2$ has been used with increasing frequency for severe CO poisoning because patient recovery appeared to exceed that expected with ambient-pressure oxygen therapy. Support for $HBO_2$ use comes from this experience (Roche et al., 1968; Goulon et al., 1969; Lamy and Hauguet, 1969; Mathieu et al., 1985; Myers et al., 1985; Norkool and Kirkpatrick, 1985; Gorman et al., 1992; Hsu and Wang, 1996). The clinical efficacy of $HBO_2$ for acute CO poisoning has been assessed in five prospective, randomized trials published in peer-reviewed journals. There is only one clinical trial that satisfies all items deemed to be necessary for the highest quality of randomized controlled trials (Weaver et al., 2002). This double-blinded, placebo-controlled clinical trial involved 152 patients. All patients received treatment with either 3 sessions of $HBO_2$ therapy or normobaric $O_2$ with sham pressurization to maintain blinding. Critically ill patients were included, with half of enrolled patients having lost consciousness and 8% requiring intubation. Neurological sequelae, defined *a priori*, were assessed by rigorous neuropsychological testing. The group treated with $HBO_2$ had a lower incidence of cognitive sequelae than the group treated with normobaric $O_2$. Post hoc subgroup analysis incorporating risk factors showed that $HBO_2$ significantly reduced cognitive sequelae in patients with any of the following: unconsciousness, COHb $\geq 25\%$, age $\geq 50$ yr , or base excess $\leq 2$ mEq/L. $HBO_2$ did not improve outcome in patients with none of these criteria.

The only other blinded, prospective, randomized trial was published in 1999 and it involved 191 patients of different severity treated with atypical protocols. One group received daily $HBO_2$ (3.0 ATA for 60 min) with intervening high-flow $O_2$ for 3 or 6 d and the second group received high-flow normobaric $O_2$ for 3 or 6 d (Scheinkestel et al., 1999). Additional $HBO_2$

treatments (up to 6 daily) were performed in patients without neurological recovery. The primary outcome measure for this trial was testing performed at completion of treatment (3–6 d), hence a time point too early to evaluate many CO poisoning victims. This study had a high rate of adverse neurological outcomes in all patients, regardless of treatment assignment. Neurological sequelae were reported in 74% in $HBO_2$-treated patients and 68% in controls. No other clinical trial has described this magnitude of neurological dysfunction. The high incidence is likely to be related to the neuropsychological assessment tool which could not discern true neurological impairments from poor test-taking related to depression (Schiltz, 2000). Suicide attempts with CO represented 69% of cases in this trial. Moreover, there were 54% of subjects lost to follow-up. Outcomes at one-month were not reported, but remarked to show no difference. Multiple statistical comparisons were reported without apparent planning or the requisite statistical correction. Both treatment arms received continuous supplemental mask $O_2$ for 3 d between their hyperbaric treatments (both true $HBO_2$ and 'sham'), resulting in greater overall $O_2$ doses than conventional therapy. Multiple flaws in the design and execution of this study are discussed in the literature, so it is impossible to draw meaningful conclusions from the data (Moon and DeLong, 1999; Hampson, 2000).

The first prospective clinical trial involving $HBO_2$ therapy did not demonstrate therapeutic benefits (Raphael et al., 1989). This study has been criticized because the authors used a low $O_2$ partial pressure (2 ATA) versus the more usual protocols with 2.5–3 ATA and because nearly half of the patients received hyperbaric treatments more than 6 h after they were discovered (Brown and Piantadosi, 1989). In 1969, a retrospective study indicated that $HBO_2$ reduced mortality and morbidity only if $HBO_2$ was administered within 6 h of CO poisoning (Goulon et al., 1969). $HBO_2$ was effective in several other prospective investigations. In a trial involving only mildly to moderately poisoned patients, 23% of patients (7/30) treated with ambient-pressure oxygen developed neurologic sequelae, whereas no patients (0/30; $p < 0.05$) treated with $HBO_2$ (2.8 ATA) developed sequelae (Thom et al., 1995). In another prospective, randomized trial, 26 patients were hospitalized within 2 h of discovery and were equally divided between 2 treatment groups: ambient-pressure $O_2$ or 2.5 ATA $O_2$ (Ducasse et al., 1995). Unfortunately, only surrogate outcome measures were utilized 3 wk later to evaluate treatment efficacy. Patients treated with $HBO_2$ had significantly fewer abnormalities on electroencephalogram and single-photon emission computed tomography (SPECT) scans showed that cerebral vessels had nearly normal reactivity to carbon dioxide, in contrast to diminished reactivity in patients treated with ambient-pressure $O_2$.

In conclusion, published clinical trials span a broad range in quality. Efficacy of $HBO_2$ for acute CO poisoning is supported in animal trials and studies provide a mechanistic basis for treatment. This is the era of evidence-based medicine and a great deal of emphasis has been placed on systematic reviews, although flaws in this analysis are increasingly recognized (Tobin and Jubran, 2008). Treatment of CO poisoning has undergone a number of systematic reviews, but the analytical fidelity has been poor. For example, profound flaws in two successive Cochrane Library Reviews have been identified but, to date, are not corrected (Logue, 2006).

Several recent reports have provided additional insight into risks for neurological sequelae post-CO poisoning and the benefit of $HBO_2$. Weaver, et al . reported on a cohort of 238 patients and found that independent risk factors for developing neurological sequelae include age $\geq$ 36 yr , exposure for 24 h or longer (with or without intermittent CO exposures), and acute complaints of memory abnormalities (Weaver et al., 2007). These conclusions were based on univariate analysis but once subjected to multivariate analysis, only age $\geq$ 36 and exposure over 24 h persist as independent risk factors. The only risk factor where $HBO_2$ demonstrated a reduction in incidence of sequelae was for the group $\geq$ 36 yr . The trial was under powered to reliably assess the benefit of HBO in those with CO exposures longer than 24 h, but none of the 5 patients in this group manifested neurological sequelae.

Hopkins et al. have shown that $HBO_2$ is only beneficial in reducing neurological sequelae among patients who do not possess the apolipoprotein ε4 allele (Hopkins et al., 2008). Because genotype is typically unknown this report does not provide treatment guidelines, but it will be important for future research. It is well established that the apolipoprotein genotype can have profound effects on risk for a variety of neuropathological events (Aamar et al., 1992; Saunders et al., 1993; Friedman et al., 1999; McCarron et al., 2000). Whether apolipoprotein e4 modifies the primary pathophysiological insults of CO, or mechanisms of $HBO_2$ are currently unknown.

As yet no objective method is available for staging the severity of CO poisoning, although preliminary reports suggest plasma markers may be used in the future (Thom et al., 2006). Psychometric screening tests have not proved reliable because abnormalities during the initial screening do not correlate with development of delayed sequelae (Thom et al., 1995). The optimal dose of $HBO_2$ (e.g. number of treatments and treatment pressure) and the time after which it is no longer effective therapy are not clearly defined. Randomized trials have treated patients as soon as possible after CO poisoning based on work suggesting the existence of a 6-h window of greatest opportunity (Goulon et al., 1969). However, it is possible that the time of potential benefit goes beyond what has been investigated for some patients. The requisite number of treatment also remains unclear.

## Carbon Monoxide Combined With Cyanide

CO and CN poisonings can occur concomitantly in victims of smoke inhalation (Wetherill, 1966; Mohler, 1975; Anderson et al., 1979; Birky et al., 1979; Clark et al., 1981; Anderson and Harland, 1982; Way, 1984; Copeland, 1985; Barillo et al., 1986; Madden et al., 1986; Silverman et al., 1988; Lundquist et al., 1989; Baud et al., 1991; Kirk et al., 1993; Barillo et al., 1994; Shusterman et al., 1996). Experimental evidence suggests that they can produce synergistic toxicity (Pace et al., 1950; Pitt et al., 1979; Norris et al., 1986; Ballantyne, 1987; Moore et al., 1987). Toxicity from CN stems from binding to cytochrome a-a$_3$ thereby inhibiting oxidative phosphorylation. Animal studies demonstrate that ambient-pressure 100% O$_2$ can enhance protection from CN toxicity (Sheehy and Way, 1968) and also can enhance CN metabolism to thiocyanate when thiosulfate is used concomitantly (Breen et al., 1995). HBO$_2$ may have direct effects on reducing CN toxicity (Ivanov, 1959; Cope, 1961; Skene et al., 1966; Isom and Way, 1974; Takano et al., 1980) or augmenting antidote treatments (Way et al., 1966; Sheehy and Way, 1968; Burrows and Way, 1977). However, not all animal studies have found HBO$_2$ improved outcome (Way et al., 1972), and clinical experience regarding CN treatment with HBO$_2$ is sparse (Anderson et al., 1973; Scolnick et al., 1993; Goodhart, 1994). In a series of smoke-inhalation victims with both toxic CO and CN levels who received both HBO$_2$ and treatment for CN involving sodium nitrite and sodium thiosulfate, 4 of 5 patients survived without apparent neurologic damage (Hart et al., 1985). Clinical case reports where HBO$_2$ was used along with standard antidote treatment of that time period (sodium nitrite plus sodium thiosulfate) for isolated CN poisonings are equivocal (Litovitz et al., 1983; Trapp and Lepawsky, 1983; Scolnick et al., 1993; Goodhart, 1994). One case showed dramatic improvement (Trapp and Lepawsky, 1983), but another showed no response (Litovitz et al., 1983). Methemoglobin formation with the standard antidote treatment involving nitrite is not thought to generate seriously high methemoglobin levels, but in the face of concomitant COHb the additional reduction of oxygen carrying capacity may pose a risk. In this regard, there is a report of a patient with 75% methemoglobin due to isobutyl nitrite poisoning who was successfully treated with toluidine-blue and HBO$_2$ (Lindenmann et al., 2006). Further research in this area is necessary. Because CN is among the most lethal poisons and toxicity is rapid, standard antidotal therapy for isolated CN poisoning is of primary importance. Hyperbaric oxygen may be an adjunct for consideration in refractory cases. Consideration for possible use of HBO therapy may change once safer antidotes such as hydroxocobalamine are available worldwide.

## Carbon Tetrachloride

Carbon tetrachloride ($CCl_4$) hepatotoxicity may be diminished by $HBO_2$. Mortality was decreased in a number of animal studies (Montani and Perret, 1967; Rapin et al., 1967; Bernacchi et al., 1984; Burk et al., 1986), and there are several case reports of patients surviving potentially lethal ingestions with HB $O_2$ therapy (Stewart et al., 1963; Larcan and Lambert, 1981; Truss and Killenberg, 1982; Zearbaugh et al., 1988). HB $O_2$ appears to inhibit the mixed-function oxidase system responsible for conversion of $CCl_4$ to hepatotoxic free radicals (Burk et al., 1984; Marzella et al., 1986). Because there are no proven antidotes for $CCl_4$ poisoning, HB $O_2$ should be considered for potentially severe $CCl_4$ exposures. However, there may be a delicate balance between oxidative processes that are therapeutic and those that mediate hepatotoxicity (Brent and Rumack, 1993). Therefore, when HB $O_2$ is being considered, it should be instituted before the onset of liver function abnormalities.

## Clostridial Myonecrosis

Successful treatment of clostridial myonecrosis (gas gangrene) depends on prompt recognition and aggressive intervention. Overall mortality rates from 11 to 52 % have been reported. Assessment of the impact of $HBO_2$ has been made in 4 retrospective comparisons and 13 case series (see citations in (Darke et al., 1977; Hart et al., 1983; Thom, 1989). Due to difficulties with comparison among patient groups, evaluation of efficacy based on mortality or 'tissue salvage' rates is difficult. Most authors comment on the clinical benefit to $HBO_2$ treatment, and share the opinion that temporal improvements of vital signs can be dramatic. As discussed above, mechanisms of action for $HBO_2$ appear to fall into several categories. Tissue hyperoxia achievable with $HBO_2$ is to the microorganisms involved, hyperoxia will diminish toxin production and the increased tissue oxygenation will improve leukocyte bacterial function (Hart et al., 1974b; Korhonen, 2000) (Mader et al., 1980). An aspect of the pathological process appears to include intravascular platelet-neutrophil aggregation with vascular occlusion (Hickey et al., 2008). Whether $HBO_2$ modifies this process requires further investigation.

## Crush Injury/Limb Reperfusion

$HBO_2$ has been used to a limited degree for acute traumatic peripheral ischemia and suturing of severed limbs. A single randomized, controlled trial (involving 36 patients) found $HBO_2$ to improve healing, reduce infection and wound dehiscence (Bouachour et al., 1996). In a case series of 23 patients,

$HBO_2$ was deemed to improve limb preservation, and changes in transcutaneous tissue $O_2$ measurements when breathing hyperbaric $O_2$ may be useful in predicting the outcome (Mathieu et al., 1990). The rationale for considering $HBO_2$ is to temporarily improve oxygenation to hypoperfused tissues, and because arterial hyperoxia will cause vasoconstriction that can diminish edema formation (Bird and Telfer, 1965; Bergofsky and Bertun, 1966). Given that neutrophil adherence to reperfused microvasculature is expected to occur, inhibition of neutrophil $\beta_2$ integrins by $HBO_2$ may also play a role.

### Osteomyelitis Refractory To Standard Management

As with many of the clinical indications for $HBO_2$, there is a theoretical/ mechanistic framework for considering its application to refractory osteomyelitis that is both attractive and in some animal models, quite compelling. Oxygen tensions in infected bone are significantly lower than in normally perfused tissues (Niinikoski and Hunt, 1972; Mader et al., 1978). $HBO_2$ causes tissue hyperoxia that augments leukocyte bacteriocidal function (Mader et al., 1980). Additionally, osteoclast activity can be enhanced by some $HBO_2$ protocols to mediate augmented bone turnover and neovascularization (Barth et al., 1990). The role of pluripotent stem/ progenitor cells in bone regeneration of adults is unclear. In culture, hyperoxia causes cells to differentiate toward an osseous phenotype, versus hypoxia that favors cartilage formation (Bassett and Herrmann, 1961). Hyperoxia also accelerates mineralization of cartilage in the bone callus (Wray and Rogers, 1968).

Utilization of $HBO_2$ is considered when surgery and antibiotic treatment have failed to resolve osteomyelitis, and when the infection is so severe that life is threatened in a patient with underlying immune compromise (such as invasive zygomycetes (fungal) infections of the skull/sinuses in diabetic patients). Except for the life-threatening cases, important patient selection issues to consider are that initial surgery adequately removed dead/infected tissue, intra-operative cultures have established the causative organism(s), persistent osteomyelitis is confirmed by surgery, bacterial cultures and/or radiological imaging, and the patient has received an adequate course of an appropriate antibiotic (usually 4–6 wk). Efficacy of $HBO_2$ in refractory osteomyelitis was described in 8 retrospective comparative trials. In most, patients were considered to serve as their own controls because they failed to heal with aggressive treatment prior to consideration of $HBO_2$ therapy (Slack et al., 1955; Depenbusch et al., 1972; Morrey et al., 1979; Davis et al., 1986; Aitasalo et al., 1998; Maynor et al., 1998; Chen et al., 2004; Kemmer and Stein, 2006). When adequate surgery is not undertaken to remove grossly infected bone, it is clear that efficacy of $HBO_2$ cannot be demonstrated

(Esterhai et al., 1987). Finally, while controlled trials are lacking, there is a growing interest in $HBO_2$ treatment of aggressive fungal infections because standard treatments often fail. A recent review described 28 published reports on this subject, with an overall survival of 75% when $HBO_2$ was included as adjunctive treatment (John et al., 2005). The mechanisms considered when treating fungal infections with $HBO_2$ include: improved leukocyte microbial killing, $O_2$-mediated suppression of microbial growth and also a possible augmentation in anti-fungal drug efficacy (Gudewicz et al., 1987; John et al., 2005).

### Progressive Necrotizing Infections

The use of $HBO_2$ for treatment of mixed aerobic-anaerobic infections such as necrotizing fasciitis and Fournier's gangrene has been reported in 7 non-randomized comparisons and 3 case series (Barzilai et al., 1985; Brown et al., 1994; Sawin et al., 1994; Shupak et al., 1995; Hollabaugh et al., 1998; Korhonen et al., 1998). As with gas gangrene, variations in time of diagnosis and clinical status on admission compromise assessments of treatment benefit. Riseman et al. reviewed their experience of 29 patients and found $HBO_2$ in addition to surgery and antibiotics reduced mortality versus surgery and antibiotics alone (Riseman et al., 1990). Brown et al. reported a multicenter experience where 30 patients received $HBO_2$, and 24 received only surgery and antibiotics (Brown et al., 1994). A trend towards increased survival was seen in the $HBO_2$ group (30% with $HBO_2$ and 42% without), but this was not statistically significant. Despite this observation, the authors stated support for continued use of $HBO_2$ because of apparent selection bias between groups. Animal trials to carefully assess the mechanisms and efficacy of $HBO_2$ have been problematic because synergistic bacterial processes are difficult to establish. Several found $HBO_2$ to potentiate antibiotic efficacy (Ollodart and Blair, 1965; Park et al., 1993).

### Thermal Burns

Some burn centers employ adjunctive $HBO_2$ to severe burns, but because controversy persists this is not a universal practice. Animal models have documented $HBO_2$ benefits as reducing conversion of partial to full thickness skin loss, of hastening epithelialization and lowering mortality (see citations in (Thom, 1989). Randomized clinical trials, albeit with small patient numbers, have reported improved rates of healing with shorter hospitalization stays and therefore reduced costs (Hart et al., 1974a; Grossman, 1978; Niu et al., 1987; Cianci et al., 1988). Uncontrolled series have also reported efficacy, but some studies have failed to find benefit (Waisbren et al., 1982; Wiseman and Grossman, 1985). Rationale for treatment has been based on reducing tissue edema, and increasing capillary angiogenesis.

## Threatened Skin Flaps and Grafts

Utilization of $HBO_2$ for graft and flap compromise was introduced in the section above on mechanisms of action. The reason to add further discussion here is because this is perhaps the one indication for $HBO_2$ where the two main categories of indications—'wound healing disorders' and 'inflammatory/reperfusion insults' may overlap. Often the first few treatments administered to a patient are at higher pressures (2.5 to 3.0 ATA) and once evidence of rapid deterioration of the flap or graft is halted, treatments at a lower pressure (2.0 to 2.4 ATA) may be extended for days or weeks to augment healing. Flaps with impaired perfusion are hypoxic and pose a risk as a nidus for necrotizing infections (Chang and Mathes, 1982; Feng et al., 1983; Gottrup et al., 1984). $HBO_2$ will improve tissue oxygenation and capillary density (Gruber et al., 1970; Manson et al., 1980). A summary of citations reporting clinical efficacy of $HBO_2$ for flaps and grafts was published in (Kindwall et al., 1991). Large clinical series also support the practice (Bowersox et al., 1986), but there is a need for more randomized trials.

## Wound Healing

Hyperbaric oxygen therapy clearly increases wound oxygen concentration, even in ischemic wounds, as long as there is some arterial inflow (Fife et al., 2002; Rollins et al., 2006). Although correction of hypoxia is one mechanism by which $HBO_2$ improves wound healing, a growing body of research makes it clear that the supra-physiologic tissue $O_2$ levels attained during hyperbaric treatment act like a drug, with multiple effects that continue after treatment termination. These mechanisms, linked to reactive species production, were reviewed in the first section of the chapter. A summary of citations reporting efficacy of $HBO_2$ for diabetic wounds and post-radiation injury were listed above.

Separate from an overall assessment on healing, $HBO_2$ efficacy has also been evaluated in terms of saving patients from major amputations. There have been three studies that included major amputations as an outcome variable when investigating efficacy of $HBO_2$ for diabetic wounds [(Faglia et al., 1996), (Abidia et al., 2003), and (Doctor et al., 1992)]. Data from these trials indicate that for every 4 patients treated with $HBO_2$ one major amputation is prevented (Kranke et al., 2004). For example, Faglia et al. (1996) performed the largest (n=70) randomized controlled trial of the efficacy of adjunctive $HBO_2$ in reducing major amputations in patients with severe, infected, ischemic diabetic foot ulcers. All subjects underwent a standardized evaluation, including angiography if the ankle brachial index was <0.9 or if the transcutaneous oxygen value was <50 mmHg, initial aggressive surgical debridement, standardized wound care, and optimized medical care. Restoration of arterial inflow by angioplasty or bypass surgery was performed if indicated and possible. Subjects in the treatment arm received HBO$_2$ at 2.4 ATA for 90 min daily. The decision to perform amputation was made by a consultant surgeon unaware of the study assignment. The treatment group underwent fewer major amputations (3/35 including 2 below the knee (BKA) and 1 above the knee (AKA) vs. control 11/33 including 7 BKA and 4 AKA; p = 0.016).

Although hypoxic wounds may occur in patients without significant peripheral arterial occlusive disease, ischemic wounds are most likely to benefit from HBO$_2$. There is one randomized controlled clinical trial that demonstrated HBO$_2$ treatment led to a more rapid decrease in size in venous leg ulcers in patients without a significant compromise of arterial blood flow (Hammarlund and Sundberg, 1994). Standard therapy of venous leg ulcers includes moist wound care, and compression, and these are successful in most patients (Robson et al., 2006). HBO$_2$ cannot replace compression therapy. It may be useful in selected patients to treat compromised skin grafts, particularly in patients with mixed venous-arterial ulcers, or in patients with uncorrected wound hypoxia despite appropriate compression and edema control.

Standard therapy of arterial insufficiency ulcers includes restoration of arterial inflow, followed by debridement and moist wound care (Hopf et al., 2006). HBO$_2$ may benefit patients in whom revascularization is unsuccessful, or is successful but fails to correct wound hypoxia, although this has only been demonstrated in diabetic patients. After revascularization, many patients develop reperfusion injuries that delay healing. Mechanistically, this is likely to be ameliorated by HBO$_2$, but additional clinical trials would be useful to evaluate efficacy.

## REFERENCES

(1999a). Consensus Development Conference on Diabetic Foot Wound Care. 7–8 April 1999, Boston, Massachusetts. American Diabetes Association. J. Am. Podiatr. Med. Assoc. 89: 475–483

(1999b). Hyperbaric oxygen therapy for wound healing. Technologica MAP Suppl.: 7–12.

Aamar, S. and A. Saada and S. Rotshenker 1992. Lesion-induced changes in the production of newly synthesized and secreted apo-E and other molecules are independent of the concomitant recruitment of blood-borne macrophages into injured peripheral nerves. J. Neurochem 59: 1287–1292.

Abidia, A. and G. Laden, G. Kuhan, B.F. Johnson, A.R. Wilkinson, P.M. Renwick, E.A. Masson and P.T. McCollum 2003. The role of hyperbaric oxygen therapy in ischaemic diabetic lower extremity ulcers: a double-blind randomised-controlled trial. Eur. J. Vasc. Endovasc Surg. 25: 513–518.

Acker, T. and H. Acker 2004. Cellular oxygen sensing need CNS function: physiological and pathological implications. J. Exp. Med. 207: 3171–3188.

Aicher, A. and C. Heeschen, C. Mildner-Rihm, C. Urbich, C. Ihling, K. Technau-Ihling, A.M. Zeiher and S. Dimmeler 2003. Essential role of endothelial nitric oxide synthase for mobilization of stem and progenitor cells. Nat. Med. 9: 1370–1376.

Aitasalo, K. and J. Niinikoski, R. Grenman and E. Virolainen 1998. A modified protocol for early treatment of osteomyelitis and osteoradionecrosis of the mandible. Head Neck 20: 411–417.

Alex, J. and G. Laden, A. Cale, S. Bennett, K. Flowers, L. Madden, E. Gardiner, P. McCollum and S. Griffin. 2005. Pretreatment with hyperbaric oxygen and its effect on neuropsychometric dysfunction and systemic inflammatory response after cardiopulmonary bypass: a prospective randomized double-blind trial. J. Thorac. Cardiovasc. Surg. 130: 1623–1630.

Allen, R. and A. Balin. 1989. Oxidative influence on development and differentiation: an overview of a free radical theory of development. Fr. Radic. Biol. Med. 6: 631–661.

Anderson, E. and R. Andelman, J. Strauch, N. Fortuin and J. Knelson. 1973. Effects of low-level carbon monoxide exposure on onset and duration of angina pectoris. Ann. Intern. Med. 79: 46–50.

Anderson, R.A. and I. Thomson and W.A. Harland 1979. The importance of cyanide and organic nitriles in fire fatalities. Fire Materials 3: 91–99.

Anderson, R.A. and W.A. Harland 1982. Fire deaths in the Glasgow area. III. The role of hydrogen cyanide. Med. Sci. Law 22: 35–40.

Annane, D. and J. Depondt, P. Aubert, M. Villart, P. Gehanno, P. Gajdos and S. Chevret. 2004. Hyperbaric oxygen therapy for radionecrosis of the jaw: a randomized, placebo-controlled, double-blind trial from the ORN96 study group. J. Clin. Oncol. 22: 4893–4900.

Aramoto, H. and J.W. Breslin, P.J. Pappas, R.W. Hobson, 2nd and W.N. Duran 2004. Vascular endothelial growth factor stimulates differential signaling pathways in in vivo microcirculation. Am J. Physiol. Heart Circ. Physiol. 287: H1590–1598.

Atochin, D. and D. Fisher, I. Demchenko and S. Thom. 2000. Neutrophil sequestration and the effect of hyperbaric oxygen in a rat model of temporary middle cerebral artery occlusion. Undersea and Hyperbaric Med. 27: 185–190.

Ballantyne, B. 1987. Hydrogen cyanide as a product of combustion and a factor in morbidity and mortality from fires. Pp. 248–291. *In*: B. Ballantyne and T. Marrs (eds.). 1987. Clinical and Experimental Toxicology of Cyanides. John Wright. Bristol, UK.

Barillo, D.J. and R. Goode, B.F. Rush, R. Lin, A. Freda and J. Anderson. 1986. Lack of correlation between carboxyhemoglobin and cyanide in smoke inhalation injury. Current. Surg. 421–423.

Barillo, D.J. and R. Goode and V. Esch. 1994. Cyanide poisoning in victims of fire: Analysis of 364 cases and review of the literature. J. Burn Care Rehabil 15: 46–57.

Barth, E. and T. Sullivan and E. Berg. 1990. Animal model for evaluating bone repair with and without adjunctive HBOT: comparing dose schedules. J. Invest. Surg. 3: 387–392.

Barzilai, A. and M. Zaaroor and C. Toledano 1985. Necrotizing fasciitis: early awareness and principles of treatment. Isr. J. Med. Sci. 21: 127–132.

Bassett, C. and I. Herrmann. 1961. Influence of oxygen concentration and mechanical factors on differentiation of connective tissues in vitro. Nature 190: 460–461.

Baud, F.J. and P. Barriot, V. Toffis, B. Riou, E. Vicaut, Y. Lecarpentier, R. Bourdon and A. Astier. 1991. Elevated blood cyanide concentrations in victims of smoke inhalation. N. Engl. J. Med. 325: 1761–1766.

Beckman, J.S. and W.H. Koppenol. 1996. Nitric oxide, superoxide, and peroxynitrite: the good, the bad, and ugly. Am. J. Physiol. 271: C1424–1437.

Bennett, M. and J. Feldmeier, N. Hampson, R. Smee and C. Milross. 2008. Hyperbaric oxygen therapy for late radiation tisue injury (Cochrane review). The Cochrane Library Issue 1.

Benson, R.M. and L.M. Minter, B.A. Osborne and E.V. Granowitz. 2003. Hyperbaric oxygen inhibits stimulus-induced proinflammatory cytokine synthesis by human blood-derived monocyte-macrophages. Clin. Exp. Immunol. 134: 57–62.

Bergofsky, E.H. and P. Bertun. 1966. Response of regional circulations to hyperoxia. J. Appl. Physiol. 21: 567–572.

Bernacchi, A. and R. Myers, B.F. Trump and L. Margello. 1984. Protection of hepatocytes with hyperoxia against carbon tetrachloride induced injury. Toxicol. Pathol. 12: 315–323.

Bert, P. 1978. Barometric Pressure (La Pression Barométrique). Undersea Medical Society. Bethesda, MD.

Bird, A.D. and A.B. Telfer. 1965. Effect of Hyperbaric Oxygen on Limb Circulation. Lancet 1: 355–356.

Birky, M.M. and M. Paabo and J.E. Brown. 1979. Correlation of autopsy data and materials in the Tennessee jail fire. Fire Safety J. 2: 17–22.

Bivalacqua, T.J. and H.C. Champion, M.F. Usta, S. Cellek, K. Chitaley, R.C. Webb, R.L. Lewis, T.M. Mills, W.J. Hellstrom and P.J. Kadowitz. 2004. RhoA/Rho-kinase suppresses endothelial nitric oxide synthase in the penis: a mechanism for diabetes-associated erectile dysfunction. Proc. Natl. Acad. Sci. USA 101: 9121–9126.

Bonomo, S. R. and J.D. Davidson, Y. Yu, Y. Xia, X. Lin and T. Mustoe. 1998. Hyperbaric oxygen as a signal transducer: upregulation of platelet derived growth factor-beta receptor in the presence of HBO2 and PDGF. Undersea Hyperb Med. 25: 211–216.

Bouachour, G. and P. Cronier, J.P. Gouello, J.L. Toulemonde, A. Talha and P. Alquier. 1996. Hyperbaric oxygen therapy in the management of crush injuries: a randomized double-blind placebo-controlled clinical trial. J. Trauma. 41: 333–339.

Bowersox, J. and M. Strauss and G. Hart 1986. Clinical experience with hyperbaric oxygen therapy in the salvage of ischemic skin flaps and grafts. J. Hyperbaric Med. 1: 141–149.

Brauchle, M. and J.O. Funk, P. Kind and S. Werner. 1996. Ultraviolet B and $H_2O_2$ are potent inducers of vascular endothelial growth factor expression in cultured keratinocytes. J. Biol. Chem. 271: 21793–21797.

Bray, P. and R.A.M. Myers and R.A. Cowley. 1983. Orogenital sex as a cause of nonfatal air embolism in pregnancy. Obstet. Gynecol. 61: 653–657.

Breen, P. H. and S.A. Isserles, J. Westley, M.F. Roizen and U.Z. Taitelman. 1995. Effect of oxygen and sodium thiosulfate during combined carbon monoxide and cyanide poisoning. Toxicol. Appl. Pharmacol. 134: 229–234.

Brem, H. and M. Tomic-Canic. 2007. Cellular and molecular basis of wound healing in diabetics. J. Clin. Invest. 117: 1219–1222.

Brennan, M.L. and W. Wu, X. Fu, Z. Shen, W. Song, H. Frost, C. Vadseth, L. Narine, E. Lenkiewicz, M.T. Borchers, A.J. Lusis, J.J. Lee, N.A. Lee, H.M. Abu-Soud, H. Ischiropoulos and S.L. Hazen. 2002. A tale of two controversies: defining both the role of peroxidases in nitrotyrosine formation in vivo using eosinophil peroxidase and myeloperoxidase-deficient mice, and the nature of peroxidase-generated reactive nitrogen species. J. Biol. Chem. 277: 17415–17427.

Brent, J.A. and B.H. Rumack 1993. Role of free radicals in toxic hepatic injury: I. Free radical biochemistry. J. Toxicol. Clin. Toxicol. 31: 173–196.

Brown, D.R. and N.L. Davis, M. Lepawsky, J. Cunningham and J. Kortbeek. 1994. A multicenter review of the treatment of major truncal necrotizing infections with and without hyperbaric oxygen therapy. Am. J. Surg. 167: 485–489.

Brown, S.D. and C.A. Piantadosi. 1989. Hyperbaric oxygen for carbon monoxide poisoning. The Lancet 1032–1033.

Brown, S.D. and C.A. Piantadosi. 1991. Recovery of energy metabolism in rat brain after carbon monoxide hypoxia. J. Clin. Invest. 89: 666–672.

Bryant, A. and C. Bayer, R. Chen, P. Guth, R. Wallace and D. Stevens. 2005. Vascular dysfunction and ischemic destruction of tissue in Streptococcus pyogenes infection: The role of streptolysin O-induced platelet/neutrophil complexes. J. Infect. Dis. 192: 1014–1022.

Bryant, A. and C. Bayer, J. Huntington and D. Stevens. 2006. Group A streptococcal myonecrosis: Increased vimentin expression after skeletal-muscle injury mediates the binding of Streptococcus pyogenes. J. Infect. Dis. 193: 1685–1692.

Bucci, M. and F. Roviezzo, V. Brancaleone, M.I. Lin, A. Di Lorenzo, C. Cicala, A. Pinto, W.C. Sessa, S. Farneti, S. Fiorucci and G.Cirino. 2004. Diabetic mouse angiopathy is linked to progressive sympathetic receptor deletion coupled to an enhanced caveolin-1 expression. Arterioscler Thromb. Vasc. Biol. 24: 721–726.

Buras, J. and D. Holt, D. Orlow, B. Belikoff, S. Pavlides and W. Reenstra. 2006. Hyperbaric oxygen protects from sepsis mortality via an interleukin-10-dependent mechanism. Crit. Care Med. 34: 2624–2629.

Burk, R.F. and J.M. Lane and K. Patel 1984. Relationship of oxygen and glutathione in protection against carbon tetrachloride-induced hepatic microsomal lipid peroxidation and covalent binding in the rat. J. Clin. Invest. 74: 1996–2001.

Burk, R.F. and R. Reiter and J.M. Land. 1986. Hyperbaric oxygen protection against carbon tetrachloride hepatotoxicity in the rat: Association with altered metabolism. Gastroenterology 90: 812–818.

Burrows, G.E. and J.L. Way. 1977. Cyanide intoxication in sheep: Therapeutic value of oxygen or colbalt. Am. J. Vet. Res. 38: 223–227.

Cabigas, B.P. and J. Su, W. Hutchins, Y. Shi, R.B. Schaefer, R.F. Recinos, V. Nilakantan, E. Kindwall, J.A. Niezgoda J.E. and Baker. 2006. Hyperoxic and hyperbaric-induced cardioprotection: role of nitric oxide synthase 3. Cardiovasc. Res. 72: 143–151.

Calabrese, V. and C. Mancuso, M. Calvani, E. Rizzarelli, D. Butterfield and A. Stella. 2007. Nitric oxide in the central nervous system: neuroprotection versus neurotoxicity. Nat. Rev. Neurosci. 8: 766–775.

Calvert, J. and J. Cahill, M. Yamaguchi-Okada and J. Zhang. 2006. Oxygen treatment after experimental hypoxia-ischemia in neonatal rats alters the expression of HIF-1a and its downstream target genes. J. Appl. Physiol. 101: 853–865.

Carmeliet, P. 2000. Mechanisms of angiogenesis and arteriogenesis. Nat. Med. 6: 389–395.

Chang, N. and S. Mathes. 1982. Comparison if the effect of bacterial inoculation in musculocutaneous and random pattern flaps. Plast Reconstr. Surg. 70: 1–10.

Chen, C. and J. Ko, T. Fu and C. Wang. 2004. Results of chronic osteomyelitis of the femur treated with hyperbaric oxygen: A preliminary report. Chang Gung Med. J. 27: 91–97.

Chen, Q. and P.D. Banick and S.R.Thom. 1996. Functional inhibition of rat polymorphonuclear leukocyte B2 integrins by hyperbaric oxygen is associated with impaired cGMP synthesis. J. Pharmacol. Expt.'l. Therap. 27: 929–933.

Chen, S.-J. and C.-T. Yu, Y.-L. Cheng, S.-Y. Yu and H.-C. Lo. 2006. Effect of hyperbaric oxygen therapy on circulating interleukin-8, nitric oxide, and insulin-like growth factors in patients with type 2 diabetes mellitus. Clin. Biochem. (In press).

Choi, S. 1983. Delayed neurologic sequelae in carbon monoxide intoxication. Arch. Neurol. 40: 433–435.

Cianci, P. and H. Lueders, H. Lee, R. Shapiro, J. Sexton, C. Williams and B. Green. 1988. Adjunctive hyperbaric oxygen reduces the need for surgery in 40–80% burns. J. Hyperbaric Med. 3: 97–101.

Clark, C.J. and D. Campbell and W.H. Reid. 1981. Blood carboxyhaemoglobin and cyanide levels in fire survivors. Lancet. 1: 1332–1335.

Clark, J. and S. Thom. Oxygen under pressure. Pp. 358–418. *In*: A.O. Brubakk and T.S. Neuman (eds.). 2003. Physiology and Medicine of Diving Saunders. Philadelphia.

Clarke, R. and C. Tenorio, J. Hussey, A. Toklu, D. Cone, J. Hinojosa, S. Desai, L. Parra, S. Rodrigues, R. Long and M. Walker 2008. Hyperbaric oxygen treatment of chronic radiation proctitis: A randomized and controlled doouble blind crossover trial with long-term follow-up. Int. J. Rad. Oncol. Biol. Phys. (In press).

Coburn, R.F. and H.J. Forman. 1987. Carbon monoxide toxicity. Pp. 439–456. *In*: A.P. Fishman, L.E. Farki and S.R. Geiger (eds.). Handbook of Physiology Williams and Wilkies, Baltimore.

Conahan, T.J. 3rd 1979. Air embolization during percutaneous Swan-Ganz catheter placement. Anesthesiology 50: 360–361.

Cope, C. 1961. The importance of oxygen in the treatment of cyanide poisoning. JAMA 175: 1061–1064.

Copeland, A.R. 1985. Accidental fire deaths: The 5-year metropolitan Dade County experience from 1979 to 1983. Z. Rechtsmed 94: 71–79.

Cramlet, S.H. and H.H. Erickson and H.A. Gorman. 1975. Ventricular function following acute carbon monoxide exposure. J. Appl. Physiol. 39: 482–486.

Darke, S.G. and A.M. King and W.K. Slack. 1977. Gas gangrene and related infection: classification, clinical features and aetiology, management and mortality. A report of 88 cases. Br. J .Surg. 64: 104–112.

Davis, J.C. and J.D. Heckman, J.C. DeLee and F.J. Buckwold. 1986. Chronic non-hematogenous osteomyelitis treated with adjuvant hyperbaric oxygen. J. Bone Joint Surg. 68: 1210–1217.

Dekleva, M. and A. Neskovic, A. Vlahovic, B. Putnikovic, B. Beleslin and M. Ostojic. 2004. Adjunctive effect of hyperbaric oxygen treatment after thrombolysis on left ventricular function in patients with acute myocardial infarction. Am. Heart J. 148: E14.

Denham, J. and M. Hauer-Jensen. 2002. The radiotherapeutic injury-a complex 'wound'. Radiother Oncol 63: 129–145.

Dennog, C. and A. Hartmann, G. Frey and G. Speit. 1996. Detection of DNA damage after hyperbaric oxygen (HBO) therapy. Mutagenesis 11: 605–609.

Dennog, C. and C. Gedik, S. Wood and G. Speit. 1999a. Analysis of oxidative DNA damage and HPRT mutations in humans after hyperbaric oxygen treatment. Mutation Res. 431: 351–359.

Dennog, C. and P. Radermacher, Y.A. Barnett and G. Speit. 1999b. Antioxidant status in humans after exposure to hyperbaric oxygen. Mutation. Res. 428: 83–89.

Depenbusch, F. and R. Thompson and G. Hart. 1972. Use of hyperbaric oxygen in the treatment of refractory osteomyelitis: a preliminary report. J. Trauma 12: 807–812.

Doctor, N. and S. Pandya and A.Supe. 1992. Hyperbaric oxygen therapy in diabetic foot. J. Postgrad. Med. 38, 112–114, 111.

Du, X.L. and D. Edelstein, S. Dimmeler, Q. Ju, C. Sui and M. Brownlee. 2001. Hyperglycemia inhibits endothelial nitric oxide synthase activity by posttranslational modification at the Akt site. J. Clin. Invest. 108: 1341–1348.

Du, X.L. and D. Edelstein, S. Obici, N. Higham, M.H. Zou and M. Brownlee. 2006. Insulin resistance reduces arterial prostacyclin synthase and eNOS activities by increasing endothelial fatty acid oxidation. J. Clin. Invest. 116: 1071–1080.

Ducasse, J.L. and P. Celsis and J.P. Marc-Vergnes. 1995. Non-comatose patients with acute carbon monoxide poisoning: hyperbaric or normobaric oxygenation? Undersea and Hyperbaric Med. 22: 9–15.

Dulak, J. and A. Jozkowicz. 2003. Regulation of vascular endothelial growth factor synthesis by nitric oxide: facts and controversies. Antioxid Redox Signal 5: 123–132.

Ema, M. and K. Hirota, J. Mimura, H. Abe, J. Yodoi, K. Sogawa, L. Poellinger and Y. Fujii-Kuriyama. 1999. Molecular mechanisms of transcription activation by HLF and HIF1alpha in response to hypoxia: Their stabilization and redox signal-induced interaction with CBP/p300. Embo. J. 18: 1905–1914.

End, E. and C.W. Long. 1942. Oxygen under pressure in carbon monoxide poisoning. J. Ind. Hyg. Toxicol. 24: 302–306.

Escobar, S. and J. Slade, T. Hunt and P. Cianci. 2005. Adjuvant hyperbaric oxygen therapy (HBO$_2$) for treatment of necrotizing fasciitis reduces mortality and amputation rate. Undersea Hyperb. Med. 32: 437–443.

Esterhai, J.L. and J. Pisarello, C.T. Brighton, R.B. Heppenstall, H. Gellman and G. Goldstein. 1987. Adjunctive hyperbaric oxygen therapy in the treatment of chronic refractory osteomyelitis. J. Trauma. 27: 763–768.

Faglia, E. and F. Favales, A. Aldeghi, P. Calia, A. Quarantiello, G. Oriani, M. Michael, P. Campagnoli and A. Morabito. 1996. Adjunctive systemic hyperbaric oxygen therapy in treatment of severe prevalently ischemic diabetic foot ulcer. Diabetes Care 19: 1338–1343.

Falanga, V. 2005. Wound healing and its impairment in the diabetic foot. Lancet. 366: 1736–1743.

Feng, L. and D. Price, D. Hohn and S. Mathes. 1983. Blood flow changes and leukocyte mobilization in infection: a comparison between ischemic and well-perfused skin. Surg.Forum. 603–604.

Fife, C.E. and C. Buyukcakir, G.H. Otto, P.J. Sheffield, R.A. Warriner, T.L. Love and J. Mader. 2002. The predictive value of transcutaneous oxygen tension measurement in diabetic lower extremity ulcers treated with hyperbaric oxygen therapy: a retrospective analysis of 1,144 patients. Wound Repair Regen 10: 198–207.

Fildissis, G. and K. Venetsanou, P. Myrianthefs, S. Karatzas, V. Zidianakis and G. Baltopoulos. 2004. Whole blood pro-inflammatory cytokines and adhesion molecules post-lipopolysaccharides exposure in hyperbaric conditions. Eur. Cytokine Netw. 15: 217–221.

Friedman, G. and P. Froom and L. Sazbon. 1999. Apolipotrotein E-epsilon 4 genotype predicts a poor outcome in survivors of traumatic brain injury. Neurology 52: 244–248.

Fukumura, D. and T. Gohongi, A. Kadambi, Y. Izumi, J. Ang, C.O. Yun, D.G. Buerk, P.L. Huang and R.K. Jain. 2001. Predominant role of endothelial nitric oxide synthase in vascular endothelial growth factor-induced angiogenesis and vascular permeability. Proc. Natl. Acad. Sci. USA 98: 2604–2609.

Gallagher, K.A. and L.J. Goldstein, D.G. Buerk, A. Nedeau, M. Xaio, H. Chen, S.R. Thom, Z.-J. Liu and O.C. Velazquez. 2007. Diabetic impairments in NO-mediated endothelial progenitor-cell mobilization and homing are reversed by hyperoxia and SDF-1a. J. Clin. Invest. 117: 1249–1259.

Ginsberg, M.D. and R.E. Myers. 1974. Experimental carbon monoxide encephalopathy in the primate. I. Physiologic and metabolic aspects. Arch. Neurol. 30: 202–208.

Goldstein, L.J. and K.A. Gallagher, S.M. Bauer, R.J. Bauer, V. Baireddy, Z.J. Liu, D.G. Buerk, S.R. Thom and O.C. Velazquez. 2006. Endothelial progenitor cell release into circulation is triggered by hyperoxia-induced increases in bone marrow nitric oxide. Stem Cells 24: 2309–2318.

Goodhart, G.L. 1994. Patient treated with antidote kit and hyperbaric oxygen survives cyanide poisoning. South. Med. J. 87: 814–816.

Gorman, D.F. and D. Clayton, J.E. Gilligan and R.K. Webb. 1992. A longitudinal study of 100 consecutive admissions for carbon monoxide poisoning to the Royal Adelaide Hospital. Anaesth. Intens. Care. 20: 311–316.

Gottrup, F. and R. Firmin, T.K. Hunt and S.J. Mathes. 1984. The dynamic properties of tissue oxygen in healing flaps. Surgery 95: 527–536.

Goulon, M. and A. Barois, M. Rapin, F. Nouailhat, S. Grosbuis and J. Labrousse. 1969. Carbon monoxide poisoning and acute anoxia due to breathing coal gas and hydrocarbons. Ann. Med. Interne. Paris. J. Hyperbaric Med. 1: 23–41, 1986. 120: 335–349.

Gregorevic, P. and G. Lynch and D. Williams. 2001. Hyperbaric oxygen modulates antioxidant enzyme activity in rat skeletal muscles. Eur. J. Appl. Physiol. 86: 24–27.

Grossman, R.A. 1978. Hyperbaric oxygen in the treatment of burns. Ann. Plastic. Surg. 1: 163–171.

Gruber, R.P. and F.B. Brinkley, J.J. Amato and J.A. Mendelson. 1970. Hyperbaric oxygen and pedicle flaps, skin grafts, and burns. Plast. Reconstr. Surg. 45: 24–30.

Gu, G.-J. and Y.-P. Li, Z.-Y. Peng, J.-J. Xu, Z.-M. Kang, W.-G. Xu, H.-Y. Tao, R. Ostrowski, J. Zhang and Z.-J. Sun. 2008. Mechanism of ischemic tolerance induced by hyperbaric oxygen preconditioning involves upregulation of hypoxia-inducible factor-1a and erythropoietin in rats. J. Appl. Physiol. 104: 1185–1191.

Gudewicz, T. and J. Mader and C. Davis. 1987. Combined effects of hyperbaric oxygen and antifungal agents on the growth of Candida albicans. Aviat. Space Environ. Med. 58: 673–678.

Hailey, D. and P. Jacobs, D. Perry, A. Chuck, A. Morrison and R. Bondreau. 2007. Technology Report: Adjunctive hyperbaric oxygen therapy for diabetic foot ulcer: An economic analysis. Canadian Agency for Drugs and Technologies in Health. Report No. 75: 1–32.

Hammarlund, C. and T. Sundberg. 1994. Hyperbaric oxygen reduced size of chronic leg ulcers: a randomized double-blind study. Plast Reconstr. Surg. 93: 829–833; discussion 834.

Hampson, N. 2000. Hyperbaric oxygen for carbon monoxide poisoning. MJA 172: 141–142.

Hart, G.B. and R.R. O'Reilly, N.D. Broussard, R.H. Cave, D.B. Goodman and R.L. Yanda. 1974a. Treatment of burns with hyperbaric oxygen. Surg. Gynec. Obst. 139: 693–696.

Hart, G.B. and R.R. O'Reilly, R.H. Cave and N.D. Broussard. 1974b. The treatment of clostridial myonecrosis with hyperbaric oxygen. J. Trauma. 14: 712–715.

Hart, G.B. and R.C. Lamb and M.B. Strauss. 1983. Gas gangrene: I. A collective review. J. Trauma. 23: 991–1000.

Hart, G.B. and M.B. Strauss, P.A. Lennon and D.D. Whitcraft. 1985. Treatment of smoke inhalation by hyperbaric oxygen. J. Emerg. Med. 3: 211–215.

Hattori, K. and S. Dias, B. Heissig, N.R. Hackett, D. Lycen, M. Tateno, D.J. Hicklin, Z. Zhu, L. Witte, R.G. Crystal, M.A. Moore and S. Rafii. 2001. Vascular endothelial growth factor and angiopoietin-1 stimulate postnatal hematopoiesis by recruitment of vasculogenic and hematopoietic stem cells. J. Exp. Med. 193: 1005–1014.

Hendricks, P. and D. Hall, W. Hunter and P. Haley. 1977. Extension of pulmonary $O_2$ tolerance in man at 2 ATA by intermittent O2 exposure. J. Appl. Physiol. 42: 593–599.

Henry, C.R. and D. Satran, B. Lindgren, C. Adkinson, C.I. Nicholson and T.D. Henry. 2006. Myocardial injury and long-term mortality following moderate to severe carbon monoxide poisoning. Jama 295: 398–402.

Hess, C.L. and M.A. Howard and C.E. Attinger. 2003. A review of mechanical adjuncts in wound healing: hydrotherapy, ultrasound, negative pressure therapy, hyperbaric oxygen, and electrostimulation. Ann. Plast. Surg. 51: 210–218.

Hickey, M. and R. Kwan, M. Awad, C. Kennedy, L. Young, P. Hall, L. Cordner, D. Lyas, J. Emmins and J. Rood. 2008. Molecular and cellular basis of microvascular perfusion deficits induced by Clostridium perfringens and Clostridium septicum. PLoS Pathog. 4, e1000045.

Hildbrand, P. and V. Cirulli, R.C. Prinsen, K.A. Smith, B.E. Torbett, D.R. Salomon and L. Crisa. 2004. The role of angiopoietins in the development of endothelial cells from cord blood CD34+ progenitors. Blood 104: 2010–2019.

Hirata, T. and Y. Cui, T. Funakoshi, Y. Mizukami, Y. Ishikawa, F. Shibasaki, M. Matsumoto and T. Sakabe. 2007. The temporal profile of genomic responses and protein synthesis in ischemic tolerance of the rat brain induced by repeated hyperbaric oxygen. Brain Res. 1130: 214–222.

Hollabaugh, R.S., Jr. and R.R. Dmochowski, W.L. Hickerson and C.E. Cox. 1998. Fournier's gangrene: therapeutic impact of hyperbaric oxygen. Plast Reconstr. Surg. 101: 94–100.

Hopf, H. and C. Uneo, R. Aslam, K. Burnand, C. Fife, L. Grant, A. Holloway, M. Iafrati, R. Mani, B. Misare, N. Rosen, D. Shapshak, B. Slade, J. West and A. Barbul. 2006. Guidelines for the treatment of arterial insufficiency ulcers. Wound Repair Regen. 14: 693–710.

Hopkins, R. and L. Weaver, K. Valentine, C. Mower, S. Churchill and J. Carlquist. 2008. Apolipoprotein E genotype and response of carbon monoxide poisoning to hyperbaric oxygen treatment. Am. J. Respir. Crit. Care. Med. 176: 1001–1006.

Hsiao. 2004. Delayed encephalopathy after carbon monoxide intoxication-long term prognosis and correlation of clinical manifestations and neuroimages. Acta Neurol., Taiwan 13: 64–70.

Hsu, L.H. and J.H. Wang. 1996. Treatment of carbon monoxide poisoning with hyperbaric oxygen. Chinese. Med. J. 58: 407–413.

Hunt, T. and R. Aslam, S. Beckert, S. Wagner, Q. Ghani, M. Hussain, S. Roy and C. Sen. 2007. Aerobically derived lactate stimulates revascularization and tissue repair via redox mechanisms. Antioxid. Redox. Signal 9: 1115–1124.

Ikeda, K.K.T. and M. Harada. 2004. Factors for PBPC collection efficiency and collection predictors. Transfus. Apher. Sci. 31: 245–259.

Isom, G.E. and J.L. Way. 1974. Effect of oxygen on cyanide intoxication. VI. Reactivation of cyanide inhibited glucose metabolism. J. Pharmacol. Exp. Ther. 189: 235–243.

Ivanov, K.P. 1959. The effect of elevated oxygen pressure on animals poisoned with potassium cyanide. Pharmacol. Toxicol. 22: 476–479.

Jamieson, D. and B. Chance, E. Cadenas and A. Boveris. 1986. The relation of free radical production to hyperoxia. Annu. Rev. Physiol. 48: 703–719.

John, B. and G. Chamilos and D. Kontoyiannis. 2005. Hyperbaric oxygen as an adjunctive treatment for zygomycosis. Clin. Microbiol. Infect. 11: 515–517.

Juttner, B. and D. Scheinichen, S. Bartsch, J. Heine, H. Ruschulte, H. Elsner, W. Franko and K.Jaeger. 2003. Lack of toxic side effects in neutrophils following hyperbaric oxygen. Undersea and Hyperbaric Med. 30: 305–311.

Kaga, S. and L. Zhan, M. Matusmoto and N. Maulik. 2005. Resveratrol enhances neovascularization in the infarcted rat myocardium through the induction of thioredoxin-1, heme oxygenase-1 and vascular endothelial growth factor. J. Mol. Cell Cardiol. 38: 813–822.

Kalns, J. and J. Lane, A. Delgado, J. Scruggs, E. Ayala, E. Gutierrez, D. Warren, D. Niemeyer, E. George Wolf and rR.A. Bowden. 2002. Hyperbaric oxygen exposure temporarily reduces Mac-1 mediated functions of human neutrophils. Immunol. Lett. 83: 125–131.

Kang, T.S. and G.K. Gorti, S.Y. Quan, M. Ho and R.J. Koch. 2004. Effect of hyperbaric oxygen on the growth factor profile of fibroblasts. Arch. Facial. Plast. Surg. 6: 31–35.

Keays, F. 1909. Compressed air illness, with a report of 3,692 cases. Dept. Med. Publ. Cornell Univ. Med. Coll. 2: 1–55.

Kemmer, A. and T. Stein. Persistent osteomyelitis. Pp. 429–449. *In*: D Mathieu (ed.). 2006. Handbook on Hyperbaric Medicine. Springer. Dodrecht, The Netherlands.

Kihara, K. and S. Ueno, M. Sakoda and T. Aikou. 2005. Effects of hyperbaric oxygen exposure on experimental hepatic ischemia reperfusion injury: relationship between its timing and neutrophil sequestration. Liver Transpl. 11: 1574–1580.

Kim, C. and H. Choi, Y. Chun, G. Kim, J. Park and M. Kim. 2001. Hyperbaric oxygenation pretreatment induces catalase and reduces infarct size in ischemic rat myocardium. Pflugers Arch. 442: 519–525.

Kindwall, E. and L. Gottlieb and D. Larson. 1991. Hyperbaric oxygen therapy in plastic surgery: a review article. Plast. Reconstr. Surg. 88: 898–908.

Kirk, M.A. and R. Gerace and K.W. Kulig. 1993. Cyanide and methemoglobin kinetics in smoke inhalation victims treated with the cyanide antidote kit. Ann. Emerg. Med. 22: 1413–1418.

Korhonen, K. 2000. Hyperbaric oxygen therapy in acute necrotizing infections with a special reference to the effects on tissue gas tensions. Ann. Chir. Gynaecol. Suppl.: 7–36.

Korhonen, K. and M. Hirn and J.Niinikoski. 1998. Hyperbaric oxygen in the treatment of Fournier's gangrene. Eur. J. Surg. 164: 251–255.

Kranke, P. and M. Bennett, I. Roeckl-Wiedmann and S. Debus. 2004. Hyperbaric oxygen therapy for chronic wounds. Cochrane Database Syst. Rev., CD004123.

Labrouche, S. and S. Javorschi, D. Leroy, G. Gbikpi-Benissan and G. Freyburger. 1999. Influence of hyperbaric oxygen on leukocyte functions and haemostasis in normal volunteer divers. Thromb. Res. 96: 309–315.

Lahat, N. and H. Bitterman, N. Yaniv, A. Kinarty and N. Bitterman. 1995. Exposure to hyperbaric oxygen induces tumor necrosis factor-alpha (TNF-alpha) secretion from rat macrophages. Clin. Exp. Immunol. 102: 655–659.

Lakshmi, V. and W. Nauseef and T. Zenser. 2005. Myeloperoxidase potentiates nitric oxide-mediated nitrosation. J. Biol. Chem. 280: 1746–1753.

Lamy, M. and M. Hauguet. 1969. Fifty patients with carbon monoxide intoxication treated with hyperbaric oxygen therapy. Acta Ahes. Belgica. 1: 49–53.

Larcan, A. and H. Lambert. 1981. Current epidemiological, clinical, biological, and therapeutic aspects of acute carbon monoxide intoxication. Bull Acad. Natl. Med. (Paris) 165, 471.

Levin, L.L. and G.J. Stewart, P.R. Lynch and A.A. Bove. 1981. Blood and blood vessel wall changes induced by decompression sickness in dogs. J. Appl. Physiol. 50: 944–949.

Li, Y. and C. Zhou, J. Calvert, A. Colohan and J. Zhang. 2005. Multiple effects of hyperbaric oxygen on the expression of HIF-1a and apoptotic genes in a global ischemia-hypotension rat model. Exp. Neurol. 191: 198–210.

Lin, S. and K.G. Shyu, C.C. Lee, B.W. Wang, C.C. Chang, Y.C. Liu, F.Y. Huang and H. Chang. 2002. Hyperbaric oxygen selectively induces angiopoietin-2 in human umbilical vein endothelial cells. Biochem. Biophys. Res. Commun. 296: 710–715.

Lindenmann, J. and V. Matzi, P. Kaufmann, P. Krisper, A. Maier, C. Porubsky and F. Smolle-Juettner. 2006. Hyperbaric oxygenation in the treatment of life-threatening isobutyl nitrite-induced methemoglobinemia-a case report. Inhalation Toxicology 18: 1047–1049.

Litovitz, T.L. and R.F. Larkin and R.A.M. Myers. 1983. Cyanide poisoning treated with hyperbaric oxygen. Am. J. Emerg. Med. 1: 94–101.

Logue, C. 2006. Hyperbaric Oxygen for Carbon Monoxide Poisoning. *In*: C. Review (ed.). 2006. Letter to the Editor <*http://www.cochranefeedback.com/cf/cda/citation.do?id=9569#9569*>.

Lundquist, P. and R. Lennart and B. Sorbo. 1989. The role of hydrogen cyanide and carbon monoxide in fire casualties: A prospective study. Forensic Sci. Int. 43, 9–14.

Madden, M.R. and J.L. Finkelstein and C.W. Goodwin. 1986. Respiratory care of the burn patient. Clin. Plast. Surg. 13: 29–38.

Mader, J.T. and J.C. Guckian, D.L. Glass, J.A. Reinarz and R.V.T.R. Volume 1978. Therapy with hyperbaric oxygen for experimental osteomyelitis due to staphylococcus aureus in rabbits. J. Infect. Dis. 138: 312–318.

Mader, J.T. and G.L. Brown, J.C. Guckian, C.H. Wells and J.A. Reinarz. 1980. A mechanism for the amelioration by hyperbaric oxygen of experimental staphylococcal osteomyelitis in rabbits. J. Infect. Dis. 142: 915–922.

Maeda, Y. and Y. Kawasaki, I. Jibiki, N. Yamaguchi, H. Matsuda and K. Hisada 1991. Effect of therapy with oxygen under high pressure on regional cerebral blood flow in the interval form of carbon monoxide poisoning: observation from subtraction of technetium-99m HMPAOSPECT brain imaging. Eur. Neurol. 31: 380–383.

Maier, A. and A. Gaggl, H. Klemen, G. Santler, U. Anegg, B. Fell, H. Karcher, F. Smolle-Juttner and G. Friehs. 2000. Review of severe osteoradionecrosis treated by surgery alone or surgery with postoperative hyperbaric oxygenation. Br. J. Oral Maxillofac Surg. 38: 173–176.

Makino, Y. and A. Kanopka, W. Wilson, H. Tanaka and L. Poellinger. 2002. Inhibitory PAS domain protein (IPAS) is a hypoxia-inducible splicing variant of the hypoxia-inducible factor-3a locus. J. Biol. Chem. 277: 32405–32408.

Manson, P.N. and M.J. Im, R.A.M. Myers and J.E. Hoopes. 1980. Improved capillaries by hyperbaric oxygen in skin flaps. Surg.Forum. 31: 56–566.

Martin, J.D. and S.R. Thom. 2002. Vascular leukocyte sequestration in decompression sickness and prophylactic hyperbaric oxygen therapy in rats. Aviat. Space Environ. Med. 73: 565–569.

Martin, M. and J. Lefaix and S. Delanian. 2000. TGF-beta1 and radiation fibrosis: A master switch and a specific therapeutic target? Int. J. Radiat. Oncol. Biol. Phys. 47: 277–290.

Marx, R.E. and R.P. Johnson and S.N. Kline. 1985. Prevention of osteoradionecrosis: a randomized prospective clinical trial of hyperbaric oxygen versus penicillin. *JADA.* 111: 49–54.

Marx, R.E. and W.J. Ehler, P. Tayapongsak and L.W. Pierce. 1990. Relationship of oxygen dose to angiogenesis induction in irradiated tissue. Am. J. Surg. 160: 519–524.

Marzella, L. and K. Muhvich and R.A.M. Myers. 1986. Effect of hyperoxia on liver necrosis induced by hepatotoxins. Virchows. Arch. 51: 497–507.

Mathieu, D. and M. Nolf, A. Durocher, F. Saulnier, P. Frimat, D. Furon and F. Wattel. 1985. Acute carbon monoxide poisoning risk of late sequelae and treatment by hyperbaric oxygen. Clin. Toxicol. 23: 315–324.

Mathieu, D. and F. Wattel, G. Bouachour, V. Billard and J.F. Defoin. 1990. Post-traumatic limb ischemia: prediction of final outcome by transcutaneous oxygen measurements in hyperbaric oxygen. J. Trauma. 30: 307–314.

Mathieu, D. and F. Wattel, M. Mathieu-Nolf, C. Durak, J.P. Tempe, G. Bouachour, and J.M. Sainty. 1996. Randomized prospective study comparing the effect of HBO versus 12 hours NBO in non-comatose CO poisoned patients. Undersea and Hyperbaric. Med. 23 (suppl.): 7.

Maulik, N. 2002. Redox signaling and angiogenesis. Antioxid Redox Signal 4: 805–815.

Maynard, M. and A. Evans, W. Shi, W. Kim, F.-F. Liu and M. Ohh. 2007. Dominant-negative HIF-3a4 suppresses VHL-null renal cell carcinoma progression. Cell Cycle 6: 2810–2816.

Maynor, M. and R. Moon, E. Camporesi, T. Fawcett, P. Fracica, H. Norvell and L. Levins. 1998. Chromic osteomyelitis of the tibia: treatment with hyperbaric oxygen and autogenous microsurgical muscle transplantation. J. South Orthop. Assoc. 7: 43–57.

Mazariegos, G. and K. O'Toole, L. Mieles, I. Dvorchik, M. Meza, G. Briassoulis, J. Arzate, G. Osorio, J. Fung and J. Reyes. 1999. Hyperbaric oxygen therapy for hepatic artery thrombosis after liver transplantation in children. Liver Transpl. Surg. 5: 429–436.

McCarron, M. and K. Muir and J. Nicoll. 2000. Prospective study of apolipoprotein E genotype and functional outcome following ischemic stroke. Arch Neurol 57: 1480–1484.

Meyer, B.C. 1928. Experimentelle erfahrungen uber die kohlenoxydverguftung des zentralnervens systems. Z. Ges. Neurol. Psychiatr. 112: 187–212.

Mileski, W.J. and R.K. Winn N.B. Vedder, T.H. Pohlman, J.M. Harlan and C.L. Rice. 1990. Inhibition of CD18-dependent neutrophil adherence reduces organ injury after hemorrhagic shock in primates. Surgery 108: 206–212.

Mileski, W.J. and P. Sikes, L. Atiles, E. Lightfoot, P. Lipsky and C. Baxter. 1993. Inhibition of leukocyte adherence and susceptibility to infection. J. Surg. Res. 54: 349–354.

Milovanova, T. and V.M. Bhopale, E.M. Sorokina, J.S. Moore, T.K. Hunt, O.C. Velazquez and S.R. Thom. 2008. Lactate stimulates vasculogenic stem cells via the thioredoxin system and engages an autocrine activation loop involving hypoxia inducible factor-1. Mol. Biol. Cell 28: 6248–6261.

Milovanova, T. and V.M. Bhopale, E.M. Sorokina, J.S. Moore, T.K. Hunt, M. Hauer-Jensen, O.C. Velazquez and S.R. Thom. 2009. Hyperbaric oxygen stimulates vasculogenic stem cell growth and differentiation in vivo. J. Appl. Physiol. 106: 711–728.

Mohler, S.R. 1975. Air crash survival: injuries and evacuation toxic hazards. Aviat. Space. Environ. Med. 46: 86–88.

Mojsilovic-Petrovic, J. and D. Callaghan, H. Cui, C. Dean, D. Stanimirovic and W. Zhang 2007. Hypoxia-inducible factor-1 is involved in the regulation of hypoxia-stimulated expression of monocyte chemoattractant protein-1 (MCP-1/CCL2) and MCP-5 (Ccl12) in astrocytes. Neuroinflammation 4: 1–15.

Montani, S. and C. Perret. 1967. Oxygenation hyperbare dans l'intoxication experimentale au tetrachlorure de carbon. Rev. Fr Etudes Clin. Biol. 12: 274–278.

Moon, R. and E. DeLong. 1999. Hyperbaric oxygen for carbon monoxide poisoning. MJA 170: 197–198.

Moore, S.J. and J.C. Norris, D.A. Walsh and A.S. Hume. 1987. Antidotal use of methemoglobin forming cyanide antagonists in concurrent carbon monoxide/cyanide intoxication. J. Pharmacol. Exp. Ther. 242: 70–73.

Morrey, B.F. and J.M. Dunn, R.D. Heimbach and J. Davis. 1979. Hyperbaric oxygen and chronic osteomyelitis. Clin. Ortho. Rel. Res. 144: 121–127.

Myers, R. and S. Snyder and T. Emhoff. 1985. Subacute sequelae of carbon monoxide poisoning. Ann. Emerg. Med. 14: 1163–1167.

Narkowicz, C.K. and J.H. Vial and P.W. McCartney. 1993. Hyperbaric oxygen therapy increases free radical levels in the blood of humans. Free Radic. Res. Commun. 19: 71–80.

Nemoto, S. and K. Takeda, Z.X. Yu, V.J. Ferrans and T. Finkel. 2000. Role for mitochondrial oxidants as regulators of cellular metabolism. Mol. Cell Biol. 20: 7311–7318.

Nie, H. and L. Xiong, N. Lao, S. Chen, N. Xu and Z. Zhu. 2006. Hyperbaric oxygen preconditioning induces tolerance against spinal cord ischemia by upregulation of antioxidant enzymes in rabbits. J. Cereb. Blood Flow Metab. 26: 666–674.

Niinikoski, J. and T.K. Hunt. 1972. Oxygen tensions in healing bone. Surg. Gynecol. Obstet. 134: 746–750.

Niu, A.K.C. and C. Yand, H.C. Lee, S.H. Chen, L.P. Chang and R.V.T.R. Volume. 1987. Burns treated with adjunctive hyperbaric oxygen therapy: a comparative study in humans. J. Hyperbaric Med. 2: 75–85.

Norkool, D.M. and J.N. Kirkpatrick. 1985. Treatment of acute carbon monoxide poisoning with hyperbaric oxygen : a review of 115 cases. Ann. Emerg. Med. 14: 1168–1171.

Norris, J.C. and S.J. Moore and A.S. Hume. 1986. Synergistic lethality induced by the combination of carbon monoxide and cyanide. Toxicology 40: 121–129.

Nossum, V. and S. Koteng and A. Brubakk. 1999. Endothelial damage by bubbles in the pulmonary artery of the pig. Undersea Hyperb. Med. 26: 1–8.

Nossum, V. and A. Hjelde and A. Brubakk. 2002. Small amounts of venous gas embolism cause delayed impairment of endothelial function and increase polymorphonuclear neutrophil infiltration. Eur. J. Appl. Physiol. 86: 209–214.

Ollodart, R. and E. Blair. 1965. High-pressure oxygen as an adjunct in experimental bacteremic shock. JAMA 191: 132–135.

Pace, N. and E. Strajman and E.L.Walker. 1950. Acceleration of carbon monoxide elimination in man by high pressure oxygen. Science 111: 652–654.

Parenti, A. and L. Morbidelli, X.L. Cui, J.G. Douglas, J.D. Hood, H.J. Granger, F. Ledda and M. Ziche. 1998. Nitric oxide is an upstream signal of vascular endothelial growth factor-induced extracellular signal-regulated kinase1/2 activation in postcapillary endothelium. J. Biol. Chem. 273: 4220–4226.

Park, M.K. and R.A.M. Myers and L. Marzella. 1993. Hyperoxia and prolongation of aminoglycoside-induced postantibiotic effect in pseudomonas aeruginosa: role of reactive oxygen species. Antimicrobial. Agents Chemotherapy 37: 120–122.

Peichev, M. and A.J. Naiyer, D. Pereira, Z. Zhu, W.J. Lane, M. Williams, M.C. Oz, D.J. Hicklin, L. Witte, M.A. Moore and S. Rafii. 2000. Expression of VEGFR-2 and AC133 by circulating human CD34(+) cells identifies a population of functional endothelial precursors. Blood 95: 952–958.

Peirce, E.C., 2nd and A. Zacharias, J.M. Alday, Jr., B.A. Hoffman J.H. and Jacobson 2nd. 1972. Carbon monoxide poisoning: experimental hypothermic and hyperbaric studies. Surgery 72: 229–237.

Perrins, D.J.D. and M.B. Cantab. 1967. Influence of hyperbaric oxygen on the survival of split skin grafts. Lancet II: 868–871.

Piantadosi, C.A. and J. Zhang, E.D. Levin, R.J. Folz and D.E. Schmechel. 1997. Apoptosis and delayed neuronal damage after carbon monoxide poisoning in the rat. Exp. Neurol. 147: 103–114.

Pitt, B.R. and E.P. Radford, G.H. Gurtner and R.J. Traystman. 1979. Interaction of carbon monoxide and cyanide on cerebral circulation and metabolism. Arch Environ Health 34: 354–359.

Platzbecker, U.B.M. and K. Zimmer, L. Lerche, C. Rutt, G. Ehninger and K. Holig. 2005. Second donation of granulocyte-colony-stimulating factor-mobilized peripheral blood progenitor cells: risk factors associated with a low yield of CD34+ cells. Transfusion 45: 11–15.

Pol, B. and T. Wattelle. 1854. Mémoire sur les effets de la compression de l'air appliquée au creusement des puits à houille. Ann Hyg Pub. Med. Leg 2: 241–279.

Raphael, J.C. and D. Elkharrat, M.C.J. Guincestre, C. Chastang, J.B. Vercken, V. Chasles and P. Gajdos. 1989. Trial of normobaric and hyperbaric oxygen for acute carbon monoxide intoxication. Lancet. 414–419.

Rapin, M. and C. Got and J.R. Le Gall. 1967. Effect de l'oxygene hyperbare sur la toxicite tetrachlorure de carbone chez le rat. Rev. Fr. Etudes Clin. Biol. 12, 594–599.

Riseman, J.A. and W.A. Zamboni, A. Curtis, D.R. Graham, H.R. Konrad and D.S.Ross. 1990. Hyperbaric oxygen therapy for necrotizing fasciitis reduces mortality and the need for debridements. Surgery 108: 847-850.

Robson, M.C. and D.M. Cooper, R. Aslam, L.J. Gould, K.G. Harding, D.J. Margolis, D.E. Ochs, T.E. Serena, R.J. Snyder, D.L. Steed, D.R. Thomas and L.Wiersma-Bryant. 2006. Guidelines for the treatment of venous ulcers. Wound Repair Regen. 14: 649–662.

Roche, L. and A. Bertoye and P. Vincent. 1968. Comparison de deux groupes de vingt intoxications oxycarbonees traitees par oxygene normobare et hyperbare. Lyon Med. 49: 1483–1499.

Rojas, A. and H. Figueroa, L. Re and M. Morales. 2006. Oxidative stress at the vascular wall. Mechanistic and pharmacological aspects. Arch. Med. Res. 37: 436–448.

Rollins, M.D. and J.J. Gibson, T.K. Hunt and H.W. Hopf. 2006. Wound oxygen levels during hyperbaric oxygen treatment in healing wounds. Undersea Hyperb. Med. 33: 17–25.

Ross, R.M. and T.A. McAllister. 1965. Protective action of hyperbaric oxygen in mice with pneumococcal sipticaemia. Lancet. 579–581.

Rothfuss, A. and P. Radermacher and G.Speit. 2001. Involvement of heme oxygenase-1 (HO-1) in the adaptive protection of human lymphocytes after hyperbaric oxygen (HBO) treatment. Carcinogenesis 22: 1979–1985.

Salceda, S. and J. Caro. 1997. Hypoxia-inducible factor 1 alpha protein is rapidly degraded by the ubiquitin-proteasome system under normoxic conditions: its stabilization by hypoxia depends upon redox-induced changes. J. Biol. Chem. 272: 22642–22647.

Sampson, J. and Y. Ye, H. Rosen and J. Beckman. 1998. Myeloperoxidase and horseradish peroxidase catalyze tyrosine nitration in proteins from nitrite and hydrogen peroxide. Arch. Biochem. Biophys. 356: 207–213.

Satran, D. and C.R. Henry, C. Adkinson, C.I. Nicholson, Y. Bracha and T.D. Henry. 2005. Cardiovascular manifestations of moderate to severe carbon monoxide poisoning. J. Am. Coll. Cardiol. 45: 1513–1516.

Saunders, A. and W. Strittmatter and D. Schmechel. 1993. Association of apolipoprotein E allele epsilon 4 with late-onset familial and sporadic Alzheimer's disease. Neurology 43: 1467–1472.

Sawin, R.S. and R.T. Schaller, D. Tapper, A. Morgan and J. Cahill. 1994. Early recognition of neonatal abdominal wall necrotizing fasciitis. Am. J. Surg. 167: 481–484.

Scheinkestel, C. and M. Bailey, P. Myles, K. Jones, D. Cooper, I. Millar and D. Tuxen. 1999. Hyperbaric or normobaric oxygen for acute carbon monoxide poisoning: a randomised controlled clinical trial. MJA 170: 203–209.

Schiltz, K.L. 2000. Failure to assess motivation, need to consider psychiatric variables, and absence of comprehensive examination: a skeptical review of neuropsychologic assessment in carbon monoxide research. Undersea Hyperb. Med. 27: 48–50.

Schroedl, C. and D.S. McClintock, G.R. Budinger and N.S. Chandel. 2002. Hypoxic but not anoxic stabilization of HIF-1alpha requires mitochondrial reactive oxygen species. Am. J. Physiol Lung Cell Mol. Physiol. 283: L922–931.

Scolnick, B. and D. Hamel and A.D. Woolf. 1993. Successful treatment of life-threatening propionitrile exposure with sodium nitrite/sodium thiosulfate followed by hyperbaric oxygen. J. Occup. Med. 35: 577–580.

Seggewiss, R. and E.C. Buss, D. Herrmann, H. Goldschmidt, A.D. Ho and S. Fruehauf. 2003. Kinetics of peripheral blood stem cell mobilization following G-CSF-supported chemotherapy. Stem Cells 21: 568–574.

Semenza, G.L. 2001. HIF-1 and mechanisms of hypoxia sensing. Current Opinion in Cell Biology 13: 167–171.

Semenza, G.L. 2003. Angiogenesis in ischemic and neoplastic disorders. Annu. Rev. Med. 54: 17–28.

Shandling, A.H. and M.H. Ellestad, G.B. Hart, R. Crump, D. Marlow, B. Van Natta, J.C. Messenger, M. Strauss and Y. Stavitsky. 1997. Hyperbaric oxygen and thrombolysis in myocardial infarction: The HOT MI pilot study. Am. Heart J. 134: 544–550.

Sharifi, M. and W. Fares, I. Abdel-Karim, D. Petrea, J.M. Koch, D. Adler and J. Sopko. 2002. Inhibition of restenosis by hyperbaric oxygen: a novel indication for an old modality. Cardiovasc. Radiat Med. 3: 124–126.

Sharifi, M. and W. Fares, I. Abdel-Karim, J.M. Koch, J. Sopko and D. Adler. 2004. Usefulness of hyperbaric oxygen therapy to inhibit restenosis after percutaneous coronary intervention for acute myocardial infarction or unstable angina pectoris. Am. J. Cardiol. 93: 1533–1535.

Sheehy, M. and J.L. Way. 1968. Effect of oxygen on cyanide intoxication: III Mithridate. J. Pharmacol. Exp. Ther. 161: 163–168.

Sheikh, A. and J. Gibson, M. Rollins, H. Hopf, Z. Hussain and T. Hunt. 2000. Effect of hyperoxia on vascular endothelial growth factor levels in a wound model. Arch. Surg. 135: 1293–1297.

Shimosegawa, E. and J. Hatazawa, K. Nagata, T. Okudera, A. Inugami, T. Ogawa, H. Fujita, H. Itoh, I. Kanno and K. Uemura. 1992. Cerebral blood flow and glucose metabolism measurements in a patient surviving one year after carbon monoxide intoxication. J. Nucl. Med. 33: 1696–1698.

Shupak, A. and O. Shoshani, I. Goldenberg, A. Barzilai, R. Moskuna and S. Bursztein. 1995. Necrotizing fasciitis: an indication for hyperbaric oxygenation therapy? Surgery 118: 873–878.

Shusterman, D. and G. Alexeeff, C. Hargis and et al. 1996. Predictors of carbon monoxide and hydrogen cyanide exposure in smoke inhalation patients. J. Toxicol. Clin. Toxicol. 34, 61–71.

Silverman, S. H.and G.F. Purdue, J.L. Hunt and et al. 1988. Cyanide toxicity in burned patients. J. Trauma 28: 171–176.

Silverman, C.S. and J. Brenner and F.R. Murtagh. 1993. Hemorrhagic necrosis and vascular injury in carbon monoxide poisoning: MR demonstration. AJNR 14: 168–170.

Skene, W.G. and J.N. Norman and G. Smith. 1966. Effect of hyperbaric oxygen in cyanide poisoning. Pp. 705–710. In: I.W. Brown and B. Cox (eds.). Proceedings of the Third International Congress on Hyperbaric Medicine. National Academy of Sciences, National Research Council. Washington, DC.

Slack, W.K. and D.A. Thomas and L.R.J. De Jode. 1955. Hyperbaric oxygen in the treatment of trauma, ischemic disease of limbs, and varicose ulceration. Pp. 621–624. In: I.W. Beacon and B.G. Cox (eds.). Proceedings Third Internation Conference on Hyperbaric Medicine Nat'l Acad. Sci., Nat'l Res. Council, Washington, DC.

Snell, E. 1896. Compressed Air Illness or So-Called Caisson Disease. H.K. Lewis. London.

Stavitsky, Y. and A.H. Shandling, M.H. Ellestad, G.B. Hart, B. Van Natta, J.C. Messenger, M. Strauss, M.N. Dekleva, J.M. Alexander, M. Mattice and D. Clarke. 1998. Hyperbaric oxygen and thrombolysis in myocardial infarction: the "HOT MI" randomized multicenter study. Cardiology 90: 131–136.

Stewart, R.D. and E.A. Boettner and R.R. Southworth. 1963. Acute carbon tetrachloride intoxication. JAMA 183: 994–997.

Stojadinovic, O. and H. Brem, C. Vouthounis, B. Lee, J. Fallon, M. Stallcup, A. Merchant, R. Galiano and M.Tomic-Canic. 2005. Molecular pathogenesis of chronic wounds: the role of beta-catenin and c-myc in the inhibition of epithelialization and wound healing. Am. J. Pathol. 167: 59–69.

Suehiro, T. and T. Shimura, K. Okamura, T. Okada, K. Okada, S. Hashimoto, Y. Mochida, H. Kuwano, S. Saitoh and F. Gotoh. 2008. The effect of hyperbaric oxygen treatment on postoperative morbidity of left lobe donor in living donor adult liver transplantation. Hepato. Gastroenterology 55: 1014–1019.

Tahepold, P. and G. Valen, J. Starkopf, C. Kairane, M. Zilmer and J. Vaage. 2001. Pretreating rats with hyperoxia attenuates inschemia-reperfusion injury of the heart. Life Sci. 68: 1629–1640.

Takano, T. and Y. Miyazaki, I. Nashimoto and K. Kobayashi. 1980. Effect of hyperbaric oxygen on cyanide intoxication: in situ changes in intracellular oxidation reduction. Undersea. Biomed. Res. 7: 191–197.

Takita, H. and W. Olszewski, G. Schimert and E.H. Lanphier. 1968. Hyperbaric treatment of cerebral air embolism as a result of open-heart surgery. J. Thorac. Cardiov. Surg. 55: 682–685.

Tepper, O.M. and J.M. Capla, R.D. Galiano, D.J. Ceradini, M.J. Callaghan, M.E. Kleinman and G.C. Gurtner. 2005. Adult vasculogenesis occurs through in situ recruitment, proliferation, and tubulization of circulating bone marrow-derived cells. Blood 105: 1068–1077.

Thom, S. and T. Milavonava. 2008a. Adult mouse stem cell mobilization and ischemic site recruitment—redox stress is good. Int'l Society for Stem Cell Research (Abstract) Abstract 453: 62.

Thom, S. and T. Milovanova. 2008d. Hyperbaric oxygen therapy increases stem cell number and HIF-1 content in diabetics. Undersea and Hyperbaric Med. 35: 280 (abstract).

Thom, S. and I. Mendiguren and D. Fisher. 2002. Smoke inhalation-induced alveolar lung injury is inhibited by hyperbaric oxygen. Undersea and Hyperbaric Med. 28: 175–180.

Thom, S. and V. Bhopale, J. Mancini and T. Milovanova. 2008. Actin S-nitrosylation inhibits neutrophil beta-2 integrin function. J. Biol. Chem. 283: 10822–10834.

Thom, S.R. 1989. Hyperbaric oxygen therapy. J. Intensive. Care Med. 4: 58–74.

Thom, S.R. 1993. Functional inhibition of leukocyte B2 integrins by hyperbaric oxygen in carbon monoxide-mediated brain injury in rats. Toxicol Appl Pharmacol 123: 248–256.

Thom, S.R. 1997. Learning dysfunction and metabolic defects in globus pallidus and hippocampus after CO poisoning in a rat model. Undersea and Hyperbaric Med. 23: 20.

Thom, S.R. and M.W. Lauermann and G.B. Hart. 1986. Intermittent hyperbaric oxygen therapy for reduction of mortality in experimental polymicrobial sepsis. J. Infect. Dis. 154: 504–510.

Thom, S.R. and R.L. Taber, I.I. Mendiguren, J.M. Clark, K.R. Hardy and A.B. Fisher. 1995. Delayed neuropsychologic sequelae after carbon monoxide poisoning: prevention by treatment with hyperbaric oxygen. Ann. Emerg. Med. 25: 474–480.

Thom, S.R. and I. Mendiguren, K. Hardy, T. Bolotin, D. Fisher, M. Nebolon and L. Kilpatrick. 1997. Inhibition of human neutrophil beta2-integrin-dependent adherence by hyperbaric $O_2$. Am. J. Physiol. 272: C770–777.

Thom, S.R. and V.M. Bhopale, D. Fisher, J. Zhang and P. Gimotty. 2004a. Delayed neuropathology after carbon monoxide poisoning is immune-mediated. Proc. Natl. Acad. Sci. USA 101: 13660–13665.

Thom, S.R. and D. Fisher, J. Zhang, V.M. Bhopale, B. Cameron and D.G. Buerk. 2004b. Neuronal nitric oxide synthase and N-methyl-D-aspartate neurons in experimental carbon monoxide poisoning. Toxicol Appl. Pharmacol 194: 280–295.

Thom, S.R. and V.M. Bhopale, S.-T. Han, J.M. Clark and K. Hardy. 2006a. Intravascular neutrophil activation due to carbon monoxide poisoning. Am. J. Respir. Crit. Care Med. 174: 1239–1248.

Thom, S.R. and V.M. Bhopale, O.C. Velazquez, L.J. Goldstein, L.H. Thom and D.G. Buerk. 2006b. Stem cell mobilization by hyperbaric oxygen. Am. J. Physiology 290: H1378–H1386.

Tobin, M. and A. Jubran. 2008. Meta-analysis under the spotlight: Focused on a meta-analysis of ventilator weaning. Crit. Care Med. 36: 1–7.

Trapp, W.G. and M. Lepawsky. 1983. 100% survival in five life-threatening acute cyanide poisoning victims treated by a therapeutic spectrum including hyperbaric oxygen. Paper presented at the First European Conference on Hyperbaric Medicine, Amsterdam, 1983.

Truss, C.D. and P.G. Killenberg. 1982. Treatment of carbon tetrachloride poisoning with hyperbaric oxygen. Gastroenterology 82: 767–769.

Ueno, H. and T. Ohya, H. Ito, Y. Kobayashi, K. Yamada and M. Sato. 2007. Chitosan application to X-ray irradiated wound in dogs. J. Plast. Reconstr. Asthetic Surg. 60: 304–310.

Ueno, S. and G. Tanabe, K. Kihara, D. Aoki, K. Arikawa, H. Dogomori and T. Aikou. 1999. Early post-operative hyperbaric oxygen therapy modifies neutrophile activation. Hepato. Gastroenterology 46: 1798–1799.

Ushio-Fukai, M. and R. Alexander. 2004. Reactive oxygen species as mediators of angiogenesis signaling. Mol. Cell Biochem. 264: 85–97.

Valko, M. and D. Leibfritz, J. Moncol, M. Cronin, M. Mazur and J. Tesler. 2007. Free radicals and antioxidants in normal physiological functions and human disease. Int. J. Biochem. Cell Biol. 39: 44–84.

Vasa, M. and S. Fichtlscherer, A. Aicher, K. Adler, C. Urbich, H. Martin, A.M. Zeiher and S. Dimmeler. 2001. Number and migratory activity of circulating endothelial progenitor cells inversely correlate with risk factors for coronary artery disease. Circ. Res. 89: E1–7.

Waisbren, B.A. and D. Schutz, G. Collentine and E. Banaszak. 1982. Hyperbaric oxygen in severe burns. Burns. 8: 176–179.

Wang, V. and D. Davis, M. Haque, L. Huang and R. Yarchoan. 2005. Differential gene-up regulation by hypoxia-inducible factor-1alpha and hypoxia-inducible factor 2alpha in HEK293T cells. Cancer Res. 65: 3299–3306.

Waterhouse, M. and W. Zamboni, R. Brown and R. Russell. 1993. The use of HBO in compromised free tissue transfer and replantation: a clinical review. Undersea and Hyperbaric Med. 20(Suppl): 64.

Way, J.L. 1984. Cyanide intoxication and its mechanism of antagonism. Annu. Rev. Pharmacol. Toxicol. 24: 451–481.

Way, J.L. and S.L. Gibbon and M. Sheehy. 1966. Effect of oxygen on cyanide intoxication. I. Prophylactic protection. J. Pharmacol. Exp. Ther. 13: 381–382.

Way, J.L. and E. End, M.H. Sheehy, P. Demiranda, U.F. Feitknecht, R. Bachand, S.L. Gibbon and G.E. Burrows. 1972. Effect of oxygen on cyanide intoxication. Toxicol. Appl. Pharmacol. 22: 415–421.

Weaver, L.K. and R.O. Hopkins, K.J. Chan, S. Churchill, C.G. Elliott, T.P. Clemmer, J.F Orme, Jr., F.O.Thomas and A.H. Morris. 2002. Hyperbaric oxygen for acute carbon monoxide poisoning. N. Engl. J. Med. 347: 1057–1067.

Weaver, L. and K. Valentine and R. Hopkins. 2007. Carbon monoxide poisoning: Risk factors for cognitive sequelae and the role of hyperbaric oxygen. Am J. Respir. Crit Care Med. 176: 491–497.

Weisz, G. and A. Lavy, Y. Adir, Y. Melamed, D. Rubin, S. Eidelman and S. Pollack. 1997. Modification in vivo and in vitro TNF-alpha, IL-1, and IL-6 secretion by circulating monocytes during hyperbaric oxygen treatment in patients with perianal Crohn's disease. J. Clin. Immunol. 17: 154–159.

Welsh, S. and W. Bellamy, M. Briehl and G. Powis. 2002. The redox protein thioredoxin-1(Trx-1) increases hypoxia-inducible factor-1alpha protein expression: Trx-1 overexpression results in increased vascular endothelial growth factor production and enhanced tumor angiogenesis. Cancer Res. 62: 5089–5095.

Wetherill, H.R. 1966. The occurrence of cyanide in the blood of fire victims. J. Forensic Sci 11: 167–173.

Wilkinson, D. and D. Doolette. 2004. Hyperbaric oxygen treatment and survival from necrotizing soft tissue infection. Arch Surg 139: 1339–1345.

Wiseman, D.H. and A.R. Grossman. 1985. Hyperbaric oxygen in the treatment of burns. Crit. Care Clin. 1: 129–145.

Wong, H.P. and W.A. Zamboni and L.L. Stephenson. 1996. Effect of hyperbaric oxygen on skeletal muscle necrosis following primary and secondary ischemia in a rat model. Surgical. Forum 705–707.

Wood, Z. 2002. Hyperbaric oxygen in the management of chronic wounds. Br. J. Nurs. 11, S16, S18–19, S22–14.

Wray, J. and L. Rogers. 1968. Effect of hyperbaric oxygenation upon fracture healing in the rat. J. Surg. Res. 8: 373–378.

Yang, B. and T. Milovanova, K. Hardy, C. Logue, V. McCarthy and S. Thom. 2007. Stem cell mobilization in diabetics—responses to hyperbaric oxygen. Undersea and Hyperbaric Med. 34: 235–236 (abstract).

Yang, Z.J. and G. Bosco, A. Montante, X.I. Ou and E.M. Camporesi. 2001. Hyperbaric O2 reduces intestinal ischemia-reperfusion-induced TNF-alpha production and lung neutrophil sequestration. Eur. J. Appl. Physiol. 85: 96–103.

Yarbrough, O. and A. Behnke. 1939. The treatment of compressed air illness using oxygen. J. Ind. Hyg. Toxicol. 21: 213–218.

Yu, S. and J. Chiu, S. Yang, H. Yu, C. Hsieh, P. Chen, W. Lui and C. Wu. 2005. Preconditioned hyperbaric oxygenation protects the liver against ischemia-reperfusion injury in rats. J. Surg. Res. 128: 28–36.

Zamboni, W.A. and A.C. Roth, R.C. Russell, B. Graham, H. Suchy and J.O. Kucan. 1993. Morphologic analysis of the microcirculation during reperfusion of ischemic skeletal muscle and the effect of hyperbaric oxygen. Plast. Reconstr. Surg. 91.

Zearbaugh, C. and D.F. Gorman and J.E. Gilligan. 1988. Carbon tetrachloride/chloroform poisoning: Case studies of hyperbaric oxygen in the treatment of lethal dose ingestion. Undersea Biomed. Res. 15: 44.

Zhang, Q. and Q. Chang, R. Cox, X. Gong and L. Gould. 2008. Hyperbaric oxygen attenuates apoptosis and decreases inflammation in an ischemic wound model. J. Invest. Dermatol. doi:10.1038/jid.2008. 53.

Zuntz, N. 1897. Zur Pathogenese und Therapie der durch rasche Luftdruckänderungen erzeugten Krankheiten. Fortschr Med. 15: 632–639.

# 21

# High Hydrostatic Pressure and Cancer

*Gerrit Blümelhuber*

## INTRODUCTION

According to the WHO, cancer was the cause of death for 7,9 million people worldwide in 2007. This includes different kinds of cancer diseases:

1,4 million of lung cancer, 866.000 of stomach cancer, 653.000 of liver cancer, 677.000 of colon cancer and 584.000 of breast cancer.

Therefore, cancer is the most common disease leading to death woldwide (WHO, 2008).

There are numerous statistics based on cancer. At a first glance, the figures mentioned in these statistics are staggering, but do not show the effective extent of the disease: who can imagine the figure of 7,9 million people? In order to compare: approximately 7,5 million people live in the city of London (excluding the outskirts), that means: worldwide, more people than the inhabitants of London die of cancer every year!

But these statistics or comparisons mentioned above are not relevant or of great interest for the people concerned because everyone of those 7,9 million people has a single twist of fate and has to be therefore considered separately.

In this chapter, how the application of hydrostatic pressure technology could offer chances or possibilities within a cancer therapy is described. Nevertheless, any single case of cancer and every single patient has to be considered separately, too. Also hydrostatic pressure technology probably will not be the universal remedy for cancer. But for some patients and some kinds of cancer there is a possiblity that hydrostatic pressure technology

Sonnenweg 16, 85084 Reichertshofen, Germany.
E-Mail: info@immobrau.com

can be implemented as a support to ease the pain or even to heal the disease. In order not to raise any unjustified hope or euphoria for a patient concerned, one has to admit that an abatement or recovery cannot be achieved for everyone. This will be shown later on.

## What is Cancer?

Over 3500 years ago the Egyptians described cancer as a disease (Davidson, 1967). This shows how long people already suffered from this disease. Whereas the number of cancer diseases appeared quite rarely in the time between antiquity and the 20th century, the number escalated in the beginning of the 20th century. Since then, in the USA alone, the number of people who died of cancer rose significantly from 50 to nearly 175 people per 100.000 residents in 1980 (Seidman et al., 1985) One reason for this is that the death rate of infectious diseases decreased due to better hygienic conditions and was also related to increased life expectancy, a lot of different cancer diseases occur within the fourth or fifth life decade (Prescott and Flexer, 1990). To summarize: cancer is not a pure disease of civilization of recent times. Moreover, it is a very old disease which became important in the last decades due to the changed living conditions.

## How does Cancer emerge?

Although a lot of influencing factors for the appearance of cancer have been identified and a lot is known about its emergence and further development stages, a complete decoding of all causes, influencing factors and phases could not be achieved up to now.

A complete description of all influencing factors would go far beyond the scope of this chapter.

Below is a short description of the theory of cancerogenesis using high pressure technology as a treatment option is given, but it is far from complete.

Cancer cells do not have the usually associated regulating mechanisms of growth, segmentation and necrosis any longer. Also, the signals of the organism transmitted to the cells cannot be identified or will be ignored. A disorder of the so-called tumor suppressor gene i.e. the formation of oncogene is responsible for this defect. The tumor suppressor gene controls a failure-free replication of the DNA. If an error in the replication of the DNA occurs, the tumor suppressor genes are responsible for either fixing the base sequence or - if not possible due to the extent of the failure- the initiation of the cell death, the so-called apoptosis. For example, a mutation of the tumor suppressor gene affected by exterior influences or an already existing gene defect, will lead to a cell defect within the next generation, proceeding on further cell segmentations accordingly. This is the emergence of a cancer cell. New cells evolving from this first cancer cell are again cancer cells. Besides the loss of their own ability for regulation of growth and segmentation, the cancer cell achieves special abilities (compared to healthy cells) caused by the gene defect which makes them dangerous. For example they have the possibility to move through the organism and displace themselves, the so-called metastasizing. In most of the cases, these metastases are the main reason for the death of a patient and not the primary tumor.

The human immune system, which normally eliminates abnormal cells, has no sufficient abilitiy to handle cancer cells.

There are various possibilities with different chances of success to medicate the cancer which arises from an abnormal development of cells resulting in cancer cells. In principle, it is possible to divide therapies into two classes: on the one hand by medicating the cancer tissue directly with drugs, exposure or by surgery, on the other hand by the so-called immune modulation - by increasing the natural immune reaction, so that the immune system eliminates the cancer cells itself.

Hydrostatic pressure technology provides alternative opportunities or approaches for both therapies.

## HYDROSTATIC PRESSURE AND CANCER

### Hydrostatic Pressure for Immune Modulation

The first studies on the effects of hydrostatic pressures on cells go back to the end of the 19th century (Hite, 1899). It was only possible to examine the effects on prokaryotes, it took until the middle of the 20th century when research focussed on eukaryotes. First examinations were implemented by Atanasiu and Basset (Atanasiu and Basset, 1953) and focussed mainly on cell necrosis and vitality after the treatment with hydrostatic pressure. Although, first studies on immunogenicity on treated cells passed, (Dubert et al., 1952) they could only be viewed superficially due to the lack of knowledge on eukaryotes cells and their metabolism in these former times.

Yamaguchi was one of the first who paid attention to hydrostatic pressure effects on tumor cells. He examined the proliferation behavior of Ehrlich ascites tumor cells. By testing on animals he proved that the tumor cells lose their ability of proliferation within a pressure of 130 MPa (Yamaguchi et al., 1997).

The final breakthrough in terms of high pressure treatment of tumor cells for immodulation was achieved by the team of Shinitzky (Shinitzky and Goldmann, 2000). They were inspired by the possibility of extracting the tumor cells by a resection or a biopsy, to separate them and finally to kill the extracted tumor cells with a high pressure treatment. Inactivating in this way these cells can be re-injected in the patient´s body and will reach all compartments of the immune system via the blood system. If this experiment is executed with non-activated tumor cells, the patient would show a multiplicity of metastases within a very short period of time, surely leading to death very soon. An inactivation of the cells can eliminate the creation of metastases.

In fact, the tumor cell is presented to the immune system on an omnipotent level which could affect an immune reaction against this inactivated cell. With this treatment an immune reaction against the cells of the primary tumor can be expected. A recrudescence is not expected any longer as the immunological memory of the organism would prevent a recurrence of the tumor cells in the organism instantly.

As one of the first research groups, the members of Shinitzky´s team followed this approach and went on working on the effects of hydrostatic pressure on tumor cells intensively. (Ramakrishna and Shinitzky, 1991; Ramakrishna et al., 1993; Eisenthal et al., 1996; Goldmann et al., 2000).

At first, Shinitzky examined murine B16-BL6 melanoma cells, which have a very low potential of immunogenicity. After a 15 min treatment with high pressure at 120 MPa and a treatment with adenosine 2', 3' dialdehyd (oxAdo) a sharp increase of the projection of MHC and retroviral –induced tumor antigenes was discovered. Also within the immunization tests conducted on mice, the immunogenicity of the treated tumor cells was noticed. Shinitzky examined six groups, with six mice each. They were vaccinated with different pressures and oxAdo treated tumor cells. One week after the vaccination the challenge took place. Within a pressure of 100 MPa 50% of the animals survived a period of 50 d, within a pressure of 120 MPa, even 100% survived. Shinitzky called this high pressure treatment process in combination with chemical agents a 'PCL' (pressure cross-linking) modification. The pressure treated cells were inactivated by x-rays before immunization (Eisenthal et al., 1993).

In a later study, Shinitzky referred to examinations with human tumor cells. Sixteen cancer patients were examined, suffering either from lung-, renal or colon cancer. The tumors removed by surgery were prepared and subsequently treated with hydrostatic pressure and oxAdo, followed by *in vitro* sensitization, showing that the high pressure treated cells caused a significantly higher immune reaction than untreated cells. Ten out of sixteen patients showed a considerably higher lymphocyte profileration as well as a decrease of IL-10 secretion of the cell cultures, whereas the IFN-secretion increased (Eisenthal et al., 1998).

Shinitzky´s team continued to improve the method of hydrostatic pressure treatment on tumor cells. They combined their PCL with NAC (N-acetyl-Lcysteine). The assumption set up by the study group was:

After the notion that reversible intercellular bridging between target and effector cells can strengthen specific immunological signals, the aim was to examine the relevance of reversible disulfide bonds to the provoking of an anti-tumor response.

Shinitzky choose NAC as a reagent for the creation of the disulphide link, trying to confirm this hypothesis by conducting an experiment on mice. A test record close to clinical studies was selected—the metastatses regression assay:

3LL-D122 cells were injected in mice, taking into account 8 d for the creation of lung micrometastases. Subsequently, the mice were medicated with PCL+NAC treated cells and NAC i.v. It was known from former studies that NAC i.v. is a T cell immuno-stimulant.

One control group was treated with unmodified cells. After 25 d, the lungs of the animals of the control group were riddled with metastases, whereas the animals immunized with modified cells showed no metastases at all. This experiment showed the great potential of hydrostatic pressure technology, particularly when combined with well-established and described methods.

The research of another team from the Technical University of Munich, suggested the application of immunmodulation with hydrostatic pressure technology (Höhn and Meyer-Pittroff, 1999a). In this paper, Höhn and Meyer-Pittroff described the possibilities of using hydrostatic pressure technology in order to create an autologous tumor vaccine in theory. This hypothesis corresponded to the approach of Shinitzky.

Furthermore, the team went into details regarding necrosis of mammalian tumor cells. The necrosis of different human tumor cell lines at different pressures were examined (Blümelhuber et al., 2003a). The team especially focussed on the programmed cell death i.e. the apoptosis, as it plays a decisive role in immune reactions (Korn et al., 2004).

The cell lines used for this study are a human histocytic lymphoma cell line (U 937), a lymphoblast-like human cell from an urkitt´s lymphoma Raji (ATCC CCL-86), a human T cell leukemia Jurkat (DSMZ ACC 282) and a B cell line from acute lymphoblastic leukemia Ball-1. These cells were exposed to pressures up to 350 MPa at intervalls of 50 MPa. Analyses of viability, exposure of PS and PI permeability as a sign for cell death, the sub-G1 DNA content and immunogenicity were executed later.

For examination of the viability, the cells were treated with hydrostatic pressure and subsequently recultivated. It was seen that the cells treated with up to 200 MPa per day showed very low impact. The viability decreased to 20% while treating with 250 MPa. More than 99% of the cells had no viability by treating with higher pressures. On the first day after the treatment, the cells suspended with more than 100MPa had a viability of 20% or less. The longer the cells were cultivated, the more they decreased in viability. On day 7, less than 2% of the cells were still present. It is amazing that cells treated with 100MPa and less showed no decrease in viability at all.

## Exposure of PS and PI Permeability as a Sign for Cell Death

An obvious indication for the apoptosis is the exposition of phosphatidyserin (PS) on the external layer of the plasma membrane. This exposition can be determined by annexin V (AxV), which is a protein linked to PS. The indication for necrotic cell death is the penetration of propidium iodide (PI) by the plasma membrane.

The cell lines U-937 and BALL-1 were examined by the above mentioned criteria after the treatment with hydrostatic pressure. Again, the cells were obviously resistant by treating with pressures up to 250 MPa, provided that the examination was done right after the treatment with hydrostatic pressure. However, the viability decreased below 10% percentage when exposed to higher pressures, this time measured by exposure of PS and PI permeability. Nevertheless, surprising results were achieved by cultivating the cells after the treatment with hydrostatic pressure. The PI permeability of the cells treated with 200 MPa i.e. 250 MPa was on a very high level, which could be a sign of necrotic dissolution of the cells. This changed within the next two days: the PI coloration decreased significantly and signs of an apoptotic dissolution came to the fore, which could be explained by degradation of cellular DNA.

This development can still be observed more clearly on higher pressures. It is interesting to note that this rapid DNA degradation performings are totally different compared to those cells damaged conventionally e.g. by heat or chemical agents (Korn et al., 2004).

## Sub-G1 DNA Content and Immunogenicity

Korn et al also examined Sub-G1 DNA content and immunogenicity (Korn et al., 2004). The DNA content of a cell provides information about its status in the cell cycle, the DNA degradation itself about the apoptosis of the cell. On apoptotic cells, the DNA content decreases whereas on necrotic cells it remains constant. It is possible to identify the DNA content of a cell by a staining method.

Cells treated with pressure of up to 100 MPa showed no change of the sub-G1 DNA content during a 6 h observation period. A significant increase of the cells containing sub-G1 DNA was detected at a pressure of 200 MPa, as well as a DNA fragmentation of the cells. Treating the cells with 400 MPa resulted in a slight increase of the sub-G1 DNA whereas the speed of the DNA degradation was considerably higher, especially as these cells showed a negrotic dissolution.

The scientists developed the theory that the DNase was not inactivated by the hydrostatic pressure treatment and effected a degradation via the permeable membrane.

This theory could be proved by experiments with EDTA, which inhibits the DNase.

To answer the question if cells treated with hydrostatic pressure are immunogenic, mice were immunized with treated Raji cells, focussing on the humoral immune response. It showed that the cells treated with hydrostatic pressure definitively caused an immune response, the hydrostatic pressure was irrelevant.

## Further Works with Mammalian Cells

Also, Islam et al focussed on the dissolution of cells under the treatment with hydrostatic pressure (Islam et al., 2002). The target of research was chondrocytes, extracted from a patient suffering from osteoarthritis. The cells were exposed to a cyclic hydrostatic pressure treatment (One Hz) at five MPa over a period of 1 h.

The treatment caused an apoptosis in the cells. The scientists proved that the treated cells initiated a heavy inter-nucleosomal DNA fragmentation, activating the enzyme Caspase-3 as well as a separation of the poly-ADP-ribose polymerase. Furthermore, an apoptosis of the cells was detected with the increase of p53, c-myc and bax-alpha (TNF-alpha). Thus, they concluded that hydrostatic pressure is an excellent tool for initiating an apoptosis in the cells.

In summary, the scientists proved with their research that hydrostatic pressure treatment is a very good tool for reliable inactivation of tumor cells without losing any of their immunogenic characteristics. Whereas Shinitzky only worked with pressures up to 110 MPa in order to provide a reliable inactivation of the cells with radiotherapy, the Munich study group focussed on the complete inactivation of the tumor cells by high levels of hydrostatic pressure. In both cases, the treated tumor cells showed a very high level of immunogenic effects, never verified with any of the conventional inactivation methods. Nevertheless, this immunogenic effect is a mandatory basic requirement for an effective autologous tumor vaccination.

The idea of an autologous tumor vaccination is surely not a new conclusion. First experiments in regard to treatment of cancer with tumor vaccination were already published in the 50s of the last century ( Weiss, 1952; Ferrero, 1954; Aswaq et al., 1964). The experiments were executed in particular with 'classical' inactivated tumor cells. The cells were inactivated with heat and/or chemical agents and subsequently re-injected to the patients. This method should draw attention of the immune system to the 'tumor cells' enabling it to fight the latter ones, which are not only the injected ones but in the end all the cells in the tumors.

The results were different. Whereas good results were achieved frequently with animal experiments, human trials showed different results.

A significantly efficient immune therapy could not be verified. The only exception was a clinical study of the autologous tumor vaccination of the renal cell carcinoma where a significant efficiency could be verified accordingly (Jocham et al., 2004).

Other approaches for the development of an autologous tumor vaccination provide the supplementary insertion of viruses as ajuvants of the treated tumor cells. The goal is to intensify the immunogenic properties of the tumor cells by addition of viruses. The results were quite successful but a concluding study of efficiency of those vaccines is still pending. The inactivation of the tumor cells was always the major problem when creating autologous tumor vaccines. In most of the cases the immunological characteristics were lost, either partly or even completely, by the inactivation of the tumor cells. Hydrostatic pressure is a further tool for inactivation or treatment.

## HYDROSTATIC PRESSURE TREATMENT OF TUMOR-DISEASED BONE TISSUE

There are more than 20 different diseases of bone neoplasm. Some of them are the most aggressive kinds of cancer we know so far. Due to their affection of dissemination and the difficulties in diagnostics, the treatment of cancer is very difficult to examine. With the virulent diseases of bone neoplasm a resection of the bone tissue is required (partially of the entire bone or even of the whole extremities). Additionally, the prognosis of a lot of bone neoplasms are extremely bad due to the aggressiveness of the tumors. This means the possibility of a dissemination is steadily on the patient's mind, alongwith suffering from a physical reduction due to a necessary resection or amputation. Besides the removal by surgery the patient is faced with medicinal treatment. Nevertheless, the prospects of success are extremely poor.

A possible alternative treatment of bone neoplasm has been published in different papers (Smith and Struhl, 1998; Sanja et al., 1997; Freiberg et al., 1992). In these studies, the affected bone is removed and sterilized by heat. Another method of sterilization was described in 1999 (Araki et al., 1999) where a part of the bone is sterilized by x-ray.

The treated part of the bone will be subsequently re-implanted. The body recognizes the tissue as an autogenetic part and consequently does not reject it. The sterilized bone tissue adopts the role of a chemical lead, new bone tissue is regenerated on the healthy edges and growth along the chemical lead. The treated part of the bone itself will not be affected any longer.

The killed cells located in that part of the bone will be disposed by the immune system. The regeneration of the bone cells results in the creation of a new part of the bone. Nevertheless, this method has the disadvantage that the sterilized bone tissue does not have the mechanical characteristics of a healthy bone. Freiberg in his study refers to a patient suffering from chondro sarcoma in the upper arm who could easily move his arm without any pain after the treatment. Also, no metastasis or rebuilding of the tumor occurred. Nevertheless, a steel nail was already inserted during the surgery in order to strengthen the bone. But as this implant caused a lot of pain for the patient, a new surgery was done six years later. Freiberg explained that this pain was caused by the intense weakness of the bone. The studies of Smith presented similar conclusions (Smith and Struhl, 1998). Smith confirmed that besides the difficulties in stability of the bones and the failure of new cell cultivation, the bone takes over the function as a chemical lead for the creation of new bone tissue.

However, this therapeutic method has not achieved acceptance so far because of the disadvantages mentioned (weakness of the mechanical characteristics of the bone). The preferred option would be a method of complete sterilization of the bone whilst preserving the mechanical characteristics up to a large extent. Höhn et al specified this method in an application for industrial property protection (Höhn and Meyer-Pittroff, 1999b): The tissue of the affected bone should be extracted and consequently treated with hydrostatic pressure. Another possibility could be the combination with heat. Höhn´s study group developed further methods. They examined the mechanical characteristics of bones treated with hydrostatic pressure and simultaneously kept an eye on the dissolution of cancer cells (Blümelhuber et al., 2003b). In regard to the mechanical examinations, they operated with cylinders from the thigh neck as well as the from condylen and corticals segments of a human femora with dimensions of 5x5x20mm. Subsequently, they treated with pressures of 300 and 600 MPa, whereas the mocks were left untreated. The following parameters were examined:

- e-modulus
- ultimate load
- breaking elongation

Only the samples of the corticalis showed a 15% reduction of the e-modulus treated with hydrostatic pressure of 600 MPa. All other examined mechanical characteristics of the treated part of the bone had no significant changes at all. This means that hydrostatic pressure treatment has no negative impact on the mechanical characteristics of the bone compared to the treatment with heat autoclaving. In order to verify the necrotic effects on cancer cells, the following cell lines were pressurized:

- human osteosarcoma cells (SAOS-2)
- ovarial carcinoma cells (Ov-MZ-6)
- leukemia cells (U937)

All cells treated with at least 350 MPa were reliably killed within 10 min. This study evidently proves that all disadvantages which arise by the use of conventional methods, are eliminated by the treatment with hydrostatic pressure and no mechanical damages of implants will occur accordingly.

The immunological impacts on the organism still need to be discussed. The autoclaving of the bones showed no negative immunological response.

Thus, one gets mechanical intact implants. This was the first and decisive assumption using hydrostatic pressure for sterilization of tumor-diseased bone tissue. Nevertheless, a lot of further influencing factors have to be considered in regard to patient treatments later. Additional research on immunological and biochemical reactions caused by the treatment with hydrostatic pressure and the re-implantation of the bone has to be examined as well as the different kinds of dissolution of healthy and tumor-affected bone cells. Another question, which should be taken into consideration is how chondral tissue, appearing mostly in combination with bone tissue, reacts on the treatment with hydrostatic pressure. In reference to the differences in dissolution of healthy and tumor-affected cells, Naal et al., examined the impact of hydrostatic pressure treatment on human chondrocytes and human chondrosarcom cells (Naal et al., 2006). They examined chondrosarcoma cells of line SW 1353 and chondrocytes treated with pressures up to 350 MPa. In order to find any differences in cell death, both cell lines were adherently cultivated as well as in a suspension. Both cell lines treated with pressures of 350 MPa were killed completely after 10 min at 37 °C, inspite of the kind of cultivation. Additionally, the conditions of cultivation at low pressures showed an influence on the dissolution under pressure. Pressures of 350 MPa were necessary in order to kill 100% of the chondrosarkoma cells growing under suspension whereas 100% of the adherently growing chondrosarcoma cells were killed already with pressures of only 250 MPa after 10 min. This effect could also be observed with chondrocytes. To kill 100% of the cells growing in suspension, pressures of 250 MPa were necessary whereas pressures of 200 MPa were sufficient on adherently growing cells.

The studies of Diehl et al. (2006) describe the research on the impacts of hydrostatic pressure treatment on chondral tissue. Osteochondral segments from bovine femurconyles were treated with pressures between 300 and 600 MPa. Subsequently, the samples were examined for their biomechanical and immunologic-histochemical characteristics. For examination of the biomechanical characteristics, the chondral samples were tested on a machine with repetitive ball indention testing. The samples treated with 300 MPa showed no significant difference compared to the untreated samples. The samples treated with 600 MPa showed a slight deterioration of their biomechanical characteristics compared to the untreated ones. In regard to immunologic-histochemical dyeing, the samples were fixed in a methanol dilution in order to descale with EDTA-dilution. Cryosections were prepared afterwards. The samples were examined in regard to proteoglycan aggrecan, link protein and collagen II. All immunologic-histochemical examinations showed no difference between untreated and pressure-treated samples.

The examinations showed that treatment with hydrostatic pressure had almost no negative impact for the chondral tissue. Correspondingly, it is in accord with the treatment of hydrostatic pressure on bones or bone pieces.

Currently, there are no published studies of the exact immunological influences in regard to re-implantation of a bone treated with hydrostatic pressure.

Therefore, the reaction of the immune system towards the presentation of cells treated with hydrostatic pressure and re-implanted in the bone piece, is currently just speculative. From the studies of thermic sterilized bone segments it is known that the autogenetic implants are very well accepted by the immune system. No publication of any rejection or objectionable reaction of the immune system due to necrotic cells is known so far by the author.

## CONNECTING AUTOLOGOUS TUMOR VACCINE WITH TREATMENT OF BONES

Considering both specified working areas, on the one hand the development of an autogenetic tumor vaccine and on the other hand the inactivation by hydrostatic pressure i.e. re-implantation of tumor-affected bone segments, it seems that both areas only have hydrostatic pressure as a new technology in common. If one continues the thread of both new areas and considers e.g. the points of the current process of bone treatment, one is confronted with an unexpected hypothesis.

With the development of an autogenetic tumor vaccine, the immune system was stimulated by the presentation of cells treated with hydrostatic pressure and thus killed accordingly. This resulted in a respective immune reaction, activating the body´s defences which again attacked and killed the active tumor cells in the organism. Also, the chance of a relapse was decreased significantly. Hypothetically, the same effect is possible with the re-implantation of bone tissue treated with hydrostatic pressure. Hydrostatic pressure treatment, which is dedicated primary for the elimination of the tumor cells in the bone tissue, could ensure at the same time that the killed tumor cells stimulate the immune system in the same way as the tumor vaccine. This could be a great advantage in regard to a possibly affected dissemination and a lower probability of a relapse.

It becomes apparent that two working areas, which are quite different at first sight, could be connected. Besides the two described working areas, no other study, dealing with hydrostatic pressure operating directly on tumors, is known by the author. There are a few more options which will be described later. First, the focus will be on studies with hydrostatic pressure not having a direct impact on tumors or the immune system.

## HIGH HYDROSTATIC PRESSURE AS A BASIC TOOL FOR UNDERSTANDING PATHWAYS

Hydrostatic pressure is a new tool for scientists worldwide in order to answer basic questions regarding the research of pathways in tumor cells and also in healthy cells. These pathways are important for the understanding of tumor genesis and also for metabolism of healthy cells. Only with the knowledge of these pathways is there a possibility to develop effective therapies and medication against cancer. In this case, hydrostatic pressure is therefore a means to an end but not the ultimate therapeutical method.

Kopakkala-Tani et al. (2004) described the impacts of a treatment with hydrostatic pressure with 15 to 30 Mpa on the activation i.e. deactivation of different pathways in chondrosarcoma cells. This treatment showed an increase of the active phosphorylated form of signal-related kinase (ERK) as well as the phosphoinositide 3-kinase (PI3K). A cyclic treatment with hydrostatic pressure could only affect a weaker increase than a permanent one. Among other things, ERK is involved in the creation of heat shock protein 70 (hsp70). If the pathway of ERK-creation is blocked by chemical effects, this will however increase the content of hsp70 after treatment with hydrostatic pressure. Also no activation of c-Jun N-terminal protein kinase occurred nor was p38 detected. These two markers refer to a gen-induced creation of hsp70 just like the ERK. Thus, research shows the creation of a protein by hydrostatic pressure treatment without activating the responsible gene. In fact, it has d been proved that stabilization of hsp70mRNA in consequence of pressure is responsible for the creation of hsp70.

Kaarniranta et al. (2003) continued their research, the object was two further stress factors, activator protein-1 (AP-1) as well as nuclear factor-kappaB (NF-kappaB). These were examined on immortalized as well as on primary chondrocytes, murine neuro-2a neuroblastoma and HELA cervical carcinoma cell lines, as well as the mRNA and the protein level. It appeared that after a treatment with 30 MPa in immortalized chondrocytes and HELa cells a significantly higher content of hsp70 mRNA was present whereas in primary chondrocaytes and neuro-2a cells no induction was observed. This is quite remarkable as these cells are well-known to be stress-sensitive. In neuro-2a cells the hsp70 mRNA could not be activated, neither by hydrostatic pressure nor by heat despite an activation of the heat shock transcription factor, after the influence with heat was observed.

Takahashi et al. (1998) examined  HCS-2/8 ells (chondrocyte-like cell line) the impacts of hydrostatic pressure treatment in regard to cytokines and expression of proteoglycan core protein . It was observed that a treatment with 50 MPa over a period of  2 h caused a creation of a interleukin 6 mRNA as well as a TNF-alpha mRNA. However, the TNF-alpha mRNA disappeared within  4 h after the treatment. The proteoglycan core protein mRNA level increased at a pressure of 1 to 5 MPa; at a treatment with 10 resp. 50 MPa the level dropped down. No increase of stromelysin and TIMPI mRNA could be noticed. This leads one to conclude that high pressures result in the creation of IL6 and TNF-alpha, which is not the case with lower pressures. These cytikines are particular  intriguing for  tumor genesis and for possible diagnostical as well as therapeutical methods.

ir Mentré et al. (1999)  described pressure-sensitiveness of the membrane of endoplasmatic reticulum and the nucleolus in tumor cells . The examined erythroleukemia cells, treated with a pressure of up to 100 MPa for 17 min at room temperature, did not lose their viability, but there was a disruption within their ultra structures. The membrane of the endoplasmatic reticulum congealed at a pressure of 50 MPa whereas the plasmatic membrane remained unchanged up to 110 MPa. Mentré et al attributed the difference in performance of both membranes to the higher level of cholesterol in the plasmatic membrane. The cholesterol makes the membrane pressure-resistant in regard to the incipient crystallization under pressure.

All listed examinations on tumor cells, as well as on healthy mammal cells, demonstrate the already mentioned new dimension of research, arising from the application with hydrostatic pressure technology. Although cells react on high pressures under natural conditions, there is a limit at 100 MPa occurring in nature. With the technical possibilities of creating higher pressure in laboratories and to expose the cells to these pressures accordingly, the cells act differently as compared to their natural conditions. Thereby extending the possibilities of exploring pathways and characteristics never appearing under 'normal' conditions.

## CONCLUSION

In general, hydrostatic pressure technology offers an exciting opportunity for medical science, in particular for  oncology, probably not even reaching any limits beyond its possibilities. There are a series of publications and a few research groups working intensely on these subjects. However, compared to other areas of science, where hydrostatic pressure is an already implemented technology, there is a lack of research in the area of oncology in combination with hydrostatic pressure. The research executed so far appears very promising, partially innovative and shows the importance of hydrostatic pressure as a resource in the field of therapy and diagnostics. In order to build a broader base of knowledge, it would be useful if more scientists would concentrate on studies in this exciting area of research. In this way, hydrostatic pressure technology could be a component for healing some types of cancer one day.

## ACKNOWLEDGMENTS

I specially thank Anja Lewinski and Stefan Schaaf. Without your help and permanent support it wouldnot have been possible for me to finish this chapter. Thank you!

## REFERENCES

Araki, N. and A. Myoui, S. Kuratsu, N. Hashimoto, T. Inoue, I. Kudawara, T. Ueda, H. Yoshikawa, N. Masaki and A. Uchida. 1999. Intraoperative extracorporeal autogenous irradiated bone grafts in tumor surgery. Clin. Orthop. Relat. Res. Nov. (368): 196–206.

Aswaq, M. and V. Richards and S. McFadden. 1964. Immunologic Response to autologous Cancer Vaccine. Arch. Surg. Sep. 89: 485–7.

Atanasiu, P. and J. Basset. 1953. Effect of elevated pressures on the virus of leukosis and the erythroblastic cells of the chick embryo; effect of same pressures on the cells of lymphoid leukosis of the mouse. Ann. Inst. Pasteur. (Paris) 85(4): 500–503.

Blümelhuber, G. and S. Fischer, U. S. Gaipl, M. Herrmann, J.R. Kalden, R.E. Voll and R. Meyer-Pittroff. Apoptosis and necrosis of mammalian cells after high pressure treatment—Prospects for a tumor vaccine. Pp. 311–314. In: R. Winter (ed.). 2003a. Advances in High Pressure Bioscience and Biotechnology II. Springer-Verlag, Berlin, Heidelberg, New York.

Blümelhuber, G. and P. Diehl, S. Fischer, B. Frey, M. Hadaller, S. van Laak, W. Mittelmeier, M. Schmitt, C. Weiss and R. Meyer-Pittroff. High pressure treatment of human bonetissue—First investigations about mechanics and biology. Pp. 307–309. In: R.Winter (ed.). 2003b. Advances in High Pressure Bioscience and Biotechnology II. Springer-Verlag, Berlin, Heidelberg, New York.

Davidson, W.R. 1967. The Egyptian medical papyri. In: D. Brothwell and A.T. Sandison. (Hrsg.) Disease in Antiquity. Thomas Springfield.

Diehl, P. and F.F. Naal, J. Schauwecker, E. Steinhauser, S. Milz, H. Gollwitzer and W. Mittelmeier. 2006. Biomechanische Eigenschaften von Gelenkknorpeln nach hydrostatischer Hochdruckbehandlung. Biomed. Tech. 51: 8–14.

Dubert, J.M. and P. Slizewicz, M. Macheboeuf and J. Basset. 1952. Action of increased hydrostatic pressures on the reaction between antigens and antibodies. Bull. Soc. Chim. Biol. (Paris) 34(3–4): 418–430

Eisenthal, A. and V. Ramakrishna, Y. Skornick and M. Shinitzky. 1993. Induction of cell mediated immunity against B16–BL6 melanoma in mice vaccinated with cells modified by hydrostatic pressure and chemical crosslinking. Cancer Immunol. Immunother. 36: 300–306.

Eisenthal, A. and A. Matsaev, A. Gelfand, P. Kahn, B. Lifschitz-Mercer, Y. Skornick and M. Shinitzky. 1996. Surface projection of murine major histocompatibility determinants induced by hydrostatic pressure and cytokines. Pathobiology 1996 64: 142–149.

Eisenthal, A. and Y. Goldman, Y. Skornick, A. Gelfand, D. Buyaner, I. Kaver, A. Yellin, H. Yehoshuaand, B. Lifschitz-Mercer, A. Gonnene and M. Shinitzky. 1998. Human tumor cells, modified by a novel pressure/crosslinking methodology, promote autologous lymphocyte proliferation and modulate cytokine secretion. Cancer Immunol. Immunother. 46: 304–310.

Ferrero, E. 1954. Vaccinations against cancer; experimental suggestions. Dia Med. Jul 5 26(45): 1205–1206.

Freiberg, A.A. and C. Saltzmann and W.S. Smith. 1992. Replantation of an autoclaved, autogenous humerus in a patient who had chondrosarcoma. Report of case with ten-year follow-up. J. Bone Joint Surge. Am. Mar. 74(3): 438–439.

Goldman, Y. and A. Peled and M. Shinitzky. 2000. Effective elimination of lung metastases induced by tumor cells treated with hydrostatic pressure and N-acetyl-L-cysteine. Cancer Res. 60: 350–358.

Hite B.H. 1899. The effect of pressure in the preservation of milk. Bulletin of the West Viginia University of Agriculture Experiment Station Morgantown 58.

Höhn, G. and R. Meyer-Pittroff. 1999a. DE000029913522U1: Vorrichtung zur Herstellung von autologen Tumorvaccinen, Gebrauchsmuster. 3.8.1999.

Höhn, G. and R. Meyer-Pittroff. 1999b. Vorrichtung zur Behandlung von Karzinomen, Malignomen, Tumoren und krankhaft wucherndem Gewebe. Gebrauchsmusterschrift Nr. 299 13 524.1 U1, 02.12.1999.

Islam, N. and T.M. Haqqi, K.J. Jepsen, M. Kraay, J.F. Welter, V.M. Goldberg and C.J. Malemud. 2002. Hydrostatic pressure induces apoptosis in human chondrocytes from osteoarthritic cartilage through up-regulation of tumor necrosis factor-alpha, inducible nitric oxide synthase, p53, c-myc, and bax-alpha, and suppression of bcl-2. J. Cell Biochem. 87(3): 266–278.

Jocham, D. and A. Richter, L. Hoffmann, K. Iwig, D. Fahlenkamp, G. Zakrzewski, E. Schmitt, T. Dannenberg, W. Lehmacher, J. von Wietersheim and C. Doehn. 2004. Adjuvant autologous renal tumour cell vaccine and risk of tumour progression in patients with renal-cell carcinoma after radical nephrectomy: phase III, randomised controlled trial. Lancet. Feb. 21,363 (9409): 594–599.

Kaarniranta, K. and M.A. Elo, R.K. Sironen, H.M. Karjalainen, H.J. Helminen and M.J. Lammi. 2003. Stress responses of mammalian cells to high hydrostatic pressure. Biorheology 40 (1–3): 87–92.

Kopakkala-Tani, M. and M.A. Elo, R.K. Sironen, H.J. Helminen and M.J. Lammi. 2004. High hydrostatic pressure induces ERK and PI3 kinase phosphorylation in human HCS-2/8 chondrosarcoma cells. Cell Mol. Biol. (Noisy-le-grand). Jun. 50 (4): 485–490.

Korn, A. and B. Frey, A. Sheriff, U.S. Gaipl, S. Franz, R. Meyer-Pittroff, M. Herrmann and G. Bluemelhuber. 2004. High Hydrostatic Pressure Inactivated Human Tumor Cells Preserve Their Immunogenicity. Cell Mol. Biol. 50(4): 469–477.

Mentré, P. and L. Hamraoui, G. Hui Bon Hoa and P. Debey. 1999. Pressure-sensitivity of endoplasmic reticulum membrane and nucleolus as revealed by electron microscopy. Cell Mol. Biol. (Noisy-le-grand). May. 45(3): 353–362.

Naal, F.D. and K. Mengele, J. Schauwecker, H. Gollwitzer, L. Gerdesmeyer, U. Reuning, W. Mittelmeier, R. Gradinger, M. Schmitt and P. Diehl. 2006. High Hydrostatic Pressure-induced Cell Death in Human Chondrocytes and Chondrosarcoma Cells. Anticancer Res. Jan-Feb. 26(1A): 85–89.

Prescott, D.M. and A.S. Flexer. 1990. Krebs-Fehlsteuerung von Zellen; Ursachen und Konsequenzen. Spektrum der Wissenschaft Verlagsgesellschaft. Heidelberg.

Ramakrishna, V. and M. Shinitzky. 1991. Potentiation of delayed-type hypersensitivity response to syngeneic tumors in mice prevaccinated with cells modified by hydrostatic pressure and crosslinking. Cancer Immunol. Immunother. 33: 1–8.

Ramakrishna, V. and A. Eisenthal, Y. Skornick and M. Shinitzky. 1993. Increased projection of MHC and tumor antigens in murine B16-BL6 melanoma induced by hydrostatic pressure and chemical crosslinking. Cancer Immunol. Immunother. 1993, 36: 293–299.

Sanja, B.K. and P.G. Moreau and D.A. Younge. 1997. Reimplantation of autoclaved tumour bone in limb salvage surgery. Int. Orthop. 21(5): 291–297.

Seidman, H. and M.H. Mushinski, S.K. Gel and E. Silverberg. 1995. Probabilities of Eventually Developing of Dying of Cancer—United States. CA—A Cancer Journal for Clinicans 35 (1985) S. 36–56.

Shinitzky, M. and Y. Goldmann. 2000. Immunotherapy of Cancer with Pressure Modified Cells. IMAJ. 2: 615–620

Smith, WS. and S. Struhl. 1998. Replantation of an Autoclaved Autogenous Segment of Bone for Treatment of Chondrosarcoma. J. Bone Joint Surge. 70-A: 70–75 Jan.

Takahashi, K. and T. Kubo, Y. Arai, I. Kitajima, M. Takigawa, J. Imanishi and Y. Hirasawa. 1998. Hydrostatic pressure induces expression of interleukin 6 and tumour necrosis factor alpha mRNAs in a chondrocyte-like cell line. Ann. Rheum. Dis. Apr. 57(4): 231–236.

Weiss, E. 1952. Treatment of malignancy with "tumor vaccine". Urol. Cutaneous Rev. Jan. 56 (1): 36–39.

WHO. 2008. http://www.who.int/mediacentre/factsheets/fs297/en/index.html.

Yamaguchi, T. and K. Kuranoshita, M. Fujii, Y. Yokokawa and E. Kimoto. 1997. Inhibition of the proliferation of Ehrlich ascites tumor cells by hydrostatic pressure. Cancer Biochem. Biophys. Jun. 15(4): 257–261.

# Working Under Pressure

*Jean-Claude Le Péchon*

## INTRODUCTION

### Why Work Under Pressure ?

In the middle of the 18th century, compressors became available, and as the principle of underwater 'diving bell' was already established, an extension to the underground space was developed, allowing to dry out wells by the application of air pressure and to use airlocks to transfer personnel into the pressurized zone. Compressed air has been extensively used to sink caissons, bore tunnels and mine coal (Walder, 1982). It has resulted in major adverse health consequences since the working conditions under pressure had physiological effects that were not understood and the incentives to work safely were very poor (Kindwall, 2003).

Salvage and military applications have been the primary reasons for men to venture underwater, one way has been the atmosphere dry submersible, the other way being diving. In the early days diving was only for salvage and for commercial purposes. Recreational diving began in the 50's when self-contained underwater breathing apparatus (SCUBA) became available. Nowadays professional diving includes a large range of situations: commercial divers work either inland in rivers, dams and harbours, or offshore mostly for the oil and gas industry.

### Specific Issues ?

Among working situations, working under pressure has a special place. Most of the times when a hazard is identified at work, safety requires that the person should be protected against it by preventing actual exposure.

Ingénieur Conseil, 94, rue de Buzenval, 75020 Paris.
E-mail: hyperbar@club-internet.fr

This is obtained by removing the hazard, erecting physical barriers like individual protective equipment or confining the source of hazard inside protective cabinets. When barriers are not suitable or possible, special working procedures may be used. Procedures are more prone to failure than physical barriers.

When one needs to work under pressure there is no physical barrier, otherwise the worker would operate from a submersible or with a remotely operated vehicle (ROV) or would dig with a Tunnelling Boring Machine (TBM). When diving or hyperbaric work is the only way to reach the workplace, the worker is exposed to pressure eventually in immersion, with no other protection against the possible physiological adverse effects of pressure and gases than working procedures. Those procedures being still subject to failure, zero accident is not warranted. Therefore, contingency plans must be established to reduce the consequences of such potential failure in order to reach a zero damage situation (Fig. 1).

In addition, trivial risks at work may still be present, and any emergency arising while under pressure, immersed or not, will make the rescue more difficult than when access to the casualty is straightforward.

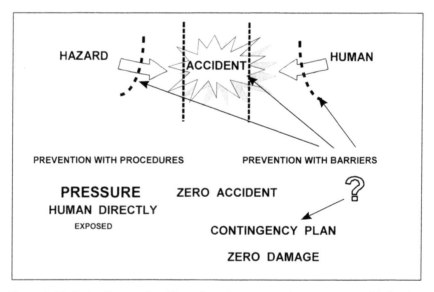

**Figure 1.** Methods of prevention: To work under pressure there are no possible barriers, only procedures protect the worker. Zero accident is only the objective, contingency plans are needed to reach 'no damage'.

## Who is Concerned?

The lower over-pressure limit for 'increased pressure environment' has been established by various national regulations and is generally accepted as a depth of 1 metre in immersion or 0.1 bar over local atmospheric pressure in the dry (Regulation, 1990a, 1990b, 1994, 1997).

### *Divers*

A professional diver is someone who dives while at work. There are many different working situations that must be considered separately as far as diving methods, organization and safety are concerned.

*Commercial Industrial Divers.* Commercial divers are involved in industrial contracts for civil engineering, construction, inspection and repair, frequently this is not very deep (0 to 40 metres), it is often referred as in-land or onshore diving. In the oil and gas industry, diving is required for construction, inspection and maintenance/repair and takes place offshore. The depth range is very wide : from the splash zone down to more than 300 metres in some deep production sites like in the Gulf of Mexico or Brazil (Imbert et al., 1990).

Most of the time divers work for commercial diving companies that are subcontracted to carry out the required underwater tasks.

Risks associated with commercial diving operations extend further than diving technology and implies all hazards resulting either from:

- the dive site natural environment,
- the diving equipment,
- the dive support vessel or structure,
- the industrial structures on which the divers work or from the tools they operate (Bevan, 2005).

*Recreational Professional Divers.* In the field of recreational diving the professionals are mostly diving instructors, members of recreational diving agencies, their depth range is most of the time limited to 50 metres sometimes less. Technical diving instructors dive much deeper. Most of the hazards involved in this type of dives are related to SCUBA diving safety, natural environment and decompression.

*Scientific Divers.* Scientists (biologists, oceanographers, archaeologists) dive to carry out observations or to collect samples or specimens. Those dives are basically SCUBA. In some exceptional cases sub-sea habitats are used (Lin and Shida, 1990; Joiner, 2001).

*Media Divers.* Divers involved in movie making (actors, cameramen, light support divers…), professional sub-sea photographers, underwater TV reporters etc. are listed as professional divers. Very few organizations or national regulations are specific (Regulation, 1997).

*Fishermen Crew.* On board fishing boats, divers may operate for collection of seafood (lobsters, abalone, sea weeds, coral …), aquarium fishes, or to salvage fishing gears (nets, traps…). In aquaculture, cage cleaning or inspection and maintenance of the structures require professional divers.

*Government Divers.* Customs, coast guards, civil security and other government agencies employ professional divers.

### Compressed Gas Workers

*Digging.* Normally compressed air workers are caisson workers. They dig shafts or tunnels in a compressed air environment established to prevent water ingress or to increase ground stability.

*Leak Testing of Pressure Vessels.* Many nuclear power plants, concrete buildings and air plane cabins are checked for leaks by pressurized operators at 1 bar –exceptionally 4.2 bar in nuclear plants- and at 0.6 bar in planes.

The risks associated with interventions in nuclear plants are irradiation (except during initial commissioning), ear barotrauma, accident while under pressure, and decompression sickness. In plane testing the only risk is ear barotrauma.

*HBO Personnel.* During medical treatment with hyperbaric oxygen in multiplace chambers, (See chapter IV-b) the personnel is exposed to compressed air (Le Péchon, 1998).

*Dry Simulated Dives.* Although scientific and technical research in deep diving is almost over, in the early days many hyperbaric centres have supported simulated deep or very deep dives (1965–1990) eventually using 'wet pots' to evaluate the effects of pressure and immersion. It has also been used to perform diver training, diving gear testing and certifications of welding procedures and welders.

### Units and Metrology

Units to be used below are only bar gauge (bar) for chamber pressures, bar$_{(a)}$ for absolute pressures, or bar for partial pressures. Depths are expressed in metres. A cross calculated table to convert values from non metric systems is available in each issue of a publication from UHMS: Undersea and Hyperbaric Medicine.

### HISTORY IN SHORT

Extensive histories of compressed air work and diving have been published many times (Haux, 1982; Lin and Shida, 1990; Jardine and McCallum, 1994; Brubakk and Neuman, 2003; Lamont, 2006) only the milestones will be reviewed here.

## Venturing into Higher Pressures, and Deeper Waters

As soon as compressed air work and diving became technically feasible, higher pressures and deeper dives have been the incentive for development, creating new technical, physiological and safety challenges for science and industry.

In diving, air breathing limits were understood in 1960's with the introduction of helium in deep diving technology to reduce gas density and narcosis, mostly with heliox mixes, until High Pressure Nervous Syndrome (HPNS) and the rate of compression became the new limiting factors inducing trials with deep Trimixes (Oxygen, Helium and Nitrogen), to try to balance HPNS and Narcosis. Then, to reduce further gas density and eventually eliminate HPNS and nitrogen narcosis, breathing hydrogen based mixtures made it possible in a Trimix cocktail to reach 70.1 bar in the dry and 534 m in actual open sea dive .

In compressed air, high pressures (4.5 bar) were achieved as early as early 1854 (Pol and Wattelle, 1854). However safe access to that range of pressure took a lot of time and the more stringent limiting factor has always been the required decompression time. Nowadays changes associated with TBMs reduced the need of such exposures, leading to the use of shorter exposure times and extended decompression using oxygen during the decompression stops. Breathing helium containing mixtures  was started only in the 90's (Hirata et al., 1992) as well as saturating the operators which is still a unique operation (Le Péchon et al., 2001). Presently (2008) several tunnels are being bored in the pressure range of 12 bar, none has yet required to expose workers at that pressure. The equipment for heliox saturation at 12 bar has been implemented in some TBMs.

### Staying Longer

For both dry and immersed situations the decompression time required to return 'safely' to normal pressure is the critical limiting factor. For long stays underwater, the other major issue has been cold water.

### *Decompression*

Models for short exposures first with air and its substitutes like Nitrox, then gas mixtures (heliox or trimixes) allowed bounce dives that remained most of time shorter than 2 h followed by long decompression stops, and required a delay of 12 h before the next exposure. To stay longer and produce more bottom time or compressed air shift time was a problem. It was only in the 60s that it was understood that the quantity of gas which the body may dissolve at a given pressure is limited according to Henry's Law and led to the introduction of saturation techniques. First the idea was to live in subsea habitats that proved to be unrealistic for industrial work, the second step was to hold the divers under pressure in a pressurized habitat built on board the support vessel. This method has been very successful and is still expanding very fast: in 2008, 22 new diving support vessels were built due to a sharp demand of powerful underwater work capacities to meet the requirements of the oil and gas industry.

With the saturation technique, decompression time is economically less important and can be extended to make it safe. Many 'tables' are available either from government agencies, commercial diving companies or navies (Le Péchon, 1981; Regulation, 1990a). Most of them cover the final decompression as well as upper or lower excursions from the habitat pressure. Offshore, heliox is the mixture of choice. In the habitat the initial nitrogen (0.8 bar) not being removed the actual mixture contains about twice as much nitrogen than oxygen (0.4 bar), but it is still considered as heliox. Trimix saturation techniques have been tried extensively to reduce HPNS, however it has always remained at the experimental level. At lower pressure in Westerschelde tunnel Trimix saturation was the choice that allowed workers to breathe air at 4 bar in the airlock during the transfer (Le Péchon et al., 2001).

### *Protection from Cold*

There are many ways to protect the diver from cold, choosing the more appropriate is a matter of the type of dive. Wet suits, dry suits, heated suits with batteries for SCUBA dives, and mostly hot water suits for commercial diving are now very well designed and altogether they provide comfort for long stays in immersion even in freezing waters. For heliox dives it has also been established that helium induces a dramatic heat loss that requires warming the breathing gases (35°C) with a hot water heat exchanger supplied in parallel with the hot water suit.

Protection from heat during warm water commercial diving is also sometimes a limiting factor and precautions must be taken to dive in water temperature above 30°C since hyperthermia gives very few signs before cardiac problems occur.

### Safer Working Conditions

Only recently were divers' and compressed air workers' safety based on physiology and general risk assessment enforced into a single regulation, the Brazilian and French (Regulation, 1983, 1990a) include both activities in a set of rules established for workers exposed to pressure, with or without immersion Norwegian (Regulation, 1990b), German (Regulation, 1994),

United Kingdom (Regulation, 1997), and US regulations, for example are separated for both activities and still address professions rather than risks. There is not yet any unified European policy in this particular field although a lobbying organization (European Diving Technology Committee) calls for a European Directive for professional divers and surprisingly for HBO personnel who are not divers...!

Despite the progress in regulation enforcement the number of fatalities is still very high with 20 deaths recorded among commercial divers in first eight months of 2008. Most of the reported cases resulting from non application of risk assessment and basic safety rules described in national rules or recommendations from IMCA (Chesshire personal communication).

## PHYSIOLOGICAL INVOLVEMENT

When one undergoes pressurization, the whole body is exposed to pressure and pressure changes. This situation induces many changes on the individual's physiology most of which are potentially hazardous. To evaluate the consequences and the required safety precautions, extensive reviews are available of diving and compressed gases, physiology and medicine Brubakk and Neuman (2003) in English or in French Broussolle et al. (2006).

Based on the physical laws of gases, the physiological impacts of pressure and pressure changes on the body can be split as shown in Fig. 2.

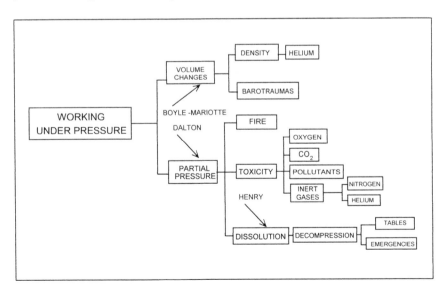

**Figure 2.** Physiological effects of pressure: The various consequences of gas laws are illustrated in a diagram that is a pathway for the risk assessment.

## Mechanical Effects on Gas Volumes

### Barotrauma

Barotrauma is caused by pressure variations and resulting volume changes in the body gas cavities: lungs, ears, sinuses, stomach, intestines and eventually teeth. The diving gear may also create facial barotrauma (mask / helmet) or whole body squeeze (suit).

*Lungs:* Lungs volume changes permanently during the respiratory cycle and must be kept at a pressure very close from ambient pressure in order that respiratory muscles can support a ventilation minute volume adapted to the workload. The lungs are compliant, the alveolar membrane is fragile and may withstand an over pressurization of only about 100 hPa (Neuman, 2003).

Underwater or when diving

Lung stress while breathing underwater remains minimal when the breathing system can deliver the right flow and volume within acceptable inhalation and exhalation pressure limits. Due to the delivery of breathing gas at local ambient pressure, the lungs keep almost their 'normal' volume. While pressure rises, the mass of gas in the lung increases according to Boyle-Mariotte's Law. Decompression will have the reverse effect and, provided the natural ventilation of the lung is maintained, the mass of gas in the lungs will adapt to the reducing pressure. Should this natural ventilation not be maintained during ascent (panic, air breakdown and free ascent, or on the occasion of submarine escape training…) it will induce an over pressurization of the lung which may disrupt the alveolar membrane and inject air into the blood flow creating the worst diving accident: Burst lung and Arterial Gas Embolism (AGE). Training, medical fitness and ascent procedures are efficient prevention methods.

Compressed gas breathing in the dry

At increased pressure, when breathing the ambient gas there is no added breathing resistance from equipment or from the effects of immersion. Inspiration flow and pressure are only modified due to gas density. There is no risk of a burst lung for medically fit persons. Natural lung ventilation will not be stopped at any time during depressurization and balancing the mass of gas is spontaneous. Except in the case of chamber burst, no decompression of a chamber will be fast enough to beat free expiration (Violette, 1955) and then there would be more serious damages due to blow, noise, cold, and eventually anoxia or decompression sickness.

*Ear*: The middle ear is a not-compliant semi-closed gas volume contained in a small bone cavity behind the tympanic membrane. During pressure changes, inside and outside pressures must be maintained equal otherwise the tympanic membrane will be stretched by differential pressure. Therefore when pressure is modified, admission or evacuation of gas in or from the middle ear is needed to keep balance. As the Eustachian tube the middle ear is connected outside. This tube opens spontaneously about once per minute on the occasion of swallowing, and pressure balance is automatically maintained during slow minor pressure changes.

Upon fast pressurization, a depression forms in the middle ear; the Eustachian Tube is flat, tight, no gas can be admitted. By willingly creating a slight over pressure from the nose side, the Eustachian tube opens, balancing the pressure on both sides of the tympanic membrane. Upon depressurization, the Eustachian tube flows out the gas with minimal resistance since the higher pressure is on the 'blowing' side.

There are various manoeuvres used by divers and compressed gas workers to equalize their ears. They have been given names following their authors, Valsalva, Frenzel. Swallowing may be sufficient if the pressure difference is kept minimal. Diving helmets are fitted with a 'nose pad' to help the diver close his nostrils.

Failure to control middle ear volumes during compression results in ear barotrauma. Training, medical fitness and careful compression procedures are efficient means of prevention.

*Sinuses*: Sinuses are non compliant gas cavities located inside the facial bones. They normally communicate freely with the oro-nasal cavities. During pressure changes, pressure inside and outside is maintained equal spontaneously. Severe pain in the frontal part of head is the sign of equalization failure. There are no manoeuvres to cure the pain and the only means is to reverse pressure. A medical fitness check is necessary when sinus pain is noticed upon compression or descent.

*Stomach and intestine*: There are significant pockets of gas in the normal stomach and intestines, their volume is variable and as pressurization or depressurization proceeds it will easily adjust. In compressed gases work, drinking sprinkling beverage while under pressure may produce a large volume of stomach gas upon depressurization. Such drinks should be avoided, however liquid beverages/water should be available while working to prevent dehydration often in relation to heat. Gases can be produced in the intestines or may move along the intestinal track during the exposure, their expansion during depressurization may create distension and pain would result. Halting depressurization and a very slow return to normal pressure will solve the problem.

*Teeth*: A normal tooth does not include any gas volume, therefore there is no risk of barotrauma. When a tooth has been fixed, some gas may be trapped behind the fillings that may become a problem when exposed to pressure variations.

## Equipment squeeze

Diving masks must be blown by the nose during descent to keep the volume constant while descending. When a mask is not balanced its volume will be reduced and a depression in the mask may induce blood to fill capillaries and eventually create haemorrhagic zones. Free flow diving helmets may produce similar situations when the descent rate is too high to allow the free flow to balance pressure.

When using a dry suit, the pressure difference between head and feet in upward position will induce a squeeze in the legs and a bubble in the upper part of the suit. The volume of the bubble should be controlled to prevent blow up, and undergarments should be worn to prevent painful skin pinching. When in head down position, the effect is reversed and buoyancy control may become a problem for poorly experienced divers.

In the dry any closed item which may not be externally pressure resistant must be open during pressurization otherwise it will be squeezed and opening becomes impossible under pressure. When an object has been opened under pressure it should be decompressed open (or with a pressure relief hole).

## Gas density

The effect of breathing dense gases under pressure is explained in details in chapter 17. Changing from air to helium containing breathing gases in diving is set at 50 metres by most national diving regulations, and for compressed gas work experience shows that above 3 to 4 bar it can be recommended to breathe trimixes.

## Partial Pressures

In hyperbaric exposures the partial pressure of the gases in the environment, change with total pressure in accordance with Dalton's law. The partial pressure of a gas species in an atmosphere is the product of the decimal concentration by the total pressure as shown in the magic triangle (Fig. 3) that solves all gas under pressure issues.

Gases may be toxic when absorbed into the body. Most of the time absorption takes place via the lungs. The gas exchange between the atmosphere and the body is directed by the alveolar partial pressure of the concerned gas at the lung level and the duration of the exposure to that partial pressure. The relative difference between ambient and alveolar partial pressures decreases as the pressure rises allowing, for increased pressure environment to consider only the ambient partial pressure to evaluate gases toxicity with an acceptable error.

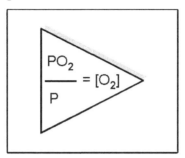

**Figure 3.** The magic triangle: $PO_2$ is the partial pressure of a gas species in bar (here oxygen), P is the total pressure in $Bar_{(a)}$, $[O_2]$ is the decimal concentration or fraction of this gas species in a mixture.

## Oxygen

Oxygen is the most important molecule to support life. Ambient oxygen partial pressure at sea level is close to 0.2 bar. The biological effects of oxygen are associated with the energy rich chemical agents resulting from the reduction of the molecular oxygen. In that process of oxygen reduction, several intermediate molecules are formed that are called 'free radicals'. Their high oxidative potential leads to the destruction of structures like cell membranes or the degradation of proteins. Protective enzymatic mechanisms keep free radicals in the acceptable range of concentration.

These mechanisms are universal among living cells, the potential resulting adverse effects of oxygen for the cells are dose related. Oxygen dose is a combination of the oxygen tension in the concerned tissue and the duration of exposure. A clear and exhaustive review of the actions and reactions to oxygen is given by Clark and Thom ( 2003).

The latent period preceding symptoms is influenced by several factors in addition to ambient $PO_2$:

- Periodic interruptions of high $PO_2$ breathing extends both pulmonary and brain tolerance. It is probably the more efficient way to gain toler-ance for medical or operational applications.
- Alveolar/arterial carbon dioxide levels reduces central nervous system tolerance.
- Several drugs may be efficient (antioxidants or free radical scavengers)…

Pulmonary effects and brain damage are the first noticeable signs in oxygen poisoning. Which one is first depends mostly on the range of $PO_2$.

*Pulmonary Signs:* For continuous exposures at moderate partial pressure of oxygen, the pulmonary signs are the first to appear and are progressive, they begin with a reduction of vital capacity, followed by coughing, then respiratory impairment, dyspnea and may ultimately be lethal when extended too far. The early signs are reversible within a few days.

This pulmonary damage is called the Lorrain-Smith effect. The oxygen dose relationship with pulmonary damage has been evaluated in detail and several means of quantification have been proposed. The most popular has been introduced at the University of Pennsylvania and specific units called Unit Pulmonary Toxic dose (UPTD), CPTD (Wright 1972) or OTU are now common practices. One UPTD is the equivalent effect of 1 min breathing pure oxygen at atmospheric pressure ($PO_2$ = 1 bar).

*Central Nervous System Signs*: In the range of oxygen partial pressure above 2 bar, the first manifestations of poisoning involve the brain. Described in detail by P. Bert and named after him 'effect Paul Bert' and extensively evaluated (Donald, 1992) for the World War II battle swimmers. The present status of brain reaction has been reviewed recently (Clark and Thom, 2003).

It is usually considered that convulsions or seizures are the symptoms of acute CNS oxygen poisoning, however this is only the end point. Most other signs occur very shortly before convulsions and they may pass unnoticed by the exposed person. Among these, changes in behaviour, visual signs like tunnel vision, bell ringing, nausea, lip twitching, eventually syncope are the more frequent signs. Restoring normoxic gas during this phase of 'pre-convulsions' is often sufficient to prevent seizures.

In trying to establish a safety policy for exposures to high partial pressures of oxygen, these discrepancies have resulted in larges ranges of recommended $PO_2$.

There is a significant difference between the risks when exposure is in the dry or in water. First warning signs are more difficult to recognize when immersed and breathing either from SCUBA or in a helmet. Second, in case of warning signs, shifting to air takes more time or is eventually not possible and in the end convulsions underwater may lead to drowning. Therefore $PO_2$ limits are adapted to exposure conditions and higher $PO_2$ values are acceptable in the dry.

The maximum duration of continuous exposure to high partial pressures of oxygen must also be specified and several tables have been published by organizations, navies, diving agencies and national regulations:

- maximum value of 1.6 bar.
- for breathing in the dry the maximum $PO_2$ value is 2.2 bar Eventually including an open bell situation when specific breathing masks are used, the divers are prevented from face immersion should convulsions or loss of consciousness is likely to occur.

- for medical purposes either for HBO or on site treatment of DCI, it is acceptable to breathe gases providing a $PO_2$ up to 2.8 bar.

*Operational applications of hyperoxia.* Breathing high $PO_2$ gases, either as pure oxygen or as mixtures is necessary to improve safety, by reducing decompression times (Nitrox, mixed gas decompression, constant $PO_2$ decompression in rebreathers or saturation, oxygen stops.) and for DCI management. Establishing safe limits for breathing hyperoxic mixtures must be based on a risk assessment. Then the benefit expected from hyperoxic breathing and the possible consequences of pulmonary or CNS damages must be balanced to recommend an acceptable procedure.

   During medical fitness checks, EEG recordings associated with light stimulation and with hyperventilation may give evidence of an early epileptic reaction, susceptible persons should not be exposed to $PO_2$ higher than 1 bar in hyperbaric operations in diving as well as in a compression chamber.

*Safe Oxygen Limits:* To prevent pulmonary damage, it has been established that 614 UPTD correspond to a minor decrease of Vital Capacity of about 2%, which is reversible (Fig. 4). An exposure to 1400 UPTD corresponds to a 10% decrease of Vital Capacity, still reversible (Wright, 1972).

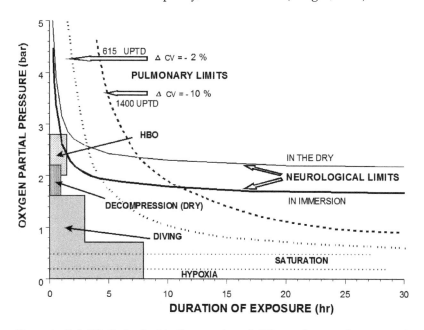

**Figure 4.** Safe $PO_2$ limits: In this diagram, the safe $PO_2$ are shown in boxes according to the condition of exposure (saturation, in water decompression, decompression in the dry or HBO applications).

In normal operations 600 UPTD in one day is safe. For medical applications the balance risk/benefit is in favour of more oxygen, the safe limit is established at 1400 UPDT. However for every day operational exposures lower values had to be enforced since it has been shown (Sterk, 1986; Hamilton and Kenyon, 1989) that daily cumulative effects are not acceptable.

### Carbon Dioxide

*Physiological Signs.* Carbon dioxide should not be considered as a pollutant since it is naturally produced by breathing. The expired gas contains a mean $PCO_2 = 0.04$ bar and the alveolar gases 0,053 bar.

The effects observed during exposure to $PCO_2$ above 0.005 bar are dose related. The dose is defined by duration of exposure and $PCO_2$.

When breathing in closed circuit in saturation habitats or from a rebreather, poor scrubbing may result in too high $PCO_2$ and the signs of intoxication may become very severe and could lead to death.

Due to the relatively large volumes of saturation habitats, the rising of $PCO_2$ is very slow and there is time to take action to fix the scrubbing capability of the environmental control unit. Due to the small volume of rebreathers counter lungs, the intoxication is acute in a few minutes should the scrubber fail.

Exposures to high partial pressures of carbon dioxide induce progressive signs and symptoms (ESA, 1992).

Long duration exposures have been evaluated extensively by the French navy for the environment of nuclear submersibles (Radziszewki, 1987). In saturation exposures, most organizations have selected 0.005 bar as maximal $PCO_2$ value, with extension to 0.01 bar for short durations.

*Synergic Effects:* It has long been known that carbon dioxide induces synergic effects with nitrogen narcosis. In the early days narcosis was even considered as an effect of high carbon dioxide partial pressure until careful evaluation, eventually using lighter inert gases demonstrated the role of nitrogen and the synergic effect (Rostain, this book chapter 18).

The direct effect of high $PCO_2$ in breathing gases has been evaluated in detail, however high arterial $PCO_2$ induced by dense gases breathing has been demonstrated to reduce tolerance to oxygen intoxication (Clark and Thom, 2003).

During compressed air work, high $PCO_2$ in the air locks may increase the incidence of DCI. An appropriate ventilation to control $PCO_2$ below 0.01 bar is easy to achieve during the decompression stops. Monitoring $PCO_2$ may be recommended but is scarcely needed.

*Scrubbing Carbon Dioxide:* Low atmospheric $PCO_2$ in chambers is obtained by natural air ventilation. In closed circuit situations a scrubbing is achieved by absorption in sodalime canisters. Sodalime is a mixture of several chemicals Soda -NA(OH)-, Hydrated lime $-Ca(OH)_2-$, some water – $H_2O$- and eventually accelerators (like potassium hydroxide –KOH-) according to the manufacturers. The chemical absorption produces heat that contributes significantly to breathing gas heating while diving with a rebreather.

Practical absorption capacity: Most soda limes can absorb 22% of carbon dioxide by weight. (1 kg of sodalime should store 0.22 kg of carbon dioxide). For a medium work load producing about 2 g/min of carbon dioxide, the overall capacity of sodalime carbon dioxide removal is about 110 min/kg. This figure is a fair approximation but is also a function of the geometry of the canister, the humidity of the gas (rebreathers and saturation chambers atmosphere are significantly different in this matter). Useful practical rules of the thumb are: 1 kg of soda lime is good for 2 h for 1 person or a 1 litre rebreather canister should support a diver for about 100 min.

### Inert Gases

Nitrogen and helium are used as breathing gases in operational hyperbaric situations, ultimately breathing hydrogen only in experimental hyperbaric exposures. The choice of a proper mixture to work at high pressure is related to operational conditions, duration and environmental situation (dry or immersion) for physiological data refer to Rostain in this book chapter 18.

### Pollutants

*Threshold Limit Values (TLV)*: At atmospheric pressure, the toxicity of volatile organic compounds is well established and maximum acceptable values have been elaborated by most national regulations and international standards. TLV are available under two different formats: ppm or $mg/m^3$. Some of these limits are established for long duration exposure (8 h/d, 5 d/wk) or for short duration exposure (15 min) allowing to escape, to don breathing masks or to restore an acceptable atmosphere.

Values in ppm are equivalent to a fraction and need correction for hyperbaric application, they have been validated for standard atmospheric pressure, air breathing, ambient temperature and humidity. However, the toxic effects controlling factor is the partial pressure at the point of breathing. For a given pollutant, the ppm value for 1 bar$_{(a)}$ corresponds to a partial pressure which can be read with the same figures in micro-bar (1 µb = 0.1 Pa = $10^{-6}$ bar). A similar calculation needs to be performed for any pollutant when a TLV in ppm is available.

A significant part of a table published in The Professional Diver's Hand Book (Bevan, 2005) translated into partial pressures for direct use at any pressure is given in Table 1.

Table 1. Threshold Limit Values in Partial Pressures.

| Main pollutants Threshold Limit Values (in µbar or $10^{-6}$ bar) | | | |
|---|---|---|---|
| *Pollutant* | *Lethal* | *15 min Exposure* | *TLV* |
| CO | >525 | 50–100 | 50 |
| $NH_3$ | | | 25 |
| $H_2S$ | >800 | 10 | 5 |
| $NO_2$ | >200 | | 5 |
| $SO_2$ | >400 | 50 | 5 |

When the TLV is in $mg/m^3$ the calculation is different, depending on measurement conditions. If the analysis is performed under pressure, the value in $mg/m^3$ applies directly. If the measurement is performed on gas samples at atmospheric pressure, the same correction as for ppm must be calculated.

*Analysis:* Very often tube analyzers are used to monitor pollutants in atmospheres.

• When a tube analyzer is used at standard atmospheric pressure to read a sample drawn from the pressurized environment, the result is a fraction. The corresponding partial pressure must be calculated to assess the working condition (figures on the tube are printed in percentage or ppm).

• When the tube is used inside the pressurized environment, the result reads on the tube in fractions (ppm) however the mass of gas that goes through the tube is under pressure and the colour line shows the partial pressure; then the limit can be read directly (in µbar instead of ppm) on the tube with no correction.

When other types of analyzers are used under pressure they must be checked to validate calibration (fraction or partial pressure ?). When used outside, the reading is always a fraction (ppm or percentage) and partial pressure must be calculated to evaluate potential toxicity.

*Risk assessment for pollutants:* In diving, the breathing gases are compressed air or mixed gases, eventually pure oxygen during decompression stops.

Pollutants control for compressed air is based on making sure where the compressor is taking the air from. Monitoring the inlet air quality is required which would reveal if carbon monoxide, carbon dioxide, hydrogen sulphide (in particular on gas production or treatment platforms) or any other expected pollutant is present in the area. The pollution resulting from the compressors must also be considered. Filters corresponding to the type of compressor must be fitted and properly maintained. Lubricated compressors use aerosol and charcoal filters. Non lubricated dry compressors, either piston compressors or screw compressors (very often used in compressed air work) need only filters for dust and water vapour. Although water vapour removal is not a matter of air pollution it is necessary in all types of compressors to prevent water condensation in high pressure storage.

For all types of compressors a schedule to monitor the quality of produced air must be established (analysis, filters checks and replacement…) and all operations carried out in the maintenance programme must be documented.

There are several standards applicable to breatheable air and diving air quality standard. The principle is that the level of pollutant must be acceptable at the maximum depth permitted for air breathing: 50 or 60 metres according to local regulations.

- Oxygen close to 21% is never a problem (at 60 ms $PO_2 = 1.47$ bar, which is always acceptable).
- Nitrogen close to 79% (at 60 m $PN_2 = 5.53$ bar) is always narcotic and the decision to dive as deep as 60 metres on air is always associated with the acceptance of the consequences of narcosis and gas density effects.
- Carbon dioxide, the maximum acceptable partial pressure is in the range of 0.005 to 0.01 bar (depending on local regulation), at a depth of 60 m that corresponds to a maximum concentration of 714 to 1428 ppm. Normal pure dry air contains from 300 to 350 ppm of carbon dioxide it is always convenient when compressor is properly located.
- Argon represents about 1% of air, this is always neglected and considered as nitrogen since argon is an inert gas and the narcotic properties that are associated with argon (Imbert et al., 1989) require higher partial pressure than 0.070 bar (air at 60 m) to be significant.

When a risk of pollutants (nitrogen oxides, hydrogen sulphide, ozone…) has been identified, it requires a calculation using the maximum threshold values locally applicable to obtain acceptable partial pressure and therefore maximum ppm content of the stored gas.

Pollutants in mixed gases are very rare, the major source could be the compressors or boosters pumps occasionally used to transfer the gases. Mostly oil vapour may be suspected from lubricated machines. Charcoal filters are systematically fitted on the exhaust of such equipment.

In saturation habitats there are few sources of contamination outside the compressed gases blown in to pressurize the chamber. Some problems have arisen from newly painted chambers that may release toxic chemicals (benzene) and also from hydrogen sulphide ($H_2S$) formed by Sulphate Reducing Bacteria activity in stored diving bells umbilicals flooded with stagnant seawater. In diving bells however, when the divers are working on polluted objects (drums of chemicals in wrecks, pipe lines contaminated with oil, drilling cuttings…) a detailed risk assessment of the potential contamination in the diving bell must be carried out to evaluate the resulting risks of atmosphere contamination. Setting the maximum threshold values and defining the required monitoring/alarm system are a must. A specific diving procedure needs to be prepared that may include divers protection with disposable overalls to be left outside the diving bell, decontamination procedure (charcoal scrubber…) if required and controls before the divers are allowed to enter back into the habitat. Monitoring the saturation habitats atmosphere may also be recommended in some instances.

In compressed air work the situation is far more complicated. All sources of possible contamination must be looked for and evaluated. The origin and quality of compressed air is to be monitored. In particular when compressors are located on board a TBM and the atmosphere of the tunnel is the primary source of fresh air, it may be contaminated by industrial activities of construction in the tunnel, by motors of trucks and civil engineering machines. To prevent that contamination in breatheable air, the onboard compressors inlets can be connected directly into the ventilation tube providing fresh air into the tunnel.

In the working chamber there are many sources of contamination, among them:

- Chemicals from the ground when it has been contaminated by human activities (Le Péchon, 2002a),
- Organic material from natural fermentation by products like methane, hydrogen sulphide…,
- Chemicals from sealing foam and sprays,
- Pollution resulting from burning, welding and grinding (carbon monoxide, metallic fumes, nitrogen oxides…),
- Grouting chemicals used as ground treatment.

The evaluation of the risks involved should take care of the nature of the pollutant, the possible partial pressures compared with TLV (expressed in partial pressure), the beneficial effect of air losses in the ground providing extra ventilation of the working site etc.

Note: Methane is not a toxic gas. It is hazardous because it is often associated with hypoxic environment, and because of the risk of explosion when mixed with air. It as been established (Cluet and Gros, 1994) that in the range of pressures concerned, methane as well as kerosene do not form explosive mixtures. Monitoring hydrocarbons in a compressed air environment can be done with the same explosivity limit values as at atmospheric pressure however toxicity of hydrocarbons should be checked.

Medical air standard EN 737-3 specifies air quality of stored air used in medical applications. HBO patients never directly breathe medical air (except ventilated patients undergoing air breaks), it is often humidified before administration (like oxygen). It is useless to blow an HBO chamber with medical air since the resulting atmosphere in the chamber will not be 'medical air' . However the quality of the stored air should be such that pollutants TLVs are not exceeded, but also that ventilating the chamber with that air will keep the atmosphere contamination low enough. Using the diving air standard is acceptable. The standard EN12021 is the reference specified in EN 14931 concerning safety of HBO chambers however, surprisingly the maximum carbon monoxide concentration acceptable in that air (15 ppm) would make it unacceptable above 3 bar since corresponding PCO is then ($4 \times 15 \times 10^{-6} = 60$ µb) which is not acceptable (Table 1)!

### Major Pollutants Symptoms

Carbon monoxide (CO)

Carbon monoxide is particularly dangerous because it is odourless, colourless and symptoms of toxicity are not readily noticeable by the potential victims.

The toxicity of carbon monoxide is depends on carbon monoxide and oxygen partial pressures in the breathing gas. For a normoxic air ($PO_2 = 0.21$ bar) and for short duration exposures, the carbon monoxide maximum acceptable partial pressure is PCO = $50 \times 10^{-6}$ bar (50 µbar).

Hydrogen sulphide ($H_2S$)

*Sources in diving*: on board platforms extracting oil or gas containing hydrogen sulphide it may contaminate the atmosphere and therefore may be pumped in diving compressed air. Sensors installed to monitor workers risk on board will trigger alarms for TLV at atmospheric pressure that will correspond to an unacceptable high partial pressure at depth.

*Sources in tunnelling*: hydrogen sulphide may be present in the ground and may evaporate in the tunnel (EPB TBMs) or at the slurry station (bentonite TBMs) or in the working chamber during the compressed air working session (Le Péchon, 2002b).

It has a very specific odour of rotten eggs which is perceived for partial pressure as low as 0.02 µbar. However detection from the odour is a pernicious indication since above 100 µbar, the sense of smell is reduced or cancelled due to anaesthesia of nose sensible cells. TLV is 5 µb.

Hydrogen sulphide is not only a very toxic gas, at concentrations between 4% and 46% in air it is explosive (unchanged with pressure).

## Gas Dissolution and Decompression

### *Principles of Decompression*

While breathing under pressure a certain amount of inert gas dissolves into the body via the blood perfusion in application of Dalton and Henry's laws, this is a gradual process until a state of saturation is achieved. Should the exposed person is decompressed back to atmospheric pressure that dissolved gas is restored as a gas phase wherever it is located in the body. As a direct consequence the forming bubbles will cause pain by stretching tissues or interfere with blood flow when they are in the capillaries. The resulting multiple mechanical and biological undesirable effects are globally described (Francis and Smith, 1990) as: Decompression Sickness or Decompression Illness (DCI).

Although Paul Bert (Bert, 1878) had understood the underlying mechanism it was Haldane (Boycott et al., 1908) who proposed a practical model for the process of safely loading and unloading the extra gas in divers and compressed air workers bodies. Nevertheless many theories have been proposed to support algorithms to calculate decompression procedures (Workman, 1965; Bülhmann, 1984; Yount, 1986; Hennessy, 1989; Vann, 1989; Wienke, 1990). Only the more significant are cited here, an extensive theoretical and practical review is available (Brubakk and Neuman, 2003, Chapter 10).

After a given exposure to pressure the choice of the best adapted decompression procedure is very large and can be made from experience, taking into account local conditions, operational constraints and level of accepted risk. The final result is always a statistical issue and no procedure is 100% safe. Therefore when working under pressure over 0.9 bar contingency plans must always be prepared to cope with DCI cases.

The procedures are called decompression tables, they should be validated by field collected data (Schreiner and Hamilton, 1989). There are many types of tables:

- Air diving decompression tables breathing air only or pure oxygen during the decompression stops,
- Surface decompression tables when a diver is brought back to the surface then recompressed in a surface decompression chamber (breathing pure oxygen),
- Tables for all kinds of mixed gases (Nitrox, Trimixes, Heliox) with in water decompression or transfer under pressure into a deck decompression chamber,
- Saturation decompression tables with upwards and downwards excursions and various mixes compositions,

- Calculation to take into account specific circumstances like altitude, repetitive exposures, aborted decompression, diving in dense media,
- Dive computers are commercially available to calculate and display on line the decompression required according to an algorithm and limiting values implemented into the machine. Most recreational dives are performed under the control of such equipment,
- For compressed air work 'diving' tables are not adapted nor recommended. Specific procedures are now used that take into account longer exposure times, dry condition of exposure, type of exposed population and work load. Oxygen breathing during decompression stops is highly recommended (Le Péchon, 1992), some regulations have even made it compulsory by law (Regulation, 1997, 1994).

### Decompression Sickness—Decompression Illness (DCI)

*Signs and symptoms:* The signs and symptoms are extremely variable due to the fact that forming bubbles may interact in almost any part of the body (Nashimoto and Lamphier, 1991). There is a significant delay between the decompression stress and the manifestation of DCI signs. In surface oriented dives and compressed gas work, the delay of symptoms may be from a minute after surfacing or reaching atmospheric pressure up to several days. Treatment is by recompression, breathing gases with high $PO_2$, and a medical adjunct to control the effects of the bubbles in blood vessels (Moon and Gorman, 2003).

*Contingency Plans and DCI.* The management of DCI is a matter of time: The shorter the elapsed time between recognition of the signs and the recompression, the better the outcome:

- First aid on site focusses on immediate oxygen breathing, water to drink, eventually 500 mg of aspirin and organization of transportation to the closest recompression chamber.
- In commercial diving offshore, the chamber is on site in most circumstances and operators are available at short notice.
- In onshore diving, the chamber is often a hospital unit. Transportation means and selection of the hospital should be done before the beginning of the operations and availability checked when not open 24 h a day.
- In compressed air work, the chamber is on site most of the time.

The protocols for treatment (that should be validated by the site medical advisor when it is carried out on site) are based on HBO ($PO_2 = 2.8$ bar) for minor signs (skin problems, bends.), and either on HBO ($PO_2 = 2.8$ bar) or on a 3 bar recompression breathing 50/50 heliox for serious cases (neurological hits, inner ear symptoms…). A consensus of the best practice is not yet there among doctors. Recompression, high $PO_2$ breathing gas and medical adjuncts are a must and should be carried out as early as possible after the first signs have been identified. Transportation in a lightweight hyperbaric stretcher (1 bar –Pure oxygen $PO_2$ = 2 bar) is also an option.

## Bone Necrosis

Many compressed air workers and some divers have experienced severe joint pain appearing a long time after the exposure. It is due to bone necrosis induced by wrong decompression procedure even when no acute signs of DCI had been noticed (Allen et al., 1974). New decompression tables, reduction of very long shifts and suppression of decanting without oxygen breathing in compressed air work as well as the use of oxygen stops have dramatically reduced the incidence of hyperbaric aseptic bone necrosis.

Radiological control of main joints is still part of the medicals, in particular for those who may have experienced DCI.

## WORKING UNDER PRESSURE SAFELY

### In Immersion

Most diving related accidents reported in a data base listing 600 such diving accidents or near misses are in direct relation with the risks derived from the immersed situation, from the diving equipment or from the gas supplies (Chesshire personal communication, 2009). There is also a series of hazards created by the diving equipment, the environment (temperature, visibility, waves, tides and current…), by the type of tasks or the tools to be operated. In addition, due to immersion the rescue of a casualty underwater in an emergency may be very difficult to achieve.

### Diving Methods and Equipment

Due to the decompression stops requiring a long time to be spent while ascending back to surface, it may be necessary to remove the diver from the water before completing the decompression time. Therefore there are three different procedures: In water decompression, Surface decompression or Transfer Under Pressure (TUP) into a compression chamber located on the diving site.

*Surface Oriented*: When the diver is decompressed, in water or with surface decompression, the procedure is called surface oriented dive. It can be performed self contained with SCUBA or surface supplied.

For safety reasons pertaining to the lack of control on the surface, the need to immerse two persons when most of the time only one is needed and the difficulty of rescue, SCUBA diving is not recommended, generally forbidden, for commercial diving. Professional divers who need to swim around during the dive or need freedom on the bottom (scientific divers, recreational professional dive masters, media divers, fishermen...) are mostly SCUBA divers.

Surface demand is the normal air diving technique in civil engineering or offshore. A dive supervisor controls all the dive parameters from a control panel, and a stand-by diver is ready to dive in case help is needed by the working diver. The dive hoses are bundled in an umbilical that includes the gas supply (air, nitrox, oxygen for in water oxygen decompression), a pressure line to read depth in surface, a communication line, a strength member, eventually hot water, light and TV cables (Fig. 5).

**Figure 5.** Typical diving helmet and umbilical.

When surface decompressing, the recompression chamber must be close to the diver recovery site: time interval in surface should be less than 3 min and no physical effort should be needed for the diver to reach that chamber (Fig. 6). After recompression at 1.2 bar, the elimination of dissolved nitrogen takes place during sessions of oxygen breathing. Some procedures reduce pressure with intermediate stops although it is more convenient and safer to keep pressure constant and decompress only at the end of the needed oxygen breathing time.

**Figure 6.** Diving basket and recompression chamber.

*Transfer Under Pressure.* TUP allows long decompression in a chamber without going through atmospheric pressure. A closed bell system is mated on board to a compression chamber (Fig. 7).

Bounce diving is the technique when divers are decompressed after each intervention, in practice it is scarcely used. When the chamber system is a pressurized habitat, divers remain in saturation and are decompressed only at the end of the job or when they are relieved by a new team. The habitat may include several chambers, possibly at different pressures during decompression phases (Fig. 8).

A hyperbaric lifeboat, permanently connected to the complex would allow divers evacuation in case of fire.

**Figure 7.** Launching a diving bell.

**Figure 8.** Mating arrangement of a saturation habitat.

### Environmental Conditions

*Nautical*: Most diving operations take place in water bodies that may be hazardous by themselves:

- Open sea—visibility, weather, tides and current, marine life, distances…
- Harbours, rivers and dams—visibility, current, naturally sucking zones, tree in grids…
- Boats or vessels from which the dive is performed—moorings, fire, propellers of dynamic positioned vessels- vessels moving around.
- Underground in caves.

*Industry*: Commercial divers are often asked to work in industrial environments that may also add to hazards:

- Structure to be worked on—platforms, pile sheets, wellheads, pipes in production, vessels hull…-
- Tanks, wells, platforms piles, moving structures,
- Contaminated waters (details below)….

*Diving in Contaminated Waters*: Being immersed implies contact between the body and the fluid. When this fluid carries hazardous substances or dangerous living organisms, adapted precautions need to be implemented. The risks associated with diving in contaminated media is highly dependant of the type and level of contamination involved.

In each case a formal evaluation of the situation is required that will have to determine how to carry out the operation safely. With the proper information at hand, a procedure can be prepared that will include at least:

- selection of the type of diving suit, the type of breathing equipment,
- description of the means of access and egress, including the stand-by diver and a rescued diver,
- precautions for the manipulation of contaminated hoses by tenders,
- cleaning procedures, controls of efficiency,
- precaution for tenders when hazardous vapours may be involved.

A contingency plan should be prepared to cope with failure of the nominal procedure whether due to the diving related incident, or caused by direct contamination (chemicals, biological, marine life, radioactive media-irradiation and contamination—contamination of diving bell and habitat atmospheres…).

### Tools

All tools operated underwater or in compressed air have their 'ordinary' risks however prevention should be more stringent since in case of a problem, first aid and ultimately rescue require specific procedures to access and rescue a casualty from the pressurized environment.

In addition specific hazards are created by pressure (fire) or immersion (electricity, poor or no visibility …). Operational procedures must take all those into account in the risk assessment carried out by a competent person.

The details on each type of tool cannot be analyzed here, a list of commonly used tools is given below for information. The Professional Diver's Hand Book (Bevan, 2005) is exhaustive in this matter.

- Mechanics, hand held tools,
- Hydraulic (jacks, grinding, impact wrench, various saws, pipe alignment frames…)
- Electricity (video…, welding, burning, non destructive testing…)
- Burning (oxy arc, ultra-thermic…),
- Welding pipelines, anodes, structures… underwater or in dry habitats
- Crane or airbag lifting, cables and pull along, very heavy lifting
- Concrete and bags, sand bagging,
- Explosives for demolition, wellhead recovery, dredging and excavating
- High pressure jetting cleaning before non destructive testing, cutting,
- Painting and polymers.

## In the Dry

Only the two main working situations under pressure in the dry will be considered: Industrial work and hyperbaric medicine. There is a specific risk associated with fire that is common for all, it will be addressed separately.

### Compressed Gas Work in the Industry

Since the introduction of TBMs to bore tunnels, the compressed gas working conditions have changed significantly.

*Digging.* Digging under pressure may concern caisson works or tunnelling.

Caisson work consists basically in digging inside a caisson which sinks as long as digging is progressing. To prevent flooding, a compressed air atmosphere is established inside the caisson and entry locks are displayed at the upper part. Removing the excavated ground is also done via airlocks that are most of the time separated from man locks (Kindwall, 2003).

Nowadays there are still several places where workers are exposed to pressure for (too) long shifts, in poor environmental conditions, with the wrong tables, poor health and safety organization resulting in many kinds of professional diseases and long term adverse effects. This could be almost totally avoided provided responsible persons make the right decisions and whatever the impact on the resulting cost. Safety is costly but rewarding in the long term.

Recently a large caisson work operation was carried out for the construction of Nagoya Bridge, all digging was mechanically done in a remotely operated mode. Only maintenance and dismantling of the excavators required manned interventions, however the pressure being above 4 bar, the 'caisson workers' were breathing Trimix gases (Hirata et al., 1992), it proved to be a safe operation.

Since the early 80's caisson works have been dramatically reduced by the introduction of Tunnelling Boring Machines (TBM) in tunnel construction, and have nowadays become almost obsolete and practically limited to very low pressure (less than 1 bar). However professionals still enter pressurized environment at work and they need specific training (Le Péchon, 2002b).

## TBM Workers

A Tunnelling Boring Machine (TBM) is a complex mobile factory constructing a tunnel. As digging progresses it requires: digging, supporting the excavated area, evacuating the ground, erecting segments to produce the tunnel, install the necessary pipelines for water, compressed air as well as wires for electricity and communication. In addition several amenities are required to support those main functions, like main and back up power supplies and controls (electric or hydraulic), coring and grouting ahead of the front for ground stabilization or behind the already laid segments, computers to control and guide the machine and to record all digging parameters, air compressors, train station, unloading and displaying segments for erection, airlock for workers, airlocks for equipment....

The selection of the various means of operation makes it possible to categorize TBMs in three major classes: Pressurized compressed air road headers, Slurry Machines and Earth Pressure Balance TBMs.

In TBMs, the hyperbaric interventions are not systematic, all efforts are made to avoid them, however technology has not entirely succeeded as yet to solve the problems of clogging, boulders to be removed, tools inspection, remote wear measurement, tools repair and major breakdowns of the cutter heads.

Teams of hyperbaric technicians are trained to carry out this work, they can be employees from the company constructing the tunnel or subcontracted among local diving companies or specialized firms in TBM hyperbaric interventions. In all cases, safety rules must be implemented in the tunnel during the interventions and the overall safety of the pressurized workers assessed by the owner (Fig. 9).

**Figure 9.** A TBM with the airlock system.

**Figure 10.** Shuttle for transfer of saturated operators (Hallandas—13 bar) Photo Hytech.

The concept of 'Compressed Air' has widened since interventions in compressed air with mixed gases as respiratory media came of age in 2000. Working under pressure in the dry includes any situation where pressure is increased while breathing any suitable gas (compressed air or breathing mixtures), without any time spent in immersion even for emergency procedures. Presently tunnel builders are considering human intervention at pressure as high as 13 bar (Fig.10).

In addition, on rare occasions actual diving in the slurry has been performed in TBMs when it could not be removed from the cutter head in relation to ground instability.

### Hyperbaric Oxygen Therapy

During HBO sessions, patients are not at work and their medical condition is the major factor. (Ref chapter 20 in this book).

The potential hazards for the attendants are related to the confinement situation, fire under pressure, ear barotrauma, decompression illness and medical equipment that they may have to operate under pressure (Mathieu, 2006). All these matters must be addressed in the hospital safety procedures. In Europe the equipment match the requirement of EN 14931.

### Fire While Under Pressure

Fire while under pressure has always been a serious concern and it may happen inside the pressurized zone as well as in the external environment.

*Outside the Pressurized Environment*: In which case the major issues are how to keep acceptable conditions inside the chamber until evacuation. It can be performed either after the decompression has been carried out or directly under pressure in an escape chamber. Fire drills must be run and documented

- During the HBO session, the choice is rather the option of decompressing, cooling the chamber and having a chamber operator with a breathing mask to control the decompression and help out the patients and attendants.
- In tunnelling operations when the decompression is very long (several hours) an escape shuttle may be the option, otherwise cooling, and smoke protection are to be implemented as a safety measure.
- During saturation exposure only TUP is possible and most Diving Support Vessels are fitted with a hyperbaric life boat.

Fire should be considered as a serious and difficult issue where all precautions should be taken to prevent it, particularly in tunnelling where any welding, burning and grinding should not be undertaken in the tunnel when the personnel are pressurized. Oxygen dumped over from the air lock during oxygen decompression stops must also be disposed off safely.

*Inside the Pressurized Zone*: In a pressurized atmosphere the situation is very different, since the gas may be compressed air, oxygen contaminated air or various types of mixes containing helium.

In compressed air, $PO_2$ is increased as total pressure and the combustive (oxygen) is more active. The flammability of most substances and the rate of combustion are increased (Harter, 1977) (Fig. 11).

**Figure 11.** Fire in a TBM airlock–1 bar, during a training session.

In case of fire, people cannot escape immediately, heat released in a restricted volume will produce burns, smoke will make the atmosphere very toxic in direct relationship with the partial pressures of the toxic vapours. Most of the fires in pressurized chambers have resulted in the death of occupants (Sheffield and Desautels, 1997).

The situation is worse when the air is contaminated with oxygen from breathing masks. As a consequence oxygen concentration in a chamber should never be allowed to exceed 23 to 25% in volume.

In heliox saturation exposures, $PO_2$ is in the range of 0.4 b however helium reduces the burning rate by quickly dissipating the heat produced by the fire (Chianta and Stoll, 1973).

In tunnelling work, the working chamber is not a place of risk from fire except during welding, arc-air burning or grinding. The airlock, during oxygen decompression, requires a specific procedure that includes cleaning of the operators, clean overalls, oxygen masks with efficient overboard dump system for expired gas, outside control of chamber ventilation rate and monitoring oxygen content of the atmosphere to detect possible oxygen leaks into the ambient.

Fire prevention requires reducing as much as possible the amount of combustible materials inside chambers by construction or by controlling all equipment which is introduced inside during operation. Electricity may be the source of ignition, but low voltage is not an efficient means of protection, proper fuses must be fitted to prevent the effects of short circuits. Sparks are not dangerous in compressed air provided the atmosphere is not contaminated with combustible vapours. Welding, burning and grinding in compressed air are common practices that never induce explosions; all burnable substances like grease must be removed around the place of burning, in particular the operator's body and hair must be protected with a full non-flammable overall and cap; a high flow water hose must be at hand close to the welder at any time during this type of work.

Fire fighting needs an efficient deluge or spray system which can be triggered from the inside as well as from the outside. A fast decompression valve (fitted with a safety locking system) is a means to stop a fire by cooling and reducing partial pressure of oxygen and evacuating the personnel. However it may lead to decompression sickness or hypoxia, but that may be a more acceptable consequence than burning.

## SAFETY AND REGULATIONS

### Risk Assessment

Risk assessment is the basis of safety organization. It consists in a thorough listing of all possible hazards associated with a given task, quantifying the probability of occurrence according to the general environment, training, experience of the people and already existing safety rules, then quantifying the consequences (on personnel, budget, delay in the programme…) of a corresponding accident. A calculation using these values and a scale of acceptable level of risk is used to define necessary added safety precautions (prevention, rescue…). A new evaluation is carried out and the process repeated until all hazards have been reduced below the acceptable scaled risk factor. All resulting safety measures must then be implemented before beginning the operation…. Audits are to be carried out at regular intervals to check the application of the rules. Near misses and accident reports must be evaluated to ensure that new safety precautions are added as required.

## Safety Management System

Working in such hostile environments as underwater or in compressed gases requires an efficient formal safety system and a formal commitment from the management regarding safety. It could be recommended that any organization involved would implement a Safety Management System as proposed by OSHAS (OHSAS 18000). Very few national regulations enforce this type of safety structure although an association (IMCA) has demonstrated with its members that it is possible and that it improves significantly the working conditions of the divers.

## Medical Fitness

The many physiological issues described above makes it necessary for any person working under pressure to be assessed as being fit to do so by a competent medical officer. A yearly control is also required by most regulations.

## Training

Most of the time the concept of training for hyperbaric works includes only two aspects: divers and diving supervisors. Training for hyperbaric works is more than that, it should include all persons involved: 'the client', the site general manager who makes the decisions, the hyperbaric work manager, the supervisors (whether divers or lock operators and life support technicians), occupational doctors, safety officers and exposed divers or compressed air workers. Each member should have received a specific training, with corresponding certification, adapted to his responsibilities and tasks, each training content being coordinated in such a manner that the whole team will work on the same safety standard (Le Péchon, 2002b). Bridges between certificates should also be recognized to help personnel to move between the professional in the fields of hyperbaric works with immersion or not.

## National Rules

Many countries have established rules for professional diving and compressed air work. Most of the time, both regulations are independent (Regulation, 1994) and eventually contradictory (UK). Only France and Brazil have a single regulation for both. The most stringent rules are from Norway and the UK in the North Sea for offshore activities. The Arab Emirates have harmonized their practices adopting IMCA recommendation as national rules. The United States have different rules in each state, reference to the US Navy practice is everywhere and poorly adapted to commercial diving activities, it is even worse for compressed air work where the rules were set in the 50's, a trend to re-evaluate the situation is in progress in states where TBMs are working (Nevada, California, Ohio…).

## IMCA

The International Marine Contractors Association (IMCA) was formed in 1995 in the UK incorporating the Dynamic Positioned Vessels Owners Association and the Association of Offshore Diving Contractors, resulting in a worldwide association devoted to the promotion of safety among the offshore industry. IMCA has developed a large number of tools and safety recommendations for the diving industry. SMS is the leading principle, IMCA members must implement such a system, and should only contract or subcontract companies with similar internal organizations. Audits are performed to monitor the efficiency and the safety records. One of the major present drawbacks is the fact that documentation, audits, training, certification exams… are solely in the English language: in many parts of the world (France, Argentina, China, West Africa, Brazil or Russia for example) members cannot find locally IMCA certified subcontractors.

## ACKNOWLEDGEMENTS

Thanks to Tim Chesshire who agreed to read the manuscript.

## REFERENCES

Allen, T.H. and J.C. Davis and C.J. Hodgson. US Air Force experience in hypobaric osteonecrosis. Pp. 17–19. *In*: Beckman and Elliott (eds.). 1974. Dysbarism-related osteonecrosis, A. symposium. Washington, DC.
Bert, P. 1878. Barometric pressure (La Pression Barométrique). Undersea Medical Society, originally published in 1878. Translated by Hitchcock M.A. and Hitchcock F.A. Bethesda Maryland.
Bevan, J. 2005. The professional diver's handbook. 2nd Edit. SIMEX ED. Gosport.
Boycott, A.E. and G.C.C. Damant and J.S. Haldane. 1908. The prevention of compressed air illness. J. Hyg., Cambridge 8: Pp. 342–443
Broussolle, B. and J.-L. Méliet and M. Coulange. 2006. Physiologie et Médecine de la Plongée, 2nd Edit. Ellipses, Paris.
Brubakk, A.O. and T.S. Neuman. 2003. Bennett and Elliott's Physiology and Medicine of Diving. 5th Edit. Saunders. Edinburgh.
Bühlmann, A.A. 1984. Decompression—Decompression sickness. Springer-Verlag. Berlin.
Chianta, M.A. and A.M. Stoll. 1973. Fire retardance of mixtures of inert gases and oxygen. Aerospace Medicine 44(2): 169–173.
Clark, M.J. and S.R. Thom. Oxygen under pressure. Pp. 358–418. *In*: A.O. Brubakk and T.S. Neuman. (eds.). 2003. Bennett and Elliott's Physiology and Medicine of Diving. 5th Edit. Saunders. Edinburgh.

Cluet, A. and P. Gros. 1994. Mise à jour par J.M. Petit. Les mélanges explosifs, Gaz, Vapeurs, Poussières, Liquides, Solides—INRS éditeur—Paris.

Donald, K. 1992. Oxygen and the diver The SPA Ltd. Worcs.

[ESA] European Space Agency. 1992. Atmosphere Quality Standards in Manned Space Vehicles 215, ESA PSS 03 401 June. ESTEC Noorwijk, The Netherlands.

Francis, T.J.R. and D.J. Smith. 1990. Describing Decompression Illness. Proceedings of 42nd UHMS Workshop, UHMS Publ. nE 79 5/15/91.

Hamilton, R.W. and D.J. Kenyon. 1989. Managing oxygen exposure when preparing decompression tables. Pp. 72–77. *In*: Proceedings of Annual Scientific Meeting EUBS 89 Eilat.

Harter, J.V. Fire at pressure. Pp. 55–80. *In*: J.C. Davis and T.K. Hunt (eds.). 1977. Hyperbaric Oxygen Therapy. Undersea Medical Society. Bethesda.

Haux, G.F. 1982. Subsea Manned Engineering. Baillière Tindall, London.

Hennessy, T.R. On the Site of Origin, Evolution and Effects of Decompression Microbubbles. Pp. 291–340. *In*: A.O. Brubakk (ed.). 1989. Supersaturation and Bubble Formation in Fluids and Organisms. B.B. Hemmingsen and G. Sundnes.

Hirata, T. and K. Takano, Y. Gotoh, I. Nashimoto and W. Sterk. Remotely controlled caisson method and its maintenance. Pp. 519–528. *In*: F.M. Jardine and R.I. McCallum (eds.). 1992. Engineering and Health in Compressed Air Work. E & F Spon. London.

Imbert, J.P. and X. Fructus, B. Gardette, C. Gortan and M. Carlioz. 1989. JASON I, Evaluation of the effects of 2 bars of argon in heliox saturation diving. Pp. 28–35. *In*: Proceedings of XVth EUBS Annual Scientific Meeting, Eilat.

Imbert, J.P. and G. Manicacci, E. Jayet and J. Lustosa. 1990. Deep diving techniques Comex experience in Brazil. Pp. 237–247. *In*: Advances in Underwater Technology, Ocean Science and Offshore Engineering, Vol 23. Subtech '89. Fitness for purpose. Kluwer Academic Publishers.

Jardine, F.M. and R.I. McCallum. (eds.). 1994 Engineering and Health in Compressed Air Work: Proceedings of International Conference, Oxford, Sept. 1992, London: E & FN Spon Publishers.

Joiner, J.T. 2001. NOAA Diving Manual: Diving for Science and Technology. 4th Edit. Best Publishing Co. Flagstaff.

Kindwall, E.P. 2003. Compressed air work. Pp. 17–28. *In*: A.O. Brubakk and T.S. Neuman (eds.). Bennett and Elliott's Physiology and Medicine of Diving. 5th Edit. Saunders. Edinburgh.

Lamont, D.R. 2006. Decompression illness and its regulation in contemporary U.K. tunnelling—an engineering perspective. Thesis, Aston University. Aston.

Le Péchon, J.Cl. 1981. Schedules for saturation decompression, establishing principles and operational results. Pp. 205–217. *In*: Proceedings 7th EUBS Annual Scientific Meeting, Cambridge.

Le Péchon, J.Cl. Oxygen decompression in tunnelling, Pp. 529–538. *In*: F.M. Jardine. and McCallum (eds.). 1992. Engineering and health in compressed air work. Proceedings of International conference. E & FN Spon publishers. Oxford.

Le Péchon, J.Cl. Personnel safety in hyperbaric medicine, general risk analysis. Pp. 111–123. *In*: D.J. Bakker, J.Cl. Le Péchon and A. Marroni (eds.). 1998. Hyperbaric Oxygen, Wound Healing, Safety, Cost Effectiveness. 1st European Workshop on hyperbaric medicine. Best Publishing Company, Belgrade.

Le Péchon, J.Cl. 2002a. Compressed air work in a kerosene contaminated soil. Pp. 153–162. *In*: Proceedings of the 2nd International Conference on Engineering and Health in Compressed Air Work. Oxford.

Le Péchon, J.Cl. 2002b. Training for compressed air work. *In*: Proceedings of the 2nd International Conference on Engineering and Health in Compressed Air Work. British Tunnelling Society Oxford.

Le Péchon, J.Cl. and W. Sterkand T.P. Van Rees Vellinga. 2001. Saturation diving for tunnelling operations. Pp. 285–292. *In*: Proceedings of AITES-IDA 2001 World Tunnel Congress, Vol.II. Milan.

Lin, Y.C. and K.K. Shida. 1990, Man in the sea. Vol. 1 and 2. Best Publishing Company, San Pedro Cal.

Mathieu, D. (ed.). 2006. Handbook on hyperbaric medicine. 1st edit. Springer. Dordrecht

Moon, E.R. and D.F. Gorman. 2003. Treatment of the decompression disorders. Pp. 600–650. *In*: A.O. Brubakk and T.S. Neuman (eds.). 2003. Bennett and Elliott's Physiology and Medicine of Diving. 5th Edit. Saunders. Edinburgh.

Nashimoto, I. and E.H. Lanphier (eds.). 1991. What is bends ? 43rd UHMS Workshop, Shimizu. Japan.

Neuman, T.S. Arterial Gas Embolism and Pulmonary Barotrauma. Pp. 557–577. *In*: A.O. Brubakk and T.S. Neuman (eds.). 2003. Bennett and Elliott's Physiology and Medicine of Diving. 5th Edit. Saunders. Edinburgh.

[OHSAS 18000] Occupational Health and Safety, 18000. International occupational health and safety management system specification.

Pol, B. and J.J. Wattelle. 1854. Mémoire sur les effets de la décompression de l'air appliquée au creusement des puits à houille, Pp. 241–279. Annales d'Hygiène publique et de médecine légale (industrielle et sociale) 1.

Radziszewski, E. 1987. Effets physiologiques chez l'homme du confinement de longue durée en atmosphère enrichie en dioxyde de carbone, Thesis University of Lyon, Lyon.

Regulation 1983. Brazilian: Trabalho sob condicoes hiperbaricas Diario official 15 set: 16024–16028.

Regulation 1990a. French-DECRET 90-277 du 28 Mars 1990 et arrêtés d'application, Brochure 1636, Travaux en milieu hyperbare—Mesures particulières de prévention, Journal Officiel République Française.

Regulation 1990b. Regulation relating to diving. Directorate of Labour Inspection, Oslo. And Regulations relating to manned underwater operations in the petroleum activities. Stipulated by the Norwegian Petroleum Directorate, Stavanger, April 30.

Regulation 1994. Novellierung der Druckluffverordnung (DLV), Aufenthalts—und Ausschleusungszeiten fur Arbeiten in Druckluft, Bek. des BMA vom 26 Oktober 1994, IIIb2-34572.2-

Regulation 1997. UK: The diving at work Regulations 1997. n° 2776, and Approved Codes of Practice. HSE Books, Her Majesty's Stationary Office. Nordwich, UK.

Schreiner, H.R. and R.W. Hamilton. 1989. Validation of decompression tables. Proceedings of 37th UHMS Workshop UHMS. Bethesda MD.

Sheffield, P.J. and D.A. Desautels. 1997. Hyperbaric and hypobaric chamber fires: a 73 year analysis.. Undersea and Hyperbaric Med. 24: 153–164.

Sterk, W. 1986. Intermittent hyperoxia in operational diving: what are the safe limits? Pp. 55–64. *In*: Proceedings XII EUBS Annual Meeting, Rotterdam.

Vann, R.D. (ed.). 1989. The physiological Basis of Decompression Proceedings 38th UHMS Workshop, Duke University, Durham, NCa.

Violette, F. 1955. Étude expérimentale et théorique de la décompression explosive et de ses effets physiologiques. Bull. Serv. Techn. du Ministère de l'Air nE 118, Paris.

Walder, D.N. The compressed air environment. Pp. 15–30. *In*: P.B. Bennett and D.H. Elliott. (eds.). 1982.The Physiology and Medicine of Diving. 3rd Edit. Ballière Tindall, London.

Wattel, F. and D. Mathieu. 2002. Traité de Médecine Hyperbare, Ellipses. Paris.

Wienke, B.R. 1990. Reduced Gradient Bubble Model, Int. J. Biomed. Comput. 26: 237–265.

Workman, R.D. 1965. Calculation of Decompression Schedules for Nitrogen-Oxygen and Helium-Oxygen Dives. Research report 6–65 US Navy Experimental Diving Unit Washington D.C.

Wright, W.B. 1972. Use of the University of Pennsylvania, Institute of Environmental Medicine procedure for calculation of cumulative pulmonary oxygen toxicity. U.S. Navy Experimental Diving Unit, rep. NEDU 2–72.

Yount, D.E. 1986. On the use of a bubble formation model to calculate diving tables. Aviat. Space Environ. Med. 57: 149–156.

## RELEVANT WEB SITES

www.edtc.org
www.eubs.org
www.hse.gov.uk/compressedair
www.hse.gov.uk/diving
www.hyperbarie.com
www.imca-int.com
www.ohsas-18001-occupational-health-and-safety.com
www.medsubhyp.com
www.uhms.org

# Index